T0327526

ADVANCED METHODS
OF BIOMEDICAL
SIGNAL PROCESSING

ADVANCED METHODS OF BIOMEDICAL SIGNAL PROCESSING

Edited by

SERGIO CERUTTI
CARLO MARCHESI

IEEE Engineering in Medicine
and Biology Society, *Sponsor*

IEEE Press Series in Biomedical Engineering
Metin Akay, *Series Editor*

IEEE Press

A JOHN WILEY & SONS, INC., PUBLICATION

This book was previously published in Italian under the title, *Metodi avanzati elaborazione dei segnali biomedici,* by Sergio Cerutti and Carlo Marchesi. © 2004 Pátron Editore, Bologna, Italy.

Copyright © 2011 by the Institute of Electrical and Electronics Engineers, Inc.

Published by John Wiley & Sons, Inc., Hoboken, New Jersey. All rights reserved.

Published simultaneously in Canada.

For general information on our other products and services please contact our Customer Care Department within the United States at (800) 762-2974, outside the United States at (317) 572-3993 or fax (317) 572-4002.

Wiley also publishes its books in a variety of electronic formats. Some content that appears in print, however, may not be available in electronic formats. For more information about Wiley products, visit our web site at www.wiley.com.

Library of Congress Cataloging-in-Publication Data is available.

ISBN: 978-0-470-42214-4

Printed in the United States of America

oBook ISBN: 978-1-118-00774-7
ePDF ISBN: 978-1-118-00772-3
ePub ISBN: 978-1-118-00773-0

10 9 8 7 6 5 4 3 2 1

CONTENTS

Part IV. Time-Frequency, Time-Scale, and Wavelet Analysis

19 Microarray Data Analysis   **473**

Gene Regulatory Networks

Riccardo Bellazzi, Silvio Bicciato, Claudio Cobelli,
Barbara Di Camillo, Fulvia Ferrazzi, Paolo Magni,
Lucia Sacchi, and Gianna Toffolo

20 Biomolecular Sequence Analysis   **489**

Linda Pattini and Sergio Cerutti

Part VII. Classification and Feature Extraction

21 Soft Computing in Signal and Data Analysis   **511**
Neural Networks, Neuro-Fuzzy Networks, and
Genetic Algorithms
Giovanni Magenes, Francesco Lunghi, and Stefano Ramat

22 Interpretation and Classification of Patient Status   **551**
Patterns
Matteo Paoletti and Carlo Marchesi

IEEE Press Series in Biomedical Engineering

PREFACE

 THIS BOOK DEALS with some of the most advanced methodological approaches in signal analysis of biomedical interest. A basic background of digital signal treatment is generally required in order to better cope with the more advanced methods. Readers who are not familiar with the basic concepts of signal (or biomedical signal) processing should read the first two sections, which cover the most important concepts at a level that will be useful for correctly understanding the other chapters. Some basic references are included for those who want to go deep into detail on the fundamentals of signal (or biomedical signal) processing, which is not the aim of the present book.

Today, the evolution of the various approaches employed for biomedical signal processing (as well as the implementation of suitable algorithms of signals and data treatment, sometimes complex and sophisticated) makes difficult the proper choice of the method to be used according to previously defined objectives. Such objectives might consist of simple signal/noise improvement; the extraction of informative parameters that are important from the clinical standpoint; diagnostic classification; patient monitoring; diagnosis or, in critical cases, the surveillance of subjects such as elderly people or neonates, who can encounter acute episodes; as well as the control of determined chronic pathologies. Hence, to know the wide range of methods that could be employed in the various contexts of signals and data processing is certainly a major qualifying issue.

On the other hand, the clinical studies that have successfully employed advanced methods of signal processing are numerous. The sixth most cited paper from *Circulation Research* (which is perhaps the most prestigious journal dealing with advanced research themes of the cardiovascular system) is one of the fundamental papers about the theme of heart rate variability (HRV) (Pagani et al., 1986). In Medline, about 12,000 papers have been indexed that deal with this theme. More remarkably, for the journal *Circulation,* the third most cited paper is the well-known Task

Force on Heart Rate Variability study (Malik et al., 1996). In these papers, parameters obtained even via nontraditional approaches of signal processing are widely reported and commented upon. These approaches are deterministic or stochastic, linear or nonlinear, monovariate or multivariate, fractal-like, and so on, and important pathophysiological correlates are suggested and documented.

Furthermore, new equipment is available in the clinical market that employs parameters derived not only from traditional signal processing approaches but also characterized by a certain methodological complexity, such as bispectral indexing or the measurement of entropy parameters in EEG signals for the monitoring of anesthesia level (manufactured by Aspect Medical Systems, Inc. and GE-Datex/Ohmeda).

Various contributions in this book will show that biomedical signal processing has to be viewed in a wider context than the one generally attributed to it, with important links to the modeling phase of the signal-generating mechanisms, so as to better comprehend the behavior of the biological system under investigation. The fundamental concept is that often the modeling phase of a biological system and the processing phase of the relevant signals are linked to each other and sometimes reciprocally and synergically contribute to improve the knowledge of the biological system under study.

This book also includes the presentation of signals and data processing methods that are homogenous and, hence, may be used to integrate in the same metric system information derived from different approaches, coming from different biological systems, on different observation scales. In this way, a more integrated and, hence, more holistic, view of data, signal, and image processing might significantly contribute to an effective improvement of the clinical knowledge of a single patient.

If it is true that any biological signal carries information relative to the system or systems that generated it, we can say that the processing of the signal has the following main objectives: (1) to enhance the useful information from the original signal, (2) to interpret the results and to validate the obtained parameters for the following decision phase, and (3) to produce innovation for the improvement of physiological knowledge, the production of new "intelligent" medical equipment and devices, and the definition of new clinical protocols for prevention, diagnosis, and therapy.

This book is a translated and updated version of a textbook written in Italian for a summer school course organized in Bressanone by the Gruppo Nazionale di Bioingegneria (GNB), the Italian Scientific Society on Bioengineering, in 2004 and published by Pátron Editore.

In Part I, some basic elements of the peculiarities of biomedical signal processing in respect to other more traditional applications of digital signal processing and their classification are introduced.

Part II presents an experimental physiologist's and cardiologist's view of central nervous system. Both note the importance of the fundamental step of information processing in biomedical signals and data.

Part III illustrates an important link between biomedical signal processing and physiological modeling. Generally, the two approaches are separated; those who do signal processing do not do modeling and vice versa. An integration between the two approaches is required in order to establish a higher level of comprehension of complex pathophysiological phenomena.

Part IV covers time-frequency, time-scale, and wavelet analysis, with which linear, quadratic, and time-varying estimation of parameters are introduced for understanding the dynamical responses of complex physiological systems. The well-known compromise between time and frequency resolutions in time-scale approach is mainly treated and various applications in biomedical systems are described.

Part V deals with advanced methods that are employed in the fascinating area of complexity measurements, from chaotic systems, to fractal geometry of biological systems behavior when described in the space–state domain, to the nonlinear phenomena that might mimic different behaviors and, hence, provide different interpretations of the physics of the biological phenomena under study.

Part VI tackles an original and innovative application field of data processing: the one that operates at the scale of genes and proteins. Computational genomics and proteomics is a growing area of investigation in the so-called postgenomic era. The challenge is to apply well-known methods of digital signal processing at the level of data sequences constituted by the series of bases constituting DNA strings, as well as of elementary amino acids constituting proteins. Noteworthy possibilities of application are foreseen in both data processing and modeling of the biological phenomena involved.

Finally, Part VII describes important methods for signal classification, such as neural networks and neuro-fuzzy and genetic algorithms.

Due to the articulated and differentiated characteristics of biological systems and their reciprocal relationships, the expert in advanced biomedical signal processing might be compared to Plato's story of a man in a cave. The man is unchained and his sight is forced toward the bottom of the cave. He must guess the nature of reality outside the cave from the pale shadows that real objects project on the bottom of the cave. Analogously, biomedical signals, which are by their nature complex and corrupted by noise from different sources, often provide only a shadow of reality and must be processed in nontrivial ways to be really informative. The good biomedical signal processing expert should try to properly use a priori information from the model of signal-generating mechanisms, their statistical characteristics, and their interactions. In order to improve the capacity of processing, in many instances the entire procedure might include a phase of integration of information on different scales, by considering also many monovariate sensors and parameters contemporaneously, as previously mentioned.

A major issue is represented by the passage from data and parameters obtained from the processing to the real production of information and, hence, to the effective application of medical care. In biomedical applications, it is not true that the more information, the better; one has to be focused on which information is really useful and which is not.

ICT (information and communication technology) instruments and devices are thought to be crucial to properly implement efficient and effective solutions in patients with chronic or acute pathological conditions and elderly and disabled subjects. On the other hand, they may have drawbacks, as has been well documented. Among these, health costs are always increasing in developed countries, due to the need to do various exams more times or in very short time intervals, due to the lack of a link among hospitals and ambulatory systems, as well as adequately distributed storage systems. We are approaching a critical situation as we lack a suitable culture to widely employ numerous low-cost tools with very simplified procedures that are available to end the patient's isolation and to encourage his participation in self-care. It is worth remembering that among the prioritary research projects at MIT in Boston as well as in EU projects (Sixth and Seventh Framework Programmes), the development of personal and wearable equipment for the detection and treatment of vital signs is included.

The problems caused by the offer of personal technologies in a variety limited only by imagination will bring attention to the topic of the cooperation among different disciplines, among users and service providers, and among researchers and equipment manufacturers. The problems connected with the ergonomics and usability of the devices as well as esthetic, functional, psychological, and environmental issues, will join the technical ones related to the efficient and effective realization of these new personal devices. In this way, the single patient/subject constitutes a node of the world information system.

According to some experts on complex systems, life as a process, not as an attribute, will always find the way to establish itself in any environment and situation, as it will always find how to adapt itself. The anthropological implication of this statement is given by technology itself. We might say that man could be the species that will realize completely life through technology.

Sergio Cerutti
Carlo Marchesi

Milan, Italy
Florence, Italy
March 2011

REFERENCES TO PREFACE

Malick, M. et al., and Task Force of the European Society of Cardiology and the North American Society of Pacing and Electrophysiology, Heart rate variability—Standards and measurement, physiological interpretation, and clinical use, *Circulation,* Vol. 93, 1043–1065, 1996.

Pagani, M., Lombardi, F., Guzzetti, S., Rimoldi, O., Furlan, R., Pizzinelli, P., Sandrone, G., Malfatto, G., Dell'Orto, S., Piccaluga, E., Turiel, M., Baselli, G., Cerutti, S., and Malliani, A., Power spectral analysis of heart rate and arterial pressure variabilities as a marker of sympatho-vagal interaction in man and conscious dog, *Circulation Research,* Vol. 59, No. 2, pp. 178–193, 1986.

This page intentionally left blank

CONTRIBUTORS

FABIO BABILONI, Institute of Human Physiology, University of Rome, "La Sapienza," Rome, Italy

RITA BALOCCHI, Institute of Clinical Physiology, National Research Council (Consiglio Nazionale delle Ricerche, CNR), Pisa, Italy

GIUSEPPE BASELLI, Department of Bioengineering, Polytechnic of Milan, Milan, Italy

RICCARDO BELLAZZI, Department of Computer Engineering and System Science, University of Pavia, Pavia, Italy,

ANNA MARIA BIANCHI, Department of Bioengineering, Polytechnic of Milan, Milan, Italy

SILVIO BICCIATO, Department of Chemical Engineering, University of Padova, Padova, Italy

GABRIELE E. M. BIELLA, Institute of Bioimaging and Molecular Physiology, National Research Council (Consiglio Nazionale delle Ricerche, CNR), Milan, Italy

PAOLO BOLZERN, Department of Electronics and Information, Polytechnic of Milan, Milan, Italy

GIOVANNI CALCAGNINI, Department of Technology and Health, National Institute of Health, Rome, Italy

FEDERICA CENSI, Department of Technology and Health, National Institute of Health, Rome, Italy

SERGIO CERUTTI, Department of Bioengineering, Polytechnic of Milan, Milan, Italy

FEBO CINCOTTI, IRCCS Foundation Santa Lucia, Rome, Italy

CLAUDIO COBELLI, Department of Information Engineering, University of Padova, Padova, Italy

GIUSEPPE DE NICOLAO, Department of Computer Engineering and System Science, University of Pavia, Pavia, Italy,

FABIO DERCOLE, Department of Electronics and Information, Polytechnic of Milan, Milan, Italy

FABRIZIO DE VICO FALLANI, Institute of Human Physiology, University Rome, La Sapienza, Rome, Italy

BARBARA DI CAMILLO, Department of Information Engineering, University of Padova, Padova, Italy

MANUELA FERRARIO, Department of Bioengineering, Polytechnic of Milan, Milan, Italy

FULVIA FERRAZZI, Department of Computer Engineering and Systems Science, University of Pavia, Pavia, Italy

LORIANO GALEOTTI, Department of Systems and Informatics, University of Florence, Florence, Italy

ALEŠ HOLOBAR, Department of Electronics, Polytechnic of Torino, Torino, Italy

MARIA TERESA LA ROVERE, Cardiology Division, IRCCS Foundation, Maugeri, Montescano, Pavia, Italy

FRANCESCO LUNGHI, Department of Computer Engineering and Systems Science, University of Pavia, Pavia, Italy

GIOVANNI MAGENES, Department of Computer Engineering and Systems Science, University of Pavia, Pavia, Italy

PAOLO MAGNI, Department of Computer Engineering and Systems Science, University of Pavia, Pavia, Italy

LUCA T. MAINARDI, Department of Bioengineering, Polytechnic of Milan, Milan, Italy

CARLO MARCHESI, Department of Systems and Informatics, University of Florence, Florence, Italy

ROBERTO MERLETTI, Department of Electronics, Polytechnic of Torino, Torino, Italy

LUCA MESIN, Department of Electronics, Polytechnic of Torino, Torino, Italy

MATTEO PAOLETTI, Department of Systems and Informatics, University of Florence, Florence, Italy

LINDA PATTINI, Department of Bioengineering, Polytechnic of Milan, Milan, Italy

GIANLUIGI PILLONETTO, Department of Information Engineering, University of Padova, Padova, Italy

ALBERTO PORTA, Department of Technologies for Health, Galeazzi Institute, University of Milan, Milan, Italy

STEFANO RAMAT, Department of Computer Engineering and Systems Science, University of Pavia, Pavia, Italy

SERGIO RINALDI, Department of Electronics and Information, Polytechnic of Milan, Milan, Italy

CARMELINA RUGGIERO, Department of Informatics, Systems and Telematics, University of Genoa, Genoa, Italy

LUCIA SACCHI, Department of Computer Engineering and Systems Science, University of Pavia, Pavia, Italy

MARIA GABRIELLA SIGNORINI, Department of Bioengineering, Polytechnic of Milan, Milan, Italy

GIOVANNI SPARACINO, Department of Information Engineering, University of Padova, Padova, Italy

GIANNA TOFFOLO, Department of Information Engineering, University of Padova, Padova, Italy

MAURIZIO VARANINI, Institute of Clinical Physiology, National Research Council (Consiglio Nazionale delle Ricerche, CNR), Pisa, Italy

MAURO URSINO, Department of Electronics and Informatics, University of Bologna, Bologna, Italy

Thispageintentionallyleftblank

FUNDAMENTALS OF BIOMEDICAL SIGNAL PROCESSING AND INTRODUCTION TO ADVANCED METHODS

Thispageintentionallyleftblank

CHAPTER 1

METHODS OF BIOMEDICAL SIGNAL PROCESSING
Multiparametric and Multidisciplinary Integration toward a Better Comprehension of Pathophysiological Mechanisms

Sergio Cerutti

1.1 INTRODUCTION

It is well known that medicine, in both research environments (particularly in physiology) and clinical applications, is becoming a more and more quantitative discipline based upon objective data obtained from the patient or the subject under examination through digital parameters, vital signs and signals, images, statistical and epidemiological indicators, and so on. Today, even clinical applications cannot be made without a more or less extended background of quantitative indicators that complement generally anamnestic parameters as well as those of the typical objective examination. Patients' medical records therefore contain more and more data, signals, images, indicators of normality, morbidity, sensitivity, and specificity, and other numerical parameters that must be properly integrated in order to help physicians to make correct decisions for diagnostic evaluations as well as in therapeutic interventions.

Actually, proper sensors or transducers are able to make biological measurements at the various parts of the human body, from the usual electrocardiographic (ECG) or electroencephalographic (EEG) signals, arterial blood pressure signals (ABP), respiration, and so on, up to metabolic signals obtainable from a proper processing of functional images from functional magnetic resonance imaging (fMRI), positron emission tomography

Advanced Methods of Biomedical Signal Processing. Edited by S. Cerutti and C. Marchesi
Copyright © 2011 the Institute of Electrical and Electronics Engineers, Inc.

(PET), and other means. Hence, the information detectable from the patient has grown by considering various organs and systems simultaneously.

Even in biology, mainly due to the very strong acceleration of research in the last few years into the genome and proteome structure sequencing, much importance has been attributed to the methods that allow the information treatment starting from sequences formed by four bases (A, C, G, T) in the genome and twenty amino acids, which are the constituents of the proteins in the proteome. Many of the methods of information treatment, in the form of biomedical signals and data, can be applied to these sequences, after a proper transformation into numerical series; and this is certainly only a single example of the possible important innovative applications of biomedical engineering to molecular biology.

This chapter and the next one illustrate the fundamentals of a modern approach to the processing and interpretation of biomedical signals by exploring the main research lines and some of the many applications that will be examined in more detail in subsequent chapters. Knowledge of these methods, their algorithms, and the techniques used for their realizations can become a sophisticated investigational tool, not only to enhance useful information from the signals (which is the more "traditional" objective of this discipline) but even to approach in a more quantitative way the study of complex biological systems, certainly with a more important impact on the improvement of the physiological knowledge as well as the clinical applications. It is fundamental to remember that the more advanced methods in the area of biomedical signal processing do integrate the more advanced processing algorithms with the necessary knowledge of the systems under study. Hence, to obtain innovative results in this research, it is necessary to achieve multidisciplinary and interdisciplinary knowledge; certainly, the biomedical engineer can be one of the main actors toward these ends.

In Section 1.2, the fundamentals on which the main methods of traditional signal analysis are based will be examined and the most important definitions provided. It is not pretended that this theme can be adequately covered in a short section; rather, it is intended to define the starting bases and provide some topical references to provide a better understanding of the following contributions in the book, even for readers who have not had the opportunity to tackle the fundamentals of signal processing. Then, a section will be dedicated to the important connection between signal processing and modeling of biological systems (which is also referred to as "model-based biomedical signal processing"), which allows one to obtain information through an approach that employs knowledge of the pathophysiological phenomenon under examination. Finally, some methods of information integration will be introduced, which come from different signals, images, systems, modalities, and observation means, constituting the basis for the comprehension of more advanced methods of information-content treatment of biological phenomena, which is the main objective of the present volume.

1.2 FUNDAMENTAL CHARACTERISTICS OF BIOMEDICAL SIGNALS AND TRADITIONAL PROCESSING APPROACHES

A biomedical signal is a phenomenon that carries information relative to one or more biological systems involved. Obviously, we may find biomedical signals at different observation scales: as an example, at the level of a functional organ (heart, brain, liver, kidneys, etc.), at a system level (cardiovascular, central nervous, endocrine–metabolic systems, etc.), but also at the level of the cell or even at a subcellular level, as indicated in the previous section, as well as in higher dimension systems, as in the case of the quantitative study of morbidity, mortality, and the mechanisms of propagation of an epidemic disease inside a certain population.

The usual objective is generally to enhance the information that "travels" on these signals or data sequences in the best suitable way, according to prefixed aims. In fact, there does not exist a unique way to process signals, but rather a plurality of approaches, even very different ones, which the expert in biomedical signal processing will choose on the bases of the objectives (*which* information am I interested in?), the experimental protocols, the available a priori information, and the cost/performance ratio to be optimized.

The fundamental biological signals, which to a certain extent are distinct from other signals, are:

- Signals that derive from dynamical systems, which are singularly characterized by very different behaviors and various degrees of complexity
- Signals most often derived from an interaction among biological systems
- Signals characterized by great variability (inter- and intraindividual)
- Signals corrupted by noise (often with low signal/noise ratio). Even superimposed noise is often a complex process with endogenous and exogenous contributions in respect to the biological system under examination.

Figure 1.1 shows a general block diagram of how biological signals contribute to medical decision making and, hence, to the traditional diagnostic and therapeutic processes. The signal is acquired from the patient or from a biological process under study and is detected by a sensor or transducer. The measurement must be obtained by means of a proper calibration operation and the usual preprocessing procedures have to be fulfilled: amplifi-

6 CHAPTER 1 METHODS OF BIOMEDICAL SIGNAL PROCESSING

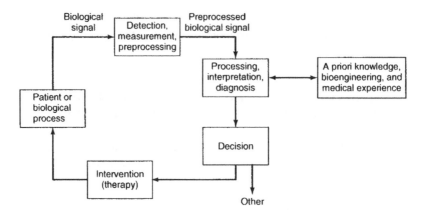

Figure 1.1. General block diagram of the operations involved in the procedures of biomedical signal processing and interpretation.

cation, antialiasing filtering (to limit the signal bandwidth), analog-to-digital (A/D) conversion of the sample (using Shannon's theorem), and quantization procedures (by taking into account the possible effects of the quantization error variance). The signal is then ready for the successive processing and interpretation phases. Here, it is important to emphasize the interaction with the block of a priori knowledge and previous experience, both from the standpoint of the bioengineer (for the development of the most suitable processing and classification algorithms) and from the medical standpoint (for a correct interpretation of the clinical data). Hence, the obtained information allows one to make a diagnosis. Then the successive action implies generally an intervention for the patient (therapy, either pharmacological, surgical, or both) which closes the ideal circuit of the scheme.

It is not necessary to go deep into details of the general elements reported above: a more extensive treatment of the topic can be found in Cohen, 1986; Webster, 1998; and Rangayyan, 2002.

1.2.1 Deterministic and Stochastic Systems and Signals

A system is said to be *deterministic* when, on the basis of the past behavior of the system, we are able to predict the trend of the system itself. A signal $y(t)$ that derives from such a system is said to be deterministic if in every instant t_k its value $y(t_k)$ may be calculated by means of a closed mathematical expression as a function of time or extrapolated by the knowledge of a certain number of preceding samples of the signal. As a consequence, the parameters of interest of a deterministic signal are directly measurable from its mathematical expression or from the graphic form of the signal itself.

A *stochastic* system does not allow one to predict its behavior from its past by means of a mathematical equation; hence, a stochastic signal can be processed only through statistical methods. In a more rigorous way, it is assumed that every sample of the signal $y(t_k)$ is intended as a random variable with probability distribution $p(y_{tk})$. Therefore, the entire sequence of the samples of the $y(t)$ signal can be intended as one of the possible realizations of the stochastic process.

It is worth remembering that de facto rigorously determinist or rigorously stochastic biological signals do not exist, whereas there are biological signals that are prevalently deterministic or prevalently stochastic.

As an example, an ECG tracing can be considered a deterministic signal as, at least in a large majority of cases, from the analysis of the past trend of the signal over a few seconds, we may estimate with good approximation the future trend and, therefore, the parameters of interest are directly detectable from the graphic form of the signal (wave durations and amplitudes, isoelectric tracts, morphology measurements of waves or wave complexes, etc.). On the other hand, a physiological beat-to-beat signal variability (and/or of the superimposed noise) could be present, which would make the signal unpredictable from a mathematical point of view. In the same way, on an EEG signal, despite its predominant nonpredictability (clearly a nondeterministic signal in the majority of cases) we could theoretically rescue some deterministic information through direct recordings in certain cortical areas (electrical activity of specific cortical structures or even of a single-neuron cell) and, hence, we could correlate this information with the pseudostochastic trend of the EEG signal. In practice, an EEG signal is generally processed with statistical methods [signal mean value and variance, besides other higher order statistical moments, and its autocorrelation function (in time domain) or its power spectral density (in frequency domain)]. These considerations support the fact that in practice an ECG signal is considered as predominantly deterministic and an EEG signal as predominantly stochastic, taking into account the methodological distinction previously made. Finally, some biological signals are called *transients* if they are correlated to an event (generally a stimulus) that happens in a determined temporal instant and their effects disappear after a time interval more or less long after the event. Examples of transient biological signals are cell action potentials and evoked (EP) or event-related sensorial potentials. In Figure 1.2, some biomedical signals are shown with their main statistical definitions.

1.2.2 Stationarity and Nonstationarity of Processes and Signals

A process is said to be *strictly stationary* (or *wide-sense stationary*) when all its statistical properties are independent of time and, hence, do not vary with time (Bendat and Piersol 2000; Papoulis, 2002). In biological

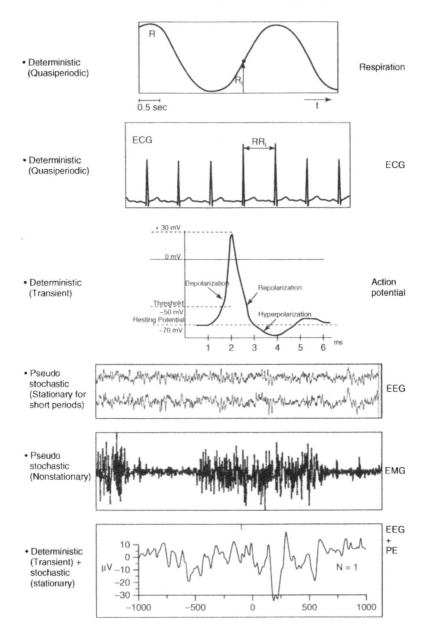

Figure 1.2. A few biomedical signals and their statistical characteristics.

processes, as in the wide majority of applications, *weak-sense stationarity* is instead said to hold, sufficient for the time invariance of the first-order ensemble mean (commonly and simply "mean") and of its autocorrelation function.

A stationary process is said to be *ergodic* if the ensemble means co-incide with the temporal means (calculated on a proper temporal time window). Therefore, the *ensemble* statistical properties may be calculated from the *temporal* statistical properties through a procedure that is obviously easier from a practical standpoint. This is the reason why it is common (and also convenient) to assume the ergodicity of the system under study. There are various stationarity and ergodicity tests that can be applied to time series with a certain statistical confidence; see, as examples, Box and Jenkins, 1970; and Ljung, 1998.

1.2.3 Gaussian and Non-Gaussian Processes

A discrete-time signal $y(t_k)$ can be intended as a realization of a *normal* or *Gaussian* process if the signal values in the various k time instants present a normal distribution. A Gaussian process is statistically determined if its mean value and variance are known. A weak-sense stationary Gaussian process maintains also the characteristics of wide-sense stationarity.

1.2.4 LTI Systems (Linear and Time-Invariant)

In the majority of the applications of signal processing, neither analog signals nor continuous-time signals are considered, but rather sampled signals, that is, sequences of samples having a constant time interval T_s, called sampling period and intended as the inverse of the sampling rate f_s. Shannon's theorem (also called sampling theorem) indicates the frequency that must be employed to correctly sample a signal without making aliasing errors and, hence, to be able to faithfully reconstruct the information contained in the signal after the sampling procedure. In particular, it is demonstrated that such a frequency (f_s) satisfies the relation $f_s \geq 2f_m$, where f_m is the maximum frequency contained in the signal spectrum, which in practice has a limited bandwidth up to f_m frequency. If the signal were not band-limited, it would be necessary to band-limit it by means of an analog low-pass filter with a cutoff frequency $\leq f_m$.

Digital signals, which, by definition, are sampled in time and quantized in amplitude, are generally employed in the processing procedures that use numerical (or digital) calculation tools (digital computers). In the following, only digital signals will be considered even if, for generality, they will be called discrete or discrete-time signals.

Linear, discrete-time signals (and, hence, also digital signals) are properly represented in their input–output relationship by the well-known *difference equation*:

$$y(k) = -a_1 y(k-1) - a_2 y(k-2) - \ldots a_N y(k-N) + b_o u(k) + b_1 u(k-1)$$
$$+ \ldots b_M u(k-M) \tag{1.1}$$

or even

$$y(k) = -\sum_{n=1}^{N} a_n y(k-n) + \sum_{m=0}^{M} b_m u(k-m) \tag{1.2}$$

where $u(k)$ and $y(k)$ are the discrete-time input and output signals, respectively.

Notations $u(k)$ or $y(k)$ imply that the signal is obtained after a constant rate sampling, that is, it is defined for $u(kT_s)$, where k is an integer number from $-\infty$ to $+\infty$. The T_s term, the inverse of which (f_s) must obviously respect Shannon's theorem, is not reported for simplicity of notation. The previous relationships demonstrate that the output of a linear discrete-time signal at instant k can be considered as the linear combination of inputs at M preceding intervals and the actual input (with b_m as time-constant coefficients), plus the linear combination of the outputs at N preceding intervals (with a_n as time-constant coefficients). N is the order of the discrete-time system. In LTI systems, coefficients a_n and b_m are obviously constant in respect to time. Another way to quantify the input–output relationships in linear system, through a finite number N of samples, is given by the following expression:

$$y(k) = \sum_{n=0}^{N} h(n) u(k-n) \tag{1.3}$$

which indicates the *convolution* of input signal $u(k)$ with the series $h(k)$ to obtain output signal $y(k)$ on $N+1$ samples in the discrete-time domain. Equation 1.3 is often indicated with the notation of the convolution product (*): $y(k) = h(k) * u(k)$. In this case, output $y(k)$ is given only by the linear combination of the delayed input $u(k-n)$ through terms $h(\cdot)$, which are the coefficients of the system response to an impulse input (*impulse response*). Many applications in the area of processing of signals relative to LTI systems (digital filtering, interpolators, predictors, etc.) employ Equations 1.2 or 1.3 for determined values of a_n, b_m, and $h(\cdot)$ in order to characterize the digital input-output relations, as generically indicated in Figure 1.3.

The preceding relations find an easy representation in the transform domain (Fourier and *z*-transform) (Cerutti, 1983; Oppenheim et al., 1999).

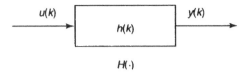

Figure 1.3. Schematic block diagram of the input–output relationship in a LTI system.

The expression $H(\cdot)$ in Figure 1.3 is generally intended as the *z-transform* of the LTI system impulse response or as the *discrete Fourier transform* (see the following section).

1.2.4.1 Fourier Analysis: Series, Fourier Integral, and Nonparametric Spectral Estimation.

Given a periodic continuous-time signal $y(t)$, it will be defined as $y(t) = y(t + kT)$ for k integer from $-\infty$ to $+\infty$, where T is the period of the periodic signal. Frequency $f_0 = 1/T$ is called the fundamental frequency of the periodic signal and $\omega_0 = 2\pi/T$ is the fundamental pulse frequency. For simplicity, the term *frequency* will be indifferently attributed to both f and ω, remembering that the two terms are related by a 2π factor (i.e., $\omega = 2\pi f$). If $y(t)$ respects Dirichelet's conditions (Ahmed and Rao, 1975), which are taken in most of the applications, we have

$$y(t) = \sum_{k=-\infty}^{+\infty} a_k e^{jk\omega_0 t}$$

The signal, under the above reported conditions, is, hence, expressed as the sum of complex exponentials (of an infinite number, in principle) and, due to Euler's theorem, of sums of sinuses and cosines. If the signal is real, the sum reduces only to real cosines (or sinuses) with a frequency multiple of k (superior-order harmonics) in respect to the fundamental frequency ω_0:

$$y(t) = a_0 + 2\sum_{k=1}^{\infty} a_k \cos(k\omega_0 t + \varphi_k)$$

The representation of these cosines is shown through the *Fourier spectrum*, the function that expresses the amplitude (a_k), the power ($P_k = a_k^2$), or the phase (φ_k) of every sinus whose sum constitutes the whole signal. Such *spectra* (of *amplitude, power,* and *phase*, respectively) are functions of frequency and are shown as "spectral rows" for a periodic signal, as illustrated in Figure 1.4. The first row refers to the DC component ($\omega = 0$), the second row to the fundamental frequency ($\omega = \omega_0$), and the others to the kth superior-order harmonics ($k\omega_0$).

The Fourier approach is also generalized even to nonperiodic sig-

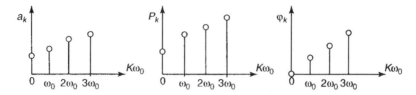

Figure 1.4. Amplitude (a_k), power (P_k), and phase (φ_k) spectra of a periodic signal. They are "row" spectra (DC component, fundamental frequency component (ω_0), and superior-order harmonics, multiple of ω_0).

nals. In this case, it is possible to start from a periodic signal and to let period T tend to infinity, thus obtaining a nonperiodic signal. It is possible to verify that in this case there is no "row spectrum" anymore; the spectrum becomes a continuous function (Fourier integral) (Oppenheim et al., 1999). It is possible to demonstrate that the spectrum becomes the envelope (continuous function) of the values of the row spectral components of the periodic signal reported above.

The Fourier integral is expressed by

$$Y(\omega) = \int_{-\infty}^{+\infty} y(t)e^{-j\omega t}\,dt$$

The preceding relation emphasizes an equivalence, an important link in nonperiodic signals when they are treated in *time domain* [$y(t)$] and in *frequency domain* [$Y(\omega)$]. $Y(\omega)$ is called the *Fourier transform* of signal $y(t)$. Such a relation is also synthetically shown as

$$y(t) \overset{F}{\leftrightarrow} Y(\omega) \tag{1.4}$$

where F indicates the Fourier transform operator. The preceding relation shows also how it is possible to obtain the *inverse Fourier transform;* that is, it is possible to pass from $Y(\omega)$ to $y(t)$ by means of a similar mathematical operation, here not shown for simplicity. Many biological signals are properly studied in the two domains indicated above.

The main problem encountered in practice in the application of Fourier theory is to consider terms that go from $-\infty$ to $+\infty$ (or from 0 to $+\infty$). Hence, there is the need to limit the number of samples to consider. As in practice we operate generally on discrete-time signals that can be represented in the frequency domain either with Fourier series (in the case of periodic signals) or with Fourier integrals (in the case of nonperiodic signals), which are discrete-frequency signals, it is possible to demonstrate that in this case (Cerutti, 1983; Oppenheim et al., 1999), the same representation is obtained in the frequency domain by introducing two different methods of limitation in the time domain (limited number of samples): (1) the signal is considered periodic from $-\infty$ to $+\infty$ with a period T equal to the finite duration of the signal; (2) the signal is non periodic (with values equal to the signal values in the considered time interval) and, for simplicity, with null (or constant) value outside the considered time interval. The two hypotheses of signal constraints for the Fourier approach lead to the same result when dealing with digital signals.

In these signals, we may obtain the discrete-frequency Fourier integral, also called *discrete Fourier transform* (DFT), which is the equivalent of Fourier transform in the previously considered case of continuous-time signals and is defined by the following expression:

$$Y(k) = \begin{cases} \displaystyle\sum_{n=0}^{N-1} y(n)e^{-jk\frac{2\pi}{N}n} & k = 0 \ldots N-1 \\ 0 & \text{elsewhere} \end{cases} \tag{1.5}$$

$Y(k)$ is the DFT (k indicates the progressive discrete-frequency term) of the discrete series $y(n)$ (n in this case indicates the discrete-time term) and N is the length of the discrete digital series (basically, N is the number of samples chosen for the signal analysis). It is possible to demonstrate that a discrete series of finite duration N has the same DFT as a periodic series of period N. Further, the DFT is periodic with period N. The DFT may have constant multiplicative terms before the sum in Equation (1.5), which depend on N and/or T_s, depending upon the magnitude assumed for $Y(k)$ terms for the successive processing (power spectrum, power spectral density, etc.), and on the opportunity to put into evidence parameters as T_s (Marple, 1987).

Without entering too much detail, with similar modalities to the ones previously described for continuous-time signals it is possible to introduce the *amplitude spectrum* [modulus of the complex value $Y(k)$], the *power spectrum,* and the *phase spectrum* by means of DFT parameters. Such spectra are obviously discrete-frequency spectra.

There are various algorithms that are applied for the spectrum estimation, either for nonparametric spectra (which basically derive from the previously introduced concept of the Fourier transform), or for parametric spectra (which are based upon a specific model of signal generation mechanism). For a deeper analysis of these topics, see Kay and Marple, 1981; Marple, 1987; and Kay, 1988.

As far as the nonparametric approach is concerned, DFT is basically considered in order to calculate the spectra (of amplitude, power, and phase). The original time series (limited in time) and the discrete-frequency series of DFT (limited in frequency) are both periodic functions of period N, due to the above considerations. As is well known, DFT calculations [and, more precisely, the values $Y(k)$] are made through the use of the algorithm of the fast Fourier transform (FFT). All the most diffused software packages of statistics or data and signal processing report the software code to calculate spectral parameters through the FFT. The so-called *periodogram* is the most common method to calculate the spectrum of a discrete signal with a nonparametric approach, as shown in the following relation (Kay and Marple, 1981):

$$P_y = \frac{1}{N} \left| \sum_{n=0}^{N-1} y(n) e^{-jk\frac{2\pi}{N}n} \right|^2 = \frac{1}{(NT_c)^2} |Y(k)^2|$$

where P_y is the power spectral density, calculated by taking the modulus values of $|Y(k)^2|$ through the FFT.

To improve the statistical significance of the periodogram, it is better to window the signal in order to obtain an efficient compromise between frequency resolution and power losses in the lateral lobes of the window. By averaging the obtained periodogram on signal-sample subsets (Bartlett and Welch methods), it is possible to obtain better spectral estimates from a statistical point of view (Oppenheim and Shafer, 1989).

It is worth remembering that in the case of a weak-sense stationary stochastic process, a relation analogous to Equation 1.4 does exist between autocorrelation function $R_y(\tau)$ of signal $y(k)$ and power spectral density $P_y(\omega)$:

$$R_y(\tau) \overset{F}{\leftrightarrow} P_y(\omega)$$

that is, the latter is the Fourier transform of the former. Even in this case, it is possible to pass from a parameter in the time domain (*autocorrelation function*) to a parameter in the frequency domain (*power spectral density*), through a Fourier transform or, vice versa, through an inverse Fourier transform.

When operating in a multivariate case [with more signals occurring contemporaneously and considering for simplicity only the bivariate case, $y_1(t)$ and $y_2(t)$], it worth mentioning that a relation analogous to Equation 1.4 does exist:

$$R_{y1,y2}(\tau) \overset{F}{\leftrightarrow} C_{y1,y2}(\omega)$$

where $R_{y1,y2}(\tau)$ is the *cross correlation* between $y_1(t)$ and $y_2(t)$ and $C_{y1,y2}(\omega)$ is the *cross spectrum* (the latter intended as the Fourier transform of the former).

Such functions are evidence of the statistical links in time and frequency domains among two (or more) different signals. Another parameter commonly employed in the multivariate analysis in the frequency domain is the *quadratic coherence k^2*, intended as the ratio between the cross-spectrum modulus and the product of the two autospectra (the coherence is basically a normalized cross spectrum). In the bivariate case, we have

$$k^2 = \frac{\left|C_{v1,v2}(\omega)\right|}{P_{v1}(\omega) \cdot P_{v2}(\omega)}$$

1.3 LINK BETWEEN PHYSIOLOGICAL MODELING AND BIOMEDICAL SIGNAL PROCESSING

One of the basic aspects of innovative approaches to biomedical signal processing is the one that strictly connects the *processing* with the *modeling* of the biological systems under study (Cohen, 1986; Bruce, 2001). It seems trivial to say that it is better for one processing a biological signal to have a more or less deep knowledge of the physiology of the system that generated the signal. However, it is possible to better formalize such a statement with the scheme reported in Figure 1.5.

In the upper part of the figure, the traditional approach of signal processing is shown. We start from the biological process under analysis, then we pass to the measurement system, which includes the signal detection, calibration, and preprocessing. The effect of noise and disturbances is indi-

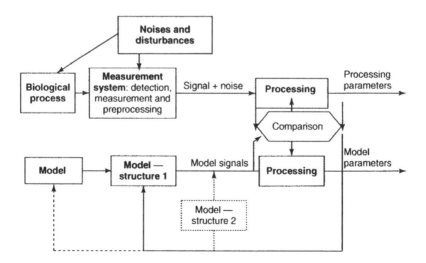

Figure 1.5. Block diagram of the possible relations existing between signal processing (upper part) and modeling of the biological system under study (lower part).

cated in the first two blocks. Then, a signal corrupted by noise is obtained, on which we can do the proper processing techniques to obtain the results (processing parameters). In the lower part of the figure, the model that is thought to be employed in this procedure is indicated. The model could be more or less complex (in the remainder of this book, different model structures will be introduced: linear, nonlinear, time-variant, time-invariant, deterministic, stochastic, chaotic, and so on, which may account for the different nuances of experimental results). The model can be defined in the first instance by using the a priori knowledge of the biological system under examination (anatomo-functional model). The model has generally a determined initial structure; hence, a comparison is made between experimental signals and signals obtained from the model, besides the comparison between signal processing parameters and model parameters. Such comparisons allow one to operate on both the process of experimental data processing and of model design (to further refine each other). A virtual feedback is therefore implemented to better define the model structure (for example, in the parameter determination or in changing the model structure or in changing also the model family itself; dotted lines in Figure 1.5). Hence, it is possible to state that signal processing can be useful for the formalization of the model and, vice versa, a proper modeling can reveal the parameters to obtain as a result of the signal processing and, hence, verify its pathophysiological relevance. It is advisable to operate with strong synergy between the two approaches.

A pertinent example to demonstrate such an important interaction is illustrated in Figures 1.6 and 1.7. Figure 1.6 shows a block diagram of the strong interconnections among autonomic regulation systems, starting from the central level (brainstem and supraspinal circuits) to the peripheral control of muscles and vessels. The figure was originally conceived by the German physiologist H. P. Koepchen (Koepchen, 1984). Figure 1.7, modified from Baselli et al. (2002), shows a scheme inspired by the preceding figure and more finalized to the localization of possible oscillatory mechanisms at low frequency (LF, around 0.1 Hz) and at high frequency (HF, the respiratory frequency in a range from 0.15 to 0.40 Hz) for the comprehension of autonomic control manifested on cardiovascular signals. With a continuous marked tract, the signals can be used to detect the information (i.e., heart rate and blood pressure variability signals and respiration). Figure 1.8 shows the block diagram used in Baselli et al. (1988) to describe the relationships among signals. In the same paper, as well as in a few of the following ones, this model has allowed investigators to determine the baroreceptive gain [modulus of the transfer function between the heart rate variability signal (t) and the one of the arterial blood pressure variability (s): H_{ts}], in a closed-loop condition, with respect to the traditional clinical measurements carried out in the "open loop" condition (such as in provocative tests, like the response to the phenylephrine bolus infusion). In

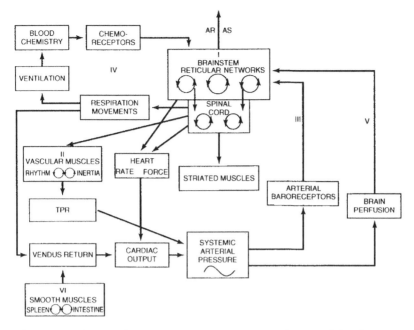

Figure 1.6. Physiological model of the complex central and peripheral interactions of autonomic control (from Koepchen, 1984. © Springer-Verlag, 1994).

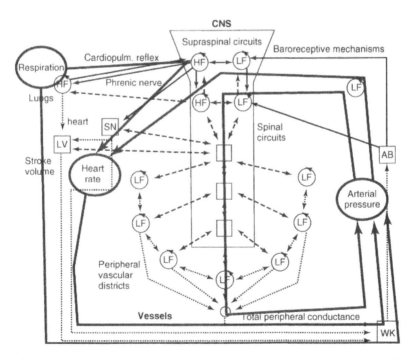

Figure 1.7. Model inspired by the structure of Figure 1.6. The biological signals from which to enhance the information for the study of cardiovascular autonomic control are evidenced by the solid line. (From Baselli et al., 2002.)

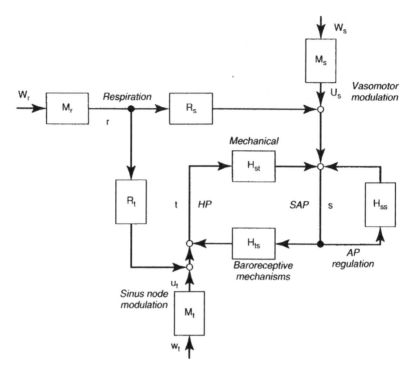

Figure 1.8. Block diagram of a double feedback loop describing the interactions among RR variability series in ECG (t), systolic blood pressure (s), and respiration (r). (From Baselli et al., 1988.)

this way, a more reliable measurement of this parameter is obtained for research as well as for clinical aims, more respectful of maintaining the physiological conditions during the test, besides allowing one to better analyze the descriptive physiological model of the complex autonomic regulation system.

The above-described case starts from a complex physiological model to get to the measurement of a parameter of a relatively "simple" phenomenon (baroreceptive gain). In the following, an opposite example will be described. We will start from a very simple (or simplified) model (even referring to the decription of baroreceptive control) to study very complex behaviors.

Kitney proposed in 1979 a model of baroreceptive control (Kitney, 1979) that accounts for some of the physiological signals introduced above. He suggested that the insertion of a delay line and a nonlinear block (a simple saturation) on a negative feedback loop circuit induced oscillations on arterial blood pressure signals, which, in normal conditions, are maintained within a certain physiological interval. Signorini and Cerutti (1997) brought some changes to the model, as reported in Figure 1.9. For simplicity, the respiration signal is schematized as a sinusoid of ε amplitude and f frequency. The vasomotor center activity is described by the second-order transfer function $G(j\omega)$ with the delay line e^{-Ts}, where y is the arterial blood pressure signal. As expected, the system, according to the values assumed by the closed-loop gain and the respiratory frequency, shows oscillations at determined values of frequency. The simple nonlinearity induced in the forward path gives rise to a very complex behavior.

Figure 1.10 shows in the plane $H_1 - f$ the various zones characterized by different ratios in the oscillation mechanisms. As an example, in the lower part of the figure (for lower values of H_1) a phenomenon of *entrainment* is noticed, that is, arterial blood pressure oscillations do synchronize with the respiratory frequency (which is the external forcing term). With increasing H_1, different behaviors are visible according to the value of respiratory frequency f. Four different frequency values are particularly evidenced (in correspondence with the four vertical notches), with passages from periodic behavior (small values of H_1) to torus, to cycle bifurcations, to torus breakdown, from period 1 to period 3. The more common range of physiological respiration rate (0.15–0.25 Hz) may present chaotic behaviors of the baroreceptive control system. Such hypothesis might also be verified by the measurement of parameters of complexity and chaoticity, carried out in an experimental protocol using recordings of respiration signal, heart rate, and arterial blood pressure. It was demonstrated (West, 1990) that chaotic behavior in biological systems very often is noticed in physiological conditions (and becomes less chaotic in pathological conditions). As a consequence, an important implication can be deduced indicating a better adaptation of a chaotic system to the various conditions to which the system itself can be subjected, besides

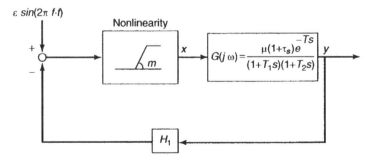

Figure 1.9. Simplified model of an arterial blood pressure signal (y) through a negative feedback loop with a nonlinearity in the forward path. (From Kitney, 1979; and Signorini and Cerutti, 1997.)

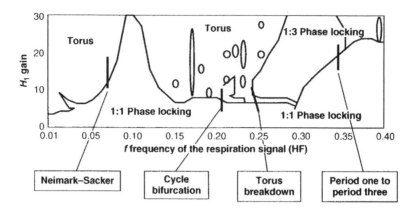

Figure 1.10. Bifurcation diagram of the model depicted in Figure 1.9, reported on axes H_1 and f. (From Signorini et al., 1997.)

satisfying more completely the "fault-tolerance" characteristics of the overall system.

In other chapters of the present volume, nonlinear characteristics of the signals will be dealt with in more detail, including fractal and chaotic properties.

1.4 THE PARADIGM OF MAXIMUM SIGNAL–SYSTEM INTEGRATION

Important developments in medicine, biology, and the life sciences tend to integrate as much as possible the information coming from different environments. The hyperspecialization in medical care as well as the marked technological development in equipment, devices, and diagnostic and therapeutic procedures, provide much and different information on the patient, sometimes much fragmented, coming from exams, reports, and different clinical experiences, which must be successively integrated for an effective comprehension of physiological phenomena under study and/or to make the most correct decision for the patient. From the standpoint of signal and image processing, that implies great attention not only to the phase of extraction and validation of the obtained parameters, but also to how these parameters can be correlated with each other, thus producing an important synergy in the information content that might be decisive in reaching a correct diagnostic decision. If, on one hand, the knowledge tools are increased, on the other hand it is fundamental to be able to integrate the obtained data in order to infer new information and to verify the effective usefulness of new data.

The fields in which it is suitable to fulfill such an integration are numerous and some examples will be reported in the following sections.

1.4.1   Integration among More Signals in the Same System

To be able to detect more signals contemporaneously from the same physiological system means in many cases to improve considerably the available information. As an example, in order to make a reliable diagnosis in cardiology in many circumstances it is necessary to make a 12-lead ECG analysis, even if from a technical point of view the information is redundant, as it is well known that the 12 standard ECG leads are not linearly independent (Webster, 1998). In ambulatory patients (as well as, for other reasons, in intensive care unit patients), the number of required leads is certainly lower, so, on the contrary, it could be necessary to have an ECG mapping analysis (with various tenths of leads) in particular forms of ischemic disease or myocardial infarction (Burnes et al., 2000). The vectorcardiogram (VCG), with its three orthogonally independent components of the electrical dipole cardiac momentum, represents certainly a better information compression, with no redundancies, even if its clinical application is actually not largely diffused.

In the same way, in EEG analysis, it is traditional to operate with the 10–20 standard leads and by considering only a few central, right, or left leads for routine analysis. In high-resolution EEG, instead, 64 or even 128 scalp leads are employed when a more detailed electrical map (with a better spatial resolution) is required (Babiloni et al., 1996).

What it is more important is to be able to calculate multivariate relations from this kind of analysis, which could not be obtainable when considering a single lead as independent from the others. As examples of these relations, we may mention *cross correlation, cross spectra, partial* or *total coherence,* as well as nonlinear parameters such as *cross entropy* or nonlinear *synchronizations.*

Further, one may also consider not only more realizations (leads) of the same biological phenomenon but even different signals relative to the same system. It is well known that in the study of various cardiac pathologies it is useful to operate on signals like ECG, arterial blood pressure, and respiration, not only by detecting informative parameters from the three single signals (in time and/or in frequency domains) but also multivariate parameters (even, in this case, coherences, entropies, and synchronizations), which allow one to get more specific information on the functioning of the autonomic nervous system and, hence, to obtain a more targeted diagnosis and therapy for pathologies such as hypertension, cardiac failure, and ischemic disease. This aspect will be properly dealt with in other chapters of the present volume.

1.4.2 Integration among Signals Relative to Different Biological Systems

It is very important from the research and clinical points of view to be able to integrate information coming from signals relative to different physiological systems. Section 1.4.1 took into consideration signals of the cardiovascular and respiratory systems together. Very significant relations can be obtained from the contemporaneous analysis of autonomic parameters together with parameters of the central nervous system. Figure 1.11 shows how it is possible to detect common rhythms in the LF frequency band (around 0.1 Hz) and HF frequency band (0.15–0.40 Hz, at the respiratory rate) in the spectra of various signals coming from different physiological systems. In many cases, a high coherence value is also obtained (not shown in figure). The top four panels in the figure refer to neurophysiology experiments in the cat and are: NA, activity of a single neuron in the bulb; SAPV, variability of systolic blood pressure; PGNA, neural efferent sympathetic activity in postganglionar fibers; and RRV, heart rate variability (on RR intervals). The bottom four panels are instead obtained from physiological experiments on patients: MSNA, mean sympathetic neural activity measured at the level of the peroneal nerve; RRV, heart rate variability (on RR intervals); SAPV, variability of systolic blood pressure; and R, respiration signal, via the thoracic belt.

Figure 1.12 refers to a patient who manifests myoclonus during sleep (stage 2 CAP). In this case, there is a precise and unexpected synchronization between the *arousal* mechanisms (on EEG signal) and the myoclonus *spike* (on the EMG signal, measured at the level of the tibial muscle). There is also a synchronous involvement of the autonomic nervous system with tachy–brady phenomena (on RR signal) and respiration modulation (on the R signal).

Actually, various branches of semiotics and clinical medicine are attributing a noticeable importance to the so-called "*dynamical diseases.*" These are those pathologies that are caused in a more specific way, by damage to some controlling system, rather than single or multiorgan damage. In these cases, important interactions among the central nervous system, autonomic nervous system, and endocrine–metabolic system may play an important role in a correct diagnosis (Mackey and Milton, 1987; Glass and Mackey, 1988). Pathologies such as certain types of hypertension, ischemic diseases, and diabetes can be caused by deficits in the cardiovascular, cardiac, and metabolic controls, with strong interconnections from the pathophysiological standpoint. In the last few years, various methods of information processing have been developed for signals coming from different biological systems, albeit interacting with one another due to the concept of dynamical disease: linear multivariate methods stay together with nonlinear approaches based upon fractal theory and deter-

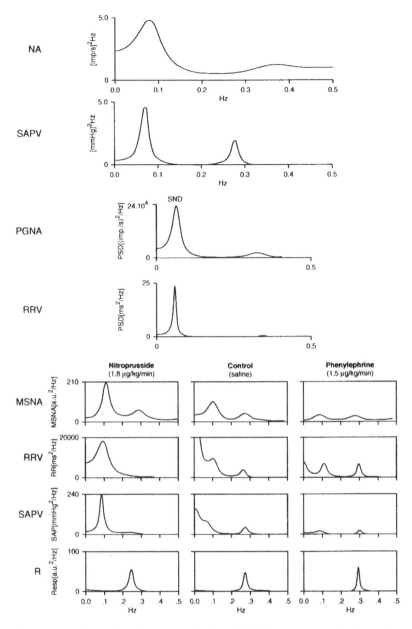

Figure 1.11. Examples of common rhythms (LF, HF) on spectra relative to signals coming from different physiological systems. The top four panels refer to a cat: NA, neural activity of a single neuron from the bulb; SAPV, systolic arterial pressure variability; PGNA, preganglion efferent neural activity; RRV, RR variability. The last four panels refer to a human case: MSNA, mean sympathetic neural activity at the peroneal nerve; RRV, RR variability; SAPV, systolic arterial pressure variability; R, respiration. (From Montano et al., 1992, 1996; Pagani et al., 1997.)

23

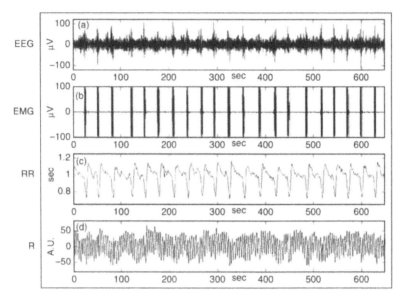

Figure 1.12. Signals recorded on a patient suffering from nocturnal myoclonus. From top to bottom: EEG (sleep stage 2 CAP, characterized by frequent pseudoperiodical arousals); EMG activity recorded at the tibial muscle; RR variability on ECG signal; respiration signal.

ministic chaos. This is a key example of how advanced concepts in medicine may find important applications when innovative methods of information processing are employed. The most important clinical trials actually designed for the diagnosis and treatment of important cardiovascular pathologies employ many of the parameters that have been validated in the course of recent years in medical and bioengineering researches, from the simple variance parameter of heart rate variability signals, to the spectral analysis in the frequency domain, to the measurement of self-similarity parameters (also called the fractal dimension), to entropy or deterministic chaos, up to the time-frequency or time-scale techniques. The present volume presents an interesting review of some of these methods.

1.4.3 Integration (or Data Fusion) from Signals and Images

A further step ahead toward a better integration of biomedical information is the fusion of signals and images. Figure 1.13 shows four bioimages detected through different modalities (multimodality): X-ray computerized tomography (CT), magnetic resonance imaging (MRI), single-photon emission tomography (SPECT), and positron emission tomography (PET).

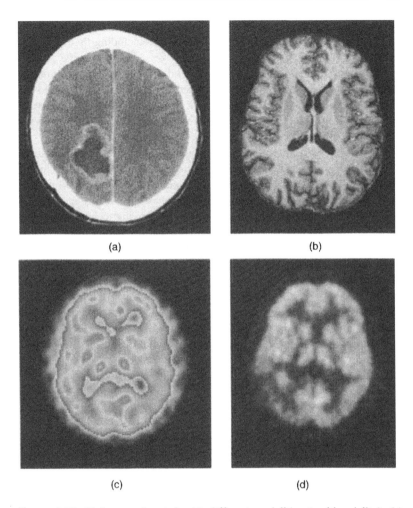

(a)   (b)

(c)   (d)

Figure 1.13. Bioimages detected with different modalities (multimodality): (a) CT, computerized X-ray tomography; (b) MRI, magnetic resonance; (c) SPECT, single-photon emission computerized tomography; (d) PET, positron emission tomography (courtesy of Scientific Institute Ospedale San Raffaele, Milan).

There are techniques that allow one to match the various obtained images, that is, to apply a geometric transformation of images in order to obtain a pixel-to-pixel (or voxel-to-voxel) correspondence of the same anatomical structure in different studies (in Figure 1.13, it is the brain), with also a possible correlation with monodimensional signals.

The *matching techniques* are based upon various approaches [external marker correlations, correspondent anatomical structure correlations, pixel-to-pixel (or voxel-to-voxel) analysis, and so on], and they can be implemented in two, three, and four dimensions (including the time variable).

Figure 1.14 shows a study of MRI and PET, originally acquired in two dimensions, matched and reconstructed in three dimensions, and successively superimposed on each other (Rizzo et al., 1994, 1997; Cerutti and Bianchi, 2001; other significant examples are given in Fazio and Valli, 1999).

The integration between biomedical signals and images is illustrated in Figure 1.15, where, starting from the functional MRI and from the potential distribution of EEG signals (32 channels) on the scalp, a unique image is obtained from the fusion between the two modalities during a spike occurrence in a focal epileptic patient. Deblurring techniques for the Laplacian solution are employed for the potential distribution of EEG signals matched on the realistic model obtained from the fMRI, thus obtaining an efficient integration between the cortical map and the geometrical structure of the cortex. In this way, the integration of information of high spatial resolution (of the MRI) with information of high temporal resolution (of the EEG mapping) is achieved. A further processing of EEG mapping may allow us to determine the potential sources and to localize their positions inside the cerebral mass on the 3-D MRI image.

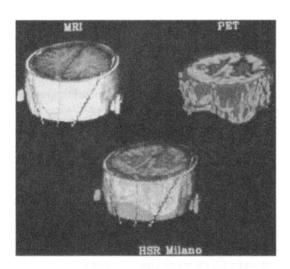

Figure 1.14. MRI and PET studies detected in two dimensions, successively reconstructed in three dimensions, and matched to each other (Rizzo et al., 1998).

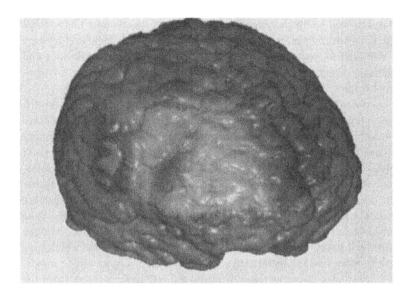

Figure 1.15. Matching of a realistic brain MRI image with the distribution of high-resolution EEG cortical potentials (32 channels) in an epileptic patient, in correspondence to an epileptic spike.

1.4.4 Integration among Different Observation Scales

Most often in modern clinical medicine, the useful information is thought to be distributed along different observational scales: from the genome sequence, to the proteic expression involved in the pathophysiological mechanisms under study, to the single-cell scale, to the entire cell set, to the organ under study, to the entire biological system, and so on. Perhaps a strong change in the physician's and biologist's cultures is indeed necessary to succeed in putting into practice a kind of Copernican revolution in which man as a whole is the focus of attention and not a disorganized set of data, reports, signals, and images. It is necessary also that biomedical engineers and scientists contribute to bring about such a revolution, which means, in this context, to be able to integrate the information obtained on different scales through the use of proper algorithms of investigation and inference.

Starting from the information that can be enhanced from the genome–proteome level, it is worth remembering that the traditional and advanced methods that actually are applied to the biomedical signals also have important applications in molecular biology and nanomedicine (see the Special Issue of *IEEE Proceedings,* Akay, 2002) with a wide variety of more or less sophisticated statistical methods. In the present book, important results will be illustrated along this direction. It is thought that such an approach will bring important benefits, not only to medicine and biology but will lead to new trends in the singling out of new biomolecules and proteins in different application areas (food farming, drug design, environmental studies, etc.).

To demonstrate how the information processing carried out on different integrative scales may bring certain benefits leading to better comprehension of the pathophysiological mechanisms, it is worth mentioning the case of the long-QT syndrome. It is a pathology detected by ECG analysis in which a dysfunctional prolongation of the ventricular repolarization phase is noticed (long QT interval). A genetic predisposition has been discovered recently; a possible mutation in some genes that are deputies, mainly to the functionality of Na^+ and K^+ ionic channels, have been observed in the population of long-QT carriers. Such a genotype manifests a strong predisposition to serious tachyarrhythmias, which may cause syncopes and sudden death (Priori et al., 2003). It would be important to find possible correlations between the mutations suspected for the genesis of this pathology (i.e., KVLQT or HERG), electrocardiographic parameters (i.e., QT variability, links with RR sinusal rhythm variability, turbulence, etc.), and ultrasound ecographic parameters (ventricular volumes, contractility, possible presence of arrhythmic substrates), as well as with the integration of patterns in cardiac SPECT (or PET or, more recently, MRI) in order to quantitatively evaluate the functioning of the ionic channels involved.

Another example is connected with the study of one of the most important degenerative pathologies of the central nervous system: Alzheimer's disease. The research has undoubtedly made noticeable progress in the fusion of EEG or EP/ERP mapping with PET/SPECT images to better localize the cerebral areas that present a greater degeneration (hippocampus, cortical associative areas, etc.). Some neural populations are more prone to the deposit of β-amyloid structures; signal and image fusion can better single out such areas and, hence, validate a possible correlation with the presence of specific genotypes (apolipoprotein E or preselinin) (Cacabelos et al., 2003). Further examples might be found for important cardiovascular pathologies (i.e., hypertension or cardiac failure) or dysmetabolic disease (i.e., diabetes) in which many important genotype–phenotype relations have been found whose implications are worthy of further investigation.

1.5 CONCLUSIONS

In conclusion, it is important to remark that biomedical signal processing requires a clearly interdisciplinary approach, as in other application areas of biomedical engineering. In fact, it is fundamental to be able to integrate knowledge of physiological modeling, data, and signal and image process-ing coming from different systems and different observational scales. this was the motivation for the slightly provocative title of the preceding sec-tion, in which a paradigm of maximum information integration was called for. Among other advantages, it is worth noting that such an approach based on advanced processing techniques could lead to noticeable im-provements from the clinical standpoint toward, paradoxically, a holistic vision of the patient and not an atomistic one. The patient is seen as a unique entity in which all the information could be properly integrated in a virtual model but, at the same time, a realistic one (based upon objective data from the patient himself) and implemented on a proper informative support, not as a set of a sometimes disordered collection of reports, data, and images.

An innovative concept of a clinical information system can be thought of, which might interact with a holistic medical record based upon the integration of objective elements obtained by means of data and signal and image processing relative to each single patient, in the various infor-mation scales. The record should be knowledge-based, with inference of data from objective protocols of evidence-based medicine and from statis-tical and epidemiological data. An important step along this direction is suggested by the Physiome Project, which is intended to define relation-ships from genome to organism and from functional behavior to gene reg-ulation. In this context, the Physiome Project includes integrated models of components of organisms, such as particular organs or cell systems, bio-hemical systems, or endocrine systems (Coatrieux et al., 2006).

Finally, there is no doubt that researchers should be able to capitalize the accessible information as much as possible to improve their knowl-edge. They must take advantage of the obtained synergies, but also identi-fy nonrelevant data and redundancies. As is well known, even in the life sciences, it is not true that to improve knowledge, "the more information, the better." This delicate passage from information to knowledge, so use-ful for physiological or clinical purposes, will be dealt with in Chapter 2.

REFERENCES

Ahmed, N., and Rao, K. R., *Orthogonal Transforms for Digital Signal Processing,* Springer-Verlag, 1975.

Akay, M., Special Issue on Bioinformatics—Part I. Advances and Challenges; Part II, Genomics and Proteomics Engineering in Medicine and Biology, *IEEE Pro-ceedings,* Vol. 90, No. 11, 2002.

Babiloni, F., Babiloni, C., Carducci, F., Fattorini, L., Onorati, P., and Urbano, A., Spline Laplacian estimate of EEG Potentials over a Realistic Magnetic Reso-nance–Constructed Scalp Surface Model, *Electroencephalogr. Clin. Neuro-physiol.,* Vol. 98, No. 4, pp. 363–373, 1996.

Baselli, G., Caiani, E., Porta, A., Montano, N., Signorini, M. G., and Cerutti, S., Biomedical Signal Processing and Modeling in Cardiovascular Systems, *Crit. Rev. Biomed. Eng.*, Vol. 30, No. 1-3, 55–84, 2002.

Baselli, G., Cerutti, S., Civardi, S., Malliani, A., and and Pagani, M., Cardiovascular Variability Signals: Towards the Identification of a Closed-Loop Model of the Neural Control Mechanisms, *IEEE Trans. Biomed. Eng.*, Vol. 35, No. 12, pp. 1033–1046, 1998.

Bendat, J. S., and Piersol, A. G., *Random Data: Analysis and Measurement Procedures*, 3rd ed., Wiley, 2000.

Box, G. E. P., and Jenkins, G. M., *Time Series Analysis: Forecasting and Control*, Holden-Day, 1970.

Bruce, E. N., *Biomedical Signal Processing and Signal Modeling*, Wiley, 2001.

Burnes, J. E., Taccardi, B., MacLeod, R. S., and Rudy, Y., Noninvasive ECG imaging of electrophysiologically abnormal substrates in infarcted hearts: A model study, *Circulation*, Vol. 8, No. 101(5), pp. 533–540, 2000.

Cacabelos, R., Fernandez-Novoa L, Lombardi V, Corzo L, Pichel V, and Kubota Y, Cerebrovascular Risk Factors in Alzheimer's Disease: Brain Hemodynamics and Pharmacogenomic Implications, *Neurol. Res.*, Vol. 25, No. 6, pp. 567–580 2003.

Cerutti, S., *Filtri Numerici per l'Elaborazione di Segnali Biologici*, CLUP ed, Milano, 1983.

Cerutti. S., and Bianchi, A. M., Metodi e Tecniche Innovative di Elaborazione di Segnali e Immagini Biomediche: Nuovi Strumenti per Migliorare le Conoscenze in Fisiologia e in Medicina, Alta Frequenza, *Rivista di Elettronica*, Vol. 13, No. 1, pp. 45–51, 2001.

Coatrieux, J. L., and Bassingthwaighte, J., Special Issue on The Physiome and Beyond, *IEEE Proceedings*, Vol. 94, No. 4, pp. 671–853, 2006.

Cohen, A., *Biomedical Signal Analysis*, Vols. I and II, CRC Press, 1986.

Fazio, F., and Valli, G. (Eds.), *Tecnologie e Metodologie per le Immagini Funzionali*, Pátron Editore, Bologna, 1999.

Glass, L., and Mackey, L. C., *From Clocks to Chaos: The Rhythms of Life*, Princeton University Press, 1988.

Kay, S.M., and Marple, S. L. M., Spectrum Analysis: A Modern Perspective, *IEEE Proc.*, Vol. 69, No. 11, pp. 1380–1419, 1981.

Kay, S. M., *Modern Spectral Estimation: Theory and Application*, Prentice-Hall, 1988.

Kitney, R. I., A Nonlinear Model for Studying Oscillations in the Blood Pressure Control System, *J. Biomed. Eng.*, Vol 1, No. 2, pp. 88–89, 1979.

Koepchen, H. P., History of Studies and Concepts of Blood Pressure Wave, in *Mechanisms of Blood Pressure Waves*, Miyakawa K., Koepchen, H. P., and Polosa C. (Eds.), pp. 3–23, Springer-Verlag, 1984.

Ljung, L., *System Identification: Theory for the User*, Prentice-Hall, 1998.

Mackey, M. C., and Milton, J. G., Dynamical Diseases, *Ann. NY Acad. Sci.*, Vol. 504, pp. 16–32, 1987.

Marchesi, C., *Tecniche Numeriche per l'Analisi di Segnali Biomedici*, Pitagora Editrice, Bologna, 1992.

Marple, S. M., *Digital Spectral Analysis with Applications*, Prentice-Hall, 1987.

Montano, N., Lombardi, F., Gnecchi Ruscone, T., Contini, M., Finocchiaro, M. L.,

Baselli, G., Porta, A., Cerutti, S., and Malliani, A., Spectral Analysis of Sympathetic Discharge, R-R Interval and Systolic Arterial Pressure in Decerebrate Cats, *J. Auton. Nerv. Syst.*, Vol. 40, No. 1, pp. 21–31, 1992.

Montano, N., Gnecchi-Ruscone, T., Porta, A., Lombardi, F., Malliani, A., and Barman, S. M., Presence of Vasomotor and Respiratory Rhythms in the Discharge of Single Medullary Neurons Involved in the Regulation of Cardiovascular System, *J. Auton. Nerv. Syst.*, Vol. 57, No. 1–2, pp. 116–122, 1996.

Oppenheim, A. V., and Shafer, R. W., *Digital Signal Processing,* Prentice-Hall, 1989.

Oppenheim, A. V., Shafer, R. W., and Buck, G. R., *Discrete-Time Signal Processing.* 2nd ed., Prentice-Hall, 1999.

Pagani, M., Montano, N., Porta, A., Malliani, A., Abboud, F. M., Birkett, C., and Somers, V. K., Relationship between Spectral Components of Cardiovascular Variabilities and Direct Measures of Muscle Sympathetic Nerve Activity in Humans, *Circulation,* Vol. 95, No. 6, pp. 1441–1448, 1997.

Papoulis, A., Probability, *Random Variables and Stochastic Processes.* McGraw-Hill, 2002.

Priori, S. G., Schwartz, P. J., Napolitano, C., Bloise, R., Ronchetti, E., Grillo, M., Vicentini, A., Spazzolini, C., Nastoli, J., Bottelli, G., Folli, R., and Cappelletti, D., Risk stratification in the Long-QT Syndrome, *N. Engl. J. Med.,* Vol. 348, No. 19, pp. 1866–1874, 2003.

Rangayyan, R., *Biomedical Signal Analysis: A Case-Study Approach,* Wiley, 2002.

Rizzo, G., Gilardi, M. C., Prinster, A., Lucignani, G., Bettinardi, V., Triulzi, F., Cardaioli, A., Cerutti, S., and Fazio, F., A Bioimaging Integration System Implemented for Neurological Applications, *J. Nucl. Biol. Med.,* Vol. 38, No. 4, pp. 579–585, 1994.

Rizzo, G., Scifo, P., Gilardi, M. C., Bettinardi, V., Grassi, F., Cerutti, S., and Fazio, F., Matching a Computerized Brain Atlas to Multimodal Medical Images, *Neuroimage,* Vol. 6, No. 1, pp. 59–69 1997.

Signorini, M. G., and Cerutti, S., Bifurcation Analysis of a Physiological Model of the Baroreceptive Control, in *Frontiers of Blood Pressure and Heart Rate Analysis,* Di Rienzo et al. (Eds.), IOS Press, Amsterdam, pp. 29–43, 1997.

Tompkins, W. J., *Biomedical Digital Signal Processing,* Prentice-Hall, 1995.

Webster, J. G., *Medical Instrumentation: Application and Design,* 3rd ed., Houghton Mifflin, 1998.

West, B. J., *Fractal Physiology and Chaos in Medicine,* World Scientific, 1990.

Thispageintentionallyleftblank

CHAPTER 2

DATA, SIGNALS, AND INFORMATION
Medical Applications of Digital Signal Processing

Carlo Marchesi, Matteo Paoletti, and Loriano Galeotti

2.1 INTRODUCTION

Digital signal processing in clinical applications provides information on the performance of the systems that constitute living organisms, particularly for obtaining a quick documentation of the onset of dangerous conditions. This processing is realized through algorithms, procedures, and autonomous devices, or incorporated in more extended systems, which are differentiated by the needs typical of the reference environment.

The efficacy and, hence, the utility of data analysis as a whole depends on the quality of technology and its development. It is possible to say that technology's impact on signal processing in medicine is improving due both to the increasing number of new procedures and their reliability, thus allowing a better analysis of known situations. It is also true that the increasing cost of healthcare (and in particular of hospitalizations), which is becoming untenable, is a critical element that contributes the need to accelerate this innovation.

The recommendations of the World Health Organization (WHO) are particularly influential and encourage the use of the home environment as a nonepisodic site to provide health assistance (OMS, 2004; Il Sole 24 Ore, 2004). Such action is accompanied by precise suggestions (i.e., the five As: Assess, Advise, Agree, Assist, Arrange), intended as guidelines for the future of health assistance development based upon patient's collaboration, though "an approach that emphasizes collaborative goal setting, patient skill building to overcome barriers, self-monitoring, personalized feedback, and systematic links to community resources."

In this chapter, we discuss the basic future trends of digital signal processing in biomedical applications, according to our experience and by recalling a selected sample of the dedicated scientific literature.

2.2   CHARACTERISTIC ASPECTS OF BIOMEDICAL SIGNAL PROCESSING

2.2.1   General Considerations Based on Actual Applications

A few characteristic elements of biomedical signal processing for clinical purposes can be summarized as follows:

- The information indicating an alteration of physiological signs in a pathology is often latent and standard stimuli are required to put it into evidence and to evaluate it properly. That extends the investigation to the developmental aspects of the pathology and to its dynamics, which have to be considered together with the statistical data (i.e., ergonometric data, ECG Holter data, and so on).
- Some pathologies also influence the structure of data and signal waveforms, which are strongly affected by the clinical information that has to be evidenced and evaluated properly. Many papers, therefore, concentrate on the analysis of single waveforms.
- Biomedical data and signal analysis has to be capable of detecting more patterns widely differentiated even in normal cases (e.g., daily patterns might be different from nightly ones). There is a further problem with respect to the false approach of considering only one "typical" normality class and excluding everything else as basically noise.
- Taking into account that we have to deal with human subjects, the preservation of performance quality must be guaranteed on one hand by a greater robustness of analysis methods and on the other by improvement in obtaining a generalization of the results. This has to be done via very rigorous methods of evaluation.
- These problems are approached through procedures that consist generally of a sequence of operations that evidence, single out, measure, transform, and classify the information contained in the data and signals in such a way as to facilitate clinicians' interpretation.
- The main operations are:
  Signal conditioning, A/D conversion, and preprocessing.
  Segmentation of processing subunits—one or more cycles if the signal is repetitive or pseudoperiodical, or one or more sections or epochs where it could have precise statistical properties (e.g., sto-

chastic and stationary)—and their characterization (parameter extraction).

Analysis of events of interest contained in one or more segments.

Classification useful for parameter extraction by the physician as an objective aid to his/her decision making. When required, signal and data information have to be integrated with the data relative to the patient and to his/her reference environment.

These operations can be executed by very different methods and can be applied to different signals for different aims, well distributed in extended fields of application (Mainardi et al., 1995).

In order to estimate the trend of themes and methods of real clinical applications, it is more convenient to refer to some texts on biomedical digital signal processing (Baura, 2002; Bruce, 2001; Rangaraj, 2002).

Without pretending to deal with the topic of automatic classification of documents, which is not convenient to go deeply into here, it is proper to remark how inadequate is the criterion of a two-parameter taxonomy (method/problem), even if well diffused in the literature. In an ongoing study, the possible advantage of a *global* classification by means of *frames* [called *ambients* (Paoletti and Marchesi, 2002)] is considered. There, a definition is proposed that uses degrees of complexity of system analysis, based upon personalized processing. Other researchers have expressed similar opinions; as an example, a recent paper proposes a classification based on five principal categories, distributed over 34 attributes (Coatrieux, 2004).

2.2.2 Some Results of the Review

It is clear that over the years, the development of sensor physics and technology, signal theory, and processing tools has happened coherently according to the level of approached medical problems, even if there is still a noticeable gap due to the slowness by which quantitative methods are successfully established in medicine. Such slowness is well motivated by the fact that a new method, algorithm, or procedure, if intended as a contribution to innovation in a clinical application, must be integrated in a system that improves the function. Performance evaluation in the laboratory is, hence, more complex but it is also more in-depth, as its reliability (meaning *dependability*) must be certified. From an examination of the literature and from our own experience, we want to stress the following observations (Baura, 2004; Bruce, 2001; Mainardi, 1995).

The traditional themes of signal processing are still active. In general, innovation concerns the continuously decreasing dimensions and costs of equipment destined to reach an increasing number of people with different pathologies, as well as people engaged in high-risk activities or sports.

In some cases, as in EEG signals, the availability of new methods and the relative implementations in packages of wide diffusion have revitalized the applications, in particular in the area of anaesthesia-level control.

Some more mature applications have found renewed interest, for example, heart rate variability, which was basically quantified as an index of dispersion of mean heart rate up to few years ago but is now the topic of sophisticated heart modeling studies, both linear and nonlinear.

A series of new methods of potential clinical interest appear continuously in the scientific environment. This book does introduce a spectrum of them and, therefore, it is advisable to recall the introductory lesson in the preceding chapter, as well as other lessons. Here, it is sufficient to cite, among the most innovative contributions, the study of complex systems and the fractal and chaotic magnitudes in medical applications; this is an important sign that a fundamental synergic rapprochement is occurring between theoretical modeling aspects and the empirical ones typical of the signal analysis approach.

Such a link may provide a path toward a unifying paradigm between the two different methods. It deals with the evolution of the error concept as a quality reference of the performance of an analysis method. In the very first applications in biomedical signal processing of about 30 years ago, it was advisable that the processing system work properly, as the implementation phase of algorithms in programs or in devices was very difficult. Quality problems were secondary due to limitations in knowledge and technological support. After the first experiences, it became very clear that the simple good functioning did not produce significant, useful results, especially for medicine.

Starting from analysis methods realized by a comparison of thresholds determined via optimality criteria, the need for certifiable quality increased. This could be made possible if an annotated database (acknowledged by independent experts) were recognized as a de facto standard. A preeminent role in this area has been played by the Physionet Project, jointly carried out by MIT and NIH, which has undertaken a few projects (Moody et al., 2001; Jager et al., 2003; Raab, 1995).

Given the effective impossibility of evaluating their utility, there is a marked trend to search for more and more sophisticated methods and procedures, with the objective of detecting the error involved and optimizing their utility according to some figure of merit, attributing less importance to the evaluation of performance. The medical environment has become more and more competent and demanding, due to the competitive pressure of using proper advanced equipment and computation means. In fact, the difference between quality and utility has been evidenced and it has been clearly stressed that a technological solution must constitute an answer to a medical question.

In particular, quality criteria expressed by the theory of signal detection might be extended from the level of algorithm performance to the global one of developing sensitivity and specificity, applied not only to signs compatible with some diseases but also to their clinical value. It is possible to sum up this issue by saying that it is advisable to have not only precise quantification tools but also accuracy in interpreting their findings.

Further important elements are:

- A culture combining systems signals with statistics (i.e., multivariate analysis and logistic regression) and information science (i.e., knowledge representation, decision theory, and data mining). Probably such an expansion of competences has many causes, among which are the deep changes in the composition of people who deal with applications in medical field. In short, we have passed from an approach of component-driven requirements to whole-system requirements. This level of change leads to the need for cultural integration.
- The generalization of analysis methods from research to clinics. This is a result obtainable by taking into account the need for quality and utility (see next section).
- Measurement errors due to an improper use of automatic equipment.
- Incompetence in physiology. In order to correctly interpret the data, it is necessary to be updated on the system knowledge under study. This could, for example, imply the generation of signals affected by high variability or noise that could mask the changes induced by a disease and, hence, mislead nonexperts.

2.3   UTILITY AND QUALITY OF APPLICATIONS

The documentation of utility of an application requires validation inside the medical environment, based upon a prefixed epidemiological protocol. The technical designer has the obligation to guarantee the quality of the application and the updating of employed methods. It is important to make aware to the stakeholders the serious problem of data quality; this last section is therefore focused on some aspects of data preprocessing and the most important variables that contribute to the quality of information processing (Jager et al., 2003; Cherkassky and Mulier, 1998; Long, 1993; Findler, 1991).

2.3.1   Input Information

First of all, it is advisable that signals be acquired in a long temporal window and in a comfortable way, that is, through wearable devices with suitable sensors. A series of operations follow that result in the transformation of the signals into a vector of parameters (*features*) whose choice is made following rational and general criteria. In some areas, heuristic parameters are preferred when they are confirmed by a long tradition of diagnostic value, and in other cases, according to a more recent technology, parameter extraction criteria are employed, which refer to a unitary criterion [i.e., singling out of the interval tachogram in ECG signals—RR beat-to-beat series—to calculate power spectral density (Esprit, 1996; Anonymous, 1999)].

We believe that the demand that new technologies decrease the number of recoveries depends upon the quality of the realizations. If each individual, as recommended by WHO, must play a primary role in his/her health management, realization quality means a reliable response of the system to the onset of dangerous symptoms. It is opportune to make a diagnosis based only upon some signals, but such a choice would obviously not be reliable; it is necessary to proceed with the analysis of data coming from different sources. In short, signal processing cannot be considered an "autonomous" function but rather must be integrated with the data characteristics of the patient. In the following, those methods suitable for this kind of global personalization will be illustrated. The following problems will be discussed:

1. Data heterogeneity
2. Most discriminative variable choice (parameters and features)
3. Reduction of the dimensions of original space parameters and of transformed variables as well
4. Graphic evaluation in a reduced space
5. Alarm generation systems

In order to provide concrete details of this problem, an example will be introduced relative to heterogeneous data in cardiac patients (Cleveland archives), by using methods illustrated in the following sections (see Figure 2.1).

2.3.2   Data Heterogeneity

This problem is approached in different ways with the objective of reaching total nonhomogeneity or partial nonhomogeneity, total representation through dichotomic variables, or making the data dichotomic (continuous variables fractionated in interval values that represent as many categories). In these three cases, many metrics have to be defined that allow the comparison and the aggregation of the patterns to be classified (Michailidis and De Leeuw, 2000).

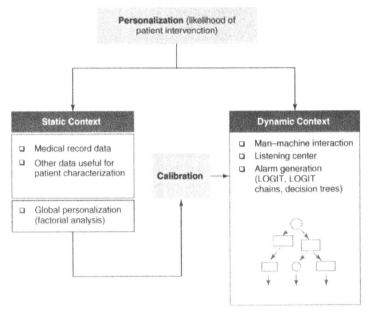

Figure 2.1. Block diagram of global personalization (Pyle, 1999; Ortega et al., 2004; Techmath, 2004). Medical record data are made homogeneous to apply a factorial analysis method (in the illustrated case, the method of multiple correspondences). Factorial analysis provides the subset of more discriminant variables. This result is used in the calibration of the portable analysis system and influences the alarm strategy.

Minkowski's metric is very often used as a convenient measure of dissimilarity for various methods of integration. Its analytical expression is the following:

$$D_n\left(x_j, x_k\right) = \left(\sum_{i=1}^{d}\left|x_{ij} - x_{ki}\right|^n\right)^{\frac{1}{n}}$$

where D_n is the distance and x_{ij} represents the ith parameter of the jth element of the set of data. When varying n, Minkovski's distance reduces to well-known particular cases: for $n = 1$, the *city-block distance* or *Manhattan's distance;* for $n = 2$, the classical Euclidean distance; and for n which tends to infinity, *Chebichev distance* or even *infinite norm* (Kaufman, 1987, Ichino, 1992).

   Let us suppose that one object is described in terms of d parameters, $X_k, k = 1, 2, \ldots, d$. Let E_k be a value assumed from a parameter X_k. The value of E_k could be an interval or a finite set, according to the type of parameter. Hence, we may represent an object by means of the Cartesian product:

$$\mathbf{E} = E_1 \times E_2 \times \cdots \times E_d \tag{2.1}$$

Let us define U_k as the domain of kth parameter and let us define with

$$\mathbf{U}^d = U_1 \times U_2 \times \cdots \times U_d$$

the *parameter space*. We indicate the d exponent in parentheses in order to distinguish it from the usual Euclidean space $\mathbf{U}^d$.

The parameters that will be dealt with in this application could be of the following types:

- *Continuous quantitative parameters.* Examples of these kinds of parameters can be the weight of a person or his/her blood pressure values. Further, the value assumed by this kind of parameter may also not be a single numerical value (i.e., $E_k = 75$ [kg] $\in U_k$) but even a close interval (i.e., $E_k = [70, 80] \subset U_k$). The U_k domain will be a finite interval of the kind $U_k = [a_k, b_k]$, where a_k and b_k are the minimum and maximum possible values for the kth parameter.

- *Discrete quantitative parameters.* This kind of parameter describes characteristics such as the number of myocardial infarctions suffered by a patient or the number of previous hospitalizations. As in the previous case, the value assumed by one of these parameters can be of the kind $E_k = 5 \in U_k$ or $E_k = [2, 8] \subset U_k$, and the domain will be an interval of the same form of the previous case.

- *Ordinal qualitative parameters.* An example of this kind of parameter can be the degree of training of a person or the seriousness degree of a certain pathology, which has been codified according to a certain scale. To treat this kind of parameter, it is sufficient to assign a numerical code (i.e., elementary school = 1, middle school = 2, highschool = 3, etc.) and we are back to the previous case.

- *Nominal qualitative parameters.* By this definition we indicate the parameters that do not possess a quantitative connotation, such as sex (male or female) or blood group (A, B, AB, O) of a person. Even here, single values or values constituted by subsets of the domain are that which are constituted by the finite set of all the possible values that can be assumed by this parameter.

- *Tree-structured parameters.* By taking a typical example from informatics, in Figure 2.2 an example of a structured domain for the parameter "microprocessor" is shown. The possible values that the parameter may assume are a terminal value or a set of these. The domain U_k is a set of all the possible terminal values.

When all the values of the characteristics of a product of the kind in Equation (2.1) satisfy the specifications of the types introduced above, we

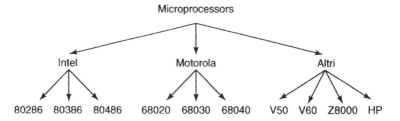

Figure 2.2. Example of a tree-structured parameter.

define the product as an *event*. When at least one component of an event is the void set $\varnothing$, it is called the *null event*. Let us now define the operator *Cartesian union*, which is associated with every ordered couple of events contained in the space of parameters of an event of $\mathbf{U}^d$ univocally determined in the following way.

Definition 1. let $\mathbf{A} = A_1 \times A_2 \times \cdots \times A_d$ and $\mathbf{B} = B_1 \times B_2 \times \cdots \times B_d$ be a couple of events belonging to $\mathbf{U}^d$. The Cartesian union $\mathbf{A} \oplus \mathbf{B}$ is defined by the Cartesian product:

$$A \oplus B = \left(A_1 \oplus B_1 \right) \times \left(A_2 \oplus B_2 \right) \times \cdots \times \left(A_d \oplus B_d \right)$$

where $A_k \oplus B_k$ is the Cartesian union of the values of the kth characteristics and is defined as follows:

1. If the kth parameter is quantitative or ordinally qualitative, $A_k \oplus B_k$ becomes a closed interval as follows:

$$A_k \oplus B_k = \left[\min \left(A_{kL}, B_{kL} \right), \max \left(A_{kU}, B_{kU} \right) \right]$$

   where A_{kL} and A_{kU} are the inferior and the superior limits, respectively, of the A_k interval.
2. If the parameter is qualitatively nominal, $A_k \oplus B_k$ becomes the union of the two parameters:

$$A_k \oplus B_k = A_k \cup B_k$$

3. If X_k is a structured parameter, the approach becomes a little more complicated. Let us indicate with $N(A_k)$ the closest parent node common to all terminal values in A_k. Then, if $N(A_k) = N(B_k)$, the union will be, similarly to the preceding case:

$$A_k \oplus B_k = A_k \cup B_k$$

whereas, in the contrary case, it will be:

$$A_k \oplus B_k = \left\{\text{all the terminal values of node } N\left(A_k \cup B_k\right)\right\}$$

Definition 2. Let $\mathbf{A} = A_1 \times A_2 \times \cdots \times A_d$ and $\mathbf{B} = B_1 \times B_2 \times \cdots \times B_d$ be a couple of events belonging to $\mathbf{U}^d$. The *Cartesian intersection* $\mathbf{A} \otimes \mathbf{B}$ is defined by the Cartesian product:

$$A \otimes B = \left(A_1 \otimes B_1\right) \times \left(A_2 \otimes B_2\right) \times \cdots \times \left(A_d \otimes B_d\right)$$

where $A_k \otimes B_k$ is the Cartesian union of the values of the kth characteristics, and is defined by the intersection of kth parameters:

$$A_k \otimes B_k = A_k \cap B_k$$

If the intersection is void even only for one characteristic, the events are disjoint and the resultant event is null. In the following, we shall indicate the mathematical model ($\mathbf{U}^{(d)}$, $\oplus$, $\otimes$) as the Cartesian spatial model (CSM) and $\Lambda(\mathbf{U}^{(d)})$ as the family of all the events contained in $\mathbf{U}^{(d)}$.

Definition 3. For a pair of events $\mathbf{A} = A_1 \times A_2 \times \cdots \times A_d$ and $\mathbf{B} = B_1 \times B_2 \times \cdots \times B_d \in \Lambda(\mathbf{U}^{(d)})$ we shall define

$$\phi\left(A_k, B_k\right) = \left|A_k \oplus B_k\right| - \left|A_k \otimes B_k\right| + \gamma\left(2\left|A_k \otimes B_k\right| - \left|A_k\right| - \left|B_k\right|\right),$$

$$k = 1, 2, \ldots, d \tag{2.2}$$

where γ is a parameter between 0 and 0.5, and $|A_k|$ is the measure of the interval A_k if the kth characteristic is continuously quantitative and, instead, the number of possible values included in the set A_k is indicated if the kth characteristic is discrete, quantitative, qualitative, or structural.

The γ parameter controls the effect of the internal extreme points in the vicinity where A_k and B_k are two intervals. Making reference to Figure 2.3, if $\gamma = 0$, the distance expression reduces to

$$\phi\left(A_k, B_k\right) = \left|A_k \oplus B_k\right| - \left|A_k \otimes B_k\right|$$

Hence, if A_k and B_k are two distinct intervals, the distance is influenced only by the vicinity of the external extreme points. Therefore, the calculated distance for the two cases (a) and (b) shown in Figure 2.3 will be the same. To the other extreme, for $\gamma = \frac{1}{2}$, the definition reduces to

$$\phi\left(A_k, B_k\right) = \left|A_k \oplus B_k\right| - \left(\left|A_k\right| + \left|B_k\right|\right)/2$$

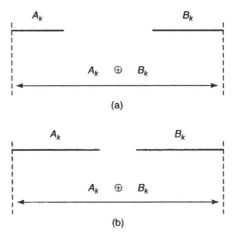

Figure 2.3. Ratio between two intervals.

which assumes instead a different value in the two cases. Obviously, for intermediate values of γ between 0 and 0.5, even the distance function shows intermediate behaviors between these two extreme values. The choice of this parameter might depend by the particular objectives of the data analysis, but for the majority of applications the solution with $\gamma = 0.5$ is the most diffused one and we have chosen it, too.

It has been demonstrated (Ichino, 1992) that the definition of distance fulfills all the axioms that define a metric function:

1. $\phi(A_k, B_k) \geq 0$ and $\phi(A_k, B_k) = 0$ if and only if $A_k = B_k$
2. $\phi(A_k, B_k) = \phi(B_k, A_k)$
3. $\phi(A_k, C_k) \leq \phi(A_k, B_k) + \phi(B_k, C_k)$

Therefore, by using Definition 3 we can define a first version of the generalized Minkovski's distance of order p, with $p \geq 1$, between two events $\mathbf{A} = A_1 \times A_2 \times \cdots \times A_d$ and $\mathbf{B} = B_1 \times B_2 \times \cdots \times B_d \in \Lambda(\mathbf{U}^{(d)})$ in the following way:

$$d_p(\mathbf{A}, \mathbf{B}) = \left[\sum_{k=1}^{d} \phi\left(A_k, B_k\right)^p \right]^{\frac{1}{p}}$$

The normalization problem is solved by

$$\psi\left(A_k, B_k\right) = \phi\left(A_k, B_k\right) / |U_k|, \quad k = 1, 2, \dots, d \qquad (2.3)$$

where notation $|U_k|$, as previously described, indicates the measure of the interval representing the parameter domain if this one is continuously quantitative and, otherwise, indicates the number of possible values contained in the domain. The metric is then

$$d_p(\mathbf{A},\mathbf{B})=\left[\sum_{k=1}^{d}\psi\left(A_k,B_k\right)^p\right]^{\frac{1}{p}}$$

If it would be necessary to differentiate the relative values of the variables, weights c_k could be employed, with $k = 1, 2, \ldots, d$. Then, a complete definition will be that of Definition 4.

Definition 4 (generalized Minkowski's metric of order p. Given two events $\mathbf{A} = A_1 \times A_2 \times \cdots \times A_d$ and $\mathbf{B} = B_1 \times B_2 \times \cdots \times B_d \in \Lambda(\mathbf{U}^{(d)})$, and a constant integer $p \geq 1$, let us define

$$d_p(\mathbf{A},\mathbf{B})=\left[\sum_{k=1}^{d}c_k\psi\left(A_k,B_k\right)^p\right]^{\frac{1}{p}}$$

where

$$c_k > 0, \quad k = 1, 2, \ldots, d, \quad \sum_{k=1}^{d}c_k = 1$$

The condition of weight normalization is useful to guarantee that $0 \leq d_p(\mathbf{A}, \mathbf{B}) \leq 1$.

2.3.3 Analysis of the Generalized Principal Components

The distance operator previously defined allows one to build "distance matrices" or "dissimilarity matrices." Some applications, for example, principal component analysis, require a representation of data intended as points in the d-dimensional space. Hence, it is possible to build a data matrix on the coordinates defined with respect to a reference chosen ad hoc. The reference is constituted from the structure centroid of the data to be analyzed. The coordinates are obtained through the comparison of couples of coordinates homogeneous between the reference and the point to be considered. The reference event is thought of as a centroid of the set of data $\mathbf{\Psi} = \{\mathbf{E}_1, \mathbf{E}_1, \ldots, \mathbf{E}_N\}$, and is defined as the individual $\mathbf{E}_r$ in $\mathbf{\Psi}$ that minimizes the sum of the distance between itself and all the other individuals:

$$\sum_{k=1}^{N}d_p\left(\mathbf{E}_r,\mathbf{E}_k\right)=\min_{1\leq i\leq N}\sum_{k=1}^{n}d_p\left(\mathbf{E}_i,\mathbf{E}_k\right)$$

In this case, the generalized Euclidean distance will be used, that is, with $p = 2$. Once the reference event has been defined, the other individuals will be obtained through Equation (2.3).

In the case in which the value of the parameter is an interval, the comparison is made between the centroids of the interval themselves.

With parameters of the qualitative nonordinal type, an arbitrary ordering is introduced in order to maintain as much information as possible.

Finally, the method of principal component analysis is employed (Joliffe, 1986).

2.3.4 Binary Variables

Binary variables may be subdivided into *symmetric* and *asymmetric*. Symmetric variables assume values that are mutually exclusive, that is, an individual could be male or female, without other possibilities. Asymmetric variables identify or contraindicate the presence of a characteristic, relatively to all the values that it could assume, generally more than 2; that is, a flower can be red or not, but not being red implies being of any other color.

If two realizations of a random vector have a binary attribute f, it is possible to obtain a table as shown in Figure 2.4., where

a = number of variables =1 for both i and j
b = number of variables f for which $x_{if} = 1$ and $x_{jf} = 0$
c = number of variables f for which $x_{if} = 0$ and $x_{jf} = 1$
d = number of variables = 0 for both i and j
$a + b + c + d = p$ = total number of variables

On the basis of this figure, it is possible to introduce a *similarity coefficient* that aims at indicating how similar are the two realizations as binary variables and, analogously, a *dissimilarity coefficient* that is its complement (Table 2.1).

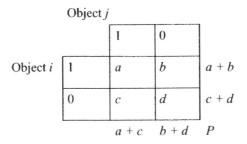

Object j

Object i	1	0	
1	a	b	$a + b$
0	c	d	$c + d$
	$a + c$	$b + d$	P

Figure 2.4

Table 2.1. Similarity and dissimilarity coefficients

	$s(i,j)$	$d(i,j)$
Jaccard coefficient for asymmetric binary variables	$\dfrac{a}{a+b+c}$	$\dfrac{b+c}{a+b+c}$
Zubin coefficient for symmetric binary variables	$\dfrac{a+b}{a+b+c+d}$	$\dfrac{b+c}{a+b+c+d}$

2.3.5 Wilson Metrics

Given two realizations of a random vector, dissimilarity distance is defined as (Wilson and Martinex, 1997)

$$d(i,j) = \frac{\sum_f \delta_{i,j}^{(f)} d_{i,j}^{(f)}}{\sum_f \delta_{i,j}^{(f)}}$$

where

$$d_{i,j}(f) = \begin{cases} 1 & \text{if } x_{i,f} \text{ and } x_{j,f} \text{ are both present} \\ 0 & \begin{cases} \text{if } x_{i,f} \text{ and } x_{j,f} \text{ are not both present, or} \\ \text{variable } f \text{ is binary asymmetric and } i,j \text{ form} \\ \text{a 0,0 matching for that variable} \end{cases} \end{cases}$$

$$d_{i,j}^{(f)} = \begin{cases} 1 & \text{if } x_{i,f} \\ 0 & \text{if } x_{i,f} = x_{j,f} \text{ and } f \text{ is binary or nominal} \\ \dfrac{\left| x_{i,f} - x_{j,f} \right|}{R_f} & \begin{cases} \text{if variable } f \text{ is linear quantitative with} \\ R_f = \max_h \left(x_{h,f} \right) - \min_h \left(x_{h,f} \right) \\ \text{and } h \text{ runs on all the missing data of } f \end{cases} \end{cases}$$

Ordinal variables are substituted by order numbers of the sequence. A Euclidean generalized metric defines a distance between two values of an attribute of a random vector. Such attribute can be quantitative, continuous, qualitative, or heterogeneous. Let x and y be such values and let us define the attribute a. Two operators on the couple (x, y) are defined in the following way:

$$\text{OVERLAP}_a(x, y) = \begin{cases} 0 & \text{if } x = y \\ 1 & \text{if } x \neq y \end{cases}$$

$$\text{RN\_DIFF}_a(x, y) = \frac{|x - y|}{\text{RANGE}(a)}$$

where $\text{RANGE}(a) = \max(a) - \min(a)$.

Then it is possibile to define the distance between x and y as follows:

$$d_a(x,y) = \begin{cases} \text{OVERLAP}_a(x,y) & \text{if } a \text{ is nominal} \\ \text{RN\_DIFF}_a(x,y) & \text{if } a \text{ is quantitative} \\ 1 & \text{if } x \text{ or } y \text{ are missing} \end{cases}$$

From the definition of RN_DIFF$_a(x, y)$, it could happen that a new value is outside the current RANGE(a) considered. This does not happen frequently and many times it could be interpreted as an error. It is possible to solve that problem by recalculating RANGE(a), taking into account the new value and by making the normalization again on all the values. Such a method could lead to an excessive thickening of most of the values, thus making less readable a graphical representation on the linear scale.

Hence, it is possible to define the Euclidean *heterogeneous overlapped metric* (HEOM) as follows:

$$\text{HEOM}(x,y) = \sqrt{\sum_{a=1}^{n} d_a(x,y)^2}$$

This metric has the advantage of an easy computation but does not use the added information given by the nominal attributes, which could help in evaluating the importance of the attribute itself.

The *values difference metric* (VDM) is a function that introduces a distance between nominal attributes. Let

$N_{a,x}$ be the number of elements inside set T that present x value for attribute a

$N_{a,x,c}$ be the number of elements inside set T that present x value for attribute a and belong to class c

$P_{a,x,c} = (N_{a,x,c}/N_{a,x})$ be the conditional probability that, given x value for attribute a, the element belongs to class c

C is the total number of classes in T

Then

$$N_{a,x} = \sum_{c=1}^{C} N_{a,x,c}$$

The VDM function is defined as follows:

$$VDM_a(x,y) = \sum_{c=1}^{C} \left| P_{a,x,c} - P_{a,y,c} \right|^q$$

where q is a constant (generally 1 or 2).

As an example, let us suppose that T is a basket of apples and oranges and we know that the conditional probability is 15% that, given red color, the element is an orange; 85% that, given red color, the element is an apple; 99% that, given orange color, the element is an orange; and 1% that, given orange color, the element is an apple. Then

$$\text{VDM}_{\text{color}}(x, y) = \left|P(\text{orange}|\text{red}) - P(\text{apple}|\text{red})\right| + \left|P(\text{orange}|\text{orange}) - P(\text{apple}|\text{orange})\right|$$
$$= \left|15 - 85\right| + \left|99 - 1\right| = 70 + 98 = 168$$

It is observed that with this definition of $P_{a,x,c}$ it could happen that a new element in T did exist with a value x for the attribute a that was not present previously. In such a situation, we will have $N_{a,x,c} = 0$; hence, the undetermined form $P_{a,x,c} = (0, 0)$. Such a problem is solved by posing $P_{a,x,c} = 0$.

Alternatively to VDM, two other distances can be defined:

$$\text{VDM2}_a = \sqrt{\sum_{c=1}^{C} \left|\frac{N_{a,x,c}}{N_{a,x}} - \frac{N_{a,y,c}}{N_{a,y}}\right|^2} = \sqrt{\sum_{c=1}^{C} \left|P_{a,x,c} - P_{a,y,c}\right|^2}$$

$$\text{VDM3}_a = \sqrt{C\sum_{c=1}^{C} \left|\frac{N_{a,x,c}}{N_{a,x}} - \frac{N_{a,y,c}}{N_{a,y}}\right|^2} = \sqrt{C\sum_{c=1}^{C} \left|P_{a,x,c} - P_{a,y,c}\right|^2}$$

It is possible to note that, using the previous definition, the distance does not take into account cases in which both probabilities $P_{a,x,c}$ and $P_{a,y,c}$ are very small or very large, that is, the cases in which the two elements are very close. In these cases, using VDM, the distance is high, while using VDM2$_a$, the distance is significantly smaller, with the disadvantage of a greater computational complexity.

2.4 GRAPHIC METHODS FOR INTERACTIVELY DETERMINING THE MOST DISCRIMINANT ORIGINAL VARIABLES

When there are few variables, trellis diagrams are very effective. Figure 2.5 shows a population divided between healthy and pathological cases (0,1). The ordinates are in arbitrary units. In order to found out which variables greatly weigh on the output (Out1), we condition the diagram with the other variables of the "Cleveland" archive (data relative to healthy/cardiopathic patients, parameters from the medical record, relative to biomedical signals such as ECG, etc.).

Let us consider the example of blood cholesterol (upper part of the diagram in Figure 2.6).

As may be noted, healthy/diseased populations are strongly depen-

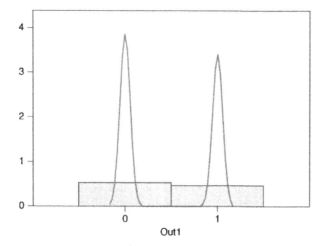

Figure 2.5. Healthy and diseased populations in the Cleveland archive.

dent on the cholesterol trend: for a value of cholesterol of about 242 (panel 3) we note the inversion of the two populations. For low values of cholesterol, we note that the population is mainly negative (healthy subjects).

Figure 2.7 is a trellis diagram relative to the population (conditioned variable) in respect to the parameter OldPeak, which represents the ST displacement during the stress-test ECG in respect to the resting ECG. Even in this case, we may note that the parameter strongly influences the output variable. In particular, a value between 1.8 and 6.8 of OldPeak divides the relative population into two very unbalanced groups. The intervals of various panels (OldPeak ranges) are calculated in such a way that there could be equal numbers in the various modules. It is also possible to use trellis diagrams that maintain equal *range* on all panels; the choice depends, obviously, on the kind of analysis we want to employ and the nature of the conditioning parameter.

Now, it is interesting to consider another parameter, maxHR, which indicates the maximum heart rate reached during the stress test (Figure 2.8).

Obviously, high values of MaxHR are found mostly in healthy subjects, whereas maximum values of heart rate around 100–125 b/m are typical of pathological cases. On the other hand, it is interesting to note that we may find some links among the various parameters (attributes) that are well shown by multipanel diagrams (multiple scatterplots), as in Figure 2.9.

2.4.1 Analysis of Homogeneity

Another method that provides useful diagrams to single out the most important variables is *correspondence analysis* (Michailidis and DeLeeuw,

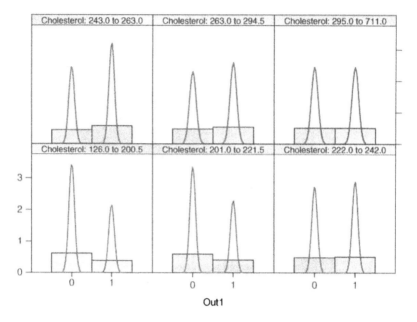

Figure 2.6. Effect of the blood cholesterol changes on the output.

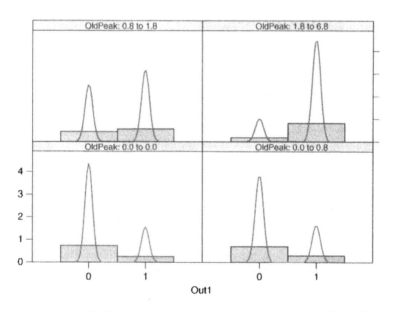

Figure 2.7. Trellis diagram relative to the conditioning variable OldPeak (ST displacement during stress-test ECG).

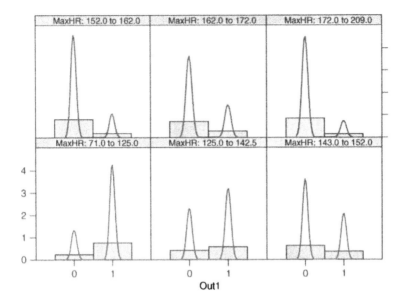

Figure 2.8. Trellis diagram relative to the conditioning variable MaxHR (maximum heart rate reached during the stress test).

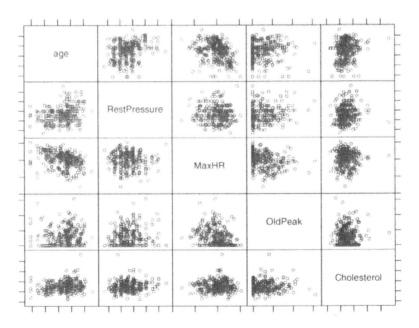

Figure 2.9. Multiple scatterplot obtained from the data contained in the Cleveland database (here only 5 of 14 available parameters are considered).

2000). Let us suppose there is a population of objects each characterized by J variables. Each object is characterized by a random J-dimensional vector, whose components may be quantitative, ordinal, or nominal according to the nature of the corresponding variable. Then there is a population of N objects homogeneous in the characteristic variables and disjoint by the values that they can reach. It is possible to comprehend the population in matricial form (data matrix; see Appendix), in which each object corresponds to a row of the table while the colums contain the distribution of the values the various variables may assume inside the population components.

According to this hypothesis, by adopting a geometrical representation, it is possible to indicate every object of the population as a point in a J-dimensional hyperspace, whose coordinate values are obtained from the parameter vector correspoinding to this subject. Then, through a metric suitable to the specific hyperspace, it would be possible to determine similarity criteria among the various objects of the population in order to subdivide them into groups, by evidencing the most important variables from their geometrical position. On the other hand, it is difficult to deal with highly dimensional hyperspaces, which require complex metrics.

For these reasons, those methods are promising which are able to condense the information into new variables that summarize most of the information content present in the data. Hence, it is possible to represent the objects of the population in hyperspace at a reduced dimensionality, yet maintain the major part of the information.

Homogeneity analysis is a method of multivariate statistical analysis that allows the processing of hetereogeneous sets of biomedical data, as it is capable of enhancing characteristics and structures present in the data.

Homogeneity analysis acts on nominal and ordinal data, that is, on discrete variables (nonquantitative). This is not restrictive, as possible quantitative variables present in the data vector of an object can be restored to an ordinal character through segmentation into classes in their variability field. The process of transformation of the quantitative variables is obviously accompanied by a loss of information, which does not appreciably impair the information associated with this variable.

On the other hand, even the way that a physician reasons is based upon a subdivision into categories of the possible numerical values that a quantitative variable might assume, rather than on attention relative to a specific detected numerical value.

In any case, if the information loss does not impair the method, the subdivision into categories of quantitative variables does not add any artifacts to the data, as happens in the codifying process of nominal variables in principal components analysis (PCA).

In practice, the homogeneity analysis makes a map of N subjects of a population, each characterized by a J-dimensional vector (*patient profile*) in a Euclidean space (IR^p) of low dimension. The Euclidean space is chosen mainly for its good geometrical characteristics as it is provided with a metric that is familiar to our visual system. Thus, is important also for a better graphical representation of the processing results.

The criterion by which the mapping is conceived is that one by which subjects with similar patient profiles are mapped in points of IR^p with a small distance between them, whereas patients with different vector parameters are mapped in points with a larger distance.

Therefore, the distance among the various points in which population patients belonging to IR^p are mapped provides a clear indication of the degree of similarity among the various patient profiles and, hence, among the various population components.

Further, the method allows, at the same time and in the same Euclidean space, even the mapping of classes (manifestations) of the J variables that constitute the patient profile. The mapping is made in such a way that a patient is close to the classes to which he/she belongs and, vice versa, the classes are close to the subjects that belong to them. As far as the rather complicated algebraic implementation is concerned, it is better to consult the cited references; here it is sufficient to provide a graphical example (see Figure 2.10).

2.5 ALARM GENERATION

In order to generate alarms or to alert the doctor/patient system when individual conditions do change suddenly or dangerous trends are evidenced, it is necessary to evaluate independent variables (predictors carefully chosen and processed by means of the above-mentioned techniques) that describe patient status and generate an index (output-dependent variable), generally a probability that is capable of singling out a potentially dangerous situation (Anonymous, 1999; Joliffe, 1986; Lieberman and Selker, 2000; Long, 1993). As an example, we shall deal with a technique that is often useful in this context and is based upon a logit model (Mitchell, 1992). The logistic regression has been chosen because it is a method that provides as output the probability of having a categorical attribute (alarm/nonalarm, criticity/normality).

Similar results have been reached in the past by the use of decisional trees.

The employment of logit models in the biomedical field has been well documented by Long (Long 1993), who described significant results on the employment of logistic regression for our aims.

This method may be used for the estimation of regression functions of a set of explicit (independent) variables with the probability of having a binary attribute (dichotomy). In fact, here the issue is that the model re-

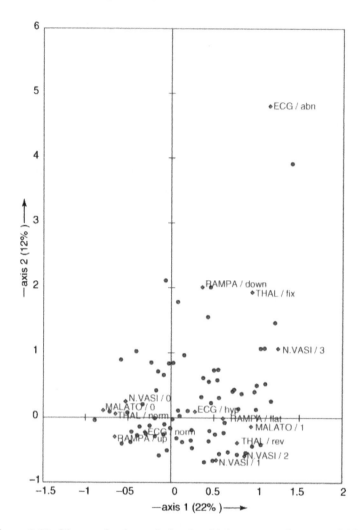

Figure 2.10. Diagram for the analysis of multiple correspondences. Individuals and categories on axis 1 and axis 2 (34%).

sponse is dichotomic (two values). In general, the cases in which dichotomic variables are involved are very common; for example, success/failure, presence/absence, and referring to a particular pathology or pathology treatment.

Techniques exist that extend the basic model of logistic regression and refer to category-dependent variables (model response) (Hosmer and Lemeshow, 1989), such as those used to evaluate three possible situations of patient health status: improvement, worsening, and stability.

The logistic regression model comes from the extension of the multiple linear regression model with continuous dependent variable Y and a set of predictors p, given by

$$Y = \alpha + \beta_1 X_1 + \beta_2 X_2 + \cdots \beta_p X_p + \varepsilon = \alpha + \sum_{j=1}^{p} \beta_j X_j + \varepsilon$$

Y is the expected value in a set of X and ε is the prediction error. In the case in which the error distribution is normal, the model is called *probit*; in the case in which the distribution is logistic, it is called *logit*.
Considering $Y = Y' + \varepsilon$, we obtain

$$E(Y|X_1,\ldots,X_p) = Y' = \alpha + \sum_{j=1}^{p} \beta_j X_j$$

We may note that Y' is function of predictors and the prediction errors. Such an error is the one connected to the measurement of the variables through the regression function. The representation with the above-described form of linear regression model, in cases of dichotomic responses, has the disadvantage of being unlimited, which is explicitly required in the initial hypotheses. In other words, when operating in this context with a linear model there is the phenomenon of additivity, which, in the case of dichotomic output, is a problem both for the measurement and the interpretation of data. For example, let us imagine calculating the success in performing a particular task as a function of two binary variables: skilfulness and sex. In this context, let us suppose that unskilful individuals who undertake the task are 50% of men and 70% of women. For the skilful men, the success probability is 90%. By supposing a linear relationship between sex and skilfulness, we get that for the skilful women the success probability is 70% + 40% or 110%, which is impossible. To overcome this unwanted effect, it is necessary to extend the linear model with unbounded output to a model in which the output can assume values between 0 and 1. Let us consider the dichotomic variable Y with values 1 or 0, where $Y = 1$ is given by $\pi = P(Y = 1)$ and $Y = 0$ is given by $1 - \pi = P(Y = 0)$, then it is possible to obtain

$$\ln\left[\frac{P(Y = 1|X_1 \ldots X_p)}{1 - P(Y = 1|X_1 \ldots X_p)}\right] = \ln\left[\frac{\pi}{1-\pi}\right] = \text{logit}(\pi) = \alpha + \beta_1 X_1 \ldots \beta_p X_p$$

In the literature, $\pi/(1 - \pi)$ is often called the odds.
For π between 0 and 1, the logit of π may vary between $-\infty$ and $+\infty$. The range of logit values is symmetric to the central value of logit 0.5 (which is equal to 0). A scale that represents this transformation is

0.3	0.4	0.6	0.7	0.8	0.9	0.95	0.99	π
−0.847	−0.405	0	0.405	0.847	1.386	2.94	4.59	logit(π)

The table above stresses that the difference between logit 0.95 and 0.99 is much higher than from 0.5 and 0.7. In fact, the logit scale is approximately linear in the center and logarithmic at the edges.

The example reported above might help in understanding such a scale. In men, the logit difference between skilful and unskilful is logit(0.5); in women it is logit(0.70) = 0.874. For unskilled men it is logit(0.9) = 2.197. Hence, we may conclude that for skilled women the success probability will be logit(0.9) + logit(0.7) = 3.044, which corresponds to a probability of 0.955.

Now, it is clear that logistic regression fits on the data through an equation of the kind

$$\log it(\pi) = \alpha + \beta_1 X_1 + \beta_2 X_2 + \cdots \beta_p X_p = \alpha + \sum_{j=1}^{p} \beta_j X_j$$

with

$$P(Y = 1 | X_1 \ldots X_p) = \frac{e^{\alpha + \sum \beta_i X_i}}{1 + e^{\alpha + \sum \beta_i X_i}} = \frac{1}{1 + e^{-\alpha - \sum \beta_i X_i}}$$

and

$$P(Y = 0 | X_1 \ldots X_p) = \frac{1}{1 + e^{\alpha + \sum \beta_i X_i}}$$

It is clear that the model of logistic regression is similar to the model of multiple linear regression, except that instead of expected variable Y we have the odd logarithm often defined as a logit transformation.

As in all regression techniques, it is necessary to choose the model parameters as well as to carry out the validation of model itself. In our case, the parameters to be estimated are constant α and β coefficients. The difference in respect to the multiple linear regression is that the criterion of coefficient choice is not the least squares but, instead, the maximum likelihood, where the probability of obtaining expected results is maximized, given regression coefficients. Therefore, the issue is not to minimize a given error but, rather, to maximize its likelihood. The likelihood of the set of data is given by

$$L = \prod_{i=1}^{n} \left[\left(\frac{e^{\alpha + \sum \beta_i X_i}}{1 + e^{\alpha + \sum \beta_i X_i}} \right)^{Y_i} \cdot \left(\frac{1}{1 + e^{\alpha + \sum \beta_i X_i}} \right)^{1 - Y_i} \right]$$

The algorithm of model determination tries to find the set of α parameters and β coefficients such that L is maximum.

It is possible to use more logit modules (logit network) in order to weigh the intervention on various groups of input parameters. In cases with few predictors, it is possible to refer to a single logit; of course, the success probability and the system reliability are bound by the initial choice of the parameters and dependent on degree of output from the parameters and from the training set used for training the module and identifying the model coefficients. Besides logit modules, it is important to mention also the decisional trees, which are much employed tools for alarm generation or reporting on patients' critical conditions.

APPENDIX

By adopting the most diffused convention, let the individuals (objects, patterns, etc.) be positioned in the rows of a matrix and every attribute (variable, magnitude, or parameter) in one column. Some useful matrices for transformations presented in the text are defined as follows:

- Data matrix. This is a rectalungar matrix $n \times p$ in which n is the number of samples under consideration and p is the number of observed variables. If the data are quantitative, every row is representated geometrically by a point in a space whose coordinated axes are the p observed variables; analogously, the column relative to a variable is represented by a point in the space defined by n statistical units.
- Frequency matrices. Let us consider n statistical units, close to which two variables are observed on any scales. The frequency matrix is $h \times m$ in which the frequencies of the joint observation of the two variables are coordinated, the first one with h modality, and the second one with m modality. Such a matrix can be considered a "profile matrix," given the fact that every row is the profile of the statistical unit set that possesses the h modality of the first variable. Analogously, the mth column can be considered the profile set of the units that possess m modality of the second variable.
- Variance–covariance matrix. Let σ_{ij} be the covariance between X_i and X_j $(i, j = 1 \ldots p)$:

$$\sigma_{ij} = E\{(X_i - \mu_i)(X_j - \mu_j)\}$$

with $i, j = 1 \ldots p$, where $E(\cdot)$ is the expected value of the argument and $\mu_j = E(X_j)$ are the variable means. The variance–covariance matrix is the squared matrix Σ of order p whose generic element is σ_i. Sample corrected covariance s_{ij} is a corrected estimator of σ_{ij}:

$$s_{ij} = \frac{1}{n-1} \sum_{h}^{n} (x_{hi} - \bar{x}_i)(x_{hj} - \bar{x}_j)$$

with converse $i, j = 1 \ldots p$, where x_{hi} is the generic element of a data matrix and $\bar{x}_i$ is the estimator of the mean μ_i:

$$\bar{x}_i = \sum_{h}^{n} x_{hi}/n$$

Hence, matrix $S = \{s_{ij}\}$ correctly estimates $\Sigma = \{\sigma_{ij}\}$.

The generic element σ_{ij} on the main diagonal represents the variance of X_i:

$$\sigma_{ii} = \sigma_i^2 = E(X_i - \mu_i)^2$$

whose correct estimator is

$$s_i^2 = \frac{1}{n-1} \sum_{h}^{n} (x_{hi} - \bar{x}_i)^2$$

Such a matrix is symmetric with respect to the main diagonal. It is defined semipositive (it has nonnegative eigenvalues). It has the same rank of data matrix X; if any linear transformation is made on the columns of the data matrix, the variance–covariance matrix of transformed data has the same rank as the matrix of nontransformed data.

• Correlation matrix. By considering the statistical standardization of observed data,

$$X_i^* = \frac{x_i - \mu_i}{\sigma_i}$$

with $i = 1 \ldots p$, where μ_i and σ_i maintain the previously assigned meaning. The data transformed in this way have null mean value and unitary variance. The covariance between the two standardized variables X_i and X_j is the correlation coefficient between the two variables:

$$r_{ij} = E(X_i^* X_j^*) = \frac{s_{ij}}{[s_{ii}s_{jj}]^{1/2}}$$

with converse $i = 1 \ldots p$, which measures the strength of the linear relation between the two variables. r_{ij} is an estimator of ρ_{ij}:

$$r_{ij} = \frac{\sum_{h}^{n}(x_{hi} - \overline{x}_i)(x_{hj} - \overline{x}_j)}{\left[\sum_{h}^{n}(x_{hi} - \overline{x}_i)^2 \sum_{h}^{n}(x_{hj} - \overline{x}_j)^2\right]^{1/2}}$$

The correlation matrix is a squared symmetric matrix of order p, in whose diagonal all ones are present and where the generic element is ρ_{ij}. Such matrices are symmetric and semipositive definite, with rank equal to that of the data matrix, invariant in respect to any linear transformation of data. The unitary values on the diagonal represent the variances of the standardized variables.

- Distance matrix. It is a squared matrix whose generic element δ_{hk} is a measure of distance between the entities h and k. Such matrices, besides being squared, are symmetric, semipositive definite, of rank equal to the data matrix, and the diagonal values are null. It is possible to choose the distances according to the kind of data defined in the preceding section (Bolasco, 1999; Fabbris, 1997).

REFERENCES

Anonymous, *Simbiosys: ECG Simulator Manual,* Critical Concepts Inc., 1999.

Baura, G. D., Listen to Your Data, *IEEE Signal Proc. Mag,* Vol. 21, No. 1, 21–25, 2004.

Baura, G. D., *System Theory and Practical Applications of Biomedical Signals,* IEEE Wiley, 2002.

Bolasco, S., *Analisi Multidimensionale dei Dati,* Carocci, 1999.

Bruce, E. N., *Biomedical Signal Processing and Signal Modeling,* Wiley, 2001.

Cherkassky, V., and Mulier, F., *Learning from Data,* Wiley, 1998.

Coatrieux, J. L., Integrative science and modelling challenge, *IEEE MBE Magazine,* Vol. 23, No. 3, 12–14, 2004.

Esprit program: Sensor Data Validation Methods, A Review, P22442 EM2S, 1996.

Fabbris, L., *Statistica Multivariata,* McGraw-Hill, 1997.

Findler, V. F., *An Artificial Intelligence Technique for Information and Fact Retrieval,* MIT Press, 1991.

Ichino, A., and Yaguchi, H., *IEEE Transactions on Systems, Man, and Cybernetics,* Vol. 24, No. 4, 638–708, 1994.

Il Sole 24 ore, 14 Gennaio, Interventi di routine, inutili 3 ricoveri su 4, 2002.

Jager, F., Taddei, A., Moody, G. B., et al., Long-term ST Database, *Med. Biol. Eng. Comput.,* Vol. 41, No. 2, 172–182, 2003.

Joliffe, I. T., *Principal Component Analysis,* Springer, 1986.

Kaufman, L., and Rousseeuw, P., Clustering by Means of Medoids, in Dodge, Y. (Ed.), *Statistical Data Analysis Based on the L-Norm,* pp. 405–416, North-Holland/Elsevier, 1987.

Lieberman, H., and Selker, T., *Out of Context: Computer that Adapt to and Learn from Context,* MIT Media Laboratory, 2000.

Long, W. J., *A Comparison of Logistic Regression to Decision-Tree Induction in a Medical Domain,* MIT Laboratory for Computer Science, 1993.

Mainardi, L. T., Bianchi, A. M., and Cerutti, S., Digital Biomedical Signal Acquisition and Processing, in Bronzino J. D. (editor), *The Biomedical Engineering Handbook,* CRC Press–IEEE Press, pp. 828–852, 1995.

Michailidis, G., and De Leeuw, J., The Gifi System of descriptive multivariate analysis, technical report, *Statistical Science,* Vol. 13, No. 4, 307–336, 1998.

Mitchell, D. C., *Logistic Regression Analysis,* Dept. of Measurement Statistcs and Evaluation, University of Maryland, 1992.

Moody, G. B., Mark, R. G., and Goldberger, A. L., Physionet: A Web-Based Resource for the Study of Physiological Signals, *IEEE Engineering in Medicine and Biology Magazine,* Vol. 20, No. 3, 70–75, 2001.

OMS, Annual Report, 2004.

Ortega, G. J., Bigun, J., and Reynolds, D., Authentication gets Personal with Biometrics, *IEEE Signal Processing Magazine,* Vol. 21, No. 2, 50–62, 2004.

Paoletti, M., and Marchesi, C., Model Based Signal Characterization for Longterm Personal Monitoring, *IEEE Computers in Cardiology;* Vol. 28, 413–416, 2001.

Provost, F., and Fawcett, T., *ROC Convex Hulls for Comparing Machine Learning Schemes,* School of IT and Engineering, Ottawa, Canada, 2007.

Pyle, D., *Data Preparation for Data Mining,* Morgan Kaufmann, 1999.

Raab, S. S., Thomas, P. A., Lenel, J. C., et al., Pathology and Probability, Likelihood Ratios and ROC Curves in the Interpretation of Bronchial Brush Specimens, *American Journal of Clinical Pathology,* Vol. 103, No. 5, 588–593, 1995.

Rangaraj, M. R., *Biomedical Signal Analysis,* IEEE–Wiley, 2001.

Techmath, *MTM-SHOP, Mass Customization,* brochure, 2004.

Wilkinson, L., *The Grammar of Graphics,* Springer, 1999.

Wilson, D. R., and Martinez, T. R., Improved Heterogeneous Distance Functions, *J. of Artif. Intell. Res.,* Vol. 6, 1–34, 1997.

PART II

POINTS OF VIEW OF THE PHYSIOLOGIST AND CLINICIAN

Thispageintentionallyleftblank

CHAPTER 3

METHODS AND NEURONS

Gabriele E. M. Biella

Nature loves concealing.—Heraclitus

3.1 WHAT IS AN OBJECT?

When applying analytic techniques to biological objects, obvious questions rise. Starting from the most immediate issues such as object naming, there is a progressive convolution in analytical and technical problems. We will go through the main problems, often intersecting with logic or philosophy, and offer a list of problems and solutions that studies on highly complex systems such as the central nervous system yield.

3.1.1 Different Perspectives

3.1.1.1 Object Limits. The inner object structure implies primarily the simplest approach to defining actual or functional boundaries. However seemingly simple, this involves complex processes of assignment to a class and goes deeply into the assumed object of this class. An object must be named before we can know or evaluate it. To avoid subtle discrepancies, we must attempt to name neural "objects" in a manner that offers a reliable definition and avoid reductionism.

  The abstract or material nature of objects are key to their cognitive accession. The first act of this cognitive appraisal relies on an ontologic process of delimitation and discrimination. Delimitation means the "giving of borders," whereas discrimination means the distinction of the object from the rest of the world. These operations require preliminary formal theories on the relations among the single parts and the whole, and among the very parts. To achieve a sense of this, in the following the whole will be indicated as the central nervous system (CNS) and the parts will be its subsets, whatever this could mean (nuclei, laminae, pathways, neurons,

Advanced Methods of Biomedical Signal Processing. Edited by S. Cerutti and C. Marchesi
Copyright © 2011 the Institute of Electrical and Electronics Engineers, Inc.

synapses, receptors, etc.). The accepted anatomical assumption describing the CNS dictates that the CNS is that structure of the nervous system contained in the skull and in the bony canal of the spinal cord.

Let us examine the problems arising from these canonical tenets. What about the topographic assignment of sensory nerve ganglia? Cranial nerve ganglia are effectively contained in the skull, but, according to the definition of sensory ganglion, they belong to the periphery. By contrast, the spinal cord ganglia are external to the bony canal but their axons, like those of their cranial counterparts, bypass the spinal cord boundary, entering the canal with final synaptic contacts in the very midst of the dorsal horn laminae, a central circuitry mass. Thus, part of the gangliar neurons are nested in central structures. These kinds of structures are neither central nor peripheral, and potentially pertain to a third condition, a kind of topographical manifold, yet to be acknowledged. So, we meet with unavoidable incongruities, a signal of more important inconsistencies concerning this part of the naming process. As a matter of fact, homologous scenarios appear unexpectedly in most observation scales.

Intriguing issues arise when going into deeper analyses. The part–whole or part–part boundaries can present either well-drawn features and universally shareable attribution of character (e.g., a neuron under a microscope) or show strictly indefinite boundaries (e.g., the somatosensory cortex). In the former case, consistent and reliable object identification implies universally shareable distinctions of its borders, a spatial criterion called by some ontologists the *bona fide* boundary. In this case, the indefiniteness character implies a preliminary abstract agreement on what we can establish as the element or object boundary. Ontologically, this a priori licence for placing a border (or limit) is called a *fiat* label (*fiat* is the latin concessive conjunctive meaning "that it is made"). Fiat labels can introduce discretional or erratic issues with no universal agreement (Smith and Varzi, 2000).

The central nervous system exhibits a collection of different boundary properties in its mixture of bona fide and fiat properties. For instance, the trigeminal ganglion, in a pouch of the rocca petrosa (literally Bony Rock, a grossly pyramidal complex region of the skull base) of the temporal bone is a capsulated element composed of the neurons whose packaging is, however, enclosed in the capsule. In defining the primary somatosensory cortex (SSI) or thalamic nuclei, for example, we can immediately recognize the arbitrariness of boundaries; it is undecidible where they actually begin (and they actually begin somewhere) and where they terminate (and they definitely terminate somewhere). It looks as if they fade off, more or less smoothly, along confluences with other regions. In addition, the repetitive responses to the same inputs or the involvement in different tasks may provide, however slightly, discrepant dynamic pictures of the same region. One can provide a microscopical or a cytological description (neurons with definite characteristics). However, no particular feature seems eligible to be unconditionally labeled a neuron component of the somatosensory cortex or the thalamic nucleus. From a functional perspective, one can delimit an area (for instance, the SSI) as the delimitable area responding to classes of sensory stimuli but, again, analytical observations fail to map the exact extension of an activation (look, for instance, at SSI maps in functional magnetic resonance experiments). Other interesting phenomena contribute to this sort of uncertainty or probabilistic picture, such as the so-called "field enlargement," an extension of the original boundaries of the cortical projection areas. This event shows that previously unresponsive sections of the cortex become sensitive and responsive to previously ineffective inputs. The aforementioned circumstance may happen, for instance, in the case of specific manual abilities: musicians display dramatic enlargement of the primary motor (and sensory) area commanding (perceiving) hand movements. It is particularly evident in "asymmetric" performers (e.g., violin players, who only use one hand in handling the instrument and in fingering) whose auditory areas and right sensorimotor areas are hugely enlarged and discrepant with the controlateral, expressing specific functional organization in the auditory and somatosensory representational cortex (Pantev et al., 2001). The complementary event, the "missing input" issue, reveals instead functional reduction and even disappearance of sensory brain areas in cases of loss of appropriate input sources (for instance, following the loss of a finger, hand, or arm). Neighboring preserved fibers progressively invade these central areas and substitute newly generating fibers. At the microscopical level, these events may be followed by neogenetic events such as the creation of new synapses or the strengthening of weakest ones and the reassignment of neuronal functional responsiveness to other hierarchical contexts. All these phenomena go by the collective name of neuronal plasticity. The neuronal task reassignment in plasticity could be so pervasive as to induce broad changes of the actual borders of involved areas. In the case of loss of peripheral sources, the cortical areas newly invaded by nearby spared fibers become, however, established. In the case of positive plasticity invasion of areas otherwise fed by normal connections (as in the example of musicians), the area enlargement fades or regresses without persistent input maintenance. In everyday experience, subtle events of plasticity, not readable by current means, are potentially shaping our brain maps with continuous labeling or task reassignment to fringe zones bordering other areas, which are also undergoing analogous adaptment stages (Spruston, 2008). Thus occurs the strange condition in which, so far as concerns the mean stability of the perceived world, individuals could be found "negligibly" different (and yet different!) from themselves at different time stages.

So when dynamically studying a brain, we are studying a structure with the potential, however slight, for ongoing changes. We are forced to use fiat boundary paradigms instead of endorsing basic geometries. Therefore, at the cellular level no clearcut principle of assignment can be certified except at the very moment of our observation. Obviously, the gross anatomy remains completely untouched even if the evanescence of borders of gross structures in the brain raises interesting questions. However, some boundaries (cortex to white matter, for instance) are easily established. As for neurons, interestingly, it is no more than fifty years that we have understood them as bona fide objects, as proposed by Ramon y Cajal at the dawn of the twentieth century with neuron theory (supporting the independence of neurons) against the reticular theory (Golgi's theory of the reticular essence of the nervous system as a continuous structure). Thus, in some way, the dispute was a bona fide versus a fiat boundary conflict. This implies, however, apparent contradictions: neurons can extend their terminals and generate new synaptic structures, profoundly changing their overall shapes, while remaining confined to the limits of their membranes. Thus, the bona fide framing is preserved even in the uncertainty of the original shape permanence or stability. We thus study areas continuously shifting between bona fide and fiat conditions.

There are additional sources of uncertainty evidenced by the tiniest dynamical markers. If studies are carried out by electrophysiological recording of neurons lying in the deep neural structures, we have no way of knowing which neuron we are recording from. The stability of the signal is produced by the sameness of the signal throughout the recording period. There is a challenge, however. Often, neurons change discharge or firing regime, and in this shift one can observe a change in the fine neurochemistry producing the electrical activity. The change is mainly due to the coordination and involvement of diverse ionic channels. One of the most impressive examples comes from the thalamic neuronal activity shifting from tonic to bursting activity and vice versa. The bursting activity shows brief clusters of spikes with increasing and decreasing amplitude seen on huge depolarization waves. The problem is due to the different amplitudes of the burst spikes that could be, wrongly, attributed to different sources in adherence to the principle of signal stability for neuronal dynamic identification. The boundary of recognition must be again changed from a bona fide to a fiat principle.

3.2   WHICH OBJECT PROPERTY IS DEFINITELY INTERESTING?

Objects under systematic observation are enriched by proper and relational features brought into play as either physical or mental properties. To attain an accurate definition of an object, one must appeal to categories of distinctive properties. The first, and most common, is correctly placing an object in the categories of universality and particularity. Universality denotes abstract properties independent of specific instantiation. For instance, we can speak of relevant electrochemical activity just to indicate the neuronal spike. Particularity is embodied in the very distinct detail of the object, for instance, the response to a stimulus of a specific neuron. Particularity entails a number of intricacies because of the richness and flexibility of attributes. For example, let us take into consideration the responses to stimuli of some central neuron. Let us choose a neuron in the visual V1 cortex. If we deliver simple stimuli (even a simple flash, slightly over the threshold of detection), we can notice a disturbing and impressive phenomenon: the neuron responses to each single stimulus deploy each of them with a "different" regime or pattern. There is no apparent class of responses but the simple universal definition of response as, let us say, increase of frequency with any further possible access (Figure 3.1). In the next subsection, we will go deeper into the logical counterpart of these biological problems.

3.2.1 A Short Introduction to Logic

We encountered the problem of single neuron response variability even to simple repeated stimuli. Just to gain some deeper insight into the logical values of this particularity, we can try to use logical tools like sense and reference, the most common translation of Frege's original *sinn* and *bedeutung* (Frege, 1892).

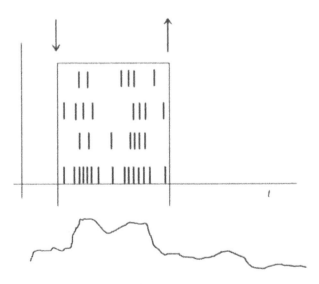

Figure 3.1.

The *reference* is the object that the expression refers to. For instance, the name cortical sensory neuron refers to a cortical neuron, a spiny stellate neuron of the somatosensory cortex. The name spiny stellate cell also refers to that neuron, as indicated by cortical sensory neuron. Hence, the two have the same reference.

The *sense* is the cognitive significance or mode of presentation of the referent (cortical sensory neuron/spiny stellate neuron). Objects with the same reference may harbor different senses.

To add an example, the thalamic transmission neuron is the target of incominig sensory inputs. It is also the target of the looping cortico-thalamic fibers. The neuron is the same in terms of reference or bedeutung (or, again, as extension or reference) but is different in terms of sinn or sense.

What about the previous problem, the variability of response to different stimuli? It is interesting to note the fact that the response is overall (we "obviously" get a response to a stimulus) and this can be the reference or bedeutung in this case. The sense is preserved in the variability. It is the mirror image of the Fregean logical version of reality identification (Frege, 1892). Poor decisions at the sinn–bedeutung level could lead to distressing problems in every formal definition. Namely, in this specific context, consider the commonly used term "pain pathways" in the central nervous system. We can recognize an object referent in the mixed collection of fibers that, as the cognitive referent (sinn, the sense of the definition), "transfer pain" from the periphery to the central structures or internally to the brain. The universal use of the terms does not recede before the lack of consistency. It seems more than obvious that fibers cannot drive any sensation, a post-hoc neuropsychological construct well differentiated from the mere signal information traveling along those pathways. Those fibers (how many are there? where are their structural borders? and how much extended into the whole fiber spectrum is their dynamic regime?) would be better described in a semantically relaxed context, that is, a kind of weakened approach asking only for a post-hoc reliance on facts: all those fibers that we can judge to be involved in the specific signal transmission will be a posteriori taken into consideration as part of the effective collection. Obviously, they do not transmit pain. They actually transmit gathered signals whose current referent is the very mass of ionic/metabolic events causing them, but the sense agrees with the theory of information. These issues obviously lead to arguments not fully relevant but open a route to a number of fragmented problems that we will discuss in the next subsection.

3.2.2 Fragments

A distinction must be made between essential and accessory object properties. As a classical example, a critical essential property of water is having two hydrogen and one oxygen atoms. Water's gaseous, liquid, or solid forms are accessory properties depending on environmental conditions. Gaseous, liquid, or solid forms do not interfere with the essential property of the molecular composition. Concerning neurons, we can agree on the fact that a continuous, electrochemically transmittable activity is their essential property of neurons (let us skip all the features distinguishing an excitable nonneuronal cell, e.g., a striate or cardiac myocite, from an actual neuron). The essential property (that of being a neuron) goes along with the accessory properties or parameters distinguishing response qualities, such as being a motor or sensory cell, thus exhibiting different shapes, frequency band discharges, discharge patterns, and so on. All variables "describing neuronal dynamics can be classified into four classes, according to their function and time scale: membrane potential, excitation variables recovery variables, and adaptation variables (Izhikevich, 2007).

The millions of different electrophysiological mechanisms of excitability and spiking, however, reduce to only four different types of bifurcations of equilibrium (Izhikevich, 2007).

Thus, in opportune frameworks we can reduce the contexts and stabilize the property field. This does not free us from all the previous constraints. If we focus on the set of neuronal properties induced by relational or environmental influences, the resultant likely simplicity must not deceive. A neuron, in perfectly equilibrated culture medium conditions in vitro, shows only "intrinsic" electrochemical features. In the proper CNS context, under the abundant input volleys from its neighbors and the regulatory trade-off at its synaptic contacts, every neuron conveys additional, previously unexpressed activities. An interesting question arises: is the new property set to be considered intrinsic and previously silent, or a shared property with the "inducing" units? However, there is no way to segregate the original from the induced, the intrinsically from the extrinsically generated activities. Also, how much induced activity changes the original activity? That is, what is the span of interference (either mutual or directed) of the two regions of activity? It seems really impossible to discretize. Thus, it is not surprising that at many inspection levels of the activity we are unable to discriminate between the constituting (proper and induced) elements. And, plainly, this interferes in distinguishing the sense–reference from the fiat–bona fide worlds.

3.2.3 Emergence

As if this were not enough, a further element increasing the degree of complexity comes into play. It can be thought of as an intersection point of the many different issues described above. This problem is the *emergence* of properties, that peculiar phenomenon of unpredictability of properties of a system starting from the mere description of the composing elements. Something new appears, not predictable by previous analyses when single elements aggregated, coalesced, and interacted, attaining a higher level of organization. Previously inexistent properties emerge from unseen properties at lower organization levels. This spontaneous property emergence generates an almost spontaneous new reality plan. Quoting John Stuart Mill, ". . . it is certain that no mere summing up of the separate actions of those elements [composing a body] will ever amount to the action of the living body itself" (Mill, 1843).

Let us start with a simple question: how is it that the problem of object characteristics turns into a problem about emergence? If we think of the object features as property collections, many of them behave like entities emerging from aggregates of more elementary entities. The emergence has a price: the newly generated entites cannot be restored to the originating elements. The emergent superelement does not preserve any individuality of the constituent subelements. A most effective example could be consciousness as neural construction; it cannot be reduced to more elementary objects.

We are faced with a crucial relationship between substructures and superstructure, where the superstructure (also called the supevenient structure) enlivens newly emergent properties depending on the entirety of substructures and in the meantime does not have an analytic relation with generating substructures (superstructure irreducibility). In some way, this is a kind of measure of emergence.

Let us go into more detail. Specifically, structure irreducibility to lower grade substructures arises when in the presence of a system S, a P property cannot be gained or inferred from the behavior of the parts composing S. Additionally, if we gather a subset of S properties we cannot gain the P property. A further feature typical of emergence is the manifestation of genuine novelties, truly new features with previosly unpredictable behaviors. It is obvious that a straight representation of the CNS refers to single neurons and neuron networks. In networks, many properties emerge, irreducible to the properties of single units. Subsets do not exploit the full properties of the major set. The immediate problem is the identification (the delimitation) of a network with emergent properties that may be considered complete (again, we encounter here in more abstract terms the bona fide–fiat duality we met at the start).

Let us refer again to the thalamo-cortical loop. The thalamo-cortico loop is a dynamic, large entity with multiple emergent, extraordinary properties (perception, consciousness, motor scheme organization, etc.), for which no true delimitation can be obtained, very much like the cortex discussed in the previous section. The conceptual limits of emergence become even harder to define because no one can effectively postulate a list of completeness for an emergent picture.

It is clear that we must build up a strategy for escaping from these traps. There are a few accessible strategies. One strategy, and the most flexible one to my mind, is a kind of programmed weakness that avoids any pretense of completeness. We must search for the minimal "reasonable incompleteness." For instance, when studying the activity of a neuron network one must rely on a set of inferences, such as the degree of homogeneity of nearby neurons, the gross homogeneity of the majority of neurons in an as small as we can admit brain area under definite experimental conditions or on an assumed range of discharge of neurons under specific experimental conditions (e.g., during the delivery of stimuli). This purposeful reduction, obviously, implies a loss of completeness but enables an inferential access to representative data. Does the openness to progressive reduction of uncertainty with an enriching analytical counterpart foregive the incompleteness? Though an impoverishing one, the strategy of approximation sanctions a conceptual adherence to more flexible analytical tool approaches such as the probabilistic world of the information theory, the algebric tensor approach to item partitiong in brain imaging, and so on. These are not insensitive to incompleteness but at least gather enough strength to open new vistas to highly promising theoretical frames. Among these frames, one of the most delicate but powerful is the fluorishing and robust idea of complexity, in spite of its incompleteness and conceptual instability.

3.2.4 Complexity

Complexity (C) is a strange item and has no conclusive definitions. It is bound to many aspects of emergence. Complexity cannot be blamed in the case of complete disorder of a system. A disordered system can be approached by number theory. A perfectly ordered system can have a complexity degree. In general, complex systems are apparently unharmful; two (or more) related parts, sets, or subsets may generate a complex system or present complexity. Interconnection or relatedness implies distinguishable and separate parts. The time consumed in describing each of the parts gives an idea of the system complexity. More complex structures will require more time to be computed and measured.

Distinguishable parts indicate variety and imply coordinated behavior. Mutual part dependence of differently behaving parts entails a loss of singularity (law of part dependence), leading to redundancy (see later).

A provisional definition of complexity C (Heylighen and Aerts, 1996) may be obtained by multiplying variety V (or entropy, as $\log_2 V$) and redundancy (a measure corresponding to the difference between variety or entropy, and maximal potential variety):

$$C = V \cdot R$$

where $R = V_{max} - V$. C goes to 0 in the case of maximal ($V = V_{max}$) or minimal ($V = 0$) variety. It reaches its maximum when redundancy and complexity are equal, $R = V$ (where V as "strange attractor" could represent the fractal dimension).

What happens when we map complexity onto the central nervous system? Let us start from the basic concepts of variety and redundancy and their relation with elementary functions such as the neuronal response to a stimulus. As we previously saw, the same stimulus repeated in sequences induces different responses in a neuron. We have a diffuse variability and some redundancy. The single neuron might be classified as a structural and functional outcome of a complex system that generats many such units. However, as a functional entity the single neuron must be considered an undivided unit, making complexity measures ineffective, as complexity is found only in the integration of heterogenic activities. Actually, even a network of a few neurons exhibits complex behaviors. The network activation does not supervene on a preconditioned path. After each stimulus, as we saw previously, the neuron expresses a diverse patterned configuration. In spite of this diversity, however, in the network topology there appear signs of redundancy, that is, different discharge patterns have the same meaning. Which features lead to redundancy and variety being inherent to the system and not adventitious applications? To produce a sufficient answer to this question, we will consider a last concept that may help to get to the end of this long list of strange analytical objects. This concept is derived from abstract algebra or group theory and logic, and is called closure. After considering closure, we will get to the very core of this chapter, the relation between technique and object.

3.2.5   Closure

As is well known, a set A is closed under a function f (or a function family F) when for each member of A, $f(a)$, is still a member of A (the set of positive real numbers is not closed under subtraction amd can have a negative outcome, whereas the set of all real numbers is closed under subtraction). In neural terms, if we want to define or delimit an overall dynamic picture of the thalamo-cortical loop involved in somatosensory input estimates, we would be compelled to analytically examine its composing elements (impossible because of the emergence contraints) but also to delimit the elements, leading to the fiat–bona fide duality. Closure could help in escaping from these tangles just by eliminating many pretenses and simply looking for a reduced set of dynamic patterns (those from the dynamic profiles of the observed neurons with inputs summated or integrated without violating the limits of pattern collection). It is obvious that pattern discrimination implies further likely problems. The gross distinctions of slow/fast frequency by arbitrary limits and the splitting between tonic and bursting patterns are strong enough criteria to lead to some inherently poor but stable result. This operation undertaken in successive refinement steps does not collapse into void nominalism but delivers a most powerful means to grasp some preliminary appraisal of CNS dynamic properties.

3.3 ARE THERE BEST TECHNIQUES?

Given the previous account of some abstract concept involved in the study of neurophysiology, let us examine concrete problems related to the chosen techniques.

3.3.1 Does a Specific Technique Influence the Data Structure?

Many techniques able to discover data have the heavy drawback of influencing the data, distorting them or reducing their inferential value. Let us see how this can happen when we decide to acquire signals from the central nervous system and try to extract the discharge coding rules. How can we quantify the information exchange as a sensory representation of the world?

Besides the coarse problems such as inappropriate technical application (e.g., signal recording with frequency undersampling), other serious drawbacks can emerge with wrong analytical techniques or wrong expectancies from the techniques. On the other hand, by using smooth instances, we can get to the very core of local problems by more universal techniques. For instance, given a time-dependent stimulus $s(t)$ one can gather data on the complete description of the temporal conditional probabilistic distribution $P[\{t_i\}|s(t)]$ measuring the probability that a spike or group of spikes is produced at definite times t_i, given a stimulus $s(t)$ under an obligatory caveat: because there is no actual quantity of finite sets of data able to determine a general description of a probability distribution, this completeness will be never achievable.

3.3.2 Coding

Only the first moments of a distribution (mean, variance, skewness, and kurtosis) are used in common probabilistic estimates. However, in the struggle for a "savant" coding we encounter a crucial problem. Coding can happen by frequency or by time. Frequency and time coding are the twin exhibitions of a unique dynamic theme and in their fringe regions they transit one to the other. The mean of conditioned distribution $P[\{t_i\}|s(t)]$ is the so-called time-dependent firing rate $r(t)$. In place of observing the single spike in a definite time t, one can choose a time window $\Delta\tau$ centered on t, gather the number of spikes under repeated stimulus presentation, and divide the final sum of intervened spikes by the number of runs (stimulus-induced responses). A decision must be made when deciding on the window width, which should be not so small as to not even contain even a single spike, and not too large. To avoid the problem of observer arbitrariness, we must choose a window small enough. To reach the very point when the probability of finding a spike is proportional to the width of the window itself, $p(t) = r(t)\Delta\tau$ (Figure 3.2). This defines the frequency or rate $r(t)$ as the probability per unit of time that a spike is present in a small temporal window around time t (formalized as a stimulus time histogram). Spike rate is, on the contrary, the mean discharge frequency of a neuron and can be computed from an individual response to a single stimulus. The time-dependent rate $r(t)$ is a collective property of spike trains and cannot be computed from single responses (Figure 3.3).

Thus, we have a conflict between the spike rate and the time-dependent rate that implies some weighting or choice when reading the codes we are most interested in. Let us briefly diverge from this problem to introduce a seemingly incoherent new subject that will help to address the problem. The concept of the *homunculus* was introduced in the practical neurosciences by the Canadian neurosurgeons Penfield and Rasmussen (Penfield and Rasmussen, 1968) to synthetize the mapping of the body onto either the sensory or motor cortical surface. It could resemble a smoother and physical Cartesian version of the pineal gland, the seat of our conscious world perception. We, as external observers of neural complexities, will assume the role of homunculi. The discrimination of the

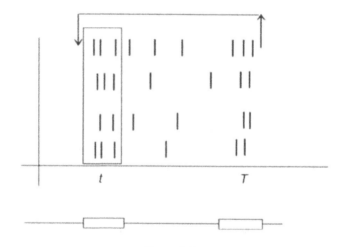

Figure 3.2.

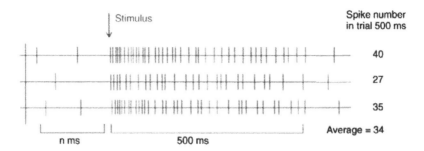

Figure 3.3.

meaning of rate will meet with the role of the homunculus by evaluating the sense and reference (intension and extension) of responses of the interacting subject. As homunculi nested in the receiving structure, we gain access to the reproduced picture of the external world. The picture is coded in the action potentials of sensory neurons. It is obvious that this implicit meaning implies a careful and "aware" decoding and this leads again to a problem of limit (what is coded? what is noise?). It is given that in natural scenarios in a multifarious and source-mixed world we do not know which input is coming from what source (of the potentially infinite sources with unknown probability distribution). Natural inputs (or stimuli) develop in time with underlying structures. They are not only time dependent but unpredictably mutually correlated because of the common deterministic principle of physical processes. Time dependency is, however, a decisive and evolutionistically sound factor, because the rapidity of responses involves the very survival of the individual. Thus, maybe we face a complex partitionable world of codings related to the urgency and importance of incoming signals. There could be a hierarchization founded on unknown criteria that have not yet been identified.

Thus, we need a typology, an evaluation criterion for those hierarchical decisions on incoming stimuli, for which an input subset may be strongly related to survival. Others, progressively less critical to primary survival, may queue up on the stack and eventually may even be lost without hampering immediate solutions. Under normal conditions, the hierarchy criticality may lack top alarms and time left to simultaneously solve less dramatic occurrences related to the richest variants in environmental adaptation, such as feeding, resting, socializing, and so on.

The homunculus must deliver a rapid interpretation of the input reference through sense, a kind of meaning assignment to the neighborhood. This accordance of sense to the world around us invokes time-dependent world features that can be processed with contextual homogeneity, that is, with time-related strategies. The content of single-spike trains becomes fundamental. The variability (and the large variance) exhibited by the system after repeated stimuli is nothing but an exuberance of the system or a kind of implicit redundancy. This feature is not acting (as a rule) in the real world, and does not allow for repetitions but goes through sinularities (single stimuli, single inputs, and so on). The homunculus response should thus suppress noise and deliver time responses, a compromise between averaging and adherence to fast requirements. Both are not soluble; there are neither enough repetitions nor the time to weigh all the input versions distributed on the spectrum of signal-carrying fibers. We can better appreciate this approach by approximation, a seemingly weak strategy for making deductions about sensory images.

In terms of programmed weakness, how many neurons are necessary to exploit this type of performance? When a stimulus is applied to the surface of the body, it involves a number of different receptors from the affected receptive field. Neurons of the same type code the same reference with different modes of representation or sense. For instance, in the visual system a retinal cell responds to light stimuli; the simple cell of area V1 responds to the light with definite positions, and responds progressively less to other positions, up to irresponsiveness for some positions. Neurons placed at higher levels of the visual pathway (e.g., the grandmother cells, so called due to the presentation of complex images as faces, such as ones grandmother's face, of the inferotemporal cortex) activate for higher traits or complex image properties. This progressive property, however, does not hold for all elements we encounter in our life. More subtle general rules are at work.

Given the previous issues, something must happen as the stimulus is transmitted to the first neurons of the chain. The access to consciousness may be dependent on the state of alertness of the individual. In highest alertness states, harming conditions or objects could gain access at highest velocities to higher stations. This would imply a simple redistribution of activated synapses or a reduced input richness.

Let us also note marginally that the random loss of neurons (natural apoptotic processes) does not damage the overall network output and the robustness of the sensory data. Were the highest level coding cells the sole interpreters of definite messages, we would respond to even the worst problems with minimal losses.

Single-neuron and distributed coding represent the two ends of a vexing debate. Horace Barlow (Barlow, 1972) wrote an impressive essay on single-neuron coding with deep insights into the problem, grounding his convictions on five dogmas. The five dogmas were extended in an attempt to address the problem of neural activity and the relation of subjective experience. On the opposite side, multiunit-based estimates of multiple stimuli flourished due to advances in techniques and analytical methods. Although Barlow's ideas were undoubtly persuasive at that time (it was 1972, and being able to focus on those kind of problems, some 40 years ago or so, shows astounding intuition), further research revealed the explosive power of synchronies and mutual correlations, strengthening the argument for distributed coding. Single-neuron coding has, however, the strong appeal of vast applicability. The diverse discharge patterns exhibited by a single neuron under different inputs undoubtly leads to a potentiality vast number of coding repertoires. These apparently similar reports are the result of profoundly different mechanisms at play in the evolutionary divergence of neuronal coding roles.

3.3.3 Do Information Estimates Generated by a Single Neuron Rely on Frequency Code?

Although it is generally agreed that the spike train output of a neuron encodes information about the inputs to that neuron, the code by which the information is transmitted remains unclear (Stevens and Zador, 1995; Ferster and Spruston, 1995). Either the mean firing rate alone encodes the signal (and variability about this mean is noise) (Shadlen and Newsome, 1994, 1995) or it is the variability itself that encodes the signal, that is, the information is encoded at the precise times at which the spikes occur (Bialek et al., 1991; Abeles et al., 1994; Softky, 1995; Rieke et al., 1997).

Some systems may preferentially code in frequency because of its immediate advanges. In the 1920s, Adrian (Adrian and Zottermann, 1926; Adrian and Bronk, 1929) observed this linear dependence of nerve spiking and muscle fiber contraction. Neuromuscular coupling is not the only such system. In rapidly moving animals, sudden obstacles must be avoided as fast as possible. The fly visual system contains neurons (H1 neurons) that employ frequency coding, an evolutionistic advantage as we will see. When sudden obstacles present, there is 20–30 ms available time to code a safely avoiding strategy. How many different neuronal states can be described in this sort of time window? If we decide on a 5 ms binning (and pay attention to the choice), how many induced neuronal states can we compute in 100 ms? In a frequency coding strategy, we can at most discriminate 20 different rates. However, if we transform the frequency code into a binary vector, estimating the probability that a spike is produced in each of the 20 5 ms length windows, it becomes possible to discriminate 2^{20} states, a conspicuous reserve of possibilities (Bialek et al., 1991). Therefore, a single neuron in this analytic context becomes a flexible tool to exploit many variables in a stimulus. However, a coordinated neuronal activity implies that this single-neuron strategy integrates with those of other involved neurons, and this obviosly implies a distributed coding. This goes under different names such as population coding, neural assembly coding, and so on. The tree types of population coding are local coding (LC), in which each neuron represents a specific feature of the object to be represented; scalar coding (SC), in which many neurons may code for the same feature; and vector coding (VC), in which each object feature is coded by neuronal populations with overlapping tuning curves in the feature space. LC is relevant in some systems, for instance, in the polyhedral, ommatidial eye of the fly (object movement speed in the visual field). VC is the most interesting (and the most expensive in terms of neuronal computation) coding. In the motor cortex, motor neuron populations code for a movement plan (Georgopulos et al., 1982). There are smooth (numerically undefinable; recall the fiat/bona fide partition) classes of different neuron responsivity, programming the whole motor scheme with strongly and weakly discharging neurons, differently involved in the partitioned coding. It is likely that most strongly discharging neurons are coherently most involved in the movement planning. A tuned vector (vector summation of the whole output) will give a more soundly weighted figure than that expressed by each single unit. This technique tries to predict the movement direction in the axis with the tuning summation vector. It is understood that the preferred direction coding of each neuron is uniformly distributed. The theory shows a fault in that spinal cord cells could decode this complex output feature before dispatching the final signals to muscles.

Other interpretations charge single neurons with lighter duties. According to these other interpretations, the motor neurons should undertake much simpler tasks; for instance, neuron populations might code for single muscles. Some authors are severe critics of VC in the motor cortex. It would be enough to choose more correct analytical tools (see the previously discussed problem of technique correctness), stabilizing variance by discharge value squaring. This common scheme delivers the surprising result that the neuronal discharge can linearly represent the movement direction. Many correlational properties in some instances would thus be solely epiphenomena and not true system features (Todorov, 2000).

The example shows the true importance of the technique, as we discussed previously. For instance, we could miss some fundamental typicality of the CNS such as temporal binding, a diffuse property of the CNS. Temporal binding refers to the simultaneous oscillating coherence that embraces many cerebral areas coding for the same object (Engel et al., 1992). The activation by the common objectual reference of many areas shows cophased firing of neuronal population units while others shift to noninphase or noncoherent simultaneous activations. This would allow, however, the multiple simultaneous representation of different integrating objects. Moving onward along the hierarchical stages of the sensory pathway, neurons progressively integrate higher and higher features. It is thought that the coincidence detection allows for this gradual extension of unit competence and reduction of population numerosity.

Neighboring neurons often belong to the same dynamic unit or the same population, but vicinity is not the only criterion for deciding on a common input. Synchronies and coincidences are evident also between units or populations placed in the two different cerebral hemispheres. Synchrony enhancement and rephasing may encode common inputs without changing the discharge frequencies (Palva et al., 2005).

As we have seen, neurons thus can code either by FC or VC. A further mode is represented by the so-called temporal coding, nested in spike trains. The relative timing between two neuron discharges could signal phase, integration, or coincidence. We have observed that differences between time and rate codes are related to synaptic integrations. For instance, the fast-acting synapses (the bouton-like synapses) could provide synchronies of coincidence detection. Coincidence detection implies the narrowest signal integration windows (Figure 3.4). Coincidence detection and timing code go together and only signal groups attaining the coincidence distributed over many neuronal lines generate the timing code. Slow synapses (e.g., the en passant synapses) can provide rate coding, enabling integration of incoming signals over long time windows. Thus, there is probably no universal solution and flexible solutions are locally chosen to provide the best outcome.

3.4 ADAPTEDNESS OF TECHNIQUES

The biological object (the brain as a whole) requires a multiplicity of approaches and adaptive strategies to respond to different demands. It is true

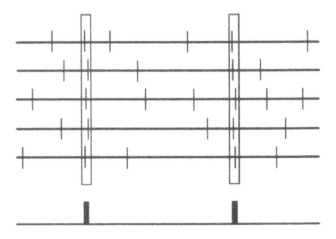

Figure 3.4.

that the application of a definite method able to extract universal properties from single particulars becomes a logical quest. Weak techniques may leave some problems untouched or unsolvable. For instance, in multiunit, multiple electrode extracellular neuronal recordings, one must ignore some classes of problems, such as the fast fluctuation of membrane potential. All the so-called low-level questions must be left unanswered. This choice has some obvious drawbacks, leading to acceptance of the operational status of ingenuous emergence, the blind acceptance of lower levels of the observation scale. In the meantime, we must pursue the right assessment of tokens, that is, of individual realizations of types or categories. In estimating specific measures, we must consider the true value of the chosen token to select truly warranted data. On a more abstract level, we must be clear on logical assumptions when we consider issues like the thalamo-cortico loop, whose definition is necessarily unsatisfactory for the principles discussed above. We must accept this implicit unsatisfactoriness as a useful tool to strengthen our approximation choices. As concerns the techniques, in summary, we must rely upon reduced versions of completeness and accept them for what they are. What we are looking for must be weighted in relation to what nature wishes to conceal.

BIBLIOGRAPHY

Abeles, M., Prut, Y., Bergman, H., and Vaadia, E., Synchronization in Neuronal Transmission and its Importance for Information Processing, *Progress in Brain Research*, Vol. 102, 395–404, 1994.

Adrian, E., and Zotterman, Y., The Impulses Produced By Sensory Nerve Endings. Part 3. Impulses Set up by Touch and Pressure. *J. Physiol.* (Lond.) Vol. 61, 465–483, 1926.

Adrian, E., and Bronk, D., The Discharge of Impulses in Motor Nerve Fibres. Part II. The Frequency of Discharge in Reflex and Voluntary Contractions. *J. Physiol.* (Lond.) Vol. 67, 119–151, 1929.

Babloyantz, A. (Ed.), Self-Organization, Emerging Properties, and Learning, *NATO Asi Series B. Physics*, Vol. 260, Plenum, 1991.

Barlow, H. B., Single Units And Sensation: A Neuron Doctrine for Perceptual Psychology? *Perception*. Vol. 1, No. 4, 371–394, 1972.

Beckermann, A., Flohr, H., and Kim, J. (Eds.), *Emergence or Reduction?* Part 1, de Gruyter, 1992.

Bialek, W., Rieke, F., de Ruyter van Steveninck, R., and Warland, D., Reading a Neural Code, *Science*, Vol. 252, 1854–1857, 1991.

Descartes, R., *Les Passions de L'âme*, Paris–Amsterdam, 1649. (English transl., *The Passions of the Soul*), in Vol. 1 of *The Philosophical Writings of Descartes*, ed. and trans. J. Cottingham, R. Stoothoff, D. Murdoch, and A. Kenny, Cambridge University Press, 1984–1991.

Descartes, R., *La Description du Corps Humain*, 1640. [English translation *The Description of the Human Body*, translation © George MacDonald Ross, 1975–1999.]

Engel, A. K., König, P., Kreiter, A. K., Schillen, T. B., and Singer W., Temporal Coding in the Visual Cortex: New Vistas on Integration in the Nervous System, *Trends Neurosci.*, Vol. 15, No. 6, 218–226, 1992.

Ferster, D., and Spruston, N., Cracking the Neuronal Code, *Science*, Vol. 270, No. 5237, 756–757, 1995.

Frege G., Über Sinn und Bedeutung, *Zeitschrift für Philosophie und Philosophische Kritik*, Vol. 100, 25–50, 1892. (Italian translation *Senso e Denotazione*, Bompiani, 1977.)

Gell-Mann, M., *The Quark and the Jaguar*, W. H. Freeman, 1994.

Georgopoulos, A. P., Kalaska, J. F., Caminiti, R., and Massey, J. T. On the Relations Between the Direction of Two-Dimensional Arm Movements and Cell Discharge in Primate Motor Cortex, *J. Neurosci.*, Vol. 2, No. 11, 1527–1537, 1982.

Hausman, D. M., Causal Relations among Tokens, Types, and Variables, University of Wisconsin, http://hypatia.ss.uci.edu/lps/psa2k/types-tokens.pdf.

Heylighen, F., The Growth of Structural and Functional Complexity during Evolution http://pespmc1.vub.ac.be/Papers/ComplexityGrowth.html.

Heylighen, F., and Aerts, D. (Eds.), *The Evolution of Complexity*, Kluwer Academic Publishers, 1996.

Izhikevich, E. M., *Dynamical Systems in Neuroscience: The Geometry of Excitability and Bursting*, MIT Press, 2007.

Mill, J. S., *A System of Logic*, Bk. III, Ch. 6, §1. University Press of the Pacific, 1998. (Originally published 1843.)

Palva J. M., Palva S., and Kaila K., Phase Synchrony among Neuronal Oscillations in the Human Cortex, *J. Neurosci.*, Vol. 25, No. 15, 3962–3972, 2005.

Pantev C., Engelien A., Candia V., and Elbert T., Representational Cortex in Musicians Plastic Alterations in Response to Musical Practice, *Annals of the New York Academy of Sciences*, Vol. 930, 300–314, 2001.

Penfield, W., and Rasmussen, T., *The Cerebral Cortex of Man: A Clinical Study of Localization of Function*, Hafner, 1968.

Rieke F., Warland D., de Ruyter van Steveninck R., and Bialek W., *Spikes: Exploring the Neural Code*, Cambridge University Press, 1997.

Shadlen, M., and Newsome, W., Noise, Neural Codes and Cortical Organization, *Current Opinion in Neurobiology*, Vol. 4, 569–579, 1994.

Shadlen, M., and Newsome, W., Is there a Signal in the Noise? [comment], *Current Opinion in Neurobiology*, Vol. 5, 248–250, 1995.

Smith B., and Varzi, A. C., Fiat and Bona Fide Boundaries, *Philosophy and Phenomenological Research*, Vol. 60, No. 2, 401–420, 2000

Softky, W., Simple Codes Versus Efficient Codes, *Current Opinion in Neurobiology*, Vol. 5, 239–247, 1995..

Spruston, N., Pyramidal Neurons: Dendritic Structure and Synaptic Integration, *Nature Reviews Neuroscience*, Vol. 9, March, 206–221, 2008.

Stephan A., Emergentism, Irreducibility and Downward Causation, *Grazer Phil. Stud.*, Vol. 65, 77–93, 2002.

Stevens, C., and Zador, A., Neural Coding: The Engima of the Brain, *Current Biology*, Vol. 12, 1370–1371, 1995.

Todorov, E., Direct Cortical Control of Muscle Activation in Voluntary Arm Movements: A Model, *Nature Neuroscience*, Vol. 3, No. 4, 391–398, 2000.

Varela, F. J., Thompson, E., and Rosch, E. (Eds.), *The Embodied Mind: Cognitive*

Science and Human Experience, Part III, MIT Press, 1991. (Italian translation, *La Via Di Mezzo Della Conoscenza,* Feltrinelli, Milan, 1992.)

Waldrop, M. M., *Complexity: the Emerging Science at the Edge of Order and Chaos,* Touchstone, 1992. (Italian translation, *Complessità: Uomini e Idee al Confine Tra Ordine e Caos,* Instar Libri, Torino 1996.)

CHAPTER 4

EVALUATION OF THE AUTONOMIC NERVOUS SYSTEM
From Algorithms to Clinical Practice

Maria Teresa La Rovere

4.1 INTRODUCTION

In the last thirty years, the development of interdisciplinary pathways between medicine and bioengineering, together with the integration of new methodologies and techniques, experimental models, and clinical realities, has led in various areas of neurosciences to new frontiers of knowledge with respect to the complexity of the nervous mechanisms that control the physiological processes and their pathological conditions, and to the definition of interpretative paradigms.

Regarding the understanding of nervous control of the cardiovascular system, biomedical engineers have provided laboratory systems able to evaluate autonomic nervous system regulation on the basis of simple cardio-respiratory measurements: heart rate, blood pressure, and respiratory activity (Malliani et al., 1991). Signal processing techniques have provided a means of analyzing those beat-to-beat fluctuations in cardiovascular signals, which have traditionally been ignored or, at best, treated as noise (Appel et al., 1989), thus offering clinicians potential evaluating tools for diagnosis and therapy. However, although the analysis of heart rate variability (HRV) using different approaches (time domain, frequency domain, and nonlinear dynamics) has been applied to many physiological and pathological conditions, its use has not yet been properly placed within clinical practice, and the initial enthusiasm aroused from early comments has cooled down due to the incapacity of the same experts to provide sound and simple enough tools to be used by nonexpert physicians (Huikuri et al., 1999). It is necessary to emphasize that, in an era of evidence-based medicine, founded on continuous defining and updating of the guidelines, the 1996 document that defined the standards regarding the measurement of heart rate variability, its physiopathological interpretation, and its clinical application has not yet been modified or revised (Task Force on HRV, 1996). For example, in the few lines dedicated to nonlinear methods they were still defined as "potential" tools for analysis of heart rate variability. The several insights provided by the theory of nonlinear dynamics in defining the abnormality of heart rate behavior and its prognostic meaning deserves extensive updating (Mäkikallio et al., 2004).

In this chapter, besides a synthesis of the clinical information provided by the analysis of the variability of the heart period, within ischemic heart disease and chronic heart failure, a clinical scenario will be defined in which these methods have been recently put to a test: the stratification of the risk of sudden death in the era of automatic implantable cardiac defibrillators.

4.2 RELATIONSHIP BETWEEN HEART RATE VARIABILITY AND MYOCARDIAL INFARCTION

Table 4.1 shows the most relevant studies mainly representing the relationship between heart rate variability and prognosis in patients after myocardial infarction.

The first clinical evidence derived from the Multicenter Post Infarction Project (MPIP) (Kleiger et al., 1987) showed that in a sample population of over 800 patients the time-domain measure, namely the standard deviation of normal-to-normal beats (SDNN) < 50 ms, was associated with a risk of death 5.3 times higher than that found in subjects with SDNN >100 ms. Despite the fact that this parameter was significantly related to heart rate, left ventricular ejection fraction, and frequency of ventricular ectopies per hour, it remained an independent predictor in the course of an average 31 month follow-up.

Subsequent analysis on the same sample population (Bigger et al., 1992) documented the value of the spectral parameters in the prognostic stratification during the postinfarction period, both in terms of all-cause mortality and mortality due to arrhythmic causes. However, in a critical analysis, the results of this study may have hindered the diffusion of these methods in clinical practice. As a matter of fact, the power of the low-frequency (LF) component, which, upon analysis of the survival curves, better separated groups with different risks of events, showed an opposite behavior with respect to the current interpretative paradigm; that is, having

Table 4.1. Relationship between heart rate variability parameters and events following acute myocardial infarction

Study	No. of patients (events)	Type and time of HRV measure	Follow-up	End points	Predictive parameter
Kleiger et al., 1987	808 (127 deaths)	24 hours, 11 ± 3 days post-MI	363 ± 243 days	All-cause mortality	SDNN < 50 ms vs. SDNN > 100 ms (RR = 5.3)
Odemuyiwa et al., 1991	385 (44 deaths)	24 hours, predischarge	151–161 days	All-cause mortality	HRV index ≤ sensitivity and 39 ms better specificity than LVEF ≤ 40%
Cripps et al., 1991	177 (17 events)	24 hours, 7 days post-MI	Median 16 months	Sudden death, sustained VT	HRV index most significant single predictor
Farrell et al., 1991	416 (24 events)	24 hours, 7 days post-MI	Mean 612 days	Sudden death, sustained arrhythmias	HRV index better univariate predictor
Bigger et al., 1992	715 (119 deaths)	24 hours, 2 weeks post-MI	Up to 4 years	All cause mortality	All spectral measures univariate predictors. Reduced ULF and VLF better predictors
Bigger et al., 1992	715 (68 arrhythmic deaths)	24 hours, 2 weeks post-MI	Up to 4 years	Arrhythmic death	All spectral measures univariate predictors. Reduced VLF better independent predictor
Zuanetti et al., 1996	567 (52 deaths)	24 hours, predischarge	1000 days	All-cause mortality	SDNN, rMSSD and pNN50 independent predictors (RR 3.0, 2.8, 3.5, respectively)
Toubol et al., 1997	471 (39 deaths)	24 hours, 10 days post-MI	Median 31 months	All-cause mortality	Nighttime mean RR interval RR < 750 ms, daytime SDNN < 100 ms

(continued)

Table 4.1. *Continued*

Study	No. of patients (events)	Type and time of HRV measure	Follow-up	End points	Predictive parameter
Zabel et al., 1998	250	24 hours, predischarge	Mean 32 months	Cardiac mortality, resuscitated VT or VF	SDNN significantly higher in subjects free of events
La Rovere et al., 1998	1284 (44 deaths, 5 nonfatal cardiac arrests)	24 hours, 15 ± 10 days post-MI	21 ± 8 months	Total cardiac mortality	SDNN < 70 ms vs. SDNN > 70 ms, independent predictor

Notes: RR: relative risk; VT: ventricular tachycardia; VF: ventricular fibrillation.

identified the frequency of the LF component as the marker of the sympathetic nervous activity (as opposed to the high-frequency component, HF, as a marker of vagal activity), the expectation was that of an increase, rather than a decrease, in those clinical situations characterized by activation of the sympathetic nervous system. The results of this study were completely overlooked for many years. These interpretative difficulties, together with some technical limitations, the absence of standardization, and a cultural approach experienced as too complex for the inexperienced user, have focused attention on the time-domain measurements, the triangular index (Odemuyiwa et al., 1991; Farrell et al., 1991), and the SDNN. With reference to Farrell and coworkers' work within a sample population of over 400 patients, the triangular index < 20 msec predicted both all-cause mortality and arrhythmic events. Depressed RR interval variability predicted more accurately arrhythmic events rather than all-cause mortality, and was independent from other well-known noninvasive arrhythmic risk indicators such as ventricular ectopies/hour and the presence of late potentials at signal-averaged ECG.

The ATRAMI study (Autonomic Tone and Reflexes After Myocardial Infarction; La Rovere et al., 1998) has generally ratified, in the postinfarction risk stratification, the role of heart rate variability and the indexes of simpatho-vagal balance by applying this analysis to a sample population of over 1200 patients. It has to be underscored that when dealing with "continuous" biological variables used to predict risk factor, the cut-off value represents a critical problem because its "optimal" value, defined in terms of sensitivity and specificity, may vary according to different pathologies and in relation to various historical periods of pharmacological treatment. As a matter of fact, both analysis of the GISSI (Zuanetti et al., 1996) study and data from the ATRAMI (La Rovere et al., 1998) study highlighted the need to use higher cut-off values with respect to the MPIP (Kleiger et al., 1987) study. Indeed, the MPIP study was carried out in the prethrombolytic era and included patients who had extended myocardial necrosis, thus with extended autonomic damage. In both GISSI and ATRAMI studies, a SDNN < 70 ms was identified in the 15th percentile of the population and contributed to a three-fold risk of cardiac death, which was independent of the left-ventricular ejection fraction and of the presence of frequent ventricular arrhythmias.

The ATRAMI study also defined the importance of the combination between autonomic parameters and left-ventricular function, and the predictive ability of SDNN in relatively old patients who represent a majority in our times. Death risk in the combined low-SDNN and depressed left-ventricular ejection fraction increased six times more in patients older than 65 years. The EMIAT (European Myocardial Infarct Amiodarone Trial, Malik et al., 2000) also supports the clinical value of combined heart rate variability with left-ventricular function. In the EMIAT study, even though mortality was not reduced by Amiodarone treatment, patients, with reduced heart rate variability and low left-ventricular function received significant benefit in the form of reduction of both arrhythmic events and overall mortality.

4.3 RELATIONSHIP BETWEEN HEART RATE VARIABILITY AND HEART FAILURE

The peculiar interest in the analysis of heart rate variability in heart failure stems from the well-recognized role of the activation of the adrenergic system as a pathogenetic factor in the progression of the disease. The quantification of the degree of adrenergic dysregulation would allow one to obtain clinical and prognostic information that could represent a guide for the choice and dosage of drugs.

Table 4.2 shows some of the numerous studies that, with different methods, have evaluated the relationship between heart rate variability and heart failure.

A number of studies have documented the significant relationship between the reduction of heart rate variability indexes in time domain, severity of clinical situation, prognosis (Binder et al., 1992; Couniham et al., 1993; Nolan et al., 1998), and neuro-hormonal activation (Buger and Aronson, 2001). The relationship with spectral decomposition parameters, with particular regard to the LF component that, in healthy subjects, prevails during adrenergic activation (Montano et al., 1994), seems to be more controversial.

In fact, although in patients with moderate heart failure in I–II class NYHA the LF component is increased (Guzzetti et al., 1995), a decrease

Table 4.2. Relationship between heart rate variability parameters and events in patients with heart failure

Study	No. of patients (events)	Etiology	Follow-up	Ejection fraction	Predictive parameter
Binder et al., 1992	61 (10)	Ischemic/ idiopathic	510 days	< 20%	Reduced SDANN
Woo et al., 1992	24 (12)	Ischemic/ idiopathic	Unknown	17%	Poincaré plot
Couniham et al., 1993	104 (10)	Hypertrophic	Unknown	Not assessed	Reduced SDANN
Brouwer et al., 1996	95 (17)	Ischemic/ idiopathic	950 days	29%	Poincaré plot
Ponikowsky et al, 1997	102 (19)	Ischemic/ idiopathic	584 days	26%	Reduced SDNN, reduced LF power
Szabo et al., 1997	159 (30)	Ischemic/ idiopathic	690 days	28%	Reduced SDNN, increased LF power
Nolan et al., 1998	433 (52)	Ischemic/ idiopathic	82 days	42%	Reduced SDNN
Galinier et al., 2000	190 (55)	Ischemic/ idiopathic	660 days	28%	Reduced SDNN, reduced day-time LF power
Lucreziotti et al., 2000	75 (11)	Ischemic/ idiopathic	340 days	22%	Reduced LF/HF in short-term recording
Makikallio et al., 2001	499 (210)	Ischemic/ idiopathic	665 days	< 35%	Fractal esponent
La Rovere et al., 2003	444 (39)	Ischemic/ idiopathic	3 years	24%	Reduced LF power in short-term recording

of all the spectral components and mainly of the LF component is observed in patients with advanced heart failure, despite a marked adrenergic activation and high levels of circulating catecholamines (Mortara et al., 1994). The data is confirmed in several series of patients (Ponikowski et al., 1997; Galinier et al., 2000; La Rovere et al., 2003) and appears to be particularly effective in predicting sudden death (Galinier et al., 2000; La

Rovere et al., 2003) with exception of Szabo and coworkers' work (Szabo et al., 1997) in which, in a sample population of 159 patients, the risk of death increased 2.5 fold in patients with increased LF frequency.

Several hypotheses have been advanced to explain the apparent paradox of a reduction of power in LF frequency in conjunction with increased sympathetic activity, including reduced responsiveness of the nodal cells during persistent adrenergic activation (Guzzetti et al., 1995), loss of oscillatory behavior during overwhelming chronic sympathetic overactivity (Mortara et al., 1994), a central abnormality in autonomic modulation (Van de Borne et al., 1997), and the effect of an altered baroreflex (Sleight et al., 1995). It is important to emphasize that in conditions in which heart rate variability is extremely reduced, the use of standardized units or the LF/HF ratio cannot highlight the relative prevalence of LF over HF. During enhanced adrenergic activation, the aspect of spectral variability is similar to that observed in transplanted hearts, with LF frequency variability nearly absent. This situation is defined as functional denervation, in which, although there may be a certain amount of variability in HF frequency due to respiratory activity, the LF/HF ratio is often less than one (Mortara et al., 1994).

4.4 RELATIONSHIP BETWEEN HEART RATE AND BLOOD PRESSURE VARIABILITY

The analysis of both spontaneous oscillations of heart rate and blood pressure and their interrelations allows one to obtain information on the activity of baroreceptive mechanisms that represent a key factor in cardiovascular regulation. The arterial baroreceptor reflex system plays a dominant role in preventing short-term, wide fluctuations in arterial blood pressure, as repeatedly shown in experiments demonstrating that, in many animal species, arterial baroreceptor denervation results in an increase in the variability of blood pressure but without a long-term change in its absolute level. Arterial baroreceptors provide the central nervous system with a continuous stream of information on changes in blood pressure (which are sensed by the stretch receptors in the wall of the carotid sinuses and aortic arch), on the basis of which efferent autonomic neural activity is dynamically modulated. Activation of arterial baroreceptors by a rise in systemic arterial pressure leads to an increase in the discharge of vagal cardioinhibitory neurons and a decrease in the discharge of sympathetic neurons both to the heart and peripheral blood vessels. This results in bradycardia, decreased cardiac contractility, decreased peripheral vascular resistance, and decreased venous return. Conversely, a decrease in systemic arterial pressure causes the deactivation of baroreceptors, with subsequent enhancement of sympathetic activity and vagal inhibition, leading to tachycardia and increase of cardiac contractility, vascular resistance, and venous return. In physiological conditions and with normal levels of arterial pressure, baroceptors are constantly active and exert a constant tonic inhibition of the sympathetic efferent activity.

Cardiovascular diseases are often accompanied by an impairment of baroreflex mechanisms, with a reduction of inhibitory activity and an imbalance in the physiological sympathetic–vagal outflow to the heart, thus resulting in a chronic adrenergic activation (Eckberg and Sleight, 1992).

In the clinical setting, the most widely used method to evaluate the sensitivity of the baroreceptor reflex and the degree of its damage involves the usage of drugs that modify systemic blood pressure, baroreflex activity, and autonomic responses. After intravenous administration of a vasoactive drug, baroreflex sensitivity is evaluated as the slope resulting from the relationship between the variation of the cardiac cycle and the variation (increase/decrease) of blood pressure induced by the utilized drug (La Rovere et al., 2000). The results of the ATRAMI study (La Rovere et al., 1998) have demonstrated that the evaluation of baroreflex sensitivity provides prognostic implications, similar to those obtained from the evaluation of heart rate variability. Indeed, the presence of depressed baroreflex sensitivity (< 3 msec/mmHg) identified a 2.8 fold risk of cardiac death in patients with preserved baroreflex function, independent of left-ventricular function and frequent ventricular ectopies.

However, the need for intravenous cannulation and the use of a drug limits the applicability of this technique. It is thus evident that the role of noninvasive methods for evaluation of biological signals is increasing for clinical evaluation of risk indicators. Based on the evidence that baroreceptors are not only activated by abrupt changes in arterial pressure but also by small variations continuously occurring during daily life, more recent computer-based techniques have allowed us to assess the baroreceptor–heart-rate reflex by analyzing spontaneous beat-to-beat fluctuations of arterial pressure and heart rate. These techniques are inherently simple, noninvasive, and low-cost, and allow a detailed assessment of the interaction between baroreflex function and the daily life modulation of cardiovascular parameters. Two basic approaches have been proposed so far: one based on time-domain and the other on frequency-domain measurements.

The sequence method, described by Parati et al. (1988), is based on the identification of three or more consecutive beats in which progressive increases/decreases in systolic blood pressure are followed by progressive lengthening/shortening in RR interval. The threshold values for including beat-to-beat systolic blood pressure and RR interval changes in a sequence are set at 1 mmHg and 6 ms, respectively.

Evaluation of baroreflex sensitivity by spectral methods is based on the concept that each spontaneous oscillation in blood pressure elicits an oscillation at the same frequency in the RR interval by the effect of arterial baroreflex activity (Cerutti et al., 2004). Two main oscillations are usually considered: one centered around 0.1 Hz within the LF band, and the other associated with respiratory activity within the HF band. Therefore, these methods allow a clear definition of the oscillatory components that contribute to baroreflex sensitivity measurement.

The prognostic value of the noninvasive methods of evaluation of baroceptive reflexes are the object of studies presently underway.

4.5 SUDDEN DEATH RISK STRATIFICATION, PROPHYLACTIC TREATMENT, AND UNRESOLVED ISSUES

Sudden cardiac death accounts for two-thirds of fatal events related to heart disease. Coronary heart disease and nonischemic cardiomyopathy are the most common causes of sudden cardiac death. A series of clinical randomized studies have consistently shown that therapy with an implantable cardioverter defibrillator (ICD) results in a significant effect on survival through a reduction in the risk of sudden death in this population.

The first evidence came from the Multicenter Automatic Defibrillator Implantation Trial (MADIT) (Moss et al., 1996), which showed that in patients with coronary artery disease, left-ventricular dysfunction with ejection fraction < 35%, spontaneous asymptomatic nonsustained ventricular tachycardia (VT), and inducible nonsuppressible VT on electrophysiologic study there was a 54% reduction in all-cause mortality in the implanted patients compared with those treated by conventional therapy.

The Multicenter Unsustained Tachycardia Trial (MUSST) (Buxton et al., 1999) included similar patients with reduced ejection fraction (≤ 40%), spontaneous asymptomatic nonsustained VT, and inducible VT, and also followed up in a registry those patients who had clinical criteria for the trial but not inducible arrhythmias. Surprisingly, the mortality rate of these patients was higher than that of inducible-ICD-treated patients. This data implied that noninducible patients with left-ventricular dysfunction and nonsustained VT may also benefit from a prophylactic ICD, and that electrophysiologic study may be an inadequate risk stratifier (Buxton et al., 2000).

The Multicenter Automatic Defibrillator Implantation Trial (MADIT-II, Moss et al., 2002) tested the hypothesis that prophylactic ICD implantation would reduce mortality in patients with ischemic cardiomyopathy without the requirement of further risk stratification. This study enrolled 1232 patients with prior myocardial infarction with ejection fraction no more than 30% and NYHA class I to III who were randomly assigned to ICD or conventional medical therapy. After a mean follow-up of 20 months, ICD patients had a 31% relative risk reduction in mortality.

Similarly, the Sudden Cardiac Death in Heart Failure Trial (SCD-

HeFT, Bardy et al., 2005) showed that in patients with either ischemic or nonischemic cardiomyopathy, an ejection fraction < 35%, no history of sustained arrhythmias, and on optimal medical therapy, there was a significant relative risk reduction of 23% with a simple shock-only ICD compared with placebo and amiodarone, and an absolute reduction of 7.2% over 5 years.

However, whether left-ventricular ejection fraction should now be the primary factor determining ICD eligibility remains controversial. Actually, although most primary prevention trials that selected ICD candidates based on reduced left-ventricular function have shown a significant reduction in sudden cardiac death and all-cause mortality with an ICD, a large number of patients in these studies did not receive any therapy, or received inappropriate therapy. For instance, in MADIT-II the probability of ICD therapy for VT/VF was 40% during a 4-year follow-up, whereas in SCD-HeFT only 21% of patients randomized to ICD therapy received appropriate treatment for ventricular tachyarrhythmias during a 45-month follow-up.

This observation together with the socioeconomic impact of the MADIT-II strategy gave way to a debate (Reynolds and Josephson, 2002; Buxton, 2005) expressing concern that the less-selective criteria for ICD implantation of MADIT-II could result in an elevated number of patients who, without receiving a real benefit from the device, could be exposed to a high unnecessary risk of complications. Editorialists also suggested that further risk stratification is needed (Buxton, 2005). The possibility of identifying, among patients with low ejection fraction, subjects with a low-risk profile for arrhythmic events and, thus, less in need of the ICD implant, could lead to a significant reduction in the number of unnecessary ICDs in spite of a minimal or null reduction in life expectancy.

4.6 THE ROLE OF AUTONOMIC MARKERS IN NONINVASIVE RISK STRATIFICATION

The information provided by autonomic markers may well be useful in the decision process for ICD implantation. Indeed, among postinfarction patients with depressed left-ventricular ejection fraction and without nonsustained spontaneous ventricular tachycardia (who could be considered at low risk following MADIT-I/MUSST), the presence or absence of an impaired baroreflex gain could identify two subgroups with significantly different two-year cardiac mortality: 18% versus 4.6% ($p = 0.01$) (La Rovere et al., 2001). Compared to the MADIT-I/MUSST strategy, the analysis of baroreflex sensitivity would extend the number of implanted ICDs to the patients with a markedly depressed autonomic balance; conversely, compared to the MADIT-II strategy (which suggests implanting the device in all subjects with depressed left-ventricular function), this approach based on the assessment of baroreflex sensitivity, by not treating the patients with well-preserved autonomic balance, could significantly reduce the number of inactive defibrillators implanted. In the ATRAMI population, we have analyzed the clinical value of baroreflex sensitivity and heart rate variability in MADIT-II-like patients (La Rovere et al., 2005). In the ATRAMI study, there was a partition into three levels: under the 15th percentile (high risk), between the 15th percentile and the median (intermediate risk), and over the median (low risk). The hypothesis to be proven was that a preserved autonomic balance (BRS, SDNN value over the median) could identify patients with a very low or null risk for sudden death, despite a severe reduction in left-ventricular function. In a population of 70 patients with an ejection fraction <30%, 28 patients were in the high-risk category for BRS (BRS < 3 ms/mmHg) and 30 for SDNN (SDNN < 70 ms), 27 in the intermediate risk for both BRS and SDNN (BRS 3–6 ms/mmHg, SDNN 70–105 ms), and 15 and 13 were in the low risk group (BRS > 6 ms/mmHg and SDNN > 105 ms). During a two-year follow-up, 11 patients died; no major arrhythmia or sudden death occurred among patients with a well-preserved BRS or SDNN. Identifying the subjects with a lower risk of cardiac events through the evaluation of autonomic balance could allow a 20% reduction in the number of ICDs used for MADIT-II-like patients.

Similar results have been obtained in a prospective study of 274 participants in the Defibrillators in Non-Ischemic Cardiomyopathy Treatment Evaluation (DEFINITE) Trial, a randomized controlled trial that evaluated the role of prophylactic ICD placement in patients with nonischemic dilated cardiomyopathy (Rashba et al., 2006). The patients underwent 24-hour Holter recording for analysis of heart rate variability. The primary heart rate variability variable was the SDNN. Patients with atrial fibrillation and frequent ventricular ectopy (>25% of beats) were excluded from analysis (23% of patients). SDNN was categorized in tertiles, and Kaplan–Meier analysis was performed to compare survival in the three tertiles and excluded patients. After three-year follow-up, significant differences in mortality rates were observed: SDNN > 113 ms: 0 (0%), SDNN 81–113 ms: 5 (7%), SDNN < 81 ms: 7 (10%), excluded patients: 11 (17%) ($p = 0.03$). There were no deaths in the tertile with SDNN > 113 ms, regardless of treatment assigned (ICD versus control). This data in patients with nonischemic dilated cardiomyopathy confirmed that subjects with preserved heart rate variability have an excellent prognosis and may not benefit from prophylactic ICD placement. Patients with severely depressed heart rate variability and patients who are excluded from heart rate variability analysis because of atrial fibrillation and frequent ventricular ectopy have the highest mortality.

This data demonstrates how the information provided by the evaluation of the neural control of cardiovascular function not only pertains to specialized laboratories, but can find application within daily clinical practice.

The time is ripe for larger multicenter clinical studies using the now available and validated noninvasive methodology, so that autonomic markers will enter routine clinical practice. This is already beginning with the NIH-sponsored 4500 patient study (VEST/PREDICTS), which aims to refine selection of patients for ICD implantation (http://clinicaltrials.gov).

REFERENCES

Appel, M. L., Berger, R. D., Saul, J. P., Smith, J. M., and Cohen, R. J., Beat-to-beat Variability in Cardiovascular Variables: Noise or Music? *J. Am. Coll. Cardiol.,* Vol. 14, No. 5, 1139–1148, 1989.

Bardy, G. H., Lee, K. L., Mark, D. B., et al., Amiodarone or an Implantable Cardioverter–Defibrillator for Congestive Heart Failure? *N. Engl. J. Med.,* Vol. 352, No. 3, 225–237, 2005.

Bigger, J. T., Jr, Fleiss, J. L., Rolnitzky, L. M., Kleiger, R. E., and Rottman, J. N., Frequency Domain Measures of Heart Period Variability and Mortality after Myocardial Infarction, *Circulation.* Vol. 85, No. 1, 164–171, 1992.

Binder, T., Frey, B., Porenta, G., et al., Prognostic Value of Heart Rate Variability in Patients Awaiting Cardiac Transplantation, *Pacing Clin. Electrophysiol.* Vol. 15, No. 11, 2215–2220, 1992.

Brouwer, J., van Veldhuisen, D. J., et al., for the Dutch Ibopamine Multicenter Trial Study Group, Prognostic Value of Heart Rate Variability During Long-term Follow-up In Patients with Mild to Moderate Heart Failure, *J. Am. Coll. Cardiol.,* Vol. 28, No. 5, 1183–1189, 1996.

Buger, A. J., and Aronson, D., Activity of the Neurohormonal System and its Relationship to Autonomic Abnormalities in Decompensated Heart Failure, *J. Card. Fail.,* Vol. 7, No. 2, 122–128, 2001.

Buxton, A. E., Lee, K. L., Fisher, J. D., et al., for the Multicenter Unsustained Ventricular Tachycardia Trial Investigators, A Randomized Study of the Prevention of Sudden Death in Patients with Coronary Artery Disease, *N. Engl. J. Med.,* Vol. 341, No. 25, 1882–1890, 1999.

Buxton, A. E., Lee, K. L., Di Carlo, L., et al., for the Multicenter Unsustained Ventricular Tachycardia Trial Investigators, Electrophysiologic Testing to Identify Patients with Coronary Artery Disease Who are at Risk For Sudden Death, *N. Engl. J. Med.,* Vol. 342, No. 27, 1937–1945, 2000.

Buxton, A. E., Should Everyone with an Ejection Fraction Less Than or Equal to 30% Receive an Implantable Cardioverter-Defibrillator? *Circulation,* Vol. 111, No., 19, 2537–2549, 2005.

Cerutti, S., Baselli, S., Bianchi, A. M., Mainardi, L. T., and Porta, A., Analysis of the Interactions Between Heart Rate and Blood Pressure Variabilities, in Malik, M., and Camm, A. J. (Eds.), *Dynamic Electrocardiography,* Blackwell Futura, pp 170–179, 2004.

Couniham, P. J., Fei, L., Bashir, Y., Farrell, T. G., Haywood, G. A., and McKenna, W. J., Assessment of Heart Rate Variability in Hypertrophic Cardiomyopathy:

Association with Clinical And Prognostic Features, *Circulation*, Vol. 88, No. 4, 1682–1690, 1993.

Cripps, T. R., Malik, M., Farrell, T. G., and Camm, A. J., Prognostic Value of Reduced Heart Rate Variability after Myocardial Infarction: Clinical Evaluation of a New Analysis Method, *Br. Heart J.*, Vol. 65, No. 1, 14–19, 1991.

Eckberg, D. L., and Sleight, P., *Human Baroreflexes in Health and Disease*, Clarendon Press, 1992.

Farrell, T. G., Bashir, Y., Cripps, T., et al., Risk Stratification for Arrhythmic Events in Postinfarction Patients Based on Heart Rate Variability, Ambulatory Electrocardiographic Variables and the Signal-Averaged Electrocardiogram, *J. Am. Coll. Cardiol.*, Vol. 18, No. 3, 687–697, 1991.

Galinier, M., Pathak, A., Fourcade, J., et al., Depressed Low Frequency Power of Heart Rate Variability as an Independent Predictor of Sudden Death in Chronic Heart Failure, *Eur. Heart J.*, Vol. 21, No. 6, 475–482, 2000.

Guzzetti, S., Cogliati, C., Turiel, M., Crema, C., Lombardi, F., and Malliani, A., Sympathetic Predominance Followed by Functional Denervation in the Progression of Chronic Heart Failure, *Eur. Heart J.*, Vol. 16, 1100–1107, 1995.

http://clinicaltrials.gov/ct2/results?term=VEST%2FPREDICTS, accessed May 19, 2008.

Huikuri, H. V., Mäkikallio, T., Airaksinen, K. E. J., et al., Measurement of Heart Rate Variability: A Clinical Tool or a Research Toy? *J. Am. Coll. Cardiol.*, Vol. 34, No. 7, 1878–1883, 1999.

Kleiger, R. E., Miller, J. P., Bigger, J. T., and Moss, A. J., and the Multicenter Post-Infarction Research Group, Decreased Heart Rate Variability and its Association with Increased Mortality after Acute Myocardial Infarction, *Am. J. Cardiol.*, Vol. 59, No. 4, 256–262, 1987.

La Rovere, M. T., Bigger, J. T., Jr., Marcus, F. I., Mortara, A., and Schwartz, P. J., for the ATRAMI Investigators, Baroreflex Sensitivity and Heart Rate Variability in Prediction of Total Cardiac Mortality after Myocardial Infarction, *Lancet*, Vol. 351, No. 9101, 478–484, 1998.

La Rovere, M. T., and Schwartz, P. J., Baroreflex Sensitivity, in *Cardiac Electrophysiology, From Cell to Bedside*, 3rd ed., Zipes, D. P., and Jalife, J. (Eds.), Saunders, pp. 771–780, 2000.

La Rovere, M. T., Pinna, G. D., Hohnloser, S. H., et al., Baroreflex Sensitivity and Heart Rate Variability in the Identification of Patients at Risk for Life-Threatening Arrhythmias: Implications for Clinical Trials, *Circulation*, Vol. 103, No. 16, 2072–2077, 2001.

La Rovere, M. T., Pinna, G. D., Maestri, R., et al., Short-Term Heart Rate Variability Strongly Predicts Sudden Cardiac Death in Chronic Heart Failure Patients, *Circulation*, Vol. 107, No. 4, 565–570, 2003.

La Rovere, M. T., and Schwartz, P. J., for the ATRAMI Investigators. Cost Concerns for Implantable Cardioverter Defibrillators Implant in Post Myocardial Infarction Patients: The Value Of Autonomic Markers, *Heart Rhythm*, 2 (Abs Suppl): S-188, 2005.

Lucreziotti, S., Gavazzi, A., Scelsi, L., et al., Five-Minute Recording of Heart Rate Variability in Severe Chronic Heart Failure: Correlates with Right Ventricular Function and Prognostic Implications, *Am. Heart J.*, Vol. 139, No. 6, 1088–1095, 2000.

Mäkikallio, T. H., Huikuri, H. V., Hintze, U., et al., Fractal Analysis and Time- and Frequency-Domain ·Measures of Heart Rate Variability as Predictors of Mortality in Patients with Heart Failure, *Am. J. Cardiol.*, Vol. 87, No. 2, 178–182, 2001.

Mäkikallio, T. H., Perkiömäki, J. S., and Huikuri, H. V., Non Linear Dynamics Of RR Interval, in *Dynamic Electrocardiography,* Malik, M., and Camm, A. J., (Eds.), Blackwell Futura, pp. 22–30, 2004.

Malik, M., Camm, A. J., Julian, D. G., et al., on behalf of the EMIAT Investigators, Depressed Heart Rate Variability Identifies Postinfarction Patients who Might Benefit From Prophylactic Treatment with Amiodarore: A Substudy of EMIAT (The European Myocardial Infarct Amiodarone Trial), *J. Am. Coll. Cardiol.*, Vol. 35, No. 5, 1263–1275, 2000.

Malliani, A., Pagani, M., Lombardi, F., and Cerutti, S., Cardiovascular Neural Regulation Explored in the Frequency Domain, *Circulation,* Vol. 84, No. 2, 482–492, 1991.

Montano, N., Ruscone, T. G., Porta, A., Lombardi, F., Pagani, M., and Malliani, A., Power Spectrum Analysis of Heart Rate Variability to Assess the Changes in Sympathovagal Balance during Graded Orthostatic Tilt, *Circulation,* Vol. 90, No. 4, 1826–1831, 1994.

Mortara, A., La Rovere, M. T., Signorini, M. G., et al., Can Power Spectral Analysis of Heart Rate Variability Identify a High Risk Subgroup of Congestive Heart Failure Patients with Excessive Sympathetic Activation? A Pilot Study Before and After Heart Transplantation, *Br. Heart J.,* Vol. 71, No. 3, 422–430, 1994.

Moss, A. J., Hall, W. J., Cannom, D. S., et al., for the Multicenter Automatic Defibrillator Implantation Trial Investigators, Improved Survival with an Implanted Defibrillator in Patients with Coronary Disease at High Risk for Ventricular Arrhythmia, *N. Engl. J. Med.,* Vol. 335, No. 26, 1933–1940, 1996.

Moss, A. J., Zareba, W., Hall, W. J., et al., for the Multicenter Automatic Defibrillator Implantation Trial Investigators, Prophylactic Implantation of a Defibrillator in Patients with Myocardial Infarction and Reduced Ejection Fraction, *N. Engl. J. Med.,* Vol. 346, No. 12, 877–883, 2002.

Nolan, J., Batin, P. D., Andrews, R., et al., Prospective Study of Heart Rate Variability and Mortality in Chronic Heart Failure: Results of the United Kingdom Heart Failure Evaluation and Assessment of Risk Trial (UK-Heart), *Circulation,* Vol. 98, No. 15, 1510–1516, 1998.

Odemuyiwa, O., Malik, M., Farrell, T., Bashir, Y., Poloniecki, J., and Camm, J., Comparison of the Predictive Characteristics of Heart Rate Variability Index and Left Ventricular Ejection Fraction in Prediction for All-Cause Mortality, Arrhythmic Events and Sudden Death after Acute Myocardial Infarction, *Am. J. Cardiol.,* Vol. 68, No. 5, 434–439, 1991.

Parati, G., Di Rienzo, M., Bertinieri, G., Pomidossi, G., Casadei, R., Groppelli, A., Pedotti, A., Zanchetti, A., and Mancia, G., Evaluation of the Baroreceptor-Heart Rate Reflex by 24-Hour Intra-Arterial Blood Pressure Monitoring in Humans, *Hypertension,* Vol. 12, No. 2, 214–222, 1988.

Ponikowski, P., Anker, S. D., Chua, T. P., et al., Depressed Heart Rate Variability as an Independent Predictor of Death in Chronic Congestive Heart Failure Secondary to Ischemic or Idiopathic Dilated Cardiomyopathy, *Am. J. Cardiol.,* Vol. 79, No. 12, 1645–1650, 1997.

Rashba, E. J., Estes, N. A., Wang, P., et al., Preserved Heart Rate Variability Identifies Low-Risk Patients with Nonischemic Dilated Cardiomyopathy: Results from the DEFINITE Trial, *Heart Rhythm,* Vol. 3, 281–286, 2006.

Reynolds, M. R., and Josephson, M. E., MADIT II Debate: Risk Stratification, Costs, and Public Policy, *Circulation,* Vol. 108, No. 13, 1779–1783, 2003.

Sleight, P., La Rovere, M. T., Mortara, A., et al., Physiology and Pathophysiology of Heart Rate and Blood Pressure Variability in Humans: Is Power Spectral Analysis Largely an Index of Baroreflex Gain? *Clin. Sci.,* Vol. 88, No. 1, 103–109, 1995.

Szabo, B. M., van Veldhuisen, D. J., van der Veer, N., et al., Prognostic Value of Heart Rate Variability in Chronic Congestive Heart Failure Secondary to Idiopathic or Ischemic Dilated Cardiomyopathy, *Am. J. Cardiol.,* Vol. 79, No. 7, 978–980, 1997.

Task Force of the European Society of Cardiology and the North American Society of Pacing and Electrophysiology, Heart Rate Variability, Standards of Measurement, Physiological Interpretation and Clinical Use, *Circulation,* Vol. 93, No. 5, 1043–1065, 1996.

Touboul, P., Andre-Fouet, X., Leizorovicz, A., et al., Risk Stratification after Myocardial Infarction: a Reappraisal in the Era of Thrombolysis; The Groupe d'Etude du Pronostic de l'Infarctus du Myocarde (GREPI), *Eur. Heart. J.,* Vol. 18, No. 1, 99–107, 1997.

Van de Borne, P., Montano, N., Pagani, M., et al., Absence of Low-Frequency Variability of Sympathetic Nerve Activity in Severe Heart Failure, *Circulation,* Vol. 95, No. 6, 1449–1454, 1997.

Woo, M. A., Stevenson, W. G., Moser, D. K., Trelase, R. B., and Harper, R. M., Patterns of Beat-to-Beat Heart Rate Variability in Advanced Heart Failure, *Am. Heart J.,* Vol. 123, No. 3, 704–710, 1992.

Zabel, M., Kligenheben, T., Franz, M. R., and Hohnloser, S. H., Assessment of QT Dispersion for Prediction of Mortality or Arrhythmic Events after Myocardial Infarction: Results of a Prospective, Long-Term Follow-up Study, *Circulation,* Vol. 97, No. 25, 2543–2550, 1998.

Zuanetti, G., Neilson, J. M. M., Latini, R., Santoro, E., Maggioni, A. P., and Ewing, D. J., Prognostic Significance of Heart Rate Variability in Post-Myocardial Infarction Patients in the Fibrinolytic Era: The GISSI-2 results, *Circulation,* Vol. 94, No. 3, 432–436, 1996.

This page intentionally left blank

PART III

MODELS AND BIOMEDICAL SIGNALS

This page intentionally left blank

CHAPTER 5

PARAMETRIC MODELS FOR THE ANALYSIS OF INTERACTIONS IN BIOMEDICAL SIGNALS

Giuseppe Baselli, Alberto Porta, and Paolo Bolzern

5.1 INTRODUCTION

In biomedical applications, dealing with several simultaneously recorded signals, with different though partially correlated information, is quite a common requirement. Hence, it is important to interpret mutual interactions and their dynamic features, beyond those of each single signal. This chapter is devoted to describing the potentials, limits, and necessary skills needed to solve this problem by means of a parametric approach (Ljung, 1999). Parametric methods permit one to exploit a priori knowledge and working hypotheses relevant to the structure of considered interactions, providing a signal processing technique based on a model of data generation mechanisms (Porta et al., 2006). The partial knowledge of underlying processes and their biological variability would not permit exact (physical) modeling; however, a completely blind or black-box analysis would lose important notions that are useful in counteracting the unfavorable signal-to-noise ratio (SNR); so a gray-box modeling often represents a flexible and efficient choice, provided that the working hypotheses are clearly stated and a posteriori validated for a correct interpretation of results.

A point of strength of a parametric approach is that causal directions are easily imposed by means of the structure of single subsystems. In some cases, a signal can be considered as exogenous to another one or to a set of other ones, hence an open-loop input–output relationship can be assumed. In many other instances, two or more signals are part of physiological control loops; hence, both causal directions should be assessed in a closed-loop identification problem. Data identification methods yield parameters specific to the single individual or experiment, which can then be used as filters to separate components in the time domain. Time domain features or transfer functions (TF) of subsystems can be easily obtained. Often, a comparison with spectral and cross-spectral plots (Kay, 1989) is carried out by means of partial or causal spectra and coherences, which account for the causal structure and the separation of independent processes performed by the gray-box identification.

Before going on to applications, some general considerations about identifiability, estimate accuracy, and possible biases induced by an incorrect model choice must be pointed out. Importantly, the theory exposed by Ljung (1999) does analyze estimate errors in the frequency domain, yielding a readily interpretable representation of the gap between the model and the true system. In this way, it is possible to underline the differences between the open-loop and closed-loop systems as to the critical choice of a correct model family. In particular, the applicability of open-loop prediction error methods to the closed-loop case (direct approach) will be shown, provided that a correct description of both the deterministic and the stochastic parts are accomplished. The aim is to provide tools for the analysis of two or more biological signals, possibly interacting in a closed-loop system.

The scope is limited to stationary and ergodic processes, that is, those presenting fluctuations and correlations with constant statistical features (at least up to the second-order moments), to be estimated from a limited time window (see Chapter 11 for a time-varying extension). Under this hypothesis, a first useful tool for the analysis of a two-dimensional discrete time signal $y(t) = |y_1(t), y_2(t)|$ is the computation of the spectral densities $\Phi_{y1}(\omega)$, $\Phi_{y2}(\omega)$ and the cross-spectral density $C_{12}(\omega)$. In a linear framework, the latter enlightens the phase $\arg[C_{12}(\omega)]$ and squared coherence $K_{12}^2(\omega) = |C_{12}|^2/[\Phi_{y1}\Phi_{y2}]$ (i.e., the degree of linear correlation) relationships at each radian frequency ω.

In the trivial case in which a clear input/output causal direction is known (e.g., $y_2 \to y_1$, from the input y_2 to the output y_1), the TF is readily obtained as $G_{12}(e^{j\omega}) = C_{12}/\Phi_{y2}$. However, a nonparametric frequency domain analysis (or a multivariate black-box parametric analysis not specifically designed) can hardly disentangle cases in which coherence can be justified both by a feed-forward path $y_2 \to y_1$ and also by a feedback one $y_1 \to y_2$. Phase information at a specific frequency is ambiguous since it can be interpreted as a time delay shorter than one period in four ways, considering the two opposite pathways and admitting either positive or negative correlations. Only in a few cases does the whole phase trend allow an immediate interpretation, for example, a phase slope indicating an overall delay in a specific direction. Nonparametric spectral factorization (Akaike, 1967) can be a solution but is not dealt with here.

A parametric approach based on the optimization of the prediction error (PEM, prediction error method) permits one to explicitly deal with the causal directions described in the model and identified in the time series. The prediction of y_1, $\hat{y}_1 = \Pi_1\{y_1(t-1), \ldots, y_1(t-p), y_2(t-1), \ldots, y_2(t-p)|\vartheta\}$ is considered as a function of past values only (according to a vector ϑ of n unknown parameters), which is, hence, based on the sole $y_2 \to y_1$ causal component of the interaction between the two signals. As presented in the following, this does not prevent one to simultaneously consider the opposite direction $y_1 \to y_2$ as well. The resulting closed-loop portrait can be analyzed starting from the joint process (y_1, y_2), or, under conditions to be specified later, the two directions can be analyzed separately by a direct approach, thus obtaining the loop details and considering the joint process only for a final validation.

The definition of causality according to Granger (1963) plays a central role in the interpretation of parametric identification results. By this very general concept, causality in the direction $y_2 \to y_1$ is recognized if the knowledge of past y_2 samples does improve y_1 prediction. Indeed, in the frequent case in which the model order is not known a priori, parameters are gradually introduced and accepted only if they improve prediction by lowering proper figures of merit (see below). So the useful introduction of nonnull parameters concerning y_2 implies causality.

The advantages of a parametric approach are well known: high data compression from the original N samples to n parameters; consequent statistical accuracy; possibility of tuning physical models (see Chapter 6) in gray-box models (Mukkamala et al., 2006); or, on the contrary, black-box description; and ease of a translation into prediction, simulation, and control models, and into filters for the separation of deterministic and stochastic components. On the other hand, the effectiveness of frequency-domain plots (rather than time-lag parameters) cannot be denied, particularly for signals commonly analyzed in this domain (e.g., EEG, cardiovascular variability, and EMG). From the parametric joint process (once the parameters are estimated from the data), a spectral and cross-spectral description is readily obtained, but it hides the causal relationships worked out, which, on the contrary, can be evidenced by the direct TF representation or by directional variants of spectra and coherences. Partial spectra exploit the lack of correlation of the identification residuals, which are seen as independent inputs to be considered one at a time in summing up the series power; hence, each partial spectrum emphasizes the power coming from a single loop side. As to coherence, which summarizes the correlation due to all interactions, factorization methods yielding a directional coherence were proposed (Saito and Harashima, 1981). A generalization to more signals stems from the concept of partial coherence (coherence between two signals, the effect of the other ones being subtracted) (Kaminski and Blinowska, 1991) and furnishes a partial directed coherence (Baccalà and

Sameshima, 2001). The causal coherence (Porta et al., 2002) of a branch is computed after setting to zero the parameters of the opposite branch (branches). It will be illustrated in detail with an example of application to cardiovascular variability. All these approaches share the property that the estimates tend to zero when no causality (in Granger's sense) is present.

In this chapter, the case of linear parametric prediction models will mainly be analyzed, which, under some circumstances (see Section 5.3), allows for the presence of nonlinearities in the other system branches. In the last part of the chapter, however, an extension of predictors in terms of conditional probability will be discussed. In this wider context, regularity, synchronization, coordination, and nonlinearity of complex interactions can be assessed by entropic indexes (Porta et al., 2000b) or by nonlinear prediction (Porta et al., 2000c).

5.2 BRIEF REVIEW OF OPEN-LOOP IDENTIFICATION

Basic identification theory assumes an input/output causal relationship; the simplest case is a single-input, single-output (SISO) system shown in Figure 5.1. The deterministic part linking y_2 to y_1 is described by the discrete time TF $G_{12}(z)$:

$$y_1(t) = G_{12}(z)y_2(t) + v_1(t)$$

where z^{-1} is the one-lag operator (the normalized radian frequency ω is found on the unit circle: $z = e^{j\omega}$).

The deterministic dependence is corrupted by an additive output noise, or equation residual $v_1(t)$, modeled in the stochastic part by a filter $H_1(z)$ fed with a white noise $\varepsilon_1(t)$ with zero mean and variance λ_1:

$$v_1(t) = H_1(z)\varepsilon_1(t)$$

$$\varepsilon_1(t) \approx WN(0, \lambda_1)$$

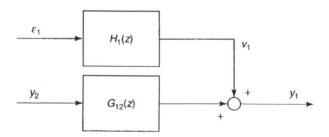

Figure 5.1. Open-loop parametric model.

Hence, the residual is colored by the spectral factor $H_1(e^{j\omega})$, that is, its spectrum is given by $\Phi_{v1}(e^{j\omega}) = |H_1|^2\lambda_1$. The transfer function $H_1(z)$ is assumed monic, that is, the filter impulse response starts at lag zero with a unit coefficient or, equivalently, $H_1(\infty) = 1$.

A prediction error method (PEM) (Ljung, 1999) considers a finite data series $Y^N = \{y(1), y(2), \ldots, y(N)\}$, where $y(t)$ is a vector with two components y_1 and y_2 and a set of models with unknown parameters (collected in a n-dimensional vector ϑ):

$$y_1(t) = G_{12}(z \mid \vartheta) y_2(t) + H_1(z \mid \vartheta) \varepsilon_1(t) \tag{5.1}$$

From this model, the one step predictor is derived as

$$\hat{y}_1(t \mid \vartheta) = H_1^{-1}(z \mid \vartheta)G_{12}(z \mid \vartheta)y_2(t) + \left[1 - H_1^{-1}(z \mid \vartheta)\right]y_1(t) \tag{5.2}$$

and the prediction error can be computed as a function of the parameters as follows:

$$\hat{\varepsilon}_1(t \mid \vartheta) = y_1(t) - \hat{y}_1(t \mid \vartheta) = H_1^{-1}(z \mid \vartheta)\left[y_1(t) - G_{12}(z \mid \vartheta)y_2(t)\right] \tag{5.3}$$

The least squares parameter estimate $\hat{\vartheta}_N$ minimizes the average of squared prediction errors:

$$\hat{\vartheta}_N = \arg\min_{\vartheta} \left[\frac{1}{N} \sum_{i=1}^{N} \hat{\varepsilon}_1^2(t \mid \vartheta)\right] \tag{5.4}$$

The optimization algorithm obviously depends on the parameterization structure of G_{12} and H_1. For instance, in an ARX model (AutoRegressive with eXogenous input) the estimated parameters can be computed in closed form by means of the normal equations. Other structures, such as models with dynamic adjustment or ARMAX, require iterative procedures; however, the implementation problem is nowadays rendered transparent to users by popular computational packages such as the Identification Toolbox of Matlab.

Least-squares estimates benefit from the following asymptotical (i.e., for $N \rightarrow \infty$) properties.

Consistency and Accuracy. If the true model is included in the considered parameterization, that is, there exists a value ϑ_0 of the parameter such that $G_{12}(z|\vartheta_0) = G_{12}(z)$ and $H_1(z|\vartheta_0) = H_1(z)$, then $\hat{\vartheta}_N \rightarrow \vartheta_0$ with probability 1 (asymptotic consistency of the estimate), provided that the input is persistently exciting (i.e., the input excites all system modes; white noise is a typical example). More precisely, the estimates of G_{12} and H_1 are asymptotically uncorrelated and their accuracy grows with N as their variance tends to zero according to:

$$\text{Cov}\hat{G}_{12}(e^{j\omega}) \approx \frac{n}{N} \frac{\Phi_{v1}(\omega)}{\Phi_{v2}(\omega)} \tag{5.5a}$$

$$\text{Cov}\hat{H}_1(e^{j\omega}) \approx \frac{n}{N} \left| H_1(e^{j\omega}) \right|^2 \tag{5.5b}$$

From Equation 5.5a, the estimate variance of G_{12} tends to zero only if all frequencies are represented in the input (persistent excitation) and it is inversely proportional to the input power compared to the output noise power. Note that parameter redundancy with high n reduces accuracy, which fosters the general principle of parametric parsimony. Although the above expressions are valid asymptotically with n (when the TF is not constrained by the particular parameterization) (Ljung, 1999), nonetheless they also provide a useful framework for the common case in which the number of parameters is limited.

Bias. If, on the contrary, the true model is not included in the considered model set, then the estimate is constrained to tend toward a biased one, which minimizes the following functional:

$$\int_{-\pi}^{\pi} \left(\left| G_{12}(e^{j\omega}) - \hat{G}_{12}(e^{j\omega} \mid \vartheta) \right|^2 \Phi_{v2}(\omega) + \Phi_{v1}(\omega) \right) \left| \hat{H}_1(e^{j\omega} \mid \vartheta) \right|^{-2} d\omega \tag{5.6}$$

Let us consider the case in which the stochastic part is arbitrarily described by $H_1(z \mid \vartheta) = H_1^*(z)$ (e.g., no parameter is assigned to the stochastic part). In this case, Equation 5.6 becomes

$$\int_{-\pi}^{\pi} \left| G_{12}(e^{j\omega}) - \hat{G}_{12}(e^{j\omega} \mid \vartheta) \right|^2 \Phi_{v2}(\omega) \left| H_1^*(e^{j\omega}) \right|^{-2} d\omega \tag{5.7}$$

$$+ \text{ terms independent of } \vartheta$$

It follows that the bias of the deterministic part G_{12} and its accuracy are independent of the noise model; hence, G_{12} can be well estimated, if correctly parameterized, even without a detailed analysis of the residual. Importantly, this result is valid only if the input y_2 is uncorrelated with the output noise v_1, but it does not hold in the closed-loop condition to be discussed below.

5.3 CLOSED-LOOP IDENTIFICATION

The important role played by feedback in physiological systems and in interfacing artificial systems to them is well known (Khoo, 2000; Xiao et al., 2005). Biological regulation processes maintaining homeostasis are based on complex closed-loop interactions, as well as the adaptation of neurosensory systems or action tuning in motor systems. Physiological systems, in turn, can be controlled from the outside for therapeutic or life-support purposes, putting an artificial system in a loop with a natural one. Finally, it is worth recalling the case of a machine driven by an operator through a human/machine interface.

This short list demonstrates the possible complexity of the systems involved and the different conditions as to the loop parts that can be known a priori or the signals that can be measured and used for the identification. Hence, the variety of approaches presented below is justified by the possibility of exploiting model and data priors, even partial, for a more pertinent and effective information extraction and signal processing.

In Figure 5.2, the structure of two signals in a closed-loop configuration is shown. In the present signal processing context, a symmetric notation has been preferred. The conversion to notation common in automatic control textbooks is trivial; however, it should be noted that the minus sign of the negative feedback, typically evidenced in the comparison with a set point, here is embedded in the loop blocks. A characteristic feature of the loop is its sensitivity function, $S(z) = (1 - G_{12}G_{21})^{-1}$, which represents the amplification from an external input to the nearest variable within the loop, for example, from v_2 to y_2. In the frequency band in which the control is effective, the sensitivity is well below unit, which, conversely, blunts oscillations that might be useful in the identification process.

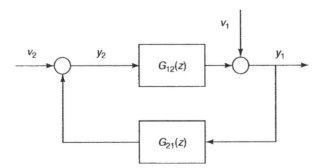

Figure 5.2. Closed-loop parametric model.

5.3.1 Joint Process, Noise Noncorrelation, and Canonical Forms

Even when the direct and indirect approaches described below are used, it is useful to start with a perspective of the closed-loop identification problem in terms of a joint process description, such as the one shown in the upper panel of Figure 5.3. Consider the stochastic bivariate process $y(t) = M(z)\boldsymbol{\varepsilon}(t)$, with $\boldsymbol{\varepsilon}(t) \approx WN(\mathbf{0}, \mathbf{\Lambda})$. The transfer functions on the diagonal of $M(z)$ are monic and the off-diagonal terms are causal (one, at least, strictly causal). The lower panel of Figure 5.3 displays the relationship with the closed-loop structure of Figure 5.2, where the unmeasured noises entering the loop are also described as a joint process $v(t) = H(z)\boldsymbol{\varepsilon}(t)$. It is clear that the problem is overparameterized and that some constraints must be introduced, since the four elements in M are now described by the four elements in H and the loop blocks G_{12} and G_{21}; that is, correlations can be established either outside or inside the loop.

A common hypothesis is to impose the nonncorrelation of noises v_1 and v_2, to be a posteriori verified from their estimates given by model residuals. This implies the diagonality of $H(z) = \mathrm{diag}[H_1 H_2]$ and also of $\mathbf{\Lambda} = \mathrm{diag}[\lambda_1 \lambda_2]$. So all correlations are explained by the loop; the consequent interpretation of results should be carefully derived in biological systems, where often links collateral to those addressed can be present and fused by the identification process into a single pathway. The full noncorrelation of input white noises described by a diagonal $\mathbf{\Lambda}$ requires description of the immediate (zero-lag) correlation of the signals through one of the loop sides allowed to be nonstrictly causal (nonnull zero-lag parameter); only one is admitted in order to avoid an algebraic loop. Hence, the canonical form of the joint process is defined by a diagonal $\mathbf{\Lambda}$ and $M(\infty)$ being upper or lower triangular with either $G_{12}(\infty) \neq 0$ or $G_{21}(\infty) \neq 0$, respectively. The choice relies on physical considerations about sampling conventions and what interactions can take place in a sampling period. Often, a black-box identification algorithm of the joint process yields the canonical form

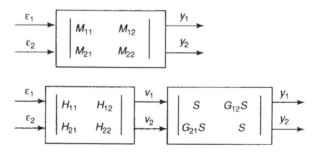

Figure 5.3. Joint process and closed-loop system.

with a nondiagonal Λ and diagonal $M(\infty)$, which requires further transformation to be projected onto the loop structure.

5.3.2 Direct Approach to Identification: Opening the Loop

In a direct approach, identification addresses a single branch at a time, for example, G_{12}, as in Figure 5.4. The dataset Y^N is processed by a standard single-output (y_1, in our example) algorithm, as if the loop was open. The least-squares PEM applies the open-loop signal model (Equation 5.1), predictor (Equation 5.2), prediction error (Equation 5.3), and estimate criterion (Equation 5.4).

Asymptotic properties, however, are different, due to the correlation between the deterministic input y_2 and the stochastic one v_1. Precisely, the spectrum of y_2 must be decomposed into the sum of two uncorrelated components: one correlated to v_1, the other correlated to v_2:

$$\Phi_{y2} = \Phi_{y2/v1} + \Phi_{y2/v2} \qquad (5.7)$$

In the fully linear scheme considered in Figure 5.4, the two components in Equation 5.7 are $|G_{21}S|^2\Phi_{v1}$ and $|S|^2\Phi_{v2}$, respectively; however, a nonlinear G_{21} can be admitted (obviously, the joint process description should be accordingly changed in this case).

Consistency and Accuracy. If the parameterized set of models is capable of adequately describing both the deterministic part G_{12} and the stochastic one H_1, then consistency is preserved and the estimate variance of the TF G_{12} follows the asymptotical law

$$\text{Cov}\hat{G}_{12} \approx \frac{n}{N}\frac{\Phi_{v1}(\omega)}{\Phi_{y2/v2}(\omega)} \qquad (5.8)$$

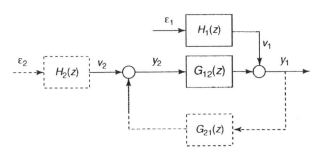

Figure 5.4. Scheme of the direct method; parts opposite to the identified pathways are dashed.

Comparing Equation 5.8 with Equation 5.5a, it can be noted that the denominator considers only the component of input y_2 which is uncorrelated with the output noise v_1 and dependent on v_2. A broad harmonic content of y_2 may give a false impression of identifiability, unless the sole v_2 dependent part is able to excite all model frequencies. Conversely, a rich behavior of v_2 may result in a poor excitation from y_2 if the loop sensitivity $S(z)$ is very low, due to an efficient loop compensation.

Bias. Differently from the open-loop case, the estimated bias of the deterministic part $B_{12}(z)$ does depend also on a correct noise description. Suppose that the G_{12} parameterization includes the true model, whereas the model of H_1 is constrained to a false value, $H_1(z|\vartheta) = H_1^*(z) \neq H_1(z)$. Then it can be shown that

$$\left|B_{12}(e^{j\omega})\right|^2 = \left|H_1(e^{j\omega}) - H_1^*(e^{j\omega})\right|^2 \frac{\lambda_1 \cdot \Phi_{v2/v1}(\omega)}{\left[\Phi_{v2}(\omega)\right]^2} \qquad (5.9)$$

So, noise estimate errors are projected onto the deterministic part proportionally to the input component correlated to the output noise. For instance, if a white residual is assumed (but this hypothesis is not verified, due to insufficient parameterization), then the spectral peaks of v_1 are aliased to the G_{12} amplification. Ultimately, oscillations external to the loop are erroneously attributed to loop resonances. Note also that the bias may overread a null G_{12}, thus invalidating a causal analysis.

In conclusion, if the system is really open loop, we may omit the estimation of the noise properties; the identification of the deterministic part will be correct even if the prediction errors will not satisfy a whiteness condition. On the contrary, an exhaustive analysis of both the deterministic and the stochastic part is required in the closed-loop case.

As already said, the direct method allows one to analyze a single branch at a time, for example, the feed-forward one alone. Nonetheless, a separate identification of the other pathway (pathways) yields as a by-product the estimate of residual v_2 and, based on its harmonic richness, permits one to verify the identifiability of the feed-forward pathway. In addition, once all branches are identified, a thorough analysis of whiteness and uncorrelation of the prediction errors ε_1 and ε_2 is possible, thus validating the identification of the joint process. Of course, if knowledge on input noises is available, either a priori or through the analysis of similar data, identification can be limited to a single branch.

5.3.3   Indirect Approach, Brief Remarks

The indirect approach is briefly introduced for its practical relevance to automatic control problems. With reference to Figure 5.2, suppose we have an artificial controller G_{21} interfaced to a biological system G_{12} (the plant), y_1 being the controlled physiological variable. Let us also suppose that $r(t)$ = $v_2(t)$ is measured, differently from the direct-approach case. Typically, $r(t)$ is a set point or reference signal. Under these assumptions, the identification of the TF from r to y_1, $F(z) = G_{12}S$, satisfies the open-loop condition and benefits from the relevant properties. Once $F(z)$ is identified, the knowledge of the controller G_{21} permits the computation of the unknown physiological part $G_{12} = F/(1 + FG_{21})$. A common application is the adaptation of the control law to slow changes in the controlled system. The simplicity of this approach is due to the knowledge of a signal *before* it enters the loop and to the knowledge of the controller; hence, nonlinearities of this block may prevent the use of linear models.

5.4 APPLICATIONS TO CARDIOVASCULAR CONTROL

Analysis of cardiovascular control can provide a helpful example of complex closed-loop interactions (Baselli et al., 2002; Porta et al., 2006). The cardiovascular variability signals (see Chapter 1) allow the exploration of the regulatory mechanisms of blood pressure and heart rate, and of the relations with the respiratory system. In this chapter, the discrete time t assumes integer values representing the progressive cardiac cycle number. The beat-to-beat series considered in the following are changes of a cardiovascular variable with respect to its mean value (or to a slow trend that is usually estimated and subtracted from the original series). In the field of cardiovascular variability analysis, the most frequently considered variables are: systolic arterial pressure (SAP), defined as the maximum of the blood pressure mostly recorded via noninvasive plethysmographic techniques; heart period, calculated as the temporal distance between two consecutive R peaks on the ECG (RR); and a respiratory signal (volume, flow, or thoracic movements), sampled once per cardiac beat. The presence of physiological rhythms in the low-frequency (LF, about 0.1 Hz) and high-frequency (HF, at the respiratory rate) bands related to autonomic and mechanical influences and to the existence of closed-loop regulatory mechanisms allow us to illustrate practical examples of the above-mentioned modeling techniques.

5.4.1 Partial Spectra

Since H is diagonal, the transfer matrix of the joint process is

$$M(z) = \begin{vmatrix} S H_1 & G_{12} S H_2 \\ G_{21} S H_1 & S H_2 \end{vmatrix} \tag{5.10}$$

The substitution of Equation 5.10 in the definition of the spectral matrix $\mathbf{\Phi}(\omega) = \mathbf{M}'(z)\mathbf{\Lambda M}(z^{-1})|_{z=e^{j\omega}}$ and the exploitation of the diagonality of $\mathbf{\Lambda}$ lead to

$$\Phi_1(\omega) = |\, S H_1(e^{j\omega})\,|^2\, \lambda_1 + |\, G_{12}\, S H_2(e^{j\omega})\,|^2\, \lambda_2 \qquad (5.11a)$$

$$\Phi_2(\omega) = |\, G_{21}\, S H_1(e^{j\omega})\,|^2\, \lambda_1 + |\, S H_2(e^{j\omega})\,|^2\, \lambda_2 \qquad (5.11b)$$

$$C_{12}(\omega) = G_{21}(e^{-j\omega})|\, S H_1(e^{j\omega})\,|^2\, \lambda_1 + G_{12}(e^{j\omega})|\, S H_2(e^{j\omega})\,|^2\, \lambda_2 \qquad (5.11c)$$

It is worth noting that the two autospectra (Equations 5.11a and b) are the sum of two parts usually referred to as partial spectra. This is the result of the noncorrelation between the two stochastic noises that allows the summation of the two partial processes seen as white noises filtered according to a transfer function depending on the sensitivity of the feedback and of the relevant feedforward pathways (from each noise to the considered signal). For example, the power spectrum of y_1 can be decomposed into two contributions related to direct perturbations linked to the presence of y_1 and to indirect influences associated with the presence of y_2. In the presence of other exogenous signals, the identification leads to the noncorrelation of the residues and to the summation of all partial spectra. Using the residue method, partial spectra can be further decomposed into spectral components relevant to the poles of the TF from each white noise to the considered signal [e.g., in Equation 5.11a TFs are SH_1 and $G_{12}SH_2$, respectively).

The analysis of the variability series of SAP and RR, using the respiratory signal as an exogenous input, allowed us to find out that the LF rhythms are related to the closed-loop regulatory mechanisms that are affected by every input: colored noises on RR, on SAP, and even respiration. In addition, we found that the LF rhythms can be related to poles relevant to the colored noises on RR and/or on SAP as well. Therefore, we suggested that both resonances of closed-loop and rhythms external to the closed-loop regulation acting directly on sinus node (these rhythms are likely to be neural central commands) or on microcirculation (these rhythms are likely to be sympathetic modulations regulating peripheral flows and resistances) contribute to the genesis of the LF oscillations. Figure 5.5 shows an example of the quantification of the spectral components in a conscious dog during bilateral carotid occlusion. In the LF band, two components were found: that at 0.05 Hz was related to the resonance of the closed-loop regulation, whereas that at 0.12 Hz is related to central commands directly affecting the sinus node.

A dynamic adjustment model such as that adopted in the previous example,

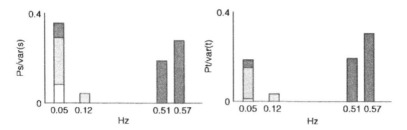

Figure 5.5. Components of the partial spectra classified according to poles of the colored noise on SAP (white), colored noise on RR (gray), and of respiration (black); and poles relevant to the closed loop (bar at 0.05 Hz) affected by all the three inputs (Baselli et al., 1997).

$$G_{12}(z) = A_{12}(z) / (1 - A_{11}(z)); \quad H_1(z) = 1 / \left[1 - A_{11}(z) \right]\left[1 - D_1(z) \right]$$
$$G_{21}(z) = A_{21}(z) / (1 - A_{22}(z)); \quad H_2(z) = 1 / \left[1 - A_{22}(z) \right]\left[1 - D_2(z) \right]$$
(5.12)

allows the definition of poles of the colored input noises independently of the network of the closed-loop interacting signals and, thus, is suitable for a classification of the rhythms similar to that above mentioned. Details about the calculation and classification of the spectral components based on multivariate dynamic adjustment (MDA) models can be found in Baselli et al. (1997). Figure 5.6 shows the general scheme of the MDA model in which the interacting variables y_{int} are $(L - X)$, whereas the exogenous signals are X.

A bivariate AR, AR(2), or, more generally, multivariate AR (MAR) model can be easily derived by imposing $D_1 = D_2 = 0$.

$$G_{12}(z) = A_{12}(z) / \left[1 - A_{11}(z) \right]; \quad H_1(z) = 1 / \left[1 - A_{11}(z) \right]$$
$$G_{21}(z) = A_{21}(z) / \left[1 - A_{22}(z) \right]; \quad H_2(z) = 1 / \left[1 - A_{22}(z) \right]$$
(5.13)

This class of models is simpler but suitable for inferring causality, as reported in the following sections.

5.4.2 Causality

The AR(2) model is defined as $y(t) = \mathbf{A}(z) \cdot y(t) + \boldsymbol{\varepsilon}(t)$, where $\mathbf{A}(z)$ is a 2×2 matrix that contains on the main diagonal the polynomials

$$A_{ii}(z) = \sum_{k=1}^{p} a_{ii}(k) \cdot z^{-k}$$

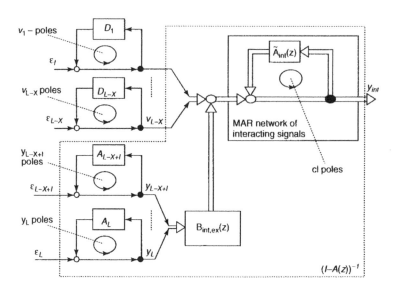

Figure 5.6. General scheme of the MDA model. The notation has been changed from the original one reported in Baselli et al. (1997) and adapted to that chosen in the present chapter.

with $i = 1, 2$ describing the dependence of $y_i(t)$ on p past values (autodependence) and out of the main diagonal the polynomials

$$A_{ij}(z) = \sum_{k=0}^{p} a_{ij}(k) \cdot z^{-k}$$

with $i, j = 1, 2$ describing the dependence of $y_i(t)$ on the current and p past values of $y_j(t)$ (cross-dependence). In order to guarantee a canonical form of the model (see Section 5.3.1), for example, triangular, $a_{12}(0) \neq 0$ and $a_{21}(0) = 0$ (i.e., y_2 can act immediately on y_1, but not vice versa).

According to the definition of causality given by Granger (Granger, 1963), y_i causes y_j (symbolized as $y_i \rightarrow y_j$ in the following) if the knowledge of y_i is helpful to predict y_j. In the AR(2) model, this means that at least one of the coefficients $a_{ji}(k)$ is significantly different from 0. According to this definition, four types of interactions between the signals y_1 and y_2 are possible:

1. $y_1 \rightarrow y_2$ but not vice versa [at least one coefficient $a_{21}(k)$ is significantly different from 0 with all the coefficients $a_{12}(k)$ equal to 0]
2. $y_2 \rightarrow y_1$ but not vice versa [at least one coefficient $a_{12}(k)$ is significantly different from 0 with all the coefficients $a_{21}(k)$ equal to 0]

3. $y_1 \rightarrow y_2$ and $y_2 \rightarrow y_1$ (symbolized as $y_1 \leftrightarrow y_2$) [both at least one coefficient $a_{21}(k)$ and $a_{12}(k)$ are significantly different from 0]
4. y_1 is not helpful to predict y_2 and vice versa [all the coefficients $a_{21}(k)$ and $a_{12}(k)$ are equal to 0]

When conditions 1 and 2 are fulfilled, the two signals y_1 and y_2 are said to be interacting in an open loop even though the two cases are dramatically different since causality is reversed. When condition 3 is satisfied, the two signals y_1 and y_2 are said to be interacting in a closed loop. When condition 4 is fulfilled, the two signals y_1 and y_2 do not interact with each other.

The relationship between the variability of RR and SAP is a convenient example of a relation between two biological signals that can result in the four above-mentioned types of interactions. A lengthening of RR induces an increased ventricular filling and, as a consequence of Starling's law, an augmentation of SAP to the next beat, but, contemporaneously, it produces a larger decrease of the diastolic pressure (Windkessel effect) and a fall of SAP to the next beat (Baselli et al., 1994). Independently of the prevalent effect, a variation of RR produces a SAP change, RR $\rightarrow$ SAP. The SAP, sensed by the baroreceptors, induces, through the involvement of the regulatory centers of the autonomic nervous system, the activation of the efferent vagal and sympathetic fibers directed to the heart and, thus, the modification of RR (Koepchen, 1984), that is, SAP $\rightarrow$ RR. Consequently, it is generally assumed that RR and SAP interact in a closed loop (RR $\leftrightarrow$ SAP) and that the opening of this loop, both along the causal path RR $\rightarrow$ SAP (usually referred to as mechanical feedforward) or along the reverse one, SAP $\rightarrow$ RR (usually referred to as baroreflex feedback), can occur solely after pharmacological or surgical interventions or through the application of an external device such as an atrial pacemaker, preventing SAP $\rightarrow$ RR, while the mechanical feedback is functioning, or a specific pumping device that regulates arterial pressure independently of the heart, thus preventing RR $\rightarrow$ SAP, while baroreflex feedback is operating. On the contrary, the conditions in which baroreflex sensitivity is evaluated from spontaneous RR and SAP variability (Laude et al., 2004) must disentangle this RR $\rightarrow$ SAP index from the more complex RR $\leftrightarrow$ SAP data structure.

5.4.3 Estimation of the Transfer Function (TF): Limitation of the Traditional Approach

Under the hypothesis of the open loop (OL) between y_2 and y_1, the TF from y_2 to y_1, is estimated as

$$\hat{G}_{12/OL}(e^{j\omega}) = \frac{C_{12}(\omega)}{\Phi_2(\omega)} \tag{5.14}$$

where $C_{12}(\omega)$ and $\Phi_2(\omega)$ are the cross spectrum between y_1 and y_2 and the autospectrum of y_2, respectively.

An interesting analysis of the sympathetic and parasympathetic influences on the sinus node has been based on the assessment of the TF from respiration to RR (Saul et al., 1991). In order to have a broadband input, respiration was randomized by exploiting a controlled breathing procedure that imposes random respiratory intervals on the subject. After nonparametric estimation of $C_{12}(\omega)$ and $\Phi_2(\omega)$, Equation 5.14 was applied, thus disregarding the closed-loop interactions RR ↔ SAP.

According to the above-reported considerations about causality in a bivariate system, let us assign y_1 = RR and y_2 = SAP. With this notation in mind, the $\hat{G}_{12/OL}$ in Equation 5.14 represents the TF from SAP to RR, the magnitude of which is traditionally utilized as a nonparametric estimate of the baroreflex gain (Robbe et al., 1987). Equation 5.14 implicitly assumes an open-loop relationship between y_1 and y_2, and a causality from y_2 to y_1 (i.e., $y_2 \rightarrow y_1$) (Porta et al., 2006). Unfortunately, in several experimental situations RR and SAP interact in a closed loop and, thus, the estimate of $\hat{G}_{12/OL}$ merges the contribution of the feedforward pathway with that of the feedback one. This observation becomes clearer when $\hat{G}_{12/OL}$ is calculated according to an AR(2) model. Indeed, Equation 5.14 assumes the form (Porta et al., 2002)

$$\hat{G}_{12/OL}(e^{j\omega}) = \frac{\left[1 - A_{22}(z)\right] \cdot A_{21}(z^{-1}) \cdot \lambda_1^2 + A_{12}(z) \cdot \left[1 - A_{11}(z^{-1})\right] \cdot \lambda_2^2}{\left|A_{21}(z)\right|^2 \cdot \lambda_1^2 + \left|\left[1 - A_{11}(z)\right]\right|^2 \cdot \lambda_2^2} \Bigg|_{z = e^{j\omega}} \quad (5.15)$$

that is, $\hat{G}_{12/OL}$ depends on both A_{21} and A_{12} describing the feedforward and feedback pathways, respectively.

Therefore, if the hypothesis of the open loop is not fulfilled, the interpretation of the TF $\hat{G}_{12/OL}$ as derived from Equation 5.14 is not easy. For example, let us consider the RR–SAP relationship and let us suppose that the aim is to estimate the baroreflex gain that actually depends solely on the A_{12}. Then Equation 5.15 provides values that depend on the gain of the mechanical feedforward (A_{21}). In addition, if the dominant causal relationship present in the real data is the reverse one ($y_1 \rightarrow y_2$) (e.g., it may occur when $\lambda_2^2 \ll \lambda_1^2$), the $\hat{G}_{12/OL}$ is not a good estimate of the true G_{12} but, conversely, it is closer to that of the opposite pathway with $\hat{G}_{12/OL} = 1/G_{21}$ (Porta et al., 2002). In this situation, the estimated baroreflex gain has nothing to do with the baroreflex feedback but is actually the inverse of the gain of the mechanical feedforward. Examples of unreliable estimates of the magnitude and gain both on simulated and real signals (RR and SAP series) can be found in Faes et al. (2004b).

5.4.4 Coherence and Causal Coherence

For a correct interpretation of the TF estimate, it is mandatory to check whether $y_2 \rightarrow y_1$. Indeed, only if $y_2 \rightarrow y_1$, and not vice versa, an open-loop causal relationship from y_2 to y_1 is actually present. An ad-hoc test has been proposed by Porta et al. (2002). It exploits the definition of squared coherence (K^2) and the AR(2) model. K^2 is defined as

$$K^2(\omega \mid \vartheta) = \frac{|C_{12}(\omega \mid \vartheta)|^2}{\Phi_1(\omega \mid \vartheta)\Phi_2(\omega \mid \vartheta)} \qquad (5.15)$$

As previously outlined, K^2 measures the amount of linear correlation (from 0 to 1) between y_1 and y_2 as a function of the frequency (Kay, 1989). The equations relevant to autospectrum and cross spectrum (Equations 5.11a, b, c) clearly indicate that K^2 cannot be utilized to infer the causal relationships between y_1 and y_2 because the feedforward and feedback pathways are combined together.

The causal coherence quantifying the degree of correlation in the causal direction from y_2 to y_1 ($y_2 \rightarrow y_1$) has been defined by Porta et al. (2002) as

$$K^2_{2\rightarrow 1}(\omega \mid \vartheta) = K^2(\omega \mid \vartheta)\Big|_{A_{21}(z)=0} \qquad (5.16)$$

The application of Equation 5.16 to the variability series of SAP and RR allowed Nollo et al. (2005) to find out that in young healthy humans the causal direction from SAP to RR (i.e., the baroreflex feedback) is less important than the causal relationship from RR to SAP (i.e., the mechanical feedback), whereas the former gains importance during head-up tilt. Classically, according to De Boer et al. (1985), values of $K^2(f) > 0.5$ are assumed to be a proof of the presence of a significant link between the variables y_1 and y_2 at the considered frequency f. Recently, a new test for the significance of K^2 has been proposed (Porta et al., 2002; Faes et al., 2004a). It is based on the construction of surrogate data derived from the original signals after destroying their cross correlation but maintaining their value distributions and autospectra (Schreiber and Schmitz, 1996; Palus, 1997). This test is more flexible, reliable, and suitable to the analysis of the cardiovascular variability series (see Figure 5.7).

5.4.4 Closed-Loop Estimation of the Baroreflex Gain

A closed-loop parametric identification based on the MDA class is more suitable to deal with the quantification of the baroreflex pathway SAP $\rightarrow$ RR. The exploitation of the AR(2) model cannot adequatly solve this issue. Indeed, since $G_{12} = A_{12}/(1 - A_{11})$ it is likely to attribute to the baroreflex block features that are pertinent to the noise in the sinus node that are described by the spectral factor $H_1 = 1/(1 - A_{11})$; the poles of H_1 must be cancelled in the G_{12} by the zeroes of A_{12}. On the other hand, as a result of the bias described by Equation 5.9, a spectral factor with a insufficient number of poles similarly leads to a projection of the spectral peaks of the

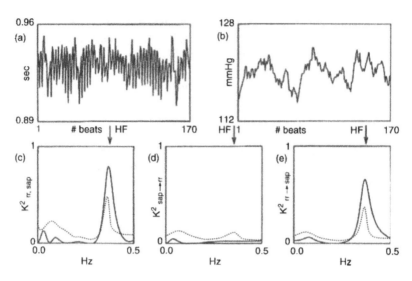

Figure 5.7. Example of beat-to-beat series of RR and SAP in a subject after recent heart transplantation (a, b). The coherence $K^2_{RR,SAP}$ (c, solid line) and its significance threshold (dotted line) are compared to the causal coherence from SAP to RR $K^2_{SAP \to RR}$ (d) and from RR to SAP $K^2_{RR \to SAP}$ (e). It is worth noting in (d) the absence of the coupling along the baroreflex feedback SAP $\to$ RR. From (Porta et al., 2002).

noise onto G_{12}. This issue can be solved by using the class of MDA models with $A_{11} = 0$. From Equation 5.12, $G_{12} = A_{12}$ and $H_1 = 1/(1 - D_{11})$, thus disentangling the parameters of the causal part from those of the stochastic parts (Baselli et al., 1988). A further possibility is to assign a very low model order to A_{11} [e.g., 1 in Porta et al. (2003)], thus preserving the possibility of deriving the dynamical properties of the sinus node while limiting the difficulties related to overparametrization of A_{11}.

A further element that is worth accounting for is the relevant effect of respiration on the sinus node due to central commands (due to the activities of respiratory centers located in the brain stem) and cardiopulmonary reflexes. The main effect on the RR variability is the so-called respiratory sinus arrhythmias that can be correlated with respiratory influences on SAP, thus challenging the hypothesis of noncorrelation relevant to unmeasurable inputs. An explicit measure of respiratory activity that can be introduced in the model as an exogenous input can solve this ambiguity and improve the estimate of the baroreflex block.

After identifying the block A_{12} linking SAP to RR, it is fed by an artificial ramp with unit slope simulating a SAP rise and the slope of the subsequent RR rise is taken as an index of the baroreflex gain (Baselli et

al., 1988). In conscious dogs, the baroreflex estimate provided by this
model (XXAR, because the loop has been opened according to a direct
approach, thus considering both respiration and SAP as exogenous signals, and the equation residue is an AR process) has been compared to a
simple linear regression (X, SAP is exogenous with respect to RR and the
equation residue is a white noise) and to a dynamic adjustment model
without respiration (XAR, SAP is an exogenous signal and the equation
residue is an AR process) (Porta et al., 2000b). As shown in Figure 5.8,
the disregard of the important influences and the underestimation of the
model complexity determine an overestimation of the barorefelx gain.
The comparison of the closed-loop estimates of the baroreflex gain with
those in open loop demonstrated that the bias of the open-loop estimates
is reduced by the baroreceptive unloading induced by active standing (Lucini et al., 2000).

5.5 NONLINEAR INTERACTIONS AND SYNCHRONIZATION

The TF estimate between two signals y_1 and y_2 carried out with both nonparametric and parametric methods, the calculation of their degree of correlation through the evaluation of the coherence, and casual coherence are
reliable only in the case of linear interactions. Unfortunately, interactions

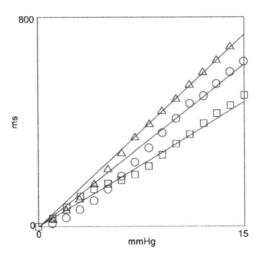

Figure 5.8. Responses of the block A_{12} to an artificial ramp with unit slope in the
case of the X (triangles), XAR (circles), and XXAR (squares) models in a dog at
control. The slope of the estimated regression line is taken as an estimate of the
baroreflex gain. From Porta et al. (2000b).

between two signals y_1 and y_2 can be nonlinear as well. Synchronization is one of the most important nonlinear interactions. Two signals are said to be synchronized if a joint pattern exists that involves the two signals and repeats itself periodically. For example, in a healthy heart an atrial contraction is followed by a ventricular contraction and, thus, atrial and ventricular activities are synchronized according to a 1:1 ratio. Some pathologies affecting atrioventricular conduction tissue (atrioventricular block) are characterized by the occurrence of a ventricular contraction every two (or, generally, n) atrial contractions, thus resulting in a synchronization with a 1:2 (or, generally, 1:n) ratio. In this case, the main harmonic of the ventricular activity is exactly half of that of the atrial activity and, in correspondence of these two frequencies, the squared coherence is insignificant even in presence of a strong coordination between the two activities. In this situation, the degree of synchronization cannot be quantified by a tool such as the coherence function. Therefore, in order to quantify the nonlinear interactions, several approaches have been proposed. They are based on the assessment of the rate of occurrence of the joint patterns and on the evaluation of their degree of repetition through the calculation of entropy (see, e.g., Hoyer et al., 1998; Palus et al., 1997, 2001; Porta et al., 1999, 2004), or predictability (see, e.g., Schiff et al., 1996; Le Van Quyen et al., 1998; Feldmann and Bhattacharya, 2004; Faes et al., 2006).

The method for the construction of the joint pattern that describes how the two signals interact each other is of capital importance. As an example, we consider the technique underlying the construction of the joint pattern proposed by Porta et al. (1999) and applied in Nollo et al. (2002, 2000c). In Porta et al. (1999) the joint pattern of length L is defined as $y_L(i) = [y_1(i), y_2(i), \ldots, y_2(i - L + 2)]$, that is, the ordered sequence of L samples formed by one sample of y_1 and $L - 1$ samples of y_2. This technique is formally equivalent to that exploited by Schiff et al. (1996) and Le Van Quyen et al. (1998) in their approach based on prediction. The normalized cross-conditional entropy is utilized to evaluate the amount of information carried by the sample $y_1(i)$ when the $L - 1$ samples of y_2, that is, $[y_2(i), \ldots, y_2(i - L + 2)]$, are known and to derive an index of coupling $\chi_{2 \rightarrow 1}$ along the causal direction $y_2 \rightarrow y_1$. Analogously, the reversal of index 2 with index 1 allows the extraction of an index of coupling $\chi_{1 \rightarrow 2}$ along the opposite causal direction ($y_1 \rightarrow y_2$). $\chi_{1,2} = \max(\chi_{2 \rightarrow 1}, \chi_{1 \rightarrow 2})$ has been proposed as an index of global synchronization (Porta et al., 1999), whereas $\chi_{2 \rightarrow 1}$ and $\chi_{1 \rightarrow 2}$ have been utilized as indexes measuring the coupling along the causal directions $y_2 \rightarrow y_1$ and $y_1 \rightarrow y_2$, respectively (Nollo et al., 2002). All the three indexes range between 0 and 1 (null and maximum degree of synchronization, respectively).

The assessment of the synchronization between RR and SAP through the application of the indexes defined in Porta et al. (1999) allowed the observation that the degree of the global synchronization between RR and SAP increases during 90° head-up tilt (a maneuver of sympathetic activation) in healthy humans (Porta et al., 2000c) and the degree of coupling along the baroreflex feedback (SAP → RR) decreases in post-myocardial infarction patients, whereas that along the mechanical feedforward (RR → SAP) increases with respect to an age-matched healthy population (Nollo, 2002). The most important disadvantage of the methods based on the calculation of entropy of the joint pattern is the reliability of the estimate of the probability of the joint pattern based on the evaluation of their rate of occurrence (i.e., sample frequency). This disadvantage is more important when these methods are applied over short series (a few hundreds samples). As a consequence, the pattern length L must be limited to a few samples (Hoyer et al., 1998; Palus et al., 1997) or, alternatively, correction terms must be introduced in the calculation of the entropy (Richman and Moorman, 2000; Porta et al., 1998, 1999, 2000c, 2007a), or of the nonlinear predictability (Porta et al., 2000a, 2007b). The use of surrogate series (Schreiber and Schmitz, 1996; Palus, 1997) constructed over the original series by destroying deterministic nonlinear interactions (e.g., by independent randomization of the phase of all the considered signals) allows the assessment of the significance of the synchronization index (see Figure 5.9). This test is very important even after identification of parameters of the multivariate parametric models, thus verifying the importance of the disregarded nonlinear features. As suggested in Porta et al. (2006), this can be done by checking for the presence of nonlinearities in the equation residues using a surrogate approach.

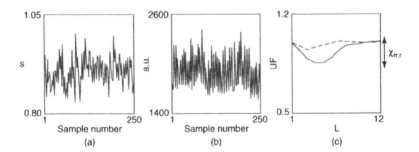

Figure 5.9. Example of synchronization analysis between respiratory signal and RR series in a healthy subject at rest derived from Porta et al. (2000c). (a) RR series; (b) respiratory signal; (c) normalized cross-conditional entropy as a function of L derived from the original series (solid line) and over a pair of surrogate series derived after independent phase randomization of the two original series (dotted line). It is worth noting that the minimum of the cross-conditional entropy, clearly visible when calculated from the original series, disappears when calculated from surrogates, thus indicating that the two original series are significantly coupled.

5.6   CONCLUSION

The various illustrated approaches share a common feature: the exploitation of past samples of a signal to predict another one through a parametric law or, more generally, through a probabilistic relationship. The adopted choices relevant to the considered interactions and their structure constitute a data-analysis model and deeply influence results and the information enlightened. For this reason, it is crucial to have tools capable of describing and assessing the followed hypotheses. This is particularly important in the biomedical field because of the complexity of the analyzed systems and the interdisciplinary approach require one to clearly understand the application limits of the adopted methods.

REFERENCES

Akaike, H., Some Problems in the Application of the Cross-Spectral Method, in B. Harris (Ed.), *Spectral Analysis of Time Series*, pp. 81–107, Wiley, 1967.

Baccalà, L. A., and Sameshima, K., Partial Directed Coherence: A New Concept in Neural Structure Determination, *Biol. Cybern.*, Vol. 84, 463–474, 2001.

Baselli, G., Caiani, E., Porta, A., Montano, N., Signorini, M. G., and Cerutti, S., Biomedical Signal Processing And Modeling In Cardiovascular Systems, *Critical Reviews in Biomed. Eng.*, Vol. 30, 57–87, 2002.

Baselli, G., Cerutti, S., Badilini, F., Biancardi, L., Porta, A., Pagani, M., Lombardi, F., Rimoldi, O., Furlan, R., and Malliani, A., Model for the Assessment of Heart Period and Arterial Pressure Variability Interactions and Respiratory Influences, *Med. Biol. Eng. Comput.*, Vol. 32, 143–152, 1994.

Baselli, G., Cerutti, S., Civardi, S., Malliani, A., and Pagani, M., Cardiovascular Variability Signals: Towards the Identification of a Closed-Loop Model of the Neural Control Mechanisms, *IEEE Trans. Biomed. Eng.*, Vol. 35, 1033–1046, 1988.

Baselli, G., Porta, A., Rimoldi, O., Pagani, M., and Cerutti, S., Spectral Decomposition in Multi-Channel Recordings Based on Multi-Variate Parametric Identification, *IEEE Trans. on Biomed. Engin.*, Vol. 44, No. 11, 1092–1101, 1997.

De Boer, R. W., Karemaker, J. M., and Strackee, J., Relationships between Short-Term Blood Pressure Fluctuations and Heart Rate Variability in Resting Subjects I: A Spectral Analysis Approach, *Med. Biol. Eng. Comput.*, Vol. 23, 352–358, 1985.

Faes, L., Cucino, R., and Nollo, G., Mixed Predictability and Cross-Validation to Assess Non-Linear Granger Causality in Short Cardiovascular Variability Series, *Biomedizinische Technik*, Vol. 51, 255–259, 2006.

Faes, L., Pinna, G. D., Porta, A., Maestri, R., and Nollo, G., Surrogate Data Analysis for Assessing the Significance of the Coherence Function, *IEEE Trans. Biomed. Eng.*, Vol. 51, 1156–1166, 2004a.

Faes, L., Porta, A., Cucino, R., Cerutti, S., Antolini, R., and Nollo, G., Causal Transfer Function to Describe Closed Loop Interactions between Cardiovascular and Cardiorespiratory Variability Signals, *Biol. Cybern.*, Vol. 90, 390–399, 2004b.

Feldmann, U., and Bhattacharya, J., Predicatability Improvement as an Asymmetrical Measure of Interdependence in Bivariate Time Series, *Int. J. Bifurc. Chaos*, Vol. 14, 505–514, 2004.

Granger, C. W. J., Economic Processes Involving Feedback, *Information and Control*, Vol. 6, 28–48, 1963.

Hoyer, D., Bauer, R., Walter, B., and Zwiener, U., Estimation of Nonlinear Couplings on the Basis of Complexity and Predictability: A New Method Applied to Cardiorespiratory Coordination, *IEEE Trans. Biomed. Eng.*, Vol. 45, 545–552, 1998.

Kaminski, M. J., and Blinowska, K. J., A New Method of the Description of the Information Flow in the Brain Structures, *Biol. Cybern.*, Vol. 65, 203–210, 1991.

Kay, S. M., *Modern Spectral Analysis: Theory and Application*, Prentice-Hall, 1989.

Khoo, C. K., *Physiological Control Systems: Analysis, Simulation, and Estimation*, IEEE Press, 2000.

Koepchen, H. P. History of Studies and Concepts of Blood Pressure Waves, in: Miyakawa, K., Polosa, C., and Koepchen, H. P. (Eds.), *Mechanisms of Blood Pressure Waves*, pp 3–23, Springer-Verlag, 1984.

Laude, D., Elghozi, J., Girard, A., Bellard, E., Bouhaddi, M., Castiglioni, P., Cerutti, C., Cividjian, A., Rienzo, M. D., Fortrat, J., Janssen, B., Karemaker, J. M., Lefthériotis, G., Parati, G., Persson, P. B., Porta, A., Quintin, L., Regnard, J., Rüdigger, H., and Stauss, H. M., Comparison of Various Techniques Used to Estimate Spontaneous Baroreflex Sensitivity (the EuroBaVar Study), *Am. J. Physiol. Regul. Integr. Comp. Physiol.*, Vol. 286, R226–R231, 2004.

Le Van Quyen, M., Adam, C., Baulac, M., Martinerie, J., and Varela, F. J., Non Linear Interdependencies of EEG Signals in Human Intracranially Recorded Temporal Lobe Seizures, *Brain Res.*, Vol. 792, 24–40, 1998.

Ljung, L., *System Identification: Theory for the User*, Prentice-Hall, 1999.

Lucini, D., Porta, A., Milani, O., Baselli, G., and Pagani, M., Assessment of Arterial Cardiopulmonary Baroreflex Gains from Simultaneous Recordings of Spontaneous Cardiovascular and Respiratory Variability, *J. Hypertension*, Vol. 18, 281–286, 2000.

Mukkamala, R., Kim, J., Li, Y., Sala-Mercado, J., Hammond, R. L., Scislo, T J., and O'Leary, D. S., Estimation of Arterial and Cardiopulmonary Total Peripheral Resistance Baroreflex Gain Values: Validation by Chronic Arterial Baroreceptor Denervation, *Am. J. Physiol. Heart Circ. Physiol.*, Vol. 290, H1830–H1836, 2006.

Nollo, G., Faes, L., Porta, A., Antolini, R., and Ravelli, F., Exploring Directionality in Spontaneous Heart Period and Systolic Arterial Pressure Variability Interactions in Humans: Implications in the Evaluation of Baroreflex Gain, *Am. J. Physiol.*, Vol. 288, H1777–H1785, 2005.

Nollo, G., Faes, L., Porta, A., Pellegrini, B., Ravelli, F., Del Greco, M., Disertori, M., and Antolini, R., Evidence of Unbalanced Regulatory Mechanism of Heart

Rate and Systolic Pressure after Acute Myocardial Infarction, *Am. J. Physiol.*, Vol. 283, H1200–H1207, 2002.

Palus, M., Detecting Phase Synchronisation in Noisy Systems, *Phys. Lett. A*, Vol. 235, 341–351, 1997.

Palus, M., Komarek, V., Hrncir, Z., and Sterbova, K., Synchronization as Adjustment of Information Rates: Detection from Bivariate Time Series, *Physical Review E*, Vol. 63, 046211, 2001.

Porta, A., Baselli, G., and Cerutti, S., Implicit and Explicit Model-Based Signal Processing for the Analysis of Short Term Cardiovascular Interactions, *Proc. IEEE*, Vol. 94, 805–818, 2006.

Porta, A., Baselli, G., Guzzetti, S., Pagani, M., Malliani, A., and Cerutti, S., Prediction of Short Cardiovascular Variability Signals Based on Conditional Distribution, *IEEE Trans. BME*, Vol. 47, 1555–1564, 2000a

Porta, A., Baselli, G., Liberati, D., Montano, N., Cogliati, C., Gnecchi-Ruscone, T., Malliani, A., and Cerutti, S., Measuring the Degree of Regularity by Means of a Corrected Conditional Entropy in Sympathetic Outflow, *Biol. Cybern.*, Vol. 78, 71–78, 1998.

Porta, A., Baselli, G., Lombardi, F., Montano, N., Malliani, A., and Cerutti, S., Conditional Entropy Approach for the Evaluation of the Coupling Strength, *Biol. Cyber.*, Vol. 81, 119–129, 1999.

Porta, A., Baselli, G., Rimoldi, O., Malliani, A., and Pagani, M., Assessing Baroreflex Gain from Spontaneous Variability in Conscious Dogs: Role of Causality and Respiration, *Am. J. Physiol.*, Vol. 279, 2558–H2567, 2000b.

Porta, A., Furlan, R., Rimoldi, O., Pagani, M., Malliani, A., and van de Borne, P., Quantifying Linear Causal Coupling in Closed Loop Interacting Cardiovascular Variability Series, *Biol. Cybern.*, Vol. 86, 241–251, 2002.

Porta, A., Gnecchi-Ruscone, T., Tobaldini, E., Guzzetti, S., Furlan, R., and Montano, N., Progressive Decrease of Heart Period Variability Entropy-Based Complexity during Graded Head-up Tilt, *J. of Appl. Physio.*, Vol. 103, 1143–1149, 2007a.

Porta, A., Guzzetti, S., Furlan, R., Gnecchi-Ruscone, T., Montano, N., and Malliani, A., Complexity and Nonlinearity in Short-Term Heart Period Variability: Comparison of Methods Based on Local Nonlinear Prediction, *IEEE Trans. Biomed. Eng.*, Vol. 54, 94–106, 2007b.

Porta, A., Guzzetti, S., Montano, N., Pagani, M., Somers, V. K., Malliani, A., Baselli, G., and Cerutti, S., Information Domain Analysis of Cardiovascular Variability Signals: Evaluation of Regularity, Synchronisation and Co-ordination, *Med. Biol. Eng. Comput.*, Vol. 38, 180–188, 2000c

Porta, A., Montano, N., Furlan, R., Cogliati, C., Guzzetti, S., Gnecchi Ruscone, T., Malliani, A., Chang, H.-S., Staras, K., and Gilbey, M. P., Automatic Classification of Interference Patterns in Driven Event Series: Application to Single Sympathetic Neuron Discharge Forced by Mechanical Ventilation, *Biol. Cybern.*, Vol. 91, 258–273, 2004.

Porta, A., Montano, N., Pagani, M., Malliani, A., Baselli, G., Somers, V. K., and van de Borne, P., Non-Invasive Model-Based Estimation of the Sinus Node Dynamic Properties from Spontaneous Cardiovascular Variability Series, *Med. Biol. Eng. Comput.*, Vol. 41, 52–61, 2003.

Richman, J. S., and Moorman, J. R., Physiological Time-Series Analysis Using

Approximate Entropy and Sample Entropy, *Am. J. Physiol.*, Vol. 278, H2039–H2049, 2000

Robbe, H. W. J., Mulder, L. J. M., Ruddel, H., Langewitz, W. A., Eldman, J. B. P., and Ider, G., Assessment of Baroreceptor Reflex Sensitivity by Means of Spectral Analysis, *Hypertension*, Vol. 10, 538–543, 1987.

Saito, Y., and Harashima, H., Tracking of Information Within Multichannel EEG Record—Causal Analysis in EEG, in Yamaguchi, N. and Fujisawa, K. (Eds.), *Recent Advances in EEG and EMG Data Processing*, pp 133–146, Elsevier, 1981.

Saul, J. P., Berger, R. D., Albrecht, P., Stein, S. P., Chen, M. H., and Cohen, R. J., Transfer Function Analysis of the Circulation: Unique Insights into Cardiovascular Regulation, *Am. J. Physiol.*, Vol. 261, H1231–H1245, 1991.

Schiff, S. J., So, P., Chang, T., Burke, R., and Sauer, T., Detecting Dynamical Interdependence and Generalized Synchrony through Mutual Prediction in a Neural Ensemble, *Phys. Rev. E*, Vol. 54, 6708–6724, 1996.

Schreiber, T., and Schmitz, A., Improved Surrogate Data for Nonlinearity Tests, *Phys. Rev. Lett.*, Vol. 77, 635–638, 1996.

Xiao, X., Mullen, T. J., and Mukkamala, R., System Identification: A Multi-Signal Approach for Probing Neural Cardiovascular Regulation, *Physiol. Meas.*, Vol. 26, R41–R71, 2005.

Thispageintentionallyleftblank

CHAPTER 6

USE OF INTERPRETATIVE MODELS IN BIOLOGICAL SIGNAL PROCESSING

Mauro Ursino

6.1 INTRODUCTION

The aim of this work is to discuss the relationships and the mutual links between mathematical modeling and signal processing techniques. Although signal processing and mathematical models have been essential aspects of biomedical engineering from the very beginning of the discipline, these two subjects evolved almost independently, without a clear connection between them. In this chapter, we emphasize how the use of mathematical models and computer simulation techniques can help in signal processing, allowing specific features of a signal to be extracted and a parsimonious representation to be achieved. Emphasis in this work will be given to mathematical models based on physiological knowledge. These are often referred to as "white-box" or "gray-box" models in the literature; in the following we will use the term "interpretative models" to signify that they provide an interpretation of the intimate nature of a system and of the mechanisms working on it. The more traditional use of "black-box" models (i.e., models based on empirical equations, without any intimate knowledge of the system) will only be mentioned where necessary, for comparison purposes.

The present chapter is organized as follows. In Section 6.2, the main relationships between mathematical models and signals will be critically evaluated and discussed, and models will be partitioned into some rough classes. This partitioning is not meant to be conclusive, but is intended to put in evidence the main properties, virtues, and limitations of the different mathematical techniques applied to signal processing. In Section 6.3, some

Advanced Methods of Biomedical Signal Processing. Edited by S. Cerutti and C. Marchesi

recent examples are presented in which interpretative mathematical models are applied to signal processing problems. Each example is derived from an advanced aspect of physiological or clinical research.

6.2   MATHEMATICAL INSTRUMENTS FOR SIGNAL PROCESSING

6.2.1   Descriptive Methods

The fundamental idea for starting our analysis is that any signal to be examined and processed requires a mathematical representation. We are often so used to processing signals with traditional techniques (such as Fourier analysis or interpolation) that we often neglect to consider the underlying mathematical representation. The traditional signal processing techniques assume, implicitly or explicitly, that this mathematical representation is completely independent of the characteristics of the system that generated the signal. Such a choice, of course, exhibits certain advantages. The most evident is that mathematical expressions and algorithms tend to be "universal" (within certain limits), that is, they may be applied to a wide class of signals independently of their effective origin. This choice, although extremely useful in several practical problems (let us consider, for instance, filtering techniques that make use of Fourier representation of signals) caused a deep separation between signal processing on one hand, and interpretative mathematical modeling on the other. This separation is still largely present in the scientific literature, although an increasing number of exceptions have appeared in recent years.

The universal nature of these mathematical models does not only imply the possible application to a wide class of signals (for instance, all signals with finite energy can be processed via the Fourier transform) but, above all, the existence of well-known processing algorithms, which can be applied automatically to data in order to produce intelligible results. Another characteristic of these methods is that they are able to account for the overall variance of the signal, that is, all original details of the signal can be reproduced and mathematically described without any loss. Let us consider, for instance, a frequency spectrum of a discrete signal. The original signal in the time domain can be perfectly reconstructed from its frequency spectrum using the inverse FFT algorithm. Of course, the restoration of the entire original information is always possible, provided some well-known constraints are satisfied (first of all, the Shannon theorem for sampling).

By way of an example, let us consider the description of a discrete time-dependent signal, $x[n]$, with finite energy, through the discrete Fourier transform (Equation 6.1) or through the discrete wavelet transform (Equation 6.2). In the first case, one can write

$$x[n] = \frac{1}{N} \sum_{k=0}^{N-1} X[k] e^{j(2\pi/N)kn} \tag{6.1}$$

where $X[k]$ represents the coefficients of the discrete Fourier transform and N is the number of available samples.

In the second case, one can write

$$x[n] = \sum_{j=1}^{\infty} \sum_{k=-\infty}^{+\infty} d_j[k] \psi[2^{-j}n - k] \tag{6.2}$$

where $d_j[k]$ represents the coefficients of the discrete wavelet transform, $\psi[n]$ is the sampled wavelet, centred at the instant $n = 0$, and j is the scale factor. In the cases of Equations 6.1 and 6.2, well-known algorithms are available in software packages such as MATLAB for the computation of coefficients from data.

However, the techniques mentioned above also exhibit important limitations. Indeed, these limitations are exactly the reason why interpretative models may be of value for the analysis of signals. The main limitations of the traditional techniques can be summarized as follows:

1. These techniques just describe the signal, by moving its representation from a domain (typically the temporal one) to another domain (such as the frequency domain in the case of the Fourier transform, or the scale domain in the case of the wavelet transform). The basic idea is that the representation in the new domain allows the user to perform a deeper signal analysis, and makes signal processing easier (for instance, filtering in the frequency domain, or detection of sudden changes in the scale domain). There is no interpretation for the genesis of the signal and no hypothesis for the mechanisms that affect the observed behavior.
2. The transformation from one domain to the other does not reduce the dimension of the problem, that is, the amount of independent data necessary to achieve a complete description of the signal remains unchanged or may even increase. Let us consider, for instance, the number of complex coefficients in the discrete Fourier transform (Equation 6.1) or the number of coefficients in the discrete wavelet transform (Equation 6.2). Of course, it is always possible to further reduce the dimension of the problem, for instance, with a filtering technique or using statistical techniques such as principal component analysis, thus reducing the overall variance of the signal.
3. This signal representation does not exhibit any predictive capacity. The processing algorithm must be applied to each new tracing.

4. Frequently in the study of physiological problems, several signals are simultaneously monitored. Of course, it is extremely important not only to represent these signals but, above all, to find a description of their mutual relationships and possible causal links.

The problems delineated above are the focus of modern techniques based on more sophisticated mathematical models.

Awareness of the limitations mentioned above spurred several authors to use different techniques for signal representation that not merely describe the signal by translating its representation from one domain to another, but also provide an interpretation of mechanisms for signal generation and a description of the mutual relationships among signals. However, these mathematical descriptions, while helping to overcome the previous limitations, also introduce new limitations and restrictions, a necessary trade-off between different and often contradictory requirements.

6.2.2   The Black-Box Models

The so-called empirical or "black-box" models are another means of signal representation. Let us assume that a signal $x[n]$ is the (sampled) input to a mathematical model and $y[n]$ is the corresponding (sampled) output. In many cases, $x[n]$ is a fictitious signal (i.e., a signal that does not exist in reality but it is just used as a mathematical phantom to generate the output). A common example of this type is the use of white noise input with zero mean value and unknown variance. In this case, the model provides just a description of the output signal $y[n]$. On the contrary, if both $x[n]$ and $y[n]$ are real signals, the model provides a relationship between them, pointing out their functional link.

Let us consider two examples, largely used in the biomedical engineering field. For the sake of generality, we will consider the case of nonlinear systems. The (simpler) case of linear systems becomes a particular case of the equations presented below, easily deducible from them.

As a first example, let us consider a mathematical model consisting of a series of functionals. This theory was originally presented by Volterra (Volterra, 1930) as a general tool for the representation of unknown analytical functionals, and subsequently included by Wiener in the context of nonlinear identification (Wiener, 1958). With reference to discrete signals, which are of interest in signal processing by a computer, the time-discrete Volterra–Wiener expansion takes the following form:

$$y[n] = k_0 + \sum_{m_1} k_1[m_1]x[n-m_1] + \sum_{m_1}\sum_{m_2} k_2[m_1,m_2]x[n-m_1]x[n-m_2] +$$

$$+ \sum_{m_1}\sum_{m_2}\sum_{m_3} k_3[m_1,m_2,m_3]x[n-m_1]x[n-m_2]x[n-m_3] + \dots \quad (6.3)$$

where k_0, $k_1[m_1]$, $k_2[m_1, m_2]$, $k_3[m_1, m_2, m_3]$, and so on are the so-called Volterra kernels, which describe system dynamics at each order of nonlinearity. k_0 represents a constant term (i.e., an offset term), whereas $k_1[m_1]$ represents the linear dynamics. In the case of a linear system, this term is the impulse response, and, as well known, its Laplace transform provides the transfer function of the system. The kernels of higher order represent the possible nonlinearities in the model, and are symmetrical functions (i.e., they do not change after a permutation of their arguments).

The practical use of the Volterra–Wiener expansion has became possible in recent years thanks to some modern techniques for kernel estimation. For instance, some authors proposed to describe the main kernels through a number of Laguerre orthogonal functions [the interested reader can find more details in Marmarelis (1994)]. However, these techniques require an onerous representation, which is often based on a great number of unknown coefficients.

An alternative and more compact choice for signal representation, still within the class of "black-box" models, can be attained with the use of NARMA models (Leontaritis and Billings, 1985). In general, a NARMA model of order L, M can be written as a multinomial difference equation:

$$P_L(\Delta)y[n] + f(x[n],...,\Delta^M x[n]; \Delta y[n],...,\Delta^L y[n]) = P_M(\Delta)x[n] \quad (6.4)$$

where $P_L()$ and $P_M()$ are polynomials in the variable Δ with degree L and M, respectively, $f()$ is a multinomial function (that is, the sum of cross products of the arguments) without linear terms, and Δ is the unit delay operator, that is, $\Delta^k x[n] = x[n - k]$.

Identification of a NARMA model requires two steps:

1. Determination of the structure of the model
2. Parameter estimation

Determination of model structure, in turn, can be split into the selection of the order of the model (that is, the values for L and M) and the selection of parameters to be incorporated in the model. In general, there are good techniques for parameter estimation if the order is known; however, selection of the optimal order and elimination of unnecessary parameters often represent difficult problems.

A third approach, conceptually similar to that used in the NARMA models, consists in the description of the input–output relationship of a system via neural networks (for instance, multilayer networks trained with the modern versions of the old back-propagation algorithm, or networks based on radial basis functions). The interested reader can find more details in canonical textbooks on neural networks (Haykin, 1994).

Mathematical representations based on black-box models allow many limitations of classic signal processing techniques to be overcome. Indeed, these techniques consent a more parsimonious representation of a signal, compared with the original amount of data. They allow a representation of the relationships among signals (when a signal is assumed as an input to the model and the other is calculated as the output) and may also exhibit some predictive capacity.

At this point, however, it must be stressed that these techniques do not represent the overall variance of the original signal. In other words, by denoting with $y[n]$ the model output and with $z[n]$ the measured output (i.e., the samples of the experimental output signal), we can write

$$z[n] = y[n] + e[n] \qquad (6.5)$$

where $e[n]$ denotes the error between the model and the measured outputs, which can be treated as a noise term. The variance of this term describes the part of the original signal that cannot be represented by the model (let us remember that, in the traditional techniques such as the Fourier representation and the wavelet decomposition, this term is identically equal to zero).

Moreover, the use of black-box models exhibits other important limitations:

1. Although parsimonious, if compared with classical techniques, the mathematical representation offered by black-box models can still be onerous and cumbersome.
2. It is difficult to find out the correct order of the model. If the order is too high, one runs the risk of "overfitting." This means that the signal is represented using an excessive number of terms (hence, an excessive number of unknown parameters, estimated from data), with the result of a poor predictive capacity (in the neural network domain, we often use the term "poor generalization," taken from psychology). If one uses an insufficient number of terms, parameter estimation results in "underfitting," that is, in a poor representation of the original signal, with an excessive increase in the noise term $e[n]$.
3. It is not easy, from the model structure, to derive a physiological/clinical interpretation of the individual terms in the model and the significance of the estimated parameters.

6.2.3 Interpretative Models

The limitations listed above can be in large measure overcome by using interpretative mathematical models, which incorporate some knowledge of the intimate structure of the system.

The basic idea is to build an "ad hoc" mathematical model, whose equations have no universal meaning (as was the case of classical equations 6.1–6.4) but aspire to reproduce the physiological process under study. We do not wish to enter here into a detailed description of knowledge that can be used to build interpretative models. Depending on the system nature, this may include physical or chemical laws, anatomical information, physiological data, or even "behavioral laws" taken from experiments on the system, which link particular quantities within the model. The fundamental feature of these models is that the equations must be individualized to the specific system under study, and one must be able to provide a clear physiological/clinical a priori meaning for equations and parameters, and for the other quantities of the model. This is the characteristic of interpretative models that distinguishes them from empirical models. To ascertain whether a model is essentially interpretative or empirical, one can try to answer these questions:

Are the equations specific to the individual system, or can they be used to describe other systems by changing just the order of the model and the values of the parameters?

Is it possible to find a meaning for the parameters, and to assign their normal range a priori, that is, even before any identification procedure? Have these parameters a clinical/physiological significance, that is, do they represent physiological/pathological mechanisms?

Is it possible to find significance for most of the other "hidden" quantities of the model?

If the response to these questions is "yes," you can consider that your model is interpretative of the process under study.

As stated above, the model contains a given number of parameters, the value of which must be assigned in order to mimic individual variability and to perform computer simulations. The procedure of parameter assignment is crucial for the problem we are considering here, and deserves particular attention. There are two fundamental ways to assign a value to the parameters of an interpretative model:

1. On the basis of previous physiological/clinical knowledge. In this case, the parameter assumes a typical or paradigmatic value that can be considered representative of an entire class of subjects (for instance, healthy subjects or subjects with a given pathology).
2. On the basis of an "a posteriori" best-fitting procedure, which minimizes a metric of the difference between model output and experimental data collected individually.

Of course, if all model parameters are given "a priori" (as assumed in method 1), the model becomes a tool to describe the behavior of a paradigmatic subject and can be used as a pure simulation instrument through which one can improve knowledge of system behavior, for instance, for scientific or didactical purposes. Indeed, this is the use of interpretative models commonly encountered in the physiological literature and, more rarely, in the clinical literature. Conversely, the use of interpretative models as a support for signal analysis requires that at least some of parameters be assigned through the modality described in method 2, that is, through a best fitting between model output and measured data. In this case, as in the case of black-box models, the model becomes an instrument to account for most of the variance of the signal, through the classic scheme illustrated in Figure 6.1. It is important to observe that not all parameters in the model must necessarily be assigned a posteriori through best fitting, but only those whose intersubject variability can explain most of the variability of the observed signals. The problem of which parameters must be assigned via the best-fitting procedure and which must be maintained at a paradigmatic value is a crucial one with a difficult solution. The latter aspect, however, is essential, since it allows a parsimonious description of a signal, even for complex models, that is, models that contain a large number of parameters.

Concerning the possibility of describing a signal in a parsimonious way, we must emphasize that the term "parsimonious" is used throughout this chapter to represent the number of unknown pieces of information that must be stored in memory to distinguish one signal from another, and to reconstruct most of its variance. Hence, "parsimonious" does not refer to

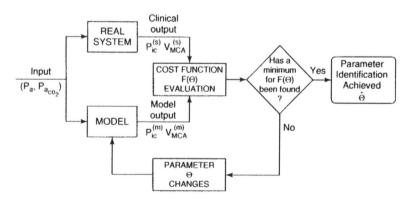

Figure 6.1. Block diagram of the procedure for parameter estimation in an interpretative model. $F(\Theta)$ represents a criterion function that quantifies the differences between model output (m) and measured signal (s). Symbols P_a, P_{CO2}, P_{ic}, and v_{MCA} are concerned with the example in Figure 6.2, and represent systemic arterial pressure, arterial carbon dioxide pressure (inputs), intracranial pressure, and blood flow velocity in the middle cerebral artery (outputs), respectively.

the complexity of equations in the model, nor to the number of mecha-nisms involved, but just to the number of estimated parameters that allows a satisfactory reproduction of individual signals, and allows the distinction of one signal from another. This is the information that cannot be deduced a priori from previous knowledge, but requires a posteriori analysis of col-lected data. This number is equal to the number of parameters estimated with the best-fitting procedure. The great complexity of some physiologi-cal models can be considered as a priori information, which does not de-pend on the individual signal (hence, not on the individual subject).

On the basis of the previous description, we can now summarize how an interpretative model can overwhelm the limitations typical of the black-box models and, even more clearly, of the purely descriptive techniques.

1. In many cases, interpretative models, including previous knowledge, can be very parsimonious. By way of comparison, let us consider that a signal $x[n]$ is described in the time domain with N samples. With Fouri-er analysis and making use of the FFT algorithm, the number of inde-pendent pieces of information is still N (we have N complex numbers to characterize the spectrum, but $N/2$ of them are complex conjugates of the others). The number of parameters is even greater if we consider the discrete wavelet decomposition. Conversely, as shown in the example of Section 6.3, the number of unknown parameters can be just a few units when a good interpretative model of a physiological process is used.

2. The problem of overfitting or underfitting can be at least in part over-come on the basis of the physiological and clinical knowledge exploited in model building. In the choice of the model and in the number of "free" parameters, the researcher can consider only those aspects of the system that can directly affect signal variability, and that can be indi-vidually estimated with sufficient accuracy. In other words, the model must not necessarily incorporate all available knowledge, and not ac-count for all existing parameters, but only for those that directly partici-pate in the genesis of the observed signals. This is, however, a complex problem even with the use of interpretative models, as will be shown in the following examples. We think that the correct choice of estimated parameters, and correct choice of the cost function to be minimized, represent the bottleneck for widespread use of interpretative models in clinical practice.

3. The significance of the estimated parameters must be a priori known. Hence, from knowledge of the parameter values, it is possible to derive information on the physiological processes that generated the signal. This may be exploited in clinics, for instance, as a first step toward a quantitative diagnosis, or to delineate the effect of a therapy in a quanti-tative way.

4. An interpretative model represents the best way to describe the relationships among signals, by explicitly considering their causal relationships and the mechanisms involved.
5. The model may exhibit interesting predictive capacities, and can be used to anticipate the effect of some maneuvers on the same patient (i.e., using the parameter set estimated from previously collected data).

Besides the previous virtues, however, one can also be aware of important difficulties that can be encountered in the use of these models, either in their building or in their use as a way to analyze individual signals. At present, these difficulties strongly restrain the routine use of these techniques, restricting them to pioneering clinical centers or to the most advanced scientific research.

First of all, as anticipated above, the numerical algorithm for parameter estimation is not universal, as it occurs with the descriptive methods and, in most cases, also with the black-box models. By way of example, if one provides the FFT algorithm or the wavelet algorithm with N samples, one always automatically finds the correct coefficients for the signal representation. Conversely, when using interpretative models, one has no a priori warranty that the minimization algorithm converges to an acceptable solution, nor that the attained solution is unique. The problem of suboptimal solutions, or solutions without a physiological meaning, is of great difficulty in the application of these models to data. Furthermore, the computation time can be elevated, thus precluding the use of these models for online solutions of problems (for instance in intensive care units, where the time for a response must be minutes or seconds). Frequently, the final solution is affected by the constraints imposed on the ranges for admissible parameter values, or on the choice of an initial guess for the parameters (any fitting procedure requires an initial guess for starting the algorithm). As a consequence of the previous problems, the application of interpretative models to real data often requires an expert user. Moreover, parameters can also be estimated with poor accuracy (a problem strictly related to the overfitting problem analyzed above), especially when signal variability can be ascribed to the convergence of two or more mechanisms working in synergism, or to the opposition of two antagonistic mechanisms. Hence, the use of these models must be accompanied with statistical techniques for assessing the accuracy of the estimated parameter values.

Of course, the model is able to account only for a portion of signal variance (by contrast, descriptive techniques provide a perfect reconstruction of the signal). The diference between the model and the signal outputs can be ascribed to measurement noise (as is usually done in the classical identification theory), but can also be a consequence of model inadequacy or of the presence of additional mechanisms voluntarily not incorporated in the model. Of course, a great difficulty in the use of interpretative models is to build a good model of the process. As presented below, however, we now have good models for many physiological systems, and improved models will certainly appear in the not-too-distant future.

In the following, we will presènt some examples, taken from different physiological fields, to illustrate possible applications of interpretative models in signal analysis and to clarify the virtues and limitations described above.

6.3   EXAMPLES

6.3.1   Mathematical Models and Signals in Intensive Care Units

As a first representative example of a possible synergic interaction between interpretative models and biomedical signals, let us consider the problem of monitoring cardiovascular quantities in intensive care units (ICUs). In particular, in this section we will focus attention on a neurosurgical ICU for the treatment of patients with severe intracranial pathologies (such as brain injury, subarachnoid haemorrhage, or rupture of an intracranial aneurism). However, similar examples could also be presented for cardiothoracic ICUs. A characteristic of these clinical settings is the simultaneous and protracted monitoring of many different physiological quantities, linked through causal relationships. One of the main problems that medical doctors must cope with during the treatment and monitoring of these patients is to extract from the plethora of monitored data and simultaneously available signals a few relevant pieces of information able to characterize the patient's status and drive the therapy. Monitored data are often insufficiently exploited because their understanding requires a comprehension of the causal relationships among the mechanisms involved and the nonlinear effects that affect the results. The problem is further complicated by the little time available to make decisions and by the large variability among individual responses. The same maneuver that may be beneficial for a subject may have a detrimental effect on another one, depending on the personal differences in the strength of the mechanisms involved.

Let us consider the signals monitored in a modern neurosurgical ICU for the treatment of brain injured patients. Among the monitored quantities, a particular clinical value can be given to the systemic arterial pressure, the end-tidal CO_2 pressure, oxygen saturation in the jugular vein, the intracranial pressure (equal to pressure in the cerebrospinal fluid), and the blood flow velocity in the middle cerebral artery (monitored with the transcranial Doppler technique as an index of cerebral blood flow). The first two quantities represent inputs that can be manipulated by clinicians (CO_2 pressure by acting on an artificial ventilator, systemic arterial pressure through pharmacological means), whereas the last three quantities represent the targets of monitoring and treatment. Some examples of signals can be found in Figure 6.2. As can be easily seen from this figure, these quantities are not independent, but strict relationships can be observed among their time changes. Actually, alterations in arterial pressure and in CO_2 pressure induce significant variations in blood flow velocity (hence, in cerebral blood flow) and in intracranial pressure. It is generally thought that these relationships contain important information on the intracranial mechanisms (such as the circulation of cerebrospinal fluid, the elasticity of the intracranial compartment, and the status of control mechanisms regulating cerebral blood flow) and on their pathological changes consequent to the head injury. Knowledge of these changes is essential for the diagnosis and the subsequent treatment of the head-injured patient. However, as specified above, the complexity of these relationships cannot be assessed in purely qualitative terms.

The use of empirical, black-box models might be useful to find out the functional link among these quantities, but without clarifying the intimate nature of the mechanisms involved. Moreover, these models provide a redundant (i.e., not parsimonious) description of the patient, based on an excessive number of parameters with poor physiological meaning.

Several interpretative models have been proposed in recent years, with different complexity, to describe intracranial dynamics and cerebral hemodynamics (Czosnyka et al., 1993; Ursino and Lodi, 1997, Ursino et al., 2000; Payne and Tarassenko, 2006; Stevens et al., 2007) but just a few of them have been used to fit real signals in neurosurgical ICUs. In the following, we will consider the model described in Ursino and Lodi (1997) (see Figure 6.2, upper panel) since it exhibits some characteristics adequate for the interpretation of intracranial signals in ICUs. The model incorporates all principal mechanisms involved in intracranial dynamics, but provides a very parsimonious description of clinical tracings, based on estimation of just a few parameters. In particular, by using mean systemic arterial pressure (i.e., arterial pressure minus the pulsatile component) and arterial CO_2 pressure as input signals, the model can correctly simulate the temporal changes in mean intracranial pressure and in mean blood flow velocity in the middle cerebral artery, by estimating just six parameters with a clear clinical meaning. Each estimated parameter contains the information on a different intracranial mechanism: the resistance to cerebrospinal fluid reabsorption, the elasticity of the craniospinal compartment, the gain and time constant of the mechanisms that maintain a constant cerebral blood flow despite pressure changes (autoregulation), and the gain and time constant of the reactivity of cerebral vessels to CO_2 changes. Some examples of fitting between model and real tracings are reported in the bottom panel of Figure 6.2, with reference to a patient with severe head injury (Ursino et al., 2000).

However, with the model it is very difficult to realize a completely automatic procedure for parameter estimation to be routinely used in a

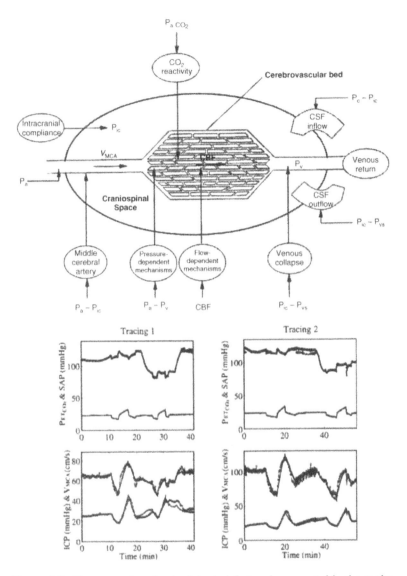

Figure 6.2. *Upper panel:* main physiological factors incorporated in the mathematical model of intracranial dynamics (Ursino and Lodi, 1997). P_a, systemic arterial pressure; P_c, capillary pressure; CBF, tissue cerebral blood flow; P_v, cerebral venous pressure; P_{vs}, sinus venous pressure; P_{ic}, intracranial pressure; CSF, cerebrospinal fluid; Pa_{CO2}, arterial carbon dioxide pressure; V_{MCA}, blood flow velocity in the middle cerebral artery. Cerebral arterioles are under the control of two regulatory mechanisms: CO_2 reactivity and autoregulation. *Lower panels:* results of two best-fitting procedures on two different patients. The upper traces represent the time patterns of the two inputs for the model (i.e., mean systemic arterial pressure and arterial CO_2 pressure); the bottom traces are the targets of the estimation procedure (intracranial pressure and blood flow velocity in the middle cerebral artery). Circles connected with thin lines are in vivo data; thick lines are model simulation results with ad hoc estimated parameters.

139

clinical setting. In some cases, in fact, the know-how of the user and his/her skills are essential to drive the fitting procedure toward an acceptable solution (for instance, by imposing a suitable choice for the initial guess, and constraints on the final solution). This limitation, which depends on the existence of multiple minima in the cost function, makes, at present, very difficult a routine use of this kind of model in daily clinical practice.

A fundamental aspect of future research will be the formulation of automatic methods for parameter estimation, without the participation of a skilled user, for instance, by automatic intelligent criteria for constraining the parameter space.

6.3.2   Mathematical Models and Cardiovascular Variability Signals

Let us consider a second example, still concerned with the cardiovascular system. It is well known that the main cardiovascular quantities (such as systemic arterial pressure and heart rate) exhibit temporal fluctuations, with a frequency lower than that of the cardiac beat. Above all, heart rate variability (HRV) has received much interest and has been the subject of intensive research in past years, in different physiological and clinical settings. The rationale for this interest is that HRV is considered an index of autonomic regulation (sympathetic and vagal) (Malliani et al., 1991); its quantitative study has been suggested as a clinical tool for the diagnosis of neurological disorders of autonomic regulation and/or for the quantification of the risk consequent on myocardial infarction (Bigger et al., 1992). Spectral analysis reveals the presence of three main components of the spectrum: a high-frequency (HF) component at the respiratory rate (about 0.2–0.5 Hz in humans) that is considered a marker of parasympathetic activity; a low-frequency (LF) component at about 0.2 Hz that is a sign of both sympathetic and parasympathetic contributions; and a component at very low frequencies (VLF) that can be ascribed to the presence of slow regulatory processes (such as termal, humoral, and vasomotor processes). Despite the disparate efforts aimed at extracting indices of clinical value from these spectra, the interpretation of cardiovascular variability signals is still very complex, due to the plethora of factors simultaneously involved.

The use of interpretative mathematical models is playing an increasing role in the study of HRV, and different models have been proposed recently, some of them with emphasis on parameter estimation. These models contain a description of the mechanisms participating in short-term cardiovascular regulation. A few parameters are estimated to mimic some aspects (in the time domain or in the frequency domain) of the measured cardiovascular signals. By way of example, Peytan et al. (2003) recently developed a model that describes the dependence of heart rate on vagal activity. The optimal parameters of the model have been obtained through a best fitting between model output and clinical data; the latter consisted of seven minutes of ECG and respiration during a block of vagal activity induced by atropine. The model is able to explain the dependence of the average heart rate and the HF component of the spectrum on atropine concentration in different subjects, as illustrated in Figure 6.3. Moreover, in six of eight examined subjects the proportion of the variance in the data explained by the model was very close to one. The authors claim that the parameter values can help in the classification of specific cardiac and neurological disorders, and that this procedure can be implemented in clinical practice in forthcoming years.

More recently, Yildez and Ider (2006) used a comprehensive cardiorespiratory model, taken from Ursino and Magosso (2003), to test the

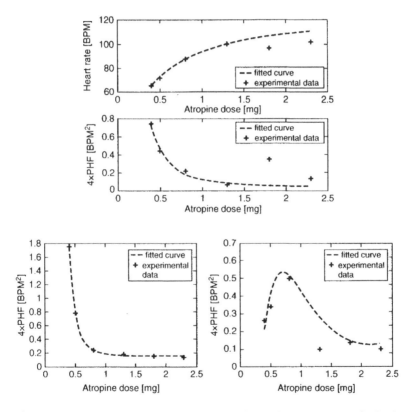

Figure 6.3. Some examples of fitting between the model by Peytan et al. (2003) and experimental data. *Upper panels:* mean heart rate and total power in the high-frequency (HF) band as a function of atropine dose in one subject. *Lower panels:* total power in the HF band in two different subjects. The model's capacity to explain both monotonic decreasing patterns (6 of 8 cases) and bimodal patterns (2 of 8 cases) is noticeable. The figures are taken from Peytan et al. (2003).

hypothesis that the LF component of heart rate variability is significantly affected by the respiratory signal. In this model, chest and abdomen circumference signals and lung volume signals are used as respiratory inputs, and validation is performed through experiments on nine volunteers. Results show that respiration not only is the major contributor to the genesis of the HF peak in the HRV power spectrum, but also plays an important role in the genesis of its LF peak. Thus, the LF/HF ratio, which is used to assess sympathovagal balance, cannot be correctly utilized in the absence of simultaneous monitoring of respiration during an HRV test. In this work, however, the comparison between model and real data is still qualitative, without using an ad hoc estimation procedure. However, in perspective this approach may be used to arrive at a best fitting between real and model HRV spectra and to provide some insight into baroreflex control in the patient.

6.3.3   Mathematical Models and EEG Signals during Epilepsy

The electroencephalographic activity during epilepsy exhibits different patterns, with specific temporal and spectral characteristics (Ebersole and Milton, 2003). Some of these patterns, measured with intracortical electrodes (deep EEG) during surgery, are shown in the right-bottom panels of Figure 6.4. The first pattern represents a normal basal EEG activity. The second consists of sporadic spikes, often observable in the period between two attacks. The third and fourth patterns represent activity with repeated sustained spikes and a slow rhythmic activity, respectively; both patterns may appear during the initial phase of an attack or, more often, during the attack itself. The fifth pattern consists in discharges with higher frequency (generally in the gamma band, 20–80 Hz) but limited amplitude. These named electrodecremental events are often encountered in the initial phase of the crisis; for this reason, they are sometimes considered as possible markers, able to anticipate the clinical symptoms. Finally, the sixth pattern consists of a slow, almost sinusoidal activity, which is generally encountered during the terminal phase of an attack.

   Although these EEG patterns can be recognized and characterized quite well from the viewpoint of signal processing, a deep understanding of their significance, etiology, and underlying neural mechanisms is still lacking. In this regard, the use of neurocomputational models, based on neural mechanisms, can be of the greatest value to deepen our understanding of these signals and help their classification on an interpretative basis.

   Various mathematical models have been proposed in the literature to describe epileptogenic phenomena, from the more detailed models at a microscopic and cellular level (Traub et al., 1997) to models at a more macroscopic level, which simulate the dynamics of entire neural groups (Lopes da Silva et al., 1976; Freeman, 1978).

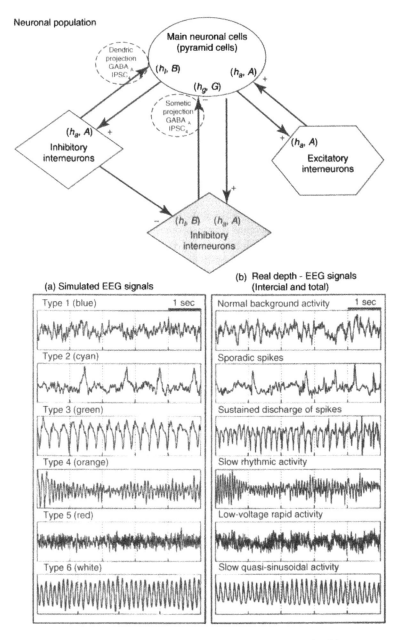

Figure 6.4. *Upper panel:* model of a neural population in the hippocampus (Wendling et al., 2002). The population consists of four subgroups: pyramidal excitatory neurons, excitatory interneurons, inhibitory interneurons with slow synaptic kinetics, and inhibitory interneurons with fast synaptic kinetics. *Bottom panels:* examples of several different patterns of electrical activity simulated with the model by varying the synaptic gains (left column) compared with cortical EEGs measured in the human hippocampus during surgery with intracerebral electrodes (right column). All figures are taken from Wendling et al. (2002).

The following example is focused on a model of the second type, recently proposed by Wendling et al. (2002) to clarify the mechanisms at the basis of EEG changes during epileptic attacks. The fundamental idea of the model is that the epileptogenic patterns can be ascribed to an imbalance between excitation and inhibition within a neural circuit. As shown in the upper panel of Figure 6.4, the model assumes that a group of pyramidal neurons (hence, excitatory) in the hippocampus can excite proximal neurons and receive back three different kinds of synapses: excitation from other excitatory neurons, slow inhibition from inhibitory interneurons with slow synaptic kinetics, and fast inhibition from other inhibitory interneurons with fast synaptic kinetics. The overall coupling is governed by three fundamental parameters represent the synaptic strength of the excitation, of slow inhibition, and of fast inhibition. Simulation results (some of which are illustrated in the left-bottom panels of Figure 6.4) show that alterations in these three parameters allow realistic simulations of the main different patterns observed during epileptic attacks. Moreover, a sensitivity analysis of parameter values allows a clear determination of the parameter changes that cause transition from one pattern to the next. For instance, a reduction in slow inhibition can explain the appearance of fast oscillations in the gamma band (pattern number 5), which represent a typical event at the beginning of the crisis (the interested reader can find more details in the paper by Wendling et al., 2002). Hence, the model provides a parsimonious description of the different EEG signals, summarizing their main aspects with just three parameters that account for the main underlying mechanisms. Subsequently, authors of the same team (Labyt et al., 2007) used this modeling approach to suggest hypotheses on the role of different types of GABAergic neurotransmission in the generation of epileptic activities in the entorhinal cortex, a brain structure largely involved in human mesiotemporal lobe epilepsy (MTLE).

Although the authors have not attempted a real fitting between model output and clinical tracings (the comparison in Figure 6.4 is just qualitative), this aspect might represent a further model application. Such a fitting, performed in limited time windows of the EEG signal, may allow the determination of nonstationary parameter changes that characterize the time evolution of the attack in the individual subject.

In this regard, in another recent work the same authors (Wendling et al., 2005) used intracerebral EEG signals recorded from the hippocampus in five patients with MTLE during four periods (interictal activity, just before seizure onset, seizure onset, and ictal activity) to identify the three main parameters of the model (related to excitation, slow dendritic inhibition, and fast somatic inhibition). The identification procedure used optimization algorithms to minimize a spectral distance between real and simulated signals. The results suggest that the transition from interictal to ictal activity cannot be simply explained by an increase in excitation and a decrease in inhibition, but requires time-varying parameter changes.

6.3.4 Mathematical Models, Electrophysiology, and Functional Neuroimaging

The possibility to study cognitive processes in men and higher order animals in quantitative terms has been rapidly increasing in the last years, especially thanks to the development of modern techniques for functional neuroimaging, such as positron emission tomography (PET) and functional nuclear magnetic resonance (fMRI), and the improvement of more traditional techniques, such as electroencephalography (EEG) and magnetoencephalography (MEG). The huge development of these techniques now allows integrated neural activity to be studied in multiple brain areas in awake subjects who execute specific voluntary tasks. However, the information so obtained is complex and variegated, and it is not simple to go back from data to the original neural mechanisms, due to a series of problems (Horwitz et al., 2000), which will be briefly summarized below.

First, the techniques for functional neuroimaging (PET and fMRI) have a high spatial resolution (on the order of 1 mm or less) but poor temporal resolution. Moreover, these techniques do not directly measure neural activity (in terms of spike frequency of membrane potential changes), but rather are sensitive to local changes in metabolism and blood flow. These latter quantities are especially affected by synaptic activity. Unfortunately, changes in the excitatory and inhibitory activity in neural groups can induce similar changes in metabolism and in PET and fMRI images, but have a completely different effect on a population of target neurons (excitation causing increasing activity and inhibition causing decreased activity). This problem has been clearly simulated by Almeida and Stetter (2002) with the use of a simple neural population model. The authors demonstrated that an increase in the imaging signal can be associated with a silent neural population, or even with a decrease in average spiking activity, in the presence of an increased inhibitory input. The reader can consult this simple and straightforward paper for more examples.

Conversely, the techniques that directly measure electrical (EEG) and magnetic (MEG) activities have a poor spatial resolution but high temporal resolution (on the order of milliseconds). However, these techniques do not directly measure neural activity. For instance, EEG can detect the changes in postsynaptic potential of a large group of pyramidal neurons, which synchronize their activity during the task.

Finally, the more traditional techniques in neurophysiology, largely used in animal experiments, make use of intracerebral electrodes, which can directly detect the activity (in spikes/s) of individual neurons.

Hence, a fundamental problem of modern cognitive neuroscience is how to link all these different pieces of information together to reach a comprehensive description of brain function, taking into account the different spatial and temporal resolution and the different physiological nature of the measured signals. Of course, the final purpose is to understand which regions of the brain participate in a given task, their role, and their dynamics. A shared common hypothesis of modern neuroscience, in fact, assumes that the brain performs cognitive or motor tasks through the participation of multiple regions, which are mutually linked and reciprocally exchange information via feedforward and feedback connections. Different tasks recruit different networks of regions, with different connectivity patterns. As a consequence, the same brain region can be involved in different tasks, and two regions can be functionally connected in a different way depending on the task they are engaged in. The critical problem is, thus, to assess the connectivity patterns among regions of interest during a given task, starting from the available data.

A classic method to evaluate the link between two signals (in our example, the connectivity between two brain areas) consists in computing their coherence. However, it is difficult to assess a connectivity network among multiple areas using coherence only, since a high coherence may be indicative of a direct or an indirect link. For instance, two signals in different regions may exhibit elevated coherence as a consequence of a direct anatomical connection, and this connection is functionally active during task execution, or the two regions may not be directly connected, but receive a common input coming from a third region. Furthermore, coherence indicates only the linear statistical link between EEG signals as a function of frequency, and so may be inadequate to represent nonlinear relationships. Alternative, nonlinear indices have been proposed, based on entropy, such as the cross mutual informative function (Jin et al., 2006).

The use of neuro-computational models may be a valid instrument to deal with the complexity of these problems, to aid in conceptualization of available data, and to establish a rational bond among data obtained with different techniques. The final purpose of such computational models is to build a realistic network of interconnected brain regions, through which to simulate the neurophysiological signals obtained with different techniques at different levels (electrophysiological, anatomical, and metabolical). However, just a few examples of this type can be found in the recent literature.

David and Friston (2003) analyzed the changes in power spectrum resulting from simple connectivity patterns among two regions of interest (ROIs), each simulated via a neural mass model of two parallel populations. This paper provides a first attempt to characterize how interactions among different ROIs are reflected in MEG/EEG oscillations, and represents a first step toward a theoretical analysis of indices for nonlinear coupling.

Sotero et al. (2007) developed a comprehensive neural-mass model to study the generation of EEG rhythms in the scalp. The model is based on an accurate anatomical description of brain connectivity, including 71 brain areas and the thalamus. The model was then used to investigate the dependence of EEG rhythms measured in the scalp on brain connectivity parameters. The main result is that anatomical data on connectivity are required to obtain physiological rhythms at the scalp. If connectivity strength is decreased, or random values are assigned to it, EEG rhythms are reduced until they eventually disappear. These results underline the strict relationship existing between connectivity and EEG rhythms. Moreover, results show that different rhythms can be produced (in the alpha, beta, delta, theta, and gamma ranges) by simply changing synaptic time constants. However, in their study the authors simulated individual rhythms separately, assigning suitable values to model parameters, without investigating conditions characterized by the simultaneous presence of different superimposed rhythms.

Ursino, Zavaglia, and coworkers, in a series of related works (Zavaglia et al., 2006; Ursino et al., 2007; Zavaglia et al., 2008) used neural-mass models of interconnected regions of interest to simulate cortical EEGs during simple cognitive and/or motor tasks. EEGs in the cortex were obtained, from multielectrode measurements on the scalp, using an inverse-propagation algorithm. Results show that the power spectral densities in the observed regions, including peaks in the alpha, beta, and gamma ranges, can be simulated with the model, using the weight of the connections among regions and the time constant of synaptic kinetics as unknown parameters. The model can be used to look for simple connectivity circuits, able to explain the main features of observed cortical power spectral densities.

A large-scale model has been proposed by Tagametz and Horwitz in a few related works starting in the late 1990s (Tagamets and Horwitz, 1998, 2000). The model intended to describe the cerebral areas involved in a specific task named a delayed match to sample. The task consists in the presentation of a visual image with a given shape, in a time delay during which the image must be maintained in the short-term memory, whereas other images are presented in a random fashion. The subject must recognize when a new image turns out equal to the original one. Hence, this task involves several brain regions devoted to shape recognition (the lateral geniculate nucleus; the primary visual cortex, V1 and V2; the occipital-temporal cortex, V4; and the interior temporal cortex, IT) and the regions devoted to maintenance of working memory (i.e., the prefrontal cortex, PF). The mathematical unit on which the model is based (see Figure 6.5, left upper panel) aspires to represent the electrical activity of a group of synchronized cortical columns, and consists of an excitatory neural population connected in feedback with an inhibitory neural population (this mathematical model, largely used to represent the dynamics of neural groups, represents the so-called Wilson–Cowan oscillator; Wilson and Cowan, 1972). Each region engaged in the task is represented by different neural groups of the same type, which differ in their reciprocal interconnections, in agreement with neuronantomical data in the primate (Figure 6.5, right upper panel). The model is able to simulate both neural activity obtained through electrophysiological measurements and metabolic activity in different brain regions obtained through PET. In the model, the latter measurement is simulated by integrating the absolute values of synaptic activity. A particular role in the model is played by the prefrontal cortex, which implements both the working memory and the decision unit. A qualitative description of the prefrontal cortex, according to this model, together with some model results, is presented in the middle and bottom panels of Figure 6.5.

In a more recent version, Winder et al. (2007) included a new learning method that adjusts the magnitude of interregional connections to match experimental results of an arbitrary functional magnetic resonance imaging (fMRI) dataset. The authors demonstrated that this method finds the appropriate connection strengths when trained starting from randomly chosen connection weights. This method expands currently available methods for estimating functional connectivity from human imaging data by including both local circuits and the main characteristics of interregional connections

6.4   CONCLUSIONS

The purpose of this brief review was to point the reader's attention to the usefulness of more stringent interrelationships between interpretative models, based on physiological knowledge, and biological signals. A strict connection among these two aspects may represent an essential aspect of modern signal processing techniques in the next decades, and may contribute to the advance of new methods for clinical investigation. To this end, an essential requisite is that the validation of interpretative mathematical models be performed on the basis of real signals, taken from individual subjects and adapted to the model through parameter estimation techniques, rather than merely on the basis of average data taken from the literature. Mathematical models in physiology have a long tradition of representing paradigmatic cases, which are instructive as examples of typical healthy individuals or typical pathological subjects of a given class. Transferring these models to the field of real signal analysis requires that the model must be tailored to the individual case. Similarly, signal processing techniques in the future should more consistently utilize interpretative models as the kernel of the mathematical procedure, rather than general or "universal" mathematical equations.

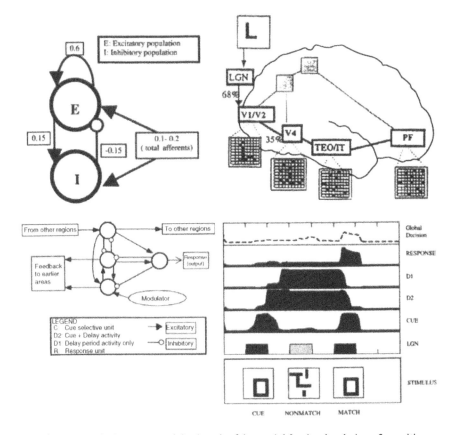

Figure 6.5. *Left upper panel:* basic unit of the model for the simulation of cognitive tasks (Tagamets and Horwitz, 1998; Tagamets and Horwitz, 2000). E represents the excitatory population, and I the inhibitory population of a generic cortical column. 60% of synapses reciprocally connect the excitatory units, 15% of synapses connect excitatory units to inhibitory units, and 15% of synapses connect the inhibitory units to the excitatory units. Just 10%–20% of synapses originate from other cortical areas. Each cortical area is described though one or more superimposed layers, each consisting of 81 basic units placed in 9 by 9 arrays. *Upper right panel:* cortical areas involved in the delayed match-to-sample task. LGN: lateral geniculate nucleus; V1,V2,V4: ventral visual areas; TEO/IT: inferotemporal cortex; PF: prefrontal cortex. The first five areas are engaged in shape recognition; the latter implements the working memory and the decision unit. The neural code becomes progressively more abstract when moving from one area to the next. The example shows the response to the letter L. *Bottom left panel:* schematic circuit of the synaptic connections among the different units in the PF area. The PF area consists of a 9 × 9 array of these circuits. *Bottom right panel:* time patterns of electrical activity in some units from the PF area. The unit "Cue" is active during the presentation of a given stimulus. The units D1 and D2 (delays) are active during the waiting period after presentation of the first stimulus. Moreover, the unit D2 is also modulated by attention. These two units excite one another through a reciprocal feedback connection and implement the working memory. Finally, the unit named "response" represents the decision unit. It displays a brief activation at the recognition of the matching by the second stimulus, and resets the delay units (from Tagamets and Horwitz, 2000).

149

This integration is still far from being satisfactorily realized and its routine use in physiological and clinical settings appears to be still at the pioneering stage, but forthcoming improvements, both of mathematical modeling and signal processing techniques, cannot ignore the benefits of such an integration. The efforts toward a stricter symbiosis between models and signals will constitute one of the most stimulating and useful research fields for biomedical engineering in future years. In this regard, the reader can note that all examples presented in this chapter are concerned with very recent studies, still in progress, whose results must be considered as preliminary. We hope that these examples can stimulate the curiosity of the reader about the potentiality of the proposed methods. These examples indicate a route to be followed. My hope is that, during the next years, these examples, and similar ones taken from the cutting edge of the scientific research, will be improved by new generations of researchers, moving toward a higher level of completeness, rigor, and simplicity, and making interpretative models routinely available in the daily practice of biomedical signal processing.

REFERENCES

Almeida, R., and Stetter, M., Modeling the Link Between Functional Imaging and Neuronal Activity: Synaptic Metabolic Demand and Spike Rates, *Neuroimage.* Vol. 17(2), 1065–1079, 2002.

Bigger, J. T., Fleiss, J. L., Steinman, R. C., Rolnitzky, L. M., Kleiger, R. E., and Rottman, J. N., (1992) Frequency Domain Measures of Heart Period Variability and Mortality after Myocardial Infarction, *Circulation.* Vol. 85, No. 1, 164–171.

Czosnyka, M., Harris, N. G., Pickard, J. D., Piechnik, S., CO_2 Cerebrovascular Reactivity as a Function of Perfusion Pressure–A Modelling Study, *Acta Neurochir (Wien).* Vol. 121, No. 3–4, 159–165, 1993.

David, O., and Friston, K. J., A Neural Mass Model For MEG/EEG: Coupling and Neuronal Dynamics, *Neuroimage.* Vol. 20(3), 1743–1755, 2003.

Ebersole, J. B., and Milton, J., The Electroencephalogram (Eeg): A Measure of Neural Synchrony, in *Epilepsy as a Dynamical Disease.* pp. 51–68, Milton, J., and Jung, P. (Eds.), Springer-Verlag. 2003.

Freeman, W. J., Models of the Dynamics of Neural Populations, *Electroencephalogr. Clin. Neurophysiol.,* Vol. 34, 9–18, 1978.

Haykin, S., *Neural Networks. A Comprehensive Foundation,* IEEE Press, 1994.

Horwitz, B., Friston, K. J., and Taylor, J. G., Neural Modeling and Functional Brain Imaging: an Overview, *Neural Networks,* Vol. 13, No. 8–9, 829–846, 2000.

Jin, S. H., Kwon, Y. J., Jeong, J. S., Kwon, S. W., and Shin, D. H., Increased Information Transmission during Scientific Hypothesis Generation: Mutual Information Analysis of Multichannel EEG, *Int. J. Psychophysiol.,* Vol. 62, No. 2, 337–44, 2006.

Labyt, E., Frogerais, P., Uva, L., Bellanger, J. J., and Wendling, F., Modeling of Entorhinal Cortex and Simulation of Epileptic Activity: Insights into the Role of Inhibition-Related Parameters, *IEEE Trans. Inf. Technol. Biomed.*, Vol. 11, No. 4, 450–461, 2007.

Leontaritis, I. J., and Billings, S. A., Input-Output Parametric Models for Non-Linear Systems. Part I: Deterministic Non-Linear Systems, *Int. J. Contr.*, Vol. 41, No. 2, 303–328, 1985.

Lopes Da Silva, F. H., Van Rotterdam, A., Barts, P., Van Heusden, E., and Burr, W., Models of Neuronal Populations: The Basic Mechanisms of Rhythmicity, *Prog. Brain Res.*, Vol. 45, 281–308, 1976.

Malliani, A., Pagani, M., Lombardi, F., and Cerutti, S., Cardiovascular Neural Regulation Explored in the Frequency Domain, *Circulation*, Vol. 84, No. 2, 482–492, 1991.

Marmarelis, V. Z., Nonlinear Modelling of Physiological Systems Using Principal Dynamic Modes, in *Advanced Methods of Physiological System Modeling*, pp. 1–27, Marmarelis, V. (Ed.), Plenum Press, 1994.

Payne, S. J., and Tarassenko, L., Combined Transfer Function Analysis and Modelling of Cerebral Autoregulation, *Ann. Biomed. Eng.*, Vol. 34, No. 5, 847–858, 2006.

Pyetan, E., Toledo, E., Zoran, O., and Akselrod, S., Parametric Description of Cardiac Vagal Control, *Autonomic Neuroscience: Basic and Clinical*, Vol. 109, No. 1–2, 42–52, 2003.

Sotero, R. C., Trujillo-Barreto, N. J., Iturria-Medina, Y., Carbonell, F., and Jimenez, J. C., Realistically Coupled Neural Mass Models Can Generate EEG Rhythms, *Neural Comput.*, Vol. 19, No. 2, 478–512, 2007.

Stevens, S. A., Thakore, N. J., Lakin, W. D., Penar, P. L., and Tranmer, B. I., A Modeling Study of Idiopathic Intracranial Hypertension: Etiology and Diagnosis, *Neurol. Res.*, Vol. 29, No. 8, 777–786, 2007.

Tagamets, M. A., and Horwitz, B., Integrating Electrophysiological and Anatomical Experimental Data to Create a Large-Scale Model that Simulates a Delayed Match-to-Sample Human Brain Imaging Study, *Cerebral Cortex*, Vol. 8, No. 4, 310–320, 1988.

Tagamets, M. A., and Horwitz, B., A Model of Working Memory: Bridging the Gap Between Electrophysiology and Human Brain Imaging, *Neural Networks*, Vol. 13, No. 8–9, 941–952, 2000.

Traub, R. D., Jefferys, J. G. R., and Whittington, M. A., Simulation of Gamma Rhythms in Networks of Interneurons and Pyramidal Cells, *J. Comput. Neurosci.*, Vol. 4, No. 2, 141–150, 1997.

Ursino, M., and Lodi, C. A., A Simple Mathematical Model of the Interaction between Intracranial Pressure and Cerebral Hemodynamics, *J. Appl. Physiol.*, Vol. 82, No. 4, 1256–1269, 1997.

Ursino, M., and Magosso, E., Role of Short-Term Cardiovascular Regulation in Heart Period Variability: A Modelling Study, *Am. J. Physiol. Heart Circ. Physiol.*, Vol. 284, No. 4, H1479–H1493, 2003.

Ursino, M., Ter Minassian, A., Lodi, C. A., and Beydon, L., Cerebral Hemodynamics During Arterial and CO_2 Pressure Changes: In Vivo Prediction by a Mathematical Model, *Am. J. Physiol. Heart Circ. Physiol.*, Vol. 279, No. 5, H2439–H2455, 2000.

Ursino, M., Zavaglia, M., Astolfi, L., and Babiloni, F., Use of a Neural Mass Model for the Analysis of Effective Connectivity among Cortical Regions Based on High Resolution EEG, *Biol Cybern.*, Vol. 96, No. 3, 351–365, 2007.

Volterra, V., *Theory of Functionals and of Integral and Integro-Differential Equations.* Dover, 1930.

Wendling, F., Hernandez, A., Bellanger, J. J., Chauvel, P., and Bartolomei, F., Interictal to Ictal Transition in Human Temporal Lobe Epilepsy: Insights from a Computational Model of Intracerebral Eeg, *J. Clin. Neurophysiol.*, Vol. 22, No. 5, 343–56, 2005.

Wendling, F., Bartolomei, F., Bellanger, J. J., and Chauvel, P., Epilepsy fast Activity can be Explained by a Model of Impaired Gabaergic Dendritic Inhibition, *Europ. J. Neurosci.*, Vol. 15, No. 9, 1499–1508, 2002.

Wiener, N., *Nonlinear Problems In Random Theory,* Wiley, 1958.

Wilson, H. R., and Cowan, J. D., Excitatory and Inhibitory Interactions in Localized Populations of Model Neurons, *Biophys. J.*, Vol. 12, No. 1, 1–24, 1972.

Winder, R., Cortes, C. R., Reggia, J. A., and Tagamets, M. A., Functional Connectivity in FMRI: A Modeling Approach for Estimation and for Relating To Local Circuits, *Neuroimage,* Vol. 34, No. 3, 1093–1107, 2006.

Yildiz, M., and Ider, Y. Z., Model Based and Experimental Investigation of Respiratory Effect on the HRV Power Spectrum, *Physiol. Meas.*, Vol. 27, No. 10, 973–988, 2006.

Zavaglia, M., Astolfi, L., Babiloni, F., and Ursino, M., A Neural Mass Model for the Simulation of Cortical Activity Estimated from High Resolution EEG During Cognitive or Motor Tasks, *J. Neurosci. Meth.*, Vol. 157, No. 2, 317–329, 2006.

Zavaglia, M., Astolfi, L., Babiloni, F., and Ursino, M., The Effect Of Connectivity on EEG Rhythms, Power Spectral Density and Coherence Among Coupled Neural Populations: Analysis With a Neural Mass Model, *IEEE Tr. Biomed. Eng.*, Vol. 55, No. 1, 69–77, 2008.

CHAPTER 7

MULTIMODAL INTEGRATION OF EEG, MEG, AND FUNCTIONAL MRI IN THE STUDY OF HUMAN BRAIN ACTIVITY

Fabio Babiloni, Fabrizio De Vico Fallani,
and Febo Cincotti

7.1 INTRODUCTION

Today, it is well understood that brain activity generates a variable electromagnetic field that can be detected quite accurately by using scalp electrodes as well as by superconductive magnetic sensors. Electroencephalography (EEG) and magnetoencephalography (MEG) are, therefore, useful techniques for the study of brain dynamics and functional cortical connectivity because of their high temporal resolution (milliseconds) (Nunez, 1981; Nunez, 1995). Electroencephalography reflects the activity of cortical generators oriented both in tangential and radial ways with respect to the scalp surface, whereas MEG reflects mainly the activity of the cortical generators oriented tangentially with respect to the magnetic sensors. However, the different electrical conductivity of the brain, skull, and scalp markedly blurs the EEG potential distributions and makes the localization of the underlying cortical generators through this technique a problematic issue. To overcome this problem, high-resolution EEG (HR-EEG) technology was introduced during the last decade and was shown to greatly improve the spatial resolution of the conventional EEG (Nunez, 1995; Gevins et al., 1991; Gevins et al., 1999). Such technology included (i) the use of realistic head models obtained from sequential magnetic resonance imaging (MRI) of the subject's head, to mathematically model the propagation of the potential from the cortex to the scalp sensors; (ii) the sampling of the spatial distribution of scalp potential with a high number of surface electrodes (64–128); (iii) the use of mathematic Laplacian (SL) operators to improve the spatial details of the recorded scalp potential distribution; and (iv) the use of an accurate model for the cortical sources that typically include 3000–5000 current dipoles.

However, the spatial resolution of the HR-EEG/MEG techniques is fundamentally limited by the intersensor distances and by the fundamental laws of electromagnetism (Nunez, 1981). Despite the lack of spatial resolution for these techniques, neural sources can be localized from HR-EEG or MEG data by making a priori hypotheses on their number and extension. When a known number of cortical sources (i.e., short-latency evoked potentials) generate the neuronal activity, the location and strength of these sources can be reliably estimated by the dipole localization technique (Scherg et al., 1984). However, with the exception of the early processing of sensory responses, event-related cortical responses include a distributed network of several unknown areas. When the distributed cortical network is supposed to be active, neural sources could be modeled by linear inverse estimation (Dale and Sereno, 1993; Dale et al., 2000). This approach implies the use of both thousands of equivalent current dipoles as a source model and realistic head models, reconstructed from magnetic resonance images, as a volume conductor medium.

The use of geometrical constraints can generally reduce the solution space (i.e., the set of all possible combinations of the cortical dipole strengths). For example, the dipoles can be disposed along the reconstruction of the cortical surface with a direction perpendicular to the local surface. An additional constraint is to force the dipoles to explain the recorded data with a minimum or a low amount of energy (minimum norm solutions) (Dale and Sereno, 1993; Hamalainen and Ilmoniemi, 1984). In addition, the use of a priori information from other neuroimaging techniques having high spatial resolution, such as functional magnetic resonance imaging (fMRI), has been suggested to improve the localization of sources from HR-EEG/MEG data (Hamalainen and Ilmoniemi, 1984; Liu et al., 1998). In fact, human neocortical processes involve temporal and spatial scales spanning several orders of magnitude, from the rapidly shifting somatosensory processes characterized by a temporal scale of milliseconds and a spatial scale of a few square millimeters to the memory processes involving periods of seconds and spatial scales of square centimeters.

Information about brain activity can be obtained by measuring different physical variables linked to brain processes, such as the increase in consumption of oxygen by the neural tissues or the variation of the electric potential over the scalp surface.

It is worth noting that all these variables have their own spatial and temporal resolution. The different neuroimaging techniques are, thus, confined to the spatiotemporal resolution offered by the measured variables. Today, no neuroimaging method provides for simultaneous spatial resolution on a millimeter scale and temporal resolution on a millisecond scale. As a consequence of the previous statement, the functional brain images obtained with the various techniques at our disposal (fMRI, HR-EEG, and MEG) are like the pieces of a puzzle, the "neuroimaging puzzle," that we have to put together to retrieve a unique picture of the underlying brain activity. Hence, it is of interest to study the possibility of integrating the information offered by the different physiological variables related to the brain functions in a unique mathematical context. This operation is called the "multimodal integration" of variables X and Y, where the X variable typically has particularly appealing spatial resolution properties (millimeter scale) and the Y variable has particularly attractive temporal properties (millisecond scale). Nevertheless, the issue of several temporal and spatial domains is critical in the study of brain functions, since different properties could become observable, depending on the spatiotemporal scales at which the brain processes are measured.

The rationale of the multimodal approach based on fMRI, MEG, and HR-EEG data to locate brain activity is that neural activity generating EEG potentials or MEG fields increases glucose and oxygen demands (Magistretti et al., 1999). This results in an increase in the local hemodynamic response that can be measured by fMRI (Grinvald et al., 1986; Puce et al., 1997). Overall, such a correlation between electrical and hemodynamic concomitants provides the basis for a spatial correspondence between fMRI responses and HR-EEG/MEG source activity. Furthermore, numerical simulations have shown that the use of fMRI priors increases the quality of the cortical current estimations for both HR-EEG and MEG recordings (Liu et al., 1998; Liu, 2000; Babiloni et al., 2003).

In the following, we first present the mathematical principle of the multimodal integration of HR-EEG with MEG data. Then we show how to integrate mathematically HR-EEG/MEG data with fMRI data. Besides these methodologies, we present some practical applications for the localization of sources responsible for intentional simple movements executed by healthy subjects.

7.2   CORTICAL ACTIVITY ESTIMATION FROM NONINVASIVE EEG AND MEG MEASUREMENTS

7.2.1   Head and Source Models

Let us assume an EEG recording performed with a set of M electrodes disposed on the scalp surface. In the following, we indicate with **b** the vector of M electrical measurements recorded from the scalp. In order to perform the estimation of the cortical activity, we also need a model of the head-volume conductor. This model is used to approximate the propagation of the potential from the modeled neural generators to the measurement sensors. The head model can be of spherical or elliptical shape or can mimic the realistic head shape. Since we have to estimate the source activity from noninvasive EEG measurements, we need a mathematical model for the neural sources. In the analysis of EEG and MEG, the commonly used mathematical model for the neural source is the current dipole. Such model is used in the literature because it approximates very well the activity of relatively small patches of cortical tissue. In this particular context, we used many current dipoles located in the brain along the entire cerebral volume or along its cortical surface. If we want to model the entire brain volume, we have to divide it into voxels and then place a triplet of orthogonal current unitary dipoles at each voxel position. Another source model takes into account only the cortical geometry, to constrain the dipoles to lie orthogonal to the modeled cortical surface. In the following, we deal with the problem of source estimation by using a representation of the cortical surface, and we use a set of dipoles disposed along such surface as a source model (see Figure 7.1). However, all the mathematical formulas presented here still hold also in the case in which the neuronal space is divided into voxels, attempting to model also the subcortical structures.

Each dipole placed inside the volume conductor model of the head has a unitary strength and different direction, according to the local cortical geometry or to the adopted reference coordinate system. There is no limitation on the number of sources placed inside the head model, which depends on the modeling capabilities of the used computational system. In the following, we indicate with N the number of dipoles whose strength is to be estimated from the M-dimensional measurement vector $\mathbf{b}$. The typical values for N are between 1000 and 7000, whereas the values for M are in the range 64–256. We indicate as x the n-dimensional vector of the unknown current strengths for the dipoles.

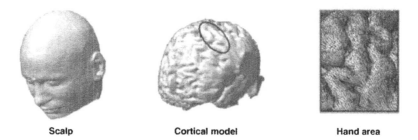

Scalp   Cortical model   Hand area

Figure 7.1. Realistic head model for linear inverse source estimation. Note the area tessellated with triangles (right figure). At the center of each triangle, a dipole with unitary strength was posed perpendicular to the triangle.

7.2.2 The Linear Inverse Problem

In the estimation of neuronal activity from noninvasive measurements, we have to use a mathematical model for the description of the propagation of the potential distribution from each modeled source to the sensor positions. In other words, we have to compute the potential distribution occurring on the set of the M sensors over the head model due to the ith unitary dipole placed at the ith cortical location. Such predictions can be made with the aid of analytical or boundary-element formulations, depending on the shape of the head model used. The equation for the potential distribution for a three-layered spherical head model can be found in literature. Appendix I contains the equations that compute the potential value due to a dipole inside a realistic head model over a point located on the scalp surface. Such equations are for the electric case; those in Appendix II are for the magnetic case.

In the following, we indicate as A_i the potential distribution over the M sensors due to the unitary ith cortical dipole. The collection of all the M-dimensional vectors A_i ($i = 1, \ldots, N$) describes how each dipole generates the potential distribution over the head model. This collection is called the lead-field matrix A.

With the definitions provided above, we can say that the estimate of the strength of the modeled dipolar source strength x from the noninvasive set of measurement b is obtained by solving the following linear system:

$$Ax = b \qquad (7.1)$$

where A is the $M \times N$ lead-field matrix, x is the N-dimensional array of the unknown cortical strengths, and b is the M-dimensional vector of the instantaneous (electrical or magnetic) measurements. This is a strongly underdetermined linear system in which the number of the unknown variables (i.e., the dimension of the vector x) is greater than the number of measurements b by about one order of magnitude. In this case, from the linear algebra we note that there are infinite solutions for the vector of dipole strengths x. All these solutions explain in the same way the data vector b. Furthermore, the linear system is ill conditioned as a result of the substantial equivalence of several columns of the electromagnetic lead-field matrix A. In fact, we know that each column of the lead-field matrix arises from the potential distribution generated by the dipolar sources that are located in similar position and orientation along the employed cortical model. Regularizing the inverse problem consists in attenuating the oscillatory modes generated by vectors associated with the smallest singular values of the lead field matrix A, introducing supplementary and a priori information on the sources to be estimated. In the following, we characterize the term "solution space," the space in which the "best" current

strength solution **x** will be found. The "measurement space" is the vectori-
al space in which the vector **b** of the gathered data is considered. The solution of the linear problem **Ax** = **b** with the variation approach is based on
the idea of selecting metrics from the solution space and the measurement
space, respectively. These two metrics are characterized by symmetric matrices and express our idea of closeness in the same spaces. With this approach, the minimization function is composed of two terms: one that
evaluates how well the solution explains the data, and another that measures the closeness of the solution to an a priori selected requires that the
solution has the minimum strength under the norm chosen for the source
space. In particular, the formulation of the problem expressed in the Equation 7.1 now becomes

$$\xi = \text{argmin} \, [\|(\mathbf{Ax} - \mathbf{b})\|_{\mathbf{Wd}}^2 + \lambda^2 \|\mathbf{x}\|_{\mathbf{Wx}}^2] \qquad (7.2)$$

where the matrices **Wd** and **Wx** are associated with the the metrics of the
measurement and source space, respectively, and λ is the Lagrangian parameter. Hence, the source estimate ξ is the source distribution that between
the infinite possible solutions to the undetermined problem described in
Equation 7.2 explains the EEG data with a minimum amount of energy
(weighted minimum norm solution). By setting the matrices **Wd** and **Wx**
to the identity, the minimum norm estimation was obtained.

The solution of the problem described in Equation 7.2 has the following the form:

$$\xi = \mathbf{Gb} \qquad (7.3)$$

where, under the hypothesis that the metrics **Wx** and **Wd** are invertible,
the pseudoinverse matrix **G** is given by

$$\mathbf{G} = \mathbf{W}_x^{-1}\mathbf{A}' \, (\mathbf{A}\mathbf{W}_x^{-1}\mathbf{A}' + \lambda \mathbf{W}_d^{-1})^{-1} \qquad (7.4)$$

Hamalainen and Ilmoniem (1984) proposed the estimation of the cortical
activity with the minimum norm solution. However, it was recognized that
in this particular application, the solutions obtained with the minimum
norm constraints were biased toward those dipoles that are located nearest
to the sensors. In fact, there is a dependence on distance in the law of potential (and magnetic field) generation and this dependence tends to increase the activity of the more superficial dipoles while depressing the activity of the dipoles far from the sensors. The solution to this bias was
obtained by taking into account a compensation factor, for each used dipole, that equalized the visibility of the dipole from the sensors. This technique, called column norm normalization, was used in the linear inverse

problem by Pascual-Marqui (1995) and then adopted largely by scientists in this field. With column norm normalization, a diagonal $N \times N$ matrix $\mathbf{W}$ was formed, whose generic ith term on the diagonal is equal to

$$\mathbf{W}_{ii} = \|\mathbf{A}i\|^{-2} \qquad (7.5)$$

representing the L2 norm of the ith column of the lead-field matrix $\mathbf{A}$. In this way, dipoles near to the sensors (with a large $\|\mathbf{A}i\|$) will be depressed in the solution of Equation 7.2, since their activations are not convenient from the point of view of the functional cost. In fact, dipoles with low visibility from the sensors (with a low $\|\mathbf{A}i\|$) have an associated cost function that is convenient to use. The use of this $\mathbf{Wx}$ matrix in the source estimation is known as the weighted minimum norm solution (Grave de Peralta et al., 1997).

Another question of interest in the solution of the linear inverse problem is the setting of the Lagrangian parameter λ that regulates the presence of the a priori information inside the solution of the problem. How it is possible to set such a parameter in an "optimal" way?

An optimal regularization of the linear system described in Equation 7.2 was obtained through the Tikhonov L-curve approach (Hansen, 1992). This curve plots the residual norm versus the solution norm at different values of the regularization parameter λ. It is worth noticing that the optimal regularization value can be selected automatically. In fact, this value is located at the "corner" of the L-curve plot. Other criteria for choosing the value of λ are the generalized cross validation (GCV) or the CRESO methods, together with the zero-crossing method and the relative minimal product.

7.2.3 Multimodal Integration of EEG and MEG Data

At a first glance, the attempt to integrate the EEG and MEG data in order to increase the quality of the source reconstruction fails when we consider that the units of the electrical and magnetic fields differ. How we can fuse these data together? One possible answer is to transform the measurements produced by the electric and magnetic sensors in terms of standard deviations of measurement noise. In this way, both the electrical and magnetic values are transformed and then reported on a common scale.

In this context, the estimation of the covariance matrix of the electrical and the magnetic noise assumes a particular importance. The estimation of such matrices requires the recording of several single sweeps of EEG and MEG data, and the possibility to determine a segment of the recorded data in which no task-related activity is present. Then, the maximum likelihood estimates for the covariance matrices of the electrical noise $\mathbf{Ne}$ and magnetic $\mathbf{Nm}$ matrices have to be computed on all the sweeps recorded for the period of interest. With the use of these matrices, we can produce the block covariance matrix of the electromagnetic measurement, called **S**, by $S = (Ne \cdot Nm)$.

The forward solution specifying the potential scalp field due to an arbitrary dipole source configuration was computed according to the following linear system:

$$\begin{bmatrix} E \\ B \end{bmatrix} [x] = \begin{bmatrix} v \\ m \end{bmatrix} \tag{7.6}$$

where **E** is the electric-lead-field matrix obtained by the boundary element technique for the realistic MR-constructed head model and **B** is the magnetic-lead-field matrix obtained for the same head model. **x** is the array of the unknown cortical dipole strengths, **v** is the array of the recorded potential values and **m** is the array of magnetic values. The lead-field matrix **E** and the array **v** were referenced consistently. In order to scale the EEG and MEG data, the rows of the lead-field matrix **E** and **B** were first normalized by the rows norm (Phillips et al., 1997). This scaling is applied in the same way to the electrical and magnetic measurements arrays, **v** and **m**. After row normalization, the linear system can be restated as

$$\underline{A}x = \underline{b} \tag{7.7}$$

where $\underline{A}$ is the matrix composed of the normalized electric and magnetic lead fields, and $\underline{b}$ is the normalized measurement array of EEG and MEG data (**v** and **m**), respectively.

As noted before, the general formulation of the linear inverse problem based on this assumption is

$$\xi = \text{argmin}(\| (\underline{A}x - \underline{b}) \|_{\underline{Wd}^2} + \lambda^2 \|x\|_{\underline{Wx}^2}) \tag{7.8}$$

where $\underline{Wd}$ is now equal to the covariance matrix **S** of the noise of the normalized EEG and MEG sensors, and $\underline{Wx}$ is the matrix that regulates how each EEG or MEG sensor is influenced by dipoles located at different depths into the source model. The covariance matrix **S** was derived from the normalized EEG and MEG data by means of the maximum likelihood estimation as described before. It is essential that the estimate of **S** be performed on both EEG and MEG data with maximum background electromagnetic noise (i.e., no event-related electromagnetic signal). The matrix $\underline{Wx}$ is a diagonal matrix in which the ith element is equal to the norm of the ith column of the normalized lead-field matrix $\underline{A}$. As said before, an optimal regularization of the linear system $\underline{A}x = \underline{b}$ was obtained by the L-curve approach (Hansen, 1992) that plots the residual norm versus the solution norm at different values of the regularization parameter λ.

In Figure 7.2, we show a representative result obtained from the linear inverse estimates of EEG, MEG, and combined EEG-MEG data. The recordings were performed during the voluntary movement of the right middle finger.

7.3 INTEGRATION OF EEG/MEG AND fMRI DATA

In this section, a proposal for the use of fMRI constraints in the source estimate by the linear inverse problem is described. In the following, we deal with the issue of integrating EEG and fMRI data, but we would like to note that a similar approach could be produced also for the fusion of the MEG with fMRI data. The methodology for the combination of all the modalities together (EEG, MEG, and fMRI) for the source estimation will be the union of the techniques presented here and in the previous section.

7.3.1 The Common Head Model

One fundamental issue in the integration of the fMRI with the EEG data is the use of a common geometrical framework to register the volume con-

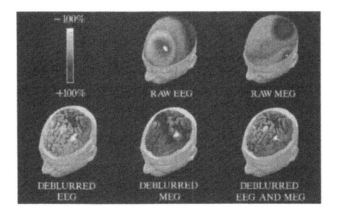

Figure 7.2. Amplitude gray scale three-dimensional maps showing linear inverse estimates of electroencephalographic (EEG), magnetoencephalographic (MEG), and combined EEG–MEG data recorded (128–150 channels, respectively) from a subject about 110 msec after the onset of electromyographic response accompanying a voluntary brisk right middle finger extension. The original movement-related EEG and MEG activity is also shown as a reference. Linear inverse estimates were mapped as electric fields forwarded over the dura mater compartment of the realistic magnetic-resonance-constructed subject's head model. Percent gray scale (256 hues) is normalized with reference to the maximum amplitude calculated for each map. Maximum negativity (–100%) is coded in white and maximum positivity (+100%) in black.

ductor of the head model and the activated voxels deriving from the fMRI images. To do that, the fMRI images have to be first coregistered with the anatomic images of the subject's head, which were obtained by the acquisition of a T1-weighted conventional spin-echo-axial-oblique sequence. The sequential MR images of the analyzed subject constituted also the base for the construction of the volume conductor of the head model to be used in the EEG analysis. In this context, the structure for the scalp, skull, and dura mater have to be recognized from the MR images by an automatic or semiautomatic segmentation procedure.

Once the different structures for scalp, skull, and dura mater have been recognized and tessellated, the use of the boundary-element modeling technique allows computation of the potential distribution over the sensors generated by a dipole inside the brain. A brief description of the analysis is described in Appendix I. Once the common geometrical reference for the head for both EEG and fMRI measurements has been established, the fusion of the information from the activated brain voxels during EEG data acquisition can be addressed.

7.3.2   Percentage Change Hemodynamic Responses

In fMRI analysis, several methods to quantify the brain hemodynamical response during a particular task have been developed. Here, we recall the statistical parametric mapping (SPM) introduced in the early 1990s by K. J. Friston, or methods relying on the use of the statistical Bonferroni-corrected t-test, or those based on the Kolmogorov–Smirnof tests. In principle, the issue of the integration between EEG and fMRI does not depend on the particular method used for the fMRI analysis. However, in order to illustrate the technique, in the following we analyze the case in which a particular fMRI quantification technique has been employed, called the percent change technique. This measure quantifies the percentage increase of the fMRI signal during the task performance with respect to a resting state. The visualization of the voxels distributed in the brain space that are statistically increased during the task condition with respect the rest is called the PC (percent change) map. The difference between the mean resting and movement-related signal intensity is generally calculated voxel by voxel. The rest-related fMRI signal intensity is obtained by averaging the premovement and recovery fMRI. The Bonferroni-corrected Student's t-test is also used to minimize the alpha inflation effects due to multiple statistical voxel-by-voxel comparisons (Type I error; $p < 0.05$). Voxels with a statistically significant PC activation have to be projected onto the modeled cortical surface. PC values of the statistically activated voxels assigned to a certain cortical triangle are summed as a measure of the movement-related cortical activation. This measure corresponded to the α_i value that must be used as a weighting function in the successive linear inverse minimization procedure.

7.3.3   EEG Linear Inverse Estimation

The forward solution specifying the potential scalp field due to an arbitrary dipole source configuration is computed according to the linear system described in Equation 7.2. The lead-field matrices **A** and the data array **b** have to be referenced consistently. To obtain a unique solution of the linear system (Equation 7.1), the variational problem for the sources **x** was posed as described in Equation 7.8. The solution of the variational problem depended on adequacy of the data and source space metrics. The metric for the data space was obtained by the use of statistics about the residual **n** = **Ax** − **b**. The metric for the source space was based on the norm of the residual vector **n**. The Mahalanobis metric for the data space has to be preferred here since it permitted us to take into account easily information from combined EEG and MEG data (see previous section). The metric for the source space has to take into account the normalization of the lead-field matrix **A** to balance the much greater visibility of the superficial cortical sources from the EEG data with respect to the deepest ones. Furthermore, in the source metric we would like to insert the a priori information that comes from the fMRI-activated voxels. To do that, we use the percentage intensity values (**α**) of the integrated fMRIs. The metric for the source space is then formed by a diagonal matrix **W**, the ith term of which was

$$\mathbf{W}_{ii} = \|\mathbf{A}_i\|^2 g(\alpha_i)^{-2} \tag{7.9}$$

where $\|\mathbf{A}_i\|$ is the norm of the ith column of the lead-field matrix **A** and $g(\alpha_i)$ is a function of the statistically significant percentage increase of the fMRI signal assigned to the ith dipole of the modeled source space. The $g(\alpha_i)$ function is expressed as

$$g(\alpha_i) = 1 + \mathbf{K}\alpha_i, \qquad \alpha_i \geq 0 \tag{7.10}$$

where the factor **K** tunes the fMRI solutions in the source space for the time-varying electromagnetic component **b**. By inspecting Equation 7.9, high **K** values (i.e., $\mathbf{K}\alpha_i \neq 10$) produce a space norm **W**, which is roughly one order of magnitude lower than the one obtained by taking into account only the column normalization. On the other hand, low **K** values (i.e., $\mathbf{K}\alpha_i \ll 1$) result in a **W** value roughly proportional to the squared column norm, which completely disregards the fMRI solution (i.e., 0% fMRI solution or 100% EEG solution).

With this metric, the pseudoinverse matrix **G** now depends on the **K** values. The problem with this optimal regulation of the **K** value is the computational effort that can be remarkable when realistic source models with thousands of dipoles are used. However, the mean computational power available on personal computers doubles each 1.5 years. Hence, the actual computational effort to compute the EEG and fMRI integration cannot represent a serious problem in the future.

Finally, it is possible to note that integration of EEG, MEG, and fMRI data can be possible by solving the linear inverse problem in Equation 7.8 with **Wd** equal to the covariance matrix of the electromagnetic noise and with **Wx** equal to that described in Equation 7.9.

In Figure 7.3, we show a representative result obtained from HR-EEG data and HR-EEG combined with fMRI data. The recordings were performed during the voluntary movement of the right middle finger.

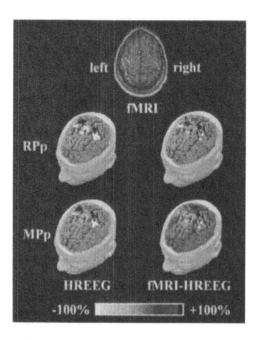

Figure 7.3. Amplitude gray-scale three-dimensional maps showing linear inverse estimates from high-resolution electroencephalographic (HREEG) and combined functional magnetic resonance image (fMRI) HREEG data computed from a subject about 50 msec before (readiness potential peak, RPp) and 20 msec after (motor potential peak, MPp) the onset of the electromyographic activity associated with self-paced right middle finger movements. A representative axial PC map of statistically significant ($p < 0.05$) fMRI signal intensity is also shown as a reference (top). The linear inverse estimate of HREEG and combined fMRI–HREEG data were mapped as electric fields forwarded over the dura mater compartment of a realistic MRI-constructed subject's head model. In fMRI data, statistically activated voxels are represented in white. Percent grey scale (256 hues) of HREEG and combined fMRI–HREEG data is normalized with reference to the maximum amplitude calculated for each map. Maximum negativity (-100%) is coded in white and maximum positivity ($+100\%$) in black.

APPENDIX I. ELECTRICAL FORWARD SOLUTION FOR A REALISTIC HEAD MODEL

Let a head model be constituted by electrically homogeneous and isotropic compartments simulating scalp, skull, and dura mater. The forward solution specifying the potential distribution (V) on these compartments S_k ($k = 1, \ldots, 3$) due to a dipole is given by the Fredholm integral equation of the second kind:

$$(\sigma_i^- + \sigma_i^+)V(\vec{r}) = 2V_0(\vec{r}) + \frac{1}{2\pi} \sum_{j=1}^{m} (\sigma_i^- - \sigma_i^+) \int_{S_j} V(\vec{r}') \, d\Omega_{\vec{r}}(\vec{r}') \quad (7.A.1)$$

with

$$d\Omega_{\vec{r}}(\vec{r}') = \frac{\vec{r}' - \vec{r}}{|\vec{r}' - \vec{r}|^3} \, d\vec{S}_j(\vec{r}') \quad (7.A.2)$$

where

$V_0(\vec{r})$ is the potential due to a dipole located in an infinite homogeneous medium

σ_i^- is the conductivity inside the surface S_j of the multicompartment head model

σ_j^+ is the conductivity outside the surface S_j

m is the total number of compartments within the head model

$d\Omega_{\vec{r}}(\vec{r}')$ is the solid angle subtended by the surface element dS located in $\vec{r}$ (point of observation $\vec{r}'$)

A numerical solution of the Fredholm integral equation can be obtained by decomposing the surfaces S_k ($k = 1, \ldots, 3$) into triangular panels and by using boundary-element techniques. With the boundary-element techniques, a discrete version of the Fredholm integral equation is given by

$$\mathbf{v} = \mathbf{g} + \mathbf{\Omega}\mathbf{v} \quad (7.A.3)$$

where the elements of matrix $\mathbf{\Omega}$, vector $\mathbf{v}$, and vector $\mathbf{g}$ are defined as follows:

$\mathbf{v}_i$ is the potential value in the center of mass of the ith triangle

$\mathbf{g}_i$ is the potential value generated by a source in the center of mass of the ith triangle

$\mathbf{\Omega}_{ij}$ is the matrix element proportional to the solid angle subtended by the jth triangle at the center of mass of the ith triangle

The numerical solution of the Fredholm integral equation can be improved using the deflation procedure. The linear system of Equation 7.A.3 is singular since the potential distribution generated on the scalp compartment by an equivalent dipole is determined up to a constant. This singularity can be removed by using a deflation procedure that yields the potential distribution (V) on the compartment surfaces S_k ($k = 1, \ldots, 3$).

APPENDIX II. MAGNETIC FORWARD SOLUTION

The magnetic field of a current dipole in a piecewise homogenous conducting medium is given by

$$B = B_0 + \frac{\mu_0}{4\pi} \sum_{j=1}^{m} \left(\sigma_j^+ - \sigma_j^- \right) \int_{S_j} V(\vec{r}) \frac{\vec{n} \times (\vec{r} - \vec{r}')}{\left| \vec{r} - \vec{r}' \right|^3} dr' \qquad (7.A.4)$$

where

B_0 is the field of the current dipole in free space with each surface integral extending to an interface between homogeneously conducting media
n is a unit vector orthogonal to the surface S_j
the summation index j runs over the surfaces

The surface integrals take into account the effects of volume currents. Notice that these currents are equivalent to layers of current dipoles orthogonal to the interfaces, with dipole moment per unit area $V(\sigma^+ - \sigma^-)$. The magnetic lead field matrix was calculated by adding, for each dipole, the free space term and the corresponding volume current term. Since a boundary element model was used, the volume currents were modeled by an array of fictitious dipoles, located at the centroids of each triangle of the reconstructed surfaces, with moment $V(\sigma^+ - \sigma^-)$ times the triangle area.

REFERENCES

Babiloni, F., Babiloni, C., Carducci, F., Romani, G. L., Rossini, P. M., Angelone, L. M., and Cincotti, F., Multimodal Integration of High Resolution EEG and Functional Magnetic Resonance Imaging Data: A Simulation Study, *Neuroimage*, Vol. 19, No. 3, 1–15, 2003.

Dale, A., Liu, A., Fischl, B., Buckner, R., Belliveau, J. W., Lewine, J., et al., Dynamic Statistical Parametric Mapping: Combining fMRI and MEG for High-Resolution Imaging of Cortical Activity, *Neuron*, Vol. 26, 55–67, 2000.

Dale, A. M., and Sereno, M., Improved Localization of Cortical Activity by Combining EEG and MEG with MRI Cortical Surface Reconstruction: A Linear Approach, *J. Cogn. Neurosci.*, Vol. 5, 162–176, 1993.

Gevins, A., Brickett, P., Reutter, B., and Desmond, J., Seeing Through the Skull: Advanced EEGs use MRIs to Accurately Measure Cortical Activity From the Scalp, *Brain Topogr.*, Vol. 4, 125–131, 1991.

Gevins, A., Le, J., Leong, H., McEvoy, L. K., and Smith, M. E., Deblurring, *J. Clin. Neurophysiol.*, Vol. 16, No. 3, 204–213, 1999.

Grave de Peralta, R., Hauk, O., Gonzalez Andino, S., Vogt, H., and Michel, C. M., Linear Inverse Solution with Optimal Resolution Kernels Applied to the Electromagnetic Tomography, *Human Brain Mapping*, Vol. 5, 454–467, 1997.

Grinvald, A., Lieke, E., Frostig, R. D., Gilbert, C. D., and Wiesel, T. N., Functional Architecture of Cortex Revealed by Optical Imaging of Intrinsic Signals, *Nature*, Vol. 324, No. 6095, 361–364, 1986.

Hamalainen, M, and Ilmoniemi, R., Interpreting Measured Magnetic Field of the Brain: Estimates of the Current Distributions, Tech Rep TKKF-A559, Espoo (Finland), Helsinki University of Technology, 1984.

Hansen, P. C., Analysis of Discrete Ill-Posed Problems by Means of the L-curve, *SIAM Rev.*, Vol. 34, 561–580, 1992.

Liu, A. K., Belliveau, J. W., and Dale, A. M., Spatiotemporal Imaging of Human Brain Activity Using Functional MRI Constrained Magnetoencephalography Data: Monte Carlo Simulations, *Proc. Natl. Acad. Sci. USA,* Vol. 95, No. 15, 8945–8950, 1998.

Liu, A. K., *Spatiotemporal Brain Imaging,* Ph.D. dissertation, Cambridge (MA): Massachusetts Institute of Technology, 2000.

Magistretti, P. J., Pellerin, L., Rothman, D. L., Shulman, R. G., Energy on Demand, *Science,* Vol. 283, No. 5401, 496–497, 1999.

Nunez, P., *Electric Fields of the Brain,* Oxford University Press, 1981.

Nunez, P. L., *Neocortical Dynamics and Human EEG Rhythms,* Oxford University Press, 1995.

Pascual-Marqui, R. D., Reply to Comments by Hamalainen, Ilmoniemi and Nunez, *ISBET Newsletter,* No. 6, December, 16–28, 1995.

Phillips, J. W., Leahy, R., and Mosher, J. C., MEG-based Imaging of Focal Neuronal Current Sources, *IEEE Trans. Med. Imag.*, Vol. 16, No. 3, pp. 338–348, 1997.

Puce, A., Allison, T., Spencer, S. S., Spencer, D. D., and McCarthy, G., Comparison of Cortical Activation Evoked by Faces Measured by Intracranial Field Potentials and Functional MRI: Two Case Studies, *Hum. Brain Mapp.*, Vol. 5, No. 4, 298–305, 1997.

Scherg, M., von Cramon, D., and Elton, M., Brain-Stem Auditory-Evoked Potentials in Post-Comatose Patients after Severe Closed Head Trauma, *J. Neurol.*, Vol. 231, No. 1, 1–5, 1984.

Thispageintentionallyleftblank

CHAPTER 8

DECONVOLUTION FOR PHYSIOLOGICAL SIGNAL ANALYSIS

Giovanni Sparacino, Gianluigi Pillonetto,
Giuseppe De Nicolao, and Claudio Cobelli

8.1. INTRODUCTION

Measuring time series of concentration in plasma is of fundamental importance both for improving the knowledge of physiological systems as well
as for diagnostic and therapeutic reasons. For instance, Figure 8.1 shows
(top) the time series of glucose and insulin plasma concentration measured
every 10 minutes for 24 hours, in spontaneous conditions, in a normal subject (left) and in a type 2 diabetic patient (right). From the picture, one can
note that, in the presence of pathology, oscillations are less regular and
more poorly synchronized. Figure 8.1 also shows (bottom left) the time series of the concentration of growth hormone (GH) in a normal patient after
the injection of a dose of growth-hormone-releasing hormone (GHRH)
(data taken from De Nicolao et al., 2000), from which the response of the
pituitary gland after a stimulus can be quantitatively assessed. Finally, Figure 8.1 (bottom right) shows the samples of the concentration of a drug
measured in the plasma for 48 hrs in a normal individual after the inges-
tion of a pill. From these data, quantitative information on how a drug is
absorbed from the gastrointestinal tract can be gathered.

   The information obtained from concentration signals can be useful in
many clinical and physiological applications, for example, for a more detailed comprehension of the inner mechanisms of physiological systems or
for a more efficient design of pharmacological therapies. In particular, a
number of ad hoc techniques (e.g., peak detection and analysis, entropy
measures) have been developed to deal with the specific feautures of con-

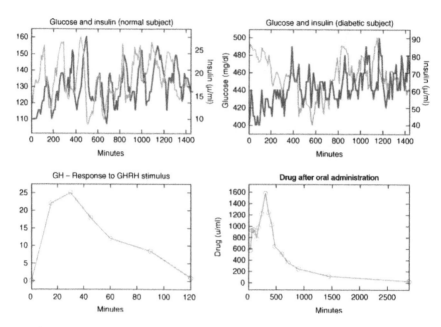

Figure 8.1. Top. Time series of insulin (thin line) and glucose concentration in plasma (thick line) measured in spontaneous conditions in a normal individual (left) and in a diabetic patient (right). Bottom. Growth hormone (GH) concentration in plasma after a GHRH stimulus (left). Concentration of a certain drug in plasma after an oral administration (right).

centration time series (concerning, e.g., number and frequency of available samples); see Veldhuis (1997) for an overview. However, concentration time series reflect not only the secretion/production process of the substance under study, but also its distribution in the circulation and metabolism, that is, the so-called kinetics of the substance. The kinetics can play an extremely important role because it can render the measured signal (concentration of the substance in plasma) much different from the unknown signal (flux of secretion/production of the substance) we want to gather information about, with obvious consequences on the informativeness of the data. In order to better illustrate the role of the kinetics, let us consider the linear case in which the blood concentration of the substance $z(t)$ and its secretion/production rate $u(t)$ are related through the following equation:

$$z(t) = \int_{-\infty}^{t} g(t, \tau)u(\tau)d\tau \qquad (8.1)$$

where $g(t, \tau)$ represents the kernel of the system, that is, the output of the system to a unitary Dirac pulse centered at time τ. Figure 8.2 (top panels)

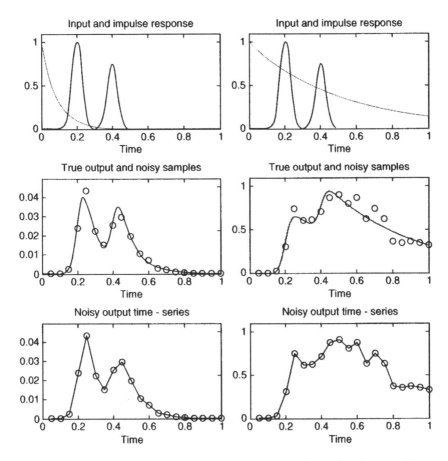

Figure 8.2. Influence of the kinetics on concentration time series. System with "fast" (left) and slow (right) dynamics. Top: input and impulse response. Middle: output and noisy samples. Bottom: noisy time series.

displays a simulation where $g(t, \tau) = g(t - \tau)$ and the input $u(t)$ is the same for two different systems, one with "fast" impulse response $g(t) = e^{-\alpha t}(\alpha = 12$, left panel), and the other with "slow" impulse response ($\alpha = 2$, right panel). The system output $z(t)$ is markedly different in the two cases (middle panel). For instance, when the "slow" impulse response is considered (right), the second peak in $z(t)$ is higher than the first one, whereas in $u(t)$ the opposite relationship holds. The existence of a common causal factor between the two time series is further masked by adding simulated noise in the measurements. For instance, two distinct peaks in the input signal may appear severely camouflaged in the time series of the noisy output samples (bottom right).

The distortion introduced by the kinetics could be removed by solving an inverse problem in which a cause (secretion/production signal) is reconstructed from the knowledge of its effect (concentration signal). In the case of linear systems, the problem amounts to solving the Fredholm integral equation of the first kind (Equation 8.1), that is, estimating $u(t)$ starting from the knowledge of $z(t)$ and the input–output description of the system given by $g(t, \tau)$. If the system is time invariant, Equation 8.1 becomes a convolution integral,

$$z(t) = \int_{-\infty}^{t} g(t - \tau)u(\tau)d\tau \qquad (8.2)$$

and the inverse problem associated with Equation 8.2 is called the deconvolution problem. In the literature, however, the term deconvolution is often used for any input estimation problem as in Equation 8.1, irrespective of the time invariance of the system.

The deconvolution problem is a classic problem in many disciplines (e.g., spectroscopy, quantum physics, image restoration, geophysics, seismology, telecommunications, astronomy, acoustics, and electromagnetism), where it is often encountered either to remove distortion from an observed signal or to indirectly measure a nonaccessible signal; see for example, Bertero (1989), De Nicolao et al. (1997), and Bertero and Boccacci (1998) for bibliographic references. In certain applications, there is the need of estimating $\hat{u}(t)$ from the samples $z(t)$ without knowing $g(t)$. This procedure is usually called blind deconvolution and, in order to avoid obvious identifiability problems, it usually requires placing rather strong hypotheses on the signals into play.

Deconvolution problems are frequently met in the study of physiological and pharmacokinetic systems. For instance, in the study of endocrine–metabolic systems, several deconvolution applications concern the indirect measurement of signals not accessible in vivo, such as hormone secretion rate or substrate production rate (Figure 8.1, top and bottom-right panels); see Sparacino et al. (2001) for a list of references. In pharmacokinetics, deconvolution is used to determine the absorption rate of a drug (see Figure 8.1, bottom left) or, in a control-system perspective, to calculate the rate $u(t)$ of the drug intravenous infusion that forces the drug plasma concentration to follow a predetermined therapeutic profile $z(t)$ (Veng-Pedersen, 2001). Deconvolution can also be used to gain insight into a black-box system via Equation 8.2, which is used to obtain the unknown impulse response $g(t)$ starting from the knowledge of system input $u(t)$ and output $z(t)$, for example, to estimate the transport function of a substance within an organ from the knowledge of its concentrations in inlet (artery) and outlet (vein) (Sparacino et al., 1998).

Other biomedical applications of deconvolution include biomechanics, confocal microscopy, blood pressure measurement, ultrasound, tracer kinetics, nuclear medicine, radiology, tomography, neurophysiology, and evoked potentials. An extended list of bibliographic references can be obtained through http://www.pubmed.com.

In recent years, deconvolution has also been employed to interpret data obtained from functional imaging techniques. For instance, in magnetic resonance imaging (MRI), hemodynamic parameters describing the physiological status of the cerebral tissue, such as volume and flow (CBF), can be estimated from the so-called residue function, the fraction of the injected exogenous contrast agent (usually gadolinium) that is still present in the system at time t. This function can be obtained by deconvolution as the impulse response $g(t)$ of a system whose input $u(t)$ and output $z(t)$ are the measured concentration of the contrast agent in artery and in tissue, respectively (Ostergaard, 2004).

8.2   DIFFICULTIES OF THE DECONVOLUTION PROBLEM

8.2.1   Ill-Posedness and Ill-Conditioning

In the present chapter, deconvolution will be addressed only from an input-estimation viewpoint, assuming that the impulse response of the system is available. In addition, for the sake of simplicity, $u(t)$ will be assumed to be causal, so that in Equation 8.1 the integral starts from zero.

In practice, in place of the continuous time output $z(t)$, only a finite number n of noisy scalar samples $\{y_k\}_{k=1,\ldots,n}$ on the sampling grid $\Omega_s = \{t_1, t_2, \ldots, t_k, \ldots, t_n\}$ is available. Assuming that measurement error is additive, the model is

$$y_k = \int_0^{t_k} g(t_k, \tau)u(\tau)d\tau + v_k = z_k + v_k \qquad k = 1, 2, \ldots, n \quad (8.3)$$

where v_k is the error affecting $z_k = z(t_k)$.

The problem of recovering the input $u(t)$ from the samples of the output is ill-posed, that is, it does not have a unique solution, as illustrated in Figure 8.3 through a simulated example, taken and adapted from a classic work by Hunt (1971). The input $u(t) = e^{-[(t-400)/75]^2} + e^{-[(t-600)/75]^2}$, $0 \leq t \leq 1025$ (panel A) drives a system with impulse response $g(t) = 1$ for $0 \leq t \leq 250$ and $g(t) = 0$ elsewhere (panel B). The output samples $z_k = z(t_k)$ are observed, without error, on the sampling grid $\Omega_s = \{kT\}$ ($T = 25$, $k = 1, 2, \ldots, 41$) (panel D). Without going into mathematical details, it is easy to see that the problem of estimating the continuous-time input $u(t)$ from the dis-

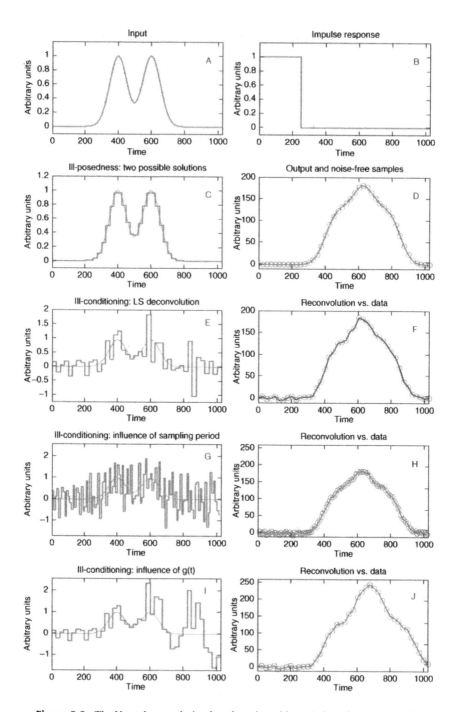

Figure 8.3. The Hunt deconvolution benchmark problem. A, Impulse response. B, True input. C, Ill-posedness of the deconvolution problem; the staircase function (thick line) perfectly explains the error-free data exactly as the true input (thin line) does. D, True continuous output with noise-free samples ($n = 41$). E, Ill-condition-

crete-time samples $\{z_k\}$ admits, in addition to the true solution, also other solutions, for example, the piecewise constant function of panel C, whose convolution with the impulse response perfectly honors the samples $\{z_k\}$.

In order to tackle ill-posedness, any deconvolution approach must somehow restrict the set of functions among which the solution of the problem is sought. For instance, in the so-called discrete deconvolution, the signal $u(t)$ is assumed to be a piecewise constant within each interval of the sampling grid Ω_s, that is, $u(t) = u_i$ for $t_{i-1} < t \le t_i$, $i = 1, 2, \ldots, n$. From Equation 8.3, it follows that

$$y_k = z_k + v_k = \sum_{i=1}^{k} u_i \int_{t_{i-1}}^{t_i} g(t_k, \tau) d\tau + v_k = \sum_{i=1}^{k} u_i \tilde{g}_{k,i} + v_k \qquad (8.4)$$

where

$$\tilde{g}_{k,i} = \int_{t_{i-1}}^{t_i} g(t_k, \tau) d\tau \qquad i = 1, \ldots, k \qquad (8.5)$$

Hence, in matrix-vector form,

$$y = z + v = Gu + v \qquad (8.6)$$

where $y = [y_1, y_2, \ldots, y_n]^T$ represents the data vector, $z = [z_1, z_2, \ldots, z_n]^T$ denotes the vector of the samples of the output $z(t)$, $v = [v_1, v_2, \ldots, v_n]^T$ is the measurement error vector, $u = [u_1, u_2, \ldots, u_n]^T$ represents the vector containing the (unknown) levels of the input $u(t)$, and G is an n-dimensional lower triangular square matrix with $[G]_{k,i} = \tilde{g}_{k,i}$ (with $i \le k$). Vector v has covariance matrix $\Sigma_v = \sigma^2 B$, where B is a known $n \times n$ positive definite matrix and σ^2 is a scale factor, possibly unknown. Usually, measurement errors are independent, so that B is diagonal. For example, assuming a constant coefficient of variation (CV) for the measurement error, one has $B = \text{diag}(y_1^2, y_2^2, \ldots, y_n^2)$ and $\sigma = CV$ (here CV is a real number, possibly unknown).

The model in Equation 8.6 describes, in matrix terms, the relationship between the unknown input u and the measured output y. Given such a model, in least-squares-parameter estimation, the Euclidean norm $\|\cdot\|_{B^{-1}}^2$

---

ing, solution provided by LS-deconvolution (thick line) from the noisy data of panel F and true input (thin line). F, Reconvolution and noisy data (SD = 3). G, Ill-conditioning (influence of an increase of the sampling rate). LS deconvolution from the noisy samples in panel H ($n = 102$). H, Reconvolution and noisy data (SD = 3). I, Ill-conditioning (influence of the impulse response). LS-deconvolution from the noisy samples in panel J (slower dynamics). J, Reconvolution and noisy data (SD = 3).

of the residual vector $r = y - G\hat{u}$ is minimized by solving the optimization problem

$$\min_{\hat{u}} \ (y - G\hat{u})^T B^{-1}(y - G\hat{u}) \qquad (8.7)$$

Notably, when measurement error is not stationary, $B \neq I_n$. Accordingly, the weighting matrix B^{-1} takes into account the possible difference in reliability of the data.

Even in presence of a pure time delay in the impulse response, G can always be reduced to a full rank matrix via a suitable shift of the time axis. Now, if G is full rank, the solution of Equation 8.7 does not depend on matrix B:

$$\hat{u} = (GB^{-1}G^T)^{-1}G^T B^{-1} y = G^{-1} y \qquad (8.8)$$

In the following, this estimate will be indicated as LS deconvolution. In this case, the so-called "reconvolution," that is, the prediction of the output $z(t)$, computed as

$$\hat{z}(t) = \int_{-\infty}^{t} g(t, \tau)\hat{u}(\tau)d\tau \qquad (8.9)$$

perfectly matches the data, and the residual vector $r = y - G\hat{u}$ is null. Notably, given the triangular structure of G, LS deconvolution can be directly obtained by the method of forward substitution. In the uniform sampling case, if the system is time invariant, it holds that $\tilde{g}_{k,i} = \tilde{g}_{k-i}$. Hence, matrix G has a Toeplitz structure and is completely specified by its first column. In addition, in this case, it is also possible to adopt a frequency domain viewpoint. In fact, since $\tilde{g}_{k,i} = \tilde{g}_{k-i}$, Equation 8.4 can be interpreted as a discrete convolution and, in the Z-transform domain, it is possible to compute LS deconvolution as $\hat{U}(z) = G^{-1}Y(z)$, with obvious meaning of notation. In practical cases, the transfer function $G(z)$ is rational, and the input estimate can be determined by exploiting a difference equation whose computational burden scales linearly with the number n of data.

LS deconvolution is appealingly simple, but, unfortunately, its performance is often poor. For instance, let us consider the simulated problem of Figure 8.3, where noisy data $\{y_k\}$ (denoted by open circles in panel F) were generated by adding white Gaussian noise (zero mean and standard deviation SD = 3). LS deconvolution returns a profile $\hat{u}$ (panel E) that exhibits wide spurious oscillations and negative swings, even if the correspondent reconvolution (continuous curve in panel F) perfectly honors the data. The lack of efficacy of LS deconvolution is due to ill-conditioning, that is, small errors in the observed data can be significantly amplified, thus yielding much larger errors in the estimate. One may conjecture that increasing the number of samples could be beneficial to the solution of the problem. On the contrary, increasing the sampling rate worsens ill-conditioning, as shown in the simulated example of Figure 8.3 (panels G and H), where, maintaining the same impulse response and measurement error variance, the sampling period was decreased from $T = 25$ to $T = 10$. In addition, the smoother the system kernel, the worse is the ill-conditioning of the deconvolution problem, as illustrated in Figure 8.3 (panels I and J), where, maintaining the same noise sequence of panel F, the duration of $g(t)$ was increased from 250 to 400. For a quantitative analysis of ill-conditioning, one may refer to De Nicolao and Liberati (1993).

8.2.2 Deconvolution of Physiological Signals

The conceptual difficulties described above made the deconvolution problem a classic topic of the engineering/mathematics/physics literature. Unfortunately, dealing with physiological signals adds to the complexity of the problem (De Nicolao et al., 1997; Sparacino et al., 2001). For instance, parsimonious sampling schemes are needed to cope with technical and cost limitations as well as with patient's comfort (e.g., each concentration determination requires drawing a separate blood sample). Consequently, concentration data are very often collected with infrequent and nonuniform sampling schedules (see Figure 8.1, bottom). Moreover, sometimes measurement noise cannot be described as a stationary process. Likewise, nonstationary properties often characterize signals such as hormone secretion after stimulus and absorption rate of a drug. Furthermore, physiological inputs, for example, a hormone secretion or a substrate production rate, are often intrinsically nonnegative, so that input estimates taking negative values due to ill-conditioning (see Figure 8.3, panels E, G, and I) appear physiologically implausible. Finally, even in short time intervals, physiological systems may be time varying; see, for example, the glucose–insulin system during a glucose perturbation.

In the literature, many methods have been developed to circumvent ill-conditioning. Broadly speaking, they can be divided into two categories. A first approach, named *parametric deconvolution,* assumes the analytic expression of the input to be known except for a small number of parameters, so that the deconvolution problem becomes a parameter estimation problem. A second class of deconvolution methods, often referred to as *nonparametric deconvolution,* does not postulate an analytic form for the input. The most known nonparametric approach is the Phillips–Tikhonov regularization method, which is described in detail in the next section in both its deterministic and stochastic formulations. Some other deconvolution approaches, both parametric and nonparametric, will be briefly reviewed in Section 8.4.

It is worth noting that, because of space constraints, in this chapter we will not consider some methodologies that are of paramount importance in signal processing, but often not applicable to physiological signals. For instance, Wiener filtering approaches are applicable only in rather restrictive circumstances (De Nicolao and Liberati, 1993), given that the occurrence of any nonuniform sampling, nonstationarity, or system time variance hinders its use. These features could be addressed by Kalman filtering methods (Commenges and Brendel, 1982), which, however, cannot deal with nonnegativity constraints. Nonetheless, it should be pointed out that, under certain circumstances and irrespective of the computational aspects, it is always possible to obtain the same estimates as for Wiener and Kalman filtering by employing the Bayes estimation approach presented in Section 8.3.2.

8.3   THE REGULARIZATION METHOD

8.3.1   Deterministic Viewpoint

The regularization method was devised by Phillips (1962) and Tikhonov (1963) in two parallel and independent works. It is probably the most commonly known deconvolution approach in the literature. The method starts from the observation that LS deconvolution is often unsatisfactory because, even if reconvolution perfectly fits the samples of the output, the estimated input profile is usually affected by high-frequency oscillations, often in contrast with expected physiological properties. The idea of the regularization method is thus to look for a solution that provides a good, instead of perfect, data fit but enjoys, at the same time, a certain degree of "smoothness," which is compatible with physiological expectations, on the unknown input signal. This is done by solving the optimization problem,

$$\min_{\hat{u}} \ (y - G\hat{u})^T B^{-1} (y - G\hat{u}) + \gamma \, \hat{u}^T F^T F \hat{u} \qquad (8.10)$$

where γ is a real nonnegative parameter and F is a $n \times n$ penalty matrix, usually chosen so as to make $F\hat{u}$ equal to the vector or the mth order differences of $\hat{u}$, that is, $F = \Delta^m$, where Δ is a square lower-triangular Toeplitz matrix (size n) whose first column is $[1, -1, 0, \ldots, 0]^T$. The parameter m is usually adjusted by trials and either $m = 1$ or $m = 2$ are normally used. Equation 8.10 is quadratic and its closed-form solution,

$$\hat{u} = (G^T B^{-1} G + \gamma F^T F)^{-1} G^T B^{-1} y \qquad (8.11)$$

linearly depends on the data vector y. Remarkably, the cost function in Equation 8.10 coincides with that of Equation 8.7, apart for the presence of the additional term $\hat{u}^T F^T F \hat{u}$, which represents the energy of the mth-order time derivative. In this way, the cost function in Equation 8.10 weights both the adherence to the data and the roughness of the solution. The relative importance given to data fit and solution regularity is governed by the so-called *regularization parameter* γ. By raising γ, the cost of roughness increases and the data match becomes relatively less important. Conversely, by decreasing the value of γ, the cost of roughness decreases and fidelity to data becomes relatively more important. The choice of the regularization parameter is a crucial problem: too large values of γ will lead to overly smooth estimates of $\hat{u}$, which may be not able to explain the data, whereas too small values of γ will lead to ill-conditioned solutions $\hat{u}$ that accurately fit the data but exhibit spurious oscillations due to their sensitivity to noise (for $\gamma \to 0$, the LS solution is approached). The regularization parameter γ must be properly tuned in each specific problem. The importance of the choice of γ is demonstrated by the deconvolution profiles reported in Figure 8.4, for the simulated problem of Figure 8.3, relative to two guess values of γ, $\gamma = 0.5$ and $\gamma = 400$. A too small value of γ leads to ill-conditioned solutions $\hat{u}$ (panel A), which well explain the data (panel B) at the cost of wide, spurious oscillations (the reconvolution $G\hat{u}$ fits both data and noise). Conversely, a too large value of γ leads to overly smooth estimates $\hat{u}$ (panel C), which may be not able to explain the data (panel D). These two extreme situations go under the name of undersmoothing and oversmoothing, respectively.

The Choice of the Regularization Parameter. In order to avoid subjectivity in the choice of the regularization parameter, several regularization criteria have been proposed, such as discrepancy, minimum risk, ordinary cross-validation (OCV), generalized cross-validation, and L-curve; see Sparacino et al. (2001) for references.

For instance, the popular discrepancy criterion (Twomey, 1965) suggests adjusting γ until the residual sum of squares equals the sum of the measurement error variances. In mathematical terms, the condition to be satisfied can be expressed as

$$WRSS = (y - G\hat{u})^T B^{-1}(y - G\hat{u}) = n\sigma^2 \qquad (8.12)$$

Since the residuals vector $r = y - G\hat{u}$ can be interpreted as an estimate of the measurement error vector v, the discrepancy criterion has a very intuitive motivation, that is, a "reasonable" estimate $\hat{u}$ should lead to weighted residuals with energy $r^T B^{-1} r$ similar to the expected value $E[v^T B^{-1} v] = n\sigma^2$.

Unfortunately, this intuitive rationale does not have a solid theoretical foundation. In particular, the discrepancy criterion (Equation 8.12) is at risk of oversmoothing, because, as will be discussed in the following, within a statistical framework it holds $E[WRSS] < n\sigma^2$. The so-called min-

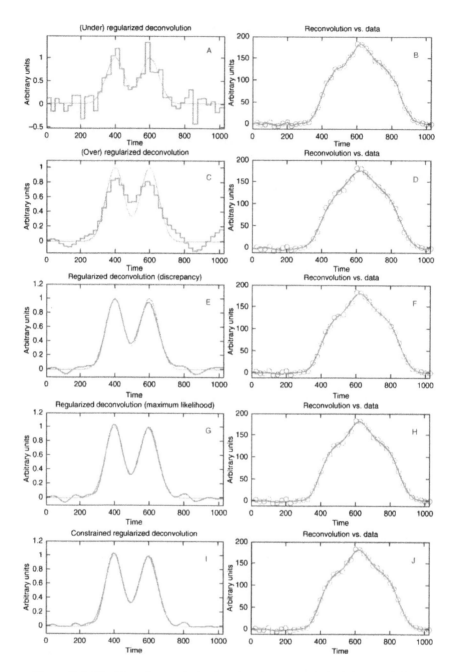

Figure 8.4. The Hunt deconvolution benchmark. A. Regularized deconvolution obtained with a too small value of the regularization parameter [$\gamma = 0.5$, $q(\gamma) = 35.12$] and true input (thin line). B, Reconvolution obtained from the input of panel A and data. C, Regularized deconvolution obtained with a too large value of the regularization parameter [$\gamma = 400$, $q(\gamma) = 11.31$]. D, Reconvolution obtained from the input of panel C and data. E, Regularized deconvolution using the discrepancy

180

imum risk criterion (Hall and Titterington, 1987) thus suggests choosing γ so as to make *WRSS* somewhat smaller than $n\sigma^2$, and, in particular, equal to

$$WRSS = \sigma^2 \left(n - trace\left[\Psi\right]\right) \tag{8.13}$$

where $\Psi = G(G^TB^{-1}G + \gamma F^TF)^{-1}G^TB^{-1}$ is the so-called hat matrix.

Both the above criteria require knowledge of σ^2. Other regularization criteria are also available that can be used when σ^2 is unavailable. For instance, OCV exploits a "leave-one-out" strategy, in which each of the output samples is, in turn, left out. Deconvolution is obtained from the remaining $n - 1$ samples, and the corresponding "leave-one-out" prediction error is computed. The best γ is the minimizer of the sum of the squared prediction errors. In order to reduce the computational burden required by OCV, an approximation called GCV is most commonly used (Wahba, 1990). Accordingly, the regularization parameter γ is selected so as to minimize GCV:

$$GCV = \frac{WRSS}{trace\left[I_n - \Psi\right]^2} \tag{8.14}$$

Finally, the so-called L-curve regularization criterion (Hansen and O'Leary, 1993) begins with noticing that, solving the deconvolution problem for increasing values of γ, the quantity $(y - G\hat{u})^TB^{-1}(y - G\hat{u})$ increases while the quantity $\hat{u}^TF^TF\hat{u}$ decreases. The loglog curve, parameterized in γ, of these two quantities often exhibits a typical L shape, whose corner is deemed a good compromise between data fit and regularity.

It is notable that, when σ^2 is available, the use of criteria that do not exploit its knowledge should be avoided, since they would induce the same amount of regularization whether σ^2 is known to be, say, 1 or 100.

The Virtual Grid. The Phillips–Tikhonov deconvolution method transforms the problem of estimating a continuous time function into that of recovering a piecewise constant function. In particular, the discrete model of Equation 8.4 was derived assuming that the unknown input is constant and equal to u_i, in the interval between t_{i-1} and t_i, no matter how long. In the

---

criterion [1 min virtual grid, $\gamma = 1.24 \times 10^6$, $q(\gamma) = 13.84$] and true input (thin line). F, Reconvolution obtained from the input of panel E and data. G, Regularized deconvolution using the maximum likelihood criterion [$\gamma = 6.18 \times 10^5$, $q(\gamma) = 15.55$]. H, Reconvolution obtained from the input of panel G and data. I, Nonnegative deconvolution, obtained with the same γ of panel G. J, Reconvolution obtained from the input of panel I and data.

nonuniform/infrequent sampling case, this often results in a poor approximation of the signal. In order to remove this assumption, a different discretization strategy can be used for the convolution integral (Equation 8.1). This strategy considers, in addition to the experimental sampling grid Ω_s, a second, finer, temporal grid $\Omega_v = \{T_1, T_2, \ldots, T_k, \ldots, T_N\}$, whose elements are close enough for the unknown input $u(t)$ to be well approximated by a piecewise constant function. The size N of the virtual grid can be much larger than the size n of Ω_s. For the sake of simplicity, it is convenient to assume that Ω_v is uniformly spaced and contains Ω_s. Apart from this, Ω_v is arbitrary and does not have an experimental counterpart. For this reason, Ω_v is called the *virtual grid*.

Now let us consider the output $z(t)$ of Equation 8.1 evaluated at the times belonging to Ω_v. In most cases, no actual measurement samples are drawn at these times (especially if $N \gg n$). Let $z_v(T_k)$ denote the (noise-free) output at the virtual sampling times T_k. Assuming that $u(t)$ is piecewise constant within each time interval of the virtual grid, in analogy to Equation 8.4, it follows that

$$z_v(T_k) = \int_0^{T_k} g(T_k, \tau)u(\tau)d\tau = \sum_{i=1}^{k} u(T_i) \int_{T_{i-1}}^{T_i} g(T_k, \tau)d\tau \qquad k = 1, 2, \ldots, N$$

$$(8.15)$$

where $T_0 = 0$. Adopting for Equation 8.15 the usual matrix notation one has $z_v = G_v u$, where z_v and u are N-dimensional vectors obtained by sampling $z(t)$ and $u(t)$ on the virtual grid, and G_v is an $N \times N$ lower triangular matrix whose (k, j) entry is given by

$$G_v(k, i) = \int_{T_{i-1}}^{T_i} g(T_k, \tau)d\tau \qquad k \leq i \qquad (8.16)$$

Times belonging to the virtual grid Ω_v but not present in the sampling grid Ω_s have no counterpart in the sampled output data. We can regard them as (virtually) missing data. By denoting by G the $n \times N$ matrix obtained by removing from G_v the $N - n$ rows that do not correspond to sampled output data, a matrix-vector model formally equal to Equation 8.6 is obtained. Hereafter, G will be of size $n \times N$. It is notable that, in presence of virtual grid, the input vector u has size N, which may be (much) larger than the size n of the measurement vector y. However, since in the optimization problem (Equation 8.10) matrix F has size $n \times N$, the solution given by Equation 8.11 can still be determined.

This method, provided that Ω_v has fine enough detail, yields a stepwise estimate that is virtually indistinguishable from a continuous profile. More precisely, it is interesting to investigate what happens when, given the virtual grid $\Omega_v = \{kT_v\}$, $k = 1, 2, \ldots, N$ (with $NT_v = t_s$), one lets $T_v \to 0$ and $N \to \infty$. Intuitively, the piecewise assumption made on the input vanishes, and the solution tends to the minimizer of

$$\min_{\hat{u}(t)} (y - \hat{z})^T B^{-1}(y - \hat{z}) + \gamma \int_0^{t_s} \hat{u}^{(m)}(\tau)^2 d\tau \tag{8.17}$$

In De Nicolao et al. (1997), it was shown that this limit estimate is continuous up to the $(2m - 2 + p)$th time derivative, where p is the relative degree of the system, that is, the difference between the degree of the denominator and that of the numerator of the Laplace transform of $g(t)$. This result offers a guideline for choosing the order of the time derivatives to penalize. For instance, for a LTI system with $p = 1$, it will be sufficient to let $m = 1$ in order to have an asymptotically continuous estimate together with its first time derivative.

8.3.2 Stochastic Viewpoint

Consider the matrix-vector model (Equation 8.6), possibly derived using the virtual grid, where G is a $n \times N$ matrix. In order to formulate the problem within a stochastic embedding, assume that u and v are zero-mean random vectors whose covariance matrices Σ_u and Σ_v are known. It is assumed that $\Sigma_v = \sigma^2 B$ and that Σ_u is factorized as $\Sigma_u = \lambda^2 (F^T F)^{-1}$ (both Σ_u and F have size $N \times N$)

In this stochastic setting, the deconvolution problem can be stated as a *linear minimum variance estimation problem:* find the estimate $\hat{u}$, linearly depending on the data vector y, such that $E[\|u - \hat{u}\|^2]$ is minimized. If u and v are uncorrelated, the linear minimum variance estimate $\hat{u}$ coincides with the solution of the optimization problem (Equation 8.10), provided that $\gamma = \gamma^0 = \sigma^2/\lambda^2$. When u and v are jointly Gaussian, the estimator (Equation 8.11) with $\gamma = \gamma^0$ has minimum error variance among all estimators, either linear or nonlinear, of u given y.

To solve the deconvolution problem in a stochastic context as a linear minimum variance estimation problem, the a priori covariance matrix of the input vector u, $\Sigma_u = \lambda^2 (F^T F)^{-1}$), is required. In practical cases, we only know that the input u is a smooth function of time. A simple a priori probabilistic model of a smooth signal on a uniformly spaced grid (possibly the virtual grid) is to describe it as the realization of a stochastic process obtained by the cascade of m integrators driven by a zero-mean white noise process $\{w_k\}$ with variance λ^2. For instance, for $m = 1$ this corresponds to a random-walk model:

$$u_k = u_{k-1} + w_k \qquad k = 1, 2, \ldots, N; u_0 = 0 \tag{8.18}$$

In a Gaussian setting, Equation 8.18 tells us that, given u_k, then u_{k+1} will be in the range $u_k \pm 3\lambda$ with probability 99.7%. Obviously, the lower λ^2, the more regular the process $\{u_k\}$. It is easily demonstrated that the covariance matrix of the random vector u whose components are obtained from m integrations of a white-noise process of variance λ^2 is given by $\Sigma_u = \lambda^2 (F^T F)^{-1}$, where $F = \Delta^m$ is the same as in the regularization method penalizing the energy of the mth-order time derivatives.

Both regularization and minimum variance estimation determine the estimate by solving Equation 8.11. This establishes an insightful analogy between the two approaches. In particular, penalizing the mth time-derivative energy in the regularization method is the same as modeling the unknown input as an $(m - 1)$-fold integrated random-walk process in the stochastic approach. In view of this analogy, $\gamma^0 = \sigma^2/\lambda^2$ represents, in some sense, the "optimal" value of the regularization parameter. Such a value is, however, unknown, since λ^2 and, possibly, also σ^2 are unknown.

Confidence Intervals. Considering Equation 8.11 for a generic value of γ in place of $\gamma^0 = \sigma^2/\lambda^2$, the estimation error $\tilde{u}$ has zero mean and covariance matrix given by

$$\mathrm{var}\left[\tilde{u}\right] = \sigma^2 \Gamma B \Gamma^T + \gamma \lambda^2 (G^T B^{-1} G + \gamma F^T F)^{-1} F^T F (G^T B^{-1} G + \gamma F^T F)^{-1} \tag{8.19}$$

where $\Gamma = (G^T B^{-1} G + \gamma F^T F)^{-1} G^T B^{-1}$. It is easily verified that the contribution of noise to the error variance, first term in the right-hand side of Equation 8.19, is a monotonically decreasing function, in matrix sense, of γ, whereas the contribution of bias, the second term in Equation 8.19, is monotonically increasing. It is worthwhile stressing the importance of taking into account both sources of error: noise and bias. Indeed, considering only the noise term would lead to the paradox of indefinitely increasing accuracy as γ grows. Not surprisingly, the minimum value of var[$\tilde{u}$] is associated with the optimal $\gamma = \gamma^0 = \sigma^2/\lambda^2$:

$$\mathrm{var}\left[\tilde{u}\right] = \sigma^2 (G^T B^{-1} G + \gamma^0 F^T F)^{-1} \tag{8.20}$$

If a reliable estimate of γ^0 is available, this covariance matrix can be used to compute confidence intervals for the entries of $\hat{u}$.

Statistically Based Choice of the Regularization Parameter. Let $WRSS = (y - G\hat{u})^T B{-}1(y - G\hat{u})$ and $WESS = \hat{u}^T F^T F \hat{u}$ denote the weighted residuals sum of squares and the weighted estimates sum of squares, respectively. Both these quantities depend on the regularized estimate $\hat{u}$ and, thus, on the value of γ. In the stochastic setting $WRSS$ and $WESS$ are random variables and, for $\gamma = \gamma^0$, their expected value can be obtained:

$$E\left[WESS\right] = \lambda^2 q(\gamma^0) \tag{8.21}$$

$$E\left[WRSS\right] = \sigma^2\{n - q(\gamma^0)\} \tag{8.22}$$

where

$$q(\gamma^0) = \text{trace}\left[G(G^T B^{-1}G + \gamma^0 F^T F)^{-1}G^T B^{-1}\right] \tag{8.23}$$

Note the analogy of Equation 8.22 with a well-known property of linear regression models: the averaged sum of squared residuals is a biased estimator of the error variance, with the bias depending on the (integer) number of degrees of freedom of the model. For this reason, $q(\gamma)$ defined by Equation 8.23 is named *equivalent degrees of freedom* associated with γ. The quantity $q(\gamma)$ is a real number varying from 0 to n: if $\gamma \to 0$, then $q(\gamma) \to n$, whereas if $\gamma \to \infty$, then $q(\gamma) \to 0$. The fact that $q(\gamma)$ is a real number is in agreement with the nature of the regularization method, for which the flexibility of the model (its degrees of freedom) can be changed with continuity through the tuning of the regularization parameter.

By dropping the expectations in Equations 8.21 and 8.22, and recalling that $\gamma^0 = \sigma^2/\lambda^2$, two "consistency" criteria can be intuitively derived, which allow the choice of γ when either λ^2 or both λ^2 and σ^2 are unknown (Sparacino et al., 1996). It is worth noting that the same two criteria can be derived on a firmer statistical ground under Gaussianity assumptions by determining necessary conditions for λ^2 and σ^2 to maximize the likelihood of the data vector y (McKay, 1992; De Nicolao et al., 1997). The two maximum likelihood (ML) criteria are formulated as follows. When λ^2 is unknown (σ^2 is assumed to be known), tune γ until

$$WESS = \sigma^2 q(\gamma)/\gamma \tag{8.24}$$

When both σ^2 and λ^2 are unknown, tune γ until

$$\frac{WRSS}{n - q(\gamma)} = \gamma \frac{WESS}{q(\gamma)} \tag{8.25}$$

and then, from Equation 8.22, estimate σ^2 as

$$\hat{\sigma}^2 = \frac{WRSS}{n - q(\gamma)} \tag{8.26}$$

Figure 8.4, panel G, shows the input profile of the Hunt deconvolution benchmark problem estimated with the virtual grid and using the regularization criterion (Equation 8.24). By comparing this profile with that of Figure 8.4, panel E, obtained using the discrepancy criterion, one notes that the latter is oversmoothed. In fact, by comparing Equation 8.12 with Equation 8.22 it is easily seen that the discrepancy criterion yields to larger values of γ.

Remark. The comparison of the value γ used in different problems does not allow the comparison of the amount of regularization employed. For instance, in the Hunt benchmark problem, $\gamma = 400$ leads to oversmoothing in Figure 8.4, panel C, but, in presence of the virtual grid, $\gamma = 6.18 \times 10^5$ leads to suitable regularization in Figure 8.4, panel G. A better indicator of the amount of regularization is the degree of freedom $q(\gamma)$, a real number varying from 0 to n. For instance, the degrees of freedom in the above two cases were $q(\gamma) = 11.31$ and $q(\gamma) = 15.55$, respectively, suggesting that less regularization (in spite of a larger value of γ) was used in the determination of the latter input estimate.

8.3.3 Numerical Aspects

In the regularization approach, the computation of the solution via Equation 8.11 would require $O(N^3)$ memory occupation and $O(N^3)$ operations. The notation $O[f(N)]$ means of the same order of magnitude as $f(N)$. This computational burden can be reduced by applying the matrix inversion lemma to Equation 8.11, obtaining $\hat{u} = F^{-1}F^{-T}G^{T}(GF^{-1}F^{-T}G^{T} + \gamma B)^{-1}y$. In this way, an $n \times n$ linear system must be solved, at the price of $O(n^3)$ operations (note that F admits an easy-to-derive inverse).

In the LTI system case with uniform sampling, matrices G and F in Equation 8.11 exhibit a Toeplitz structure so that only their first column need be stored. In addition, in this case, particularly efficient numerical techniques can be used to compute the regularized estimate. In particular, Equation 8.10 can be solved by the iterative conjugate gradient (CG) method, whose basic iteration can be performed in $O(N \log N)$ operations through the use of fast Fourier transform (Commenges, 1984) or in $O(N)$ operations by the use of recursive difference equations (De Nicolao et al., 1997). The Toeplitz structure of the matrices can also be exploited in order to devise suitable "preconditioners" that improve the rate of convergence of the algorithm (Commenges, 1984). Since theory guarantees the convergence of the CG algorithm in N iterations at most, the overall complexity of the algorithm is $O(N^2 \log N)$. Note that, in the presence of time-varying systems, these numerically efficient methods are not applicable because matrix G does not have a Toeplitz structure.

However, the bottleneck of numerical algorithms for deconvolution is due to the need for computing several trial solutions of Equation 8.10. In fact, the tuning of the regularization parameter γ (according to any criterion) requires a trial-and-error procedure. De Nicolao et al. (1997) proposed a strategy which, through singular value decomposition (SVD), first puts Equation 8.6 in diagonal form in $O(n^3)$ operations, then dramatically speeds up the trial-and-error procedure for the determination of the optimal regularization parameter, since it requires only $O(n)$ scalar operations to compute $\hat{u}$ for each trial value of λ. The strategy also allows one to compute the confidence intervals with $O(N^2)$ complexity, a significant improvement over the use of Equation 8.11, which would require $O(N^3)$ operations. The overall complexity of the algorithm is $O(n^3N)$ which, at least when $n \ll N$, is better than the above-mentioned method based on the conjugate gradient. Note that, since the SVD procedure is unaffected by the Toeplitz structure of G, this numerical strategy applies also to the time-varying case. An alternative numerical strategy that takes into account the need for calculating Equation 8.11 for several values of the regularization parameters is the QR factorization approach proposed by Hanke and Hansen (1993).

For large values of n, further refinements of the numerical algorithms can be obtained when the sampling is uniform and the system is time invariant. In particular, a spectral factorization can be exploited to calculate, with a computational burden not dependent on the complexity of the problem, the degrees of freedom $q(\gamma)$. Some explicit formulas to compute $q(\gamma)$, under appropriate assumptions, can also be obtained (De Nicolao et al., 2000). In the unconstrained case, using state-space methods it has been shown (De Nicolao and Ferrari-Trecate, 2003) that the regularized estimate $\hat{u}_k(t)$ can be seen as a regularization network whose output is the weighted sum of N suitable basis functions. Interestingly, the neural network weights are computable in $O(n)$ operations via a Kalman filtering approach (De Nicolao and Ferrari-Trecate, 2001). In absence of nonnegativity constraints, this approach is particularly convenient when deconvolution must be performed on several datasets that share the same statistical parameters and impulse response (for instance, a population model). In fact, the basis functions remain unchanged across the datasets, and one has only to compute the individual weights.

8.3.4 Nonnegativity Constraints

In a number of physiological systems, the input $u(t)$ is known to be intrinsically nonnegative (e.g., hormone secretion rates and drug absorption rates). Nevertheless, due to measurement errors and impulse response model mismatch, the solution provided by Equation 8.11 may take on negative values (see Figure 8.4, panels E and G). To obtain nonnegative estimates, the regularization method can be reformulated by solving the optimization problem (Equation 8.10) in the subspace $\hat{u} \geq 0$ (i.e., $\hat{u}_k \leq 0, \forall\, k$). The constrained problem does not admit a closed-form solution and must be solved by an iterative method such as the constrained conjugate gradient algorithm (Commenges, 1984; De Nicolao et al., 1997). Remarkably, the incorporation of nonnegativity constraints makes the estimator nonlinear. This impairs the use of some of the regularization criteria previously described, for example, GCV and ML, and forces the adoption of empirical strategies. For instance, Figure 8.4 panel I shows, for the Hunt benchmark problem, the input profile estimated with the nonnegativity constraint using the same γ adopted for the unconstrained estimate displayed in panel G. In addition, nonnegativity contradicts Gaussianity, so that the computation of confidence intervals by exploiting analytic approaches (see Equation 8.20), is no longer possible. An empirical approach to compute confidence intervals is based an a Monte Carlo approach involving the solution, by the constrained technique, of a large number of artificially perturbed problems (Sparacino et al., 1996).

The regularization approach to deconvolution was implemented, through the numerical algorithms described in Sections 8.3.3 and 8.3.4, in an interactive software called WINSTODEC. This program, developed in MATLAB, has a graphical user interface that eases its use by experimenters without specific deconvolution expertise (Sparacino et al., 2002).

8.4   OTHER DECONVOLUTION METHODS

A number of deconvolution techniques, often referred to as *parametric deconvolution* methods, circumvent ill-conditioning by making functional assumptions on the input. For example, proposed functional forms for $u(t) = u(t, \theta)$ include polynomials and linear combinations of either exponentials or Gaussians. In these methods, deconvolution turns into the problem of estimating, via LS techniques, the unknown parameter vector θ (whose size is considerably less than n). This guarantees uniqueness (at least locally) and regularity of the solution. The heavy assumptions on the shape of the unknown input constitute, in general, a major drawback for parametric deconvolution methods. However, they are still rather popular in pharmacokinetics, in spite of several open methodological problems, such as choice of model order (a too high order will lead to ill-conditioned input estimates) and determination of input confidence intervals. From an algorithmic point of view, the intrinsic nonlinearity of parametric methods raises the problem of correctly selecting the initial guess of the parameters in order to avoid local minima in parameter estimation. Recent work addressing some of these problems using a Bayesian approach is reported in Johnson (2003).

Another method proposed in pharmacokinetics relies on *regression splines* (Verotta, 1993). In this approach, the input $u(t)$ is a the linear combination, with weights $\{\mu_i\}$, of M functions $B_{i,\tau}$, where $B_{i,\tau}$ represents the ith normalized component of the (cubic) B-spline basis, τ being the knot position vector. Having fixed number and location of the spline knots, that is, having chosen M and τ, the reconvolution vector linearly depends on the M weights $\{\mu_i\}$, which can thus be estimated by least squares. The major problems with the use of regression splines in deconvolution are related to the choice of the model order M and the knot location vector τ. The problem of selecting M involves the usual fit-versus-smoothness trade-off. The problem of locating the spline knots is more complex. Intuitively, the density of knots should be higher where fast changes in the input are expected. An appealing feature of the regression splines method is the possibility of incorporating monotonicity and nonnegativity constraints on the input by adding suitable inequality constraints on the weights vector μ. However, these constraints further complicate the choice of the number and position of the spline knots.

Turning back to nonparametric methods, it is worth pointing out that, in addition to regularization, there are also other approaches to deconvolution. These include: truncated singular values decomposition (TSVD), conjugate gradient regularization (CGR), and maximum entropy (ME).

TSVD methods first perform a singular value expansion of matrix G in Equation 8.6 as $G = W\Lambda V^T$, where V and W are unitary, that is, such that $WW^T = VV^T = I_n$, and matrix Λ is diagonal, $\Lambda = \mathrm{diag}\,(\lambda_1, \lambda_2, \ldots, \lambda_n)$, with $\lambda_1 \geq \lambda_{i+1}$. Since $G^{-1} = V\Lambda^{-1}W^T$, the LS deconvolution can be expressed as $\hat{u} = G^{-1}y = V\Lambda^{-1}W^Ty$, from which

$$\hat{u} = \sum_{i=1}^{k} \frac{w_i^T y}{\lambda_i} v_i \qquad (8.27)$$

with $k = n$. In Equation 8.27, $\{w_i\}$ and $\{v_i\}$ $(i = 1, 2, \ldots, n)$ are the n column vectors of W and V, respectively. From Equation 8.27 it follows that the smallest eigenvalues of G are the most responsible for ill-conditioning. The idea of TSVD deconvolution methods is to obtain an (unconstrained) estimate by truncating the expansion (Equation 8.27) before the small singular values start to amplify noise. The (integer) number k of the eigenvectors that are left in the expansion determines the regularity of the estimate, similarly to the regularization parameter (Hansen, 1987).

CGR is based on the fact that, when the CG algorithm is used to solve the system $y = Gu$, the low-frequency components of the solution tend to converge faster than the high-frequency ones. Hence, the CG has some inherent regularization effect where the number of CG iterations tunes the regularity of the solution (Van der Sluis and Van der Vorst, 1990). Another iterative method based on the same principle is based on an algorithm developed by Landweber for the solution of algebraic systems, a modified version being available in order to deal with nonnegativity constraints (Bertero and Boccacci, 1998).

Finally, ME methods (Donoho et al., 1992) can be viewed as a variant of the regularization method in which the term $u^T \log u$ replaces the quadratic term $u^T F^T F u$ in Equation 8.10. In this way, for a fixed γ, the ME estimator provides (strictly) positive signals but it is no more linear in the data, so that a closed-form solution does not exist and an iterative method is required to compute the estimate. The structure of the cost function makes ME methods particularly suitable to solve problems in which the unknown input is essentially zero in the vast majority of its domain, for example, a nearly black image.

8.5 A STOCHASTIC NONLINEAR METHOD FOR CONSTRAINED PROBLEMS

Often, in the presence of severe ill-conditioning, the stochastic linear approach described in Section 8.3 does not yield nonnegative input estimates (see Figure 8.4, panel G). In other cases, the reconstructed input is always nonnegative, but confidence intervals computed by using Equation 8.20 are not credible since they include negative values. This situation is depicted in Figure 8.5 (top), where one can see the secretion rate of luteinizing hormone (LH) in a healthy subject and its confidence interval (left), esti-

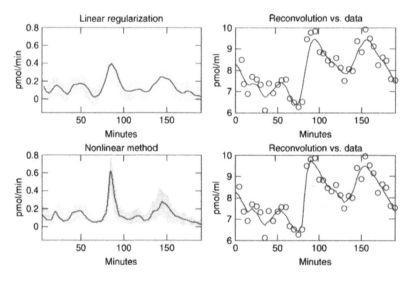

Figure 8.5. Reconstruction of secretion rate of luteinizing hormone under spontaneous conditions. Panel A, regularization using maximum-likelihood criterion (thick line) with confidence interval (shadowed area). Panel B, reconvolution obtained using the input in panel A and measurements. Panel C, deconvolution computed by the nonlinear stochastic approach (thick line) with confidence interval (shadowed area). Panel D, reconvolution obtained using the input in panel C and measurements.

mated starting from samples of its concentration in plasma collected every 5 min under spontaneous conditions (right). To take into account the fact that the secretion is not equal to zero before the first available observation, the initial estimation time is chosen at −55 min, as in De Nicolao et al. (1997). It is apparent that the confidence interval includes a negative region.

Still working in a stochastic setting, it is possible to take into account nonnegativity of the input by resorting to a log-normal prior. In particular, in Pillonetto et al. (2002) entries of the unknown input u in the measurements model (Equation 8.6) are the exponential of the entries of the vector $u_{\log}$, the latter being an m-fold integrated white Gaussian noise $\{w_k\}$. In this way, the covariance matrix of $u_{\log}$ is $\Sigma_{u\log} = \lambda^2(F^TF)^{-1}$. In this new description of the a priori information on u, the hyper-parameter λ^2 reflects the smoothness of the logarithm of the signal of interest. If m is equal to 1, it is easy to derive the following stochastic relationship between two adjacent components of the input:

$$\frac{u_k - u_{k-1}}{u_{k-1}} = e^{w_k} - 1 \qquad u(o) = u_o;\, k = 1, 2, \ldots, N \qquad (8.28)$$

where $u(o)$ is known. Equation 8.28 shows that the percent increase between two consecutive input levels on the virtual grid is a white and stationary stochastic process having a finite variance. Notice that such process is nonnormal, as opposed to that of Equation 8.18. Once the prior for $u_{\log}$ is specified and the statistical description for v in Equation 8.6 is given, it is possible to obtain the conditional density $u_{\log}|y$ (the so-called posterior density), whose optimum defines the maximum a posteriori (MAP) estimate. By substituting $u = \exp(u_{\log})$ in the measurements model (Equation 8.6), we find that the MAP estimate of $u_{\log}$, denoted by $\hat{u}_{\log}$, is the solution of the following problem:

$$\min_{u_{\log}} \left[y - G\exp(u_{\log}) \right]^T B^{-1} \left[y - G\exp(u_{\log}) \right] + \gamma u_{\log}{}^T F^T F u_{\log} \qquad (8.29)$$

where $\gamma = \sigma^2/\lambda^2$. In Pillonetto et al. (2002), it is shown how to obtain an approximated expression for the covariance matrix of the estimation error affecting $\hat{u}_{\log}$. Having obtained $\hat{u}_{\log}$, it is possible to compute the nonnegative estimate $\hat{u}$.

In practice, instead of considering Equation 8.29, it is useful to resort to a Markov chain Monte Carlo (MCMC) approach similar to that presented to solve deconvolution problems in Bellazzi et al. (1997), Magni et al. (1999), and Pillonetto and Bell (2007). The MCMC approach jointly addresses the problem of determining confidence intervals (analytically intractable due to the presence of an exponential transformation in Equation 8.29) and the tuning of the regularization parameter γ. Details can be found in Pillonetto et al. (2002). Briefly, the estimation problem is reformulated in a fully Bayesian setting, where also λ^2 and σ^2 are random variables with suitable a priori densities. The algorithm generates a Markov chain that converges in a distribution to the posterior of the unknown variables present in the problem, that is, $u_{\log}$, γ, and σ^2. Once such a posteriori solution is obtained in sampled form, Monte Carlo integration is performed to obtain the minimum variance estimate of $u_{\log}$ and the confidence intervals of interest.

Results obtained using the above-described approach, applied to the estimation of LH secretion, are depicted in Figure 8.5 (bottom). For what concerns λ^2 and σ^2, poorly informative priors have been used with support restricted to the positive axis. Reconstruction in sampled form of the posterior of u has been obtained by generating 11,000 realizations. As seen in Figure 8.5, the 95% confidence intervals are now realistic. The posterior mean of u exhibits a peak around 90 min whose amplitude is larger than that obtained from the linear estimator, thus introducing less bias in the residuals (see right panels).

In Figure 8.6, one can see results relative to the Hunt benchmark problem obtained by using the linear stochastic approach (top) and the nonlinear one (bottom), with the posterior reconstructed using 8000 realizations. In this case, one can appreciate the improved performance in

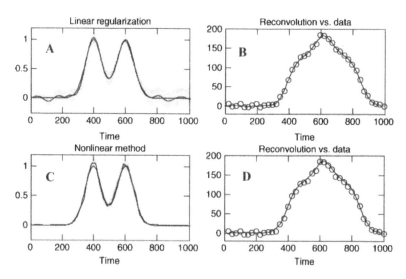

Figure 8.6. The Hunt deconvolution benchmark. A, regularization using the linear stochastic approach (thick line) with confidence intervals (shadowed area) and true profile (dashed line). B, reconvolution and measurements. C, deconvolution via the nonlinear stochastic approach (thick line) with confidence intervals (shadowed area) and true profile (dashed line). D, reconvolution and measurements.

terms of spurious oscillations. In addition, the confidence intervals are now not only more credible but also much narrower.

In conclusion, in Figure 8.7 we show some results obtained by applying the nonlinear stochastic approach for estimating hemodynamic parameters from MRI perfusion images (Zanderigo et al., 2007). In the top panel, for a generic pixel, we show measurements $\{y_k\}$ of tissue concentration, the arterial input $u(t)$ (left), and two estimates of the residue function (right) obtained using the TSVD approach described in Section 8.4 (thin line) and using the nonlinear stochastic approach (thick line) with 10,000 realizations drawn from the posterior. TSVD is the state-of-the- art approach in MRI even if it often does not provide, as in this case, physiological estimates of the residue function due to the large oscillations and the presence of negative values in the reconstructed profile. The nonlinear stochastic method, instead, leads to a more physiological profile. Applying this approach pixel by pixel, it is then possible to identify the maximum of

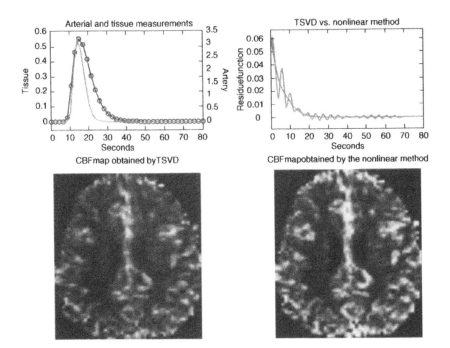

Figure 8.7. Top left, arterial input function (thick line) and time series of tissue concentrations (thin line). Top right, estimate of the residue function by TSVD (thin line) and by the nonlinear stochastic method (thick line). Bottom, CBF map reconstructed starting from the estimated residue functions using TSVD (left) and the nonlinear stochastic method (right). Lighter shades correspond to larger CBF values.

the residue function, reconstructing the cerebral blood flow CBF in the subject under study. In the bottom panels, we show the CBF map obtained using TSVD (left) and the new nonlinear approach (right). From these results, it is apparent, at least at a qualitative level, to see that the new technique yields a clearer image with improved anatomic contrast.

8.6   CONCLUSIONS AND DEVELOPMENTS

Deconvolution problems arise in several physiological systems when time series of plasma concentrations are used in order to reconstruct signals that are not directly accessible, such as rates of hormone secretion, substrate production, and drug appearance in plasma. The deconvolution problem, which suffers from intrinsic difficulties due to ill-positioning and ill-conditioning, has been extensively investigated as to its mathematical, physical, and engineering aspects. Nevertheless, physiological systems often involve additional difficulties that hamper the direct application of standard methodologies. Among these difficulties, one may mention infrequent or nonuniform sampling, time-varying nature of the underlying system, and nonnegativity constraints. In this chapter, a critical review of the existing literature has been provided, focusing on the Phillips–Tikhonov regularization approach, whose algorithmic and statistical aspects have been discussed, highlighting the connections with Bayesian estimation. An extension of the method has been presented that solves the deconvolution problem subject to nonnegativity constraints, preserving the statistical interpretation of the estimate and its confidence intervals.

Among recent research issues, there is the development of algorithms that explicitly account for the uncertainty in the impulse response, which has been neglected in the present chapter but is an important feature, given that $g(t, \tau)$ is often uncertain, being either identified from experimental data or an average population model that ignores interindividual variability. In such a case, an analytic solution being unavailable, a first approach relies on Monte Carlo techniques (Sparacino and Cobelli, 1996; De Nicolao et al., 1997), but it is also possible to pursue a fully Bayesian approach via MCMC methods (Bellazzi et al., 1997; Magni et al., 1998) at the cost of some increase in algorithmic complexity.

Another significant problem, especially in pharmacokinetics, is to embed the deconvolution problem in a population modeling context in which M individuals undergo the same experiment whose constraints (e.g., several subjects but few measurements per subject) prevent one from obtaining acceptable estimates. In such situations, the M subjects can be simultaneously analyzed so as to improve each individual estimate by taking advantage of the relevant information in the other $M - 1$ subjects. A first attempt in this direction is the semiparametric, spline-based approach proposed by Fattinger and Verotta (1995), in which the input to each individual is seen as the distortion of a template common to all the population that is estimated exploiting the data of all subjects. In a nonparametric context, it is natural to adopt a Bayesian paradigm and model the inputs to each individual as realizations of stochastic processes sharing the same prior distribution typical of the population. The two key points are the definition of the most suitable prior distribution and the development of efficient algorithms for the derivation of the posterior distribution of each individual input conditional on all the M datasets. These problems have already been successfully addressed in the function reconstruction problem (Neve et al., 2007, 2008), which can be seen as a particular deconvolution problem having a Dirac delta as impulse response. In particular, the computational burden involved by the need of jointly processing the M datasets has been substantially reduced by exploiting sampling grids common to all subjects (De Nicolao et al., 2007) or using coincident sampling times among subjects (Pillonetto et al., 2008).

On the computational side, recent research has focused on the development of numerically efficient algorithms that solve the regularization problem without any discretization grid and express the estimated input as the linear combination of continuous-time basis functions (De Nicolao and Ferrari-Trecate, 2003; Bell and Pillonetto, 2004). There are also numerically efficient MCMC algorithms that yield discretization-free continuous-time solutions (Pillonetto and Bell, 2007).

Finally, although the present chapter focused mostly on estimating the unknown input of a linear system, deconvolution can also be used to estimate the impulse response of a linear system from input and output measurements. Many considerations and techniques of this chapter apply equally well to this case, with the possible important exception of the choice of the regularization penalty, or, equivalently, the prior distribution of the unknown impulse response. In fact, imposing only regularity on the derivatives of the signal does not account for the possible exponential decay of the impulse response. A recent investigation has demonstrated the benefit of incorporating a stability constraint within the statistical prior for the impulse response (De Nicolao et al., 2008), a result that could find important applications in the context of physiological systems

REFERENCES

Bell, B., and Pillonetto, G., Estimating Parameters and Stochastic Functions of One Variable Using Nonlinear Measurement Models, *Inverse Problems*, Vol. 20, 627–646, 2004.

Bellazzi, R., Magni, P., and De Nicolao, G., Dynamic Probabilistic Networks for Modelling and Identifying Dynamics Systems, *Intelligent Data Analysis*, Vol. 1, 245–262, 1997.

Bertero, M., and Boccacci, P., *Introduction to Inverse Problems in Imaging*, IOP Publishing, 1998.

Bertero, M., Linear Inverse Problems and Ill-Posed Problems, *Advances in Electronics and Electron Phys.*, Vol. 75, 1–120, 1989.

Commenges, D., The Deconvolution Problem: Fast Algorithms Including the Preconditioned Conjugate-Gradient to Compute a MAP Estimator, *IEEE Trans. Automatic Control*, Vol. 29, 229–243, 1984.

Commenges, D., and Brendel, A. J., A Deconvolution Program for Processing Radiotracer Dilution Curves, *Computer Programs in Biomedicine*, Vol. 14, 271–276, 1982.

De Nicolao, G., Liberati, D., and Sartorio, A., Stimulated Secretion of Pituitary Hormones in Normal Humans: A Novel Direct Assessment from Blood Concentrations, *Ann. Biomed. Eng.*, Vol. 28, 1136–1145, 2000.

De Nicolao, G., and Liberati, D., Linear and Nonlinear Techniques for the Deconvolution of Hormone Time-Series, *IEEE Trans. Biomed. Eng.*, Vol. 40, 440–455, 1993.

De Nicolao, G., Sparacino, G., and Cobelli, C., Nonparametric Input Estimation in Physiological Systems: Problems, Methods, Case Studies, *Automatica*, Vol. 33, 851–870, 1997.

De Nicolao G., Ferrari Trecate, G., and Sparacino, G., Fast Spline Smoothing via Spectral Factorization Concepts, *Automatica*, Vol. 36, 1733–1739, 2000.

De Nicolao, G., and Ferrari Trecate, G., Regularization Networks: Fast Weight Calculation via Kalman Filtering, *IEEE Trans. on Neural Networks*, Vol. 12, 228–235, 2001.

De Nicolao, G., and Ferrari Trecate, G., Regularization Networks for Inverse Problems: A State-Space Approach, *Automatica*, Vol. 39, 669–676, 2003.

De Nicolao, G., Pillonetto, G., Chierici, M., and Cobelli, C., Efficient Nonparametric Population Modeling for Large Data Sets, in *Proceedings of 2007 American Control Conference*, pp. 2921–2926, New York, July 11–13, 2007.

De Nicolao, G., and Pillonetto, G., A New Kernel-Based Approach for System Identification, in *Proceedings of 2008 American Control Conference*, Seattle, WA, USA, June 11–13, 2008.

Donoho, D., Johnstone, I. M., Hoch, J., and Stern, A., Maximum Entropy and the Nearly Black Object, *J. R. Statist. Soc., Ser. B*, Vol. 54, 41–81, 1992.

Fattinger, K. E., and Verotta, D., A Nonparametric Subject-Specific Population Method for Deconvolution: I. Description, Internal Validation, and Real Data Examples, *J. Pharmacokinet. Biopharm.*, Vol. 23, 581–610, 1995.

Hall, P., and Titterington, D. M., Common Structure of Techniques for Choosing Smoothing Parameters in Regression Problems, *J. Roy. Statist. Soc., Ser. B*, Vol. 49, 184–198, 1987.

Hanke, M., and Hansen, P. C., Regularization Methods for Large Scale Problems, *Surv. Math. Ind.*, Vol. 3, 253–315, 1993.

Hansen, P. C., The Truncated SVD as a Method for Regularization, *Numerical Mathematics*, Vol. 27, 534–553, 1987.

Hansen, P. C., and O'Leary, D. P., The Use of The L-curve in the Regularization of Discrete Ill-Posed Problems, *SIAM J. Sci. Comp.*, Vol. 14, 1487–1503, 1993.

Hunt, B. R., Biased Estimation for Nonparametric Identification of Linear Systems, *Math. Biosci.*, Vol. 10, 215–237, 1971.

Johnson, T. D., Bayesian Deconvolution Analysis of Pulsatile Hormone Concentration Profiles, *Biometrics*, Vol. 59, 650–60, 2003.

MacKay, D. J. C., Bayesian Interpolation, *Neural Comp.*, Vol. 4, 415–447, 1992.

Magni P., Bellazzi, R., and De Nicolao, G., Bayesian Function Learning Using MCMC Methods, *IEEE Trans. Pattern Analysis and Machine Intelligence*, Vol. 20, 1319–1331, 1998.

McNally, J. G., Karpova, T., Cooper, J., and Conchello, J. A., Three-Dimensional Imaging by Deconvolution Microscopy, *Methods*, Vol. 19, 373–385, 1999.

Neve, M., De Nicolao, G., and Marchesi, L., Nonparametric Identification of Population Models via Gaussian Processes, *Automatica*, Vol. 43, 1134–1144, 2007.

Neve, M., De Nicolao, G., and Marchesi, L., Nonparametric Identification of Population Models: An MCMC Approach, *IEEE Trans. on Biomed. Eng.*, Vol. 1, 41–50, 2008.

Ostergaard, L., Cerebral Perfusion Imaging by Bolus Tracking, *Magn. Reson. Imaging*, Vol. 15, 3–9, 2004.

Phillips, D. L., A Technique for the Numerical Solution of Certain Integral Equations of the First Kind, *J. Ass. Comput. Mach.*, Vol. 9, 97–101, 1962.

Pillonetto, G., and Bell, B., Bayes and Empirical Bayes Semi-Blind Deconvolution Using Eigenfunctions of a Prior Covariance, *Automatica*, Vol. 43, No. 10, 1698–1712, 2007.

Pillonetto, G., and Bell, B., Bayesian Deconvolution of Functions in Reproducing Kernel Hilbert Spaces Using MCMC Techniques, in *Proceedings of the 21st IFIP TC 7 Conference on System Modeling and Optimization*, Sophia Antipolis, France, July 21–25, 2003.

Pillonetto, G., Dinuzzo, F., and De Nicolao, G., Bayesian Online Multi-Task Learning Using Regularization Networks, in *Proceedings of 2008 American Control Conference*, Seattle, WA, USA, June 11–13, 2008.

Pillonetto, G., Sparacino, G., and Cobelli, C., Handling Non-Negativity in Deconvolution of Physiological Signals: A Nonlinear Stochastic Approach, *Ann. Biomed. Eng.*, Vol. 30, 1077–1087, 2002.

Sparacino, G., and Cobelli, C., A Stochastic Deconvolution Method to Reconstruct Insulin Secretion Rate After a Glucose Stimulus, *IEEE Trans. on Biom. Eng.*, Vol. 43, 512–529, 1996.

Sparacino, G., Bonadonna, R., Steinberg, H., Baron, A., and Cobelli, C., Estimation of Organ Transport Function from Recirculating Indicator Dilution Curves, *Ann. Biomed. Eng.*, Vol. 26, 128–135, 1998.

Sparacino, G., De Nicolao, G., and Cobelli, C., Deconvolution, in *Modeling Methodology for Physiology and Medicine (Biomedical Engineering Series)*, Carson, E., and Cobelli, C. (Eds.), pp. 45–76, Academic Press, San Diego, USA, 2001.

Sparacino, G., Pillonetto, G., Capello, M., De Nicolao, G., and Cobelli, C., Winstodec: A Stochastic Deconvolution Interactive Program for Physiological And Pharmacokinetic Systems, *Comp. Meth. Prog. in Biomed.*, Vol. 67, 67–77, 2002.

Tikhonov, A. N., Solution of Incorrectly Formulated Problems and the Regularization Method, *Soviet. Math. Dokl.*, Vol. 4, 1624., 1963

Twomey, S., The Application of Numerical Filtering to the Solution of Integral

Equations of the First Kind Encountered in Indirect Sensing Measurements, *J. Franklin Inst.*, Vol. 279, 95–109, 1965.

Van der Sluis, A., and Van der Vorst, H. A., SIRT and CG Type Methods for Iterative Solutions of Sparse Linear Least-Squares Problems, *Lin. Alg. Appl.*, Vol. 130, 257–302, 1990.

Veldhuis, J. D., Novel Modalities for Appraising Individual and Coordinate Pulsatile Hormone Secretion: The Paradigm of Luteinizing Hormone and Testosterone Release in the Aging Male, *Mol. Psychiatry,* Vol. 2, 70–80, 1997.

Veng-Pedersen, P., Noncompartmentally-Based Pharmacokinetic Modeling, *Adv. Drug Deliv. Rev.*, Vol. 11, 265–300, 2001.

Verotta, D., Estimation and Model Selection in Constrained Deconvolution, *Annals of Biomedical Eng.*, Vol. 21, 605–620, 1993.

Wahba, G., *Splines Models for Observational Data*, CBMS-NFS Regional Conference Series, SIAM, Philadelphia, 1990.

Zanderigo F., Bertoldo, A., Peruzzo, D., Pillonetto, G., Cosottini, M., and Cobelli, C., Assessment on Clinical Data of Nonlinear Stochastic Deconvolution Versus SVD and Block-Circulant SVD Methods for Quantitative DSC-MRI, in *Proceedings of the 14th Scientific Meeting and Exhibition of the International Society for Magnetic Resonance in Medicine*, Berlin, Germany, 19–25 May, 2007.

PART IV

TIME-FREQUENCY, TIME-SCALE, AND WAVELET ANALYSIS

Thispageintentionallyleftblank

CHAPTER 9

LINEAR TIME-FREQUENCY REPRESENTATION

Maurizio Varanini

9.1 INTRODUCTION

The aim of signal analysis is to extract interesting information from the mixture of components that form signal variability. This extraction can be carried out by looking for a representation of the signal in which these components are well separated.

Let us suppose we are observing a landscape of mountains in the background from a position in which we cannot distinguish the ranges since we are observing only the projection on a plane. When the mountain ranges are parallel to each other and orthogonal to the direction of our view, the best position for distinguishing the ranges will be along the axis of the valley in between, that is, on a direction perpendicular to the previous one. However, when we are looking at a set of more or less isolated hills, the best viewpoint will be the bird's-eye view.

Applying this concept to our signal, consider, for example, an electroencephalographic signal affected by noise at the frequency of 60 Hz. Although, in the time-domain view, both components—information and noise—are overlapping and not separable, the signal decomposition in sinusoidal oscillations can isolate information of interest, as the basic elements involved in the components of EEG origin are disjoint from those involved in the disturbance component. Full duality is seen in the case of the electrocardiograph signal when we are interested in studying the ventricular depolarization whose manifestation on the ECG consists of the Q, R, and S waves (QRS complex). Now the information is concentrated in the time domain view of the QRS complex, while the other changes (T and P waves and baseline movements), as related to other phenomena, are considered noise. So it is in the domain of time that exists the separation between information

Advanced Methods of Biomedical Signal Processing. Edited by S. Cerutti and C. Marchesi
Copyright © 2011 the Institute of Electrical and Electronics Engineers, Inc.

and noise, whereas, in sinusoidal wave decomposition, the two sets that form the QRS and other signal variations are superimposed. These two examples are extreme cases. In practice, even if information and noise are of the oscillatory kind so that we can think about their characterization in frequency, the unknown systems are often not stationary and, thus, the basic elements in which information and noise can be decomposed change over time. It is, therefore, necessary to account simultaneously for both time and frequency domains and then observe our mountains from an airplane. However, this analogy does not fit perfectly. In the case of a signal, in fact, we have only the time and the other dimensions must be determined. The decision to decompose the signal into sinusoidal oscillations leads to the frequency dimension and to a view on a plane orthogonal to the time dimension. But in the case of oscillatory components varying in time, from the frequency point of view we can have overlap between information and noise. The decomposition into stationary oscillatory components cannot be suitable to represent the physical reality that is the basis of the signal. In fact, although the feature of many physical systems is to oscillate, the amplitude and frequency of their oscillations can only rarely be considered constant over time. Then, our basic elements should, indeed, be of an oscillatory nature but of limited duration. We will see that there are different possibilities to select combinations of these basic functions into which to decompose the signal. Usually, the decomposition of concern will be that more close to physical reality of the problem and should take into account the particular purpose that we want to accomplish by analysis of the signal. Consider, for example, the signal

$$(1 + a e^{j\omega_1 t}) e^{j\omega_0 t} = e^{j\omega_0 t} + a e^{j(\omega_0 + \omega_1)t}$$

with $\omega_1 < \omega_0$ and $a < 1$. It can be seen as an oscillation at frequency ω_0, modulated in amplitude (and phase) at frequency ω_1 if we think of it as generated by an oscillator with inconstant amplitude, or it can be considered as the sum of two components, one at frequency ω_0 and the other at frequency $\omega_0 + \omega_1$ when we are inclined to imagine two oscillators at two stable frequencies close together. In the first case, the decomposition of the signal requires a good resolution in time; in the second, a good resolution in frequency.

More generally, one way to obtain a significant representation for the separation of the components of interest (information) from the others (noise), is to decompose the signal into a set of basic signals, so that the information and noise are made up of elements belonging to distinct subsets. It is clear that this will be possible when the basic signals are close to the systems that create the physical components of information and noise. The decomposition of a signal as the sum of sinusoidal functions is particularly important for two reasons: (1) many physical systems are characterized by oscillations, and (2) sinusoidal functions are eigenfunctions of linear systems. In fact, although in nature no systems are strictly linear, we may consider them linear by approximating the behavior of the systems for small variations in the neighborhood of stable points. It is, therefore, natural to approximate real signals as the sum of sinusoidal waves rather than as the sum of rectangular waves. In addition, the oscillation produced by a system characterizes in a certain way the system itself; hence, the decomposition of a signal into a sum of oscillations is a first step to separate information in different systems.

The present chapter is a short tutorial on linear time-frequency representation (TFR) and is organized as follows. The first two sections introduce the short-time Fourier transform (STFT), the concept of time-frequency resolution, and the uncertainty principle (Heisenberg inequality). The following three sections deal with the multiresolution approach, the wavelets transform, and a generalization of STFT. A separate section then highlights the connections between discrete wavelet transforms and filter banks, and another briefly describes signal decomposition with a redundant dictionary of functions by the matching pursuit algorithm. Finally, some examples of application of TFR to biomedical signals are described, accompanied by some useful references to biomedical applications in the literature.

9.2   THE SHORT-TIME FOURIER TRANSFORM

The Fourier series for periodic signals and, more generally, the Fourier transform (FT) decomposes a signal into sinusoidal components invariant over time. Considering a signal $x(t)$, its Fourier transform is

$$FT_x(f) = \int_{-\infty}^{\infty} x(t) e^{-j2\pi f t} dt \qquad (9.1)$$

The amplitude of the complex value $FT_x(f)$ represents the strength of the oscillatory component at frequency f contained in the signal $x(t)$; however, no information is given on the time localization of such component.

The STFT (Allen and Rabiner, 1977; Portnoff, 1980; Crochiere and Rabiner, 1983) introduces a temporal dependence, applying the FT not to all of the signal but to the portion of it contained in an interval moving in the time

$$STFT_{x,w}(t,f) = \int_{-\infty}^{\infty} x(\tau) w^*(\tau - t) e^{-j2\pi f \tau} d\tau \qquad (9.2)$$

At each time instant t, we get a spectral decomposition obtained by applying the FT to the portion of signal $x(\tau)$ viewed through the window

$w^*(\tau - t)$ centered at the time t (Figure 9.1). This $w(\tau)$ is a function of limited duration, such as to select the signal belonging to an analysis interval centered around the time t and deleting parts outside the window. The STFT is, therefore, made up of those spectral components relative to a portion of the signal around the time instant t.

In order to preserve energy and to get the energy distribution in the time-frequency plane, the window $w^*(\tau - t)$ should be normalized to unitary energy. The STFT is a linear operator with properties similar to those of the FT:

- Invariance for time shifting apart from the phase factor:

$$\tilde{x}(t) = x(t - t_0) \quad \Rightarrow \quad STFT_{\tilde{x},w}(t,f) = STFT_{x,w}(t - t_0, f)e^{-j2\pi t_0 f}$$

- Invariance for frequency shifting:

$$\tilde{x}(t) = x(t)e^{j2\pi f_0 t} \quad \Rightarrow \quad STFT_{\tilde{x},w}(t,f) = STFT_{x,w}(t, f - f_0)$$

The expression (Equation 9.2) of the STFT can be interpreted as a convolution and then as the output of a filter (Hlawatsch and Boudreaux-Bartels, 1992). In particular, we can consider the STFT as frequency shifting the signal $x(t)$ by $-f$, followed by a low-pass filtering given by convolution with the function $w(-t)$:

$$STFT_{x,w}(t,f) = \int_{-\infty}^{\infty} \left[x(\tau)e^{-j2\pi f\tau} \right] w(\tau - t)d\tau \qquad (9.3)$$

(See Figure 9.2.) Otherwise, the STFT can be considered as a band-pass filter, filtering the signal $x(t)$ around the frequency f, obtained by convolution with the function $w(-t)e^{j2\pi ft}$, followed by a shift in frequency by $-f$ (Figure 9.3):

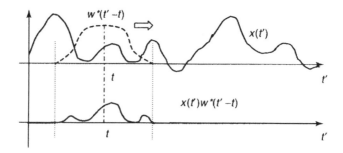

Figure 9.1. The moving analysis window $w(t)$ selects a part of signal around the analysis time t. The STFT is the FT of this selected signal.

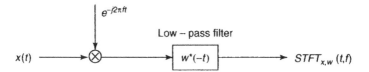

Figure 9.2. Low-pass interpretation of the STFT.

$$STFT_{x,w}\left(t,f\right)=e^{-j2\pi ft}\int_{-\infty}^{\infty}x\left(\tau\right)\left[w\left(\tau-t\right)e^{-j2\pi f(\tau-t)}\right]d\tau \qquad (9.4)$$

It should be noted that the filter impulse response is merely given by the window function modulated at the frequency f.

In addition, the convolution between $x(t)$ and $w(-t)e^{j2\pi ft}$ can be written as an inverse transform of the product $X(v)W^*(v-f)$, where $W(f)$ is the transform of the window function $w(t)$:

$$STFT_{x,w}\left(t,f\right)=e^{-j2\pi tf}\int_{-\infty}^{\infty}X\left(v\right)W*\left(v-f\right)e^{j2\pi tv}dv \qquad (9.5)$$

This expression reinforces the interpretation of the STFT as a filter bank. Indeed, the product $X(v)W^*(v-f)$ represents the transform of the output of a filter with a frequency response given by $W^*(v-f)$, which is a band-pass filter centered at frequency f, obtained by shifting the frequency of the response of the low-pass filter $W(v)$ (Figure 9.4).

The STFT provides a "local spectrum" at time t by selecting an interval of the signal around the time instant t with the analysis window $w(t)$; therefore, in order to get a good resolution in time, a narrow window $w(t)$ is required. On the other hand, the interpretation of STFT as a filter bank highlights that a good resolution in frequency requires a narrow band. These two requirements are the antithetical as the bandwidth is given by the extent of $W(f)$, which is related to the time window by

$$\Im[w(at)]=\frac{1}{|a|}W\left(\frac{1}{a}\right)$$

Hence, the use of a parameter $a > 1$ leads to a narrower time window $w(t)$, but the transformed $W(f)$ becomes more extensive.

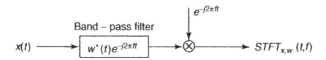

Figure 9.3. Band-pass interpretation of the STFT.

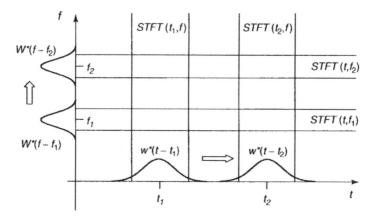

Figure 9.4. Schematic representation of the STFT in the time-frequency plane. At each analysis time t_i, the $STFT(t_i, f)$ is the FT of the signal interval selected by the window $w^*(t - t_i)$, whereas at each frequency f_i, the $STFT(t, f_i)$ is the inverse Fourier transform of the signal spectrum selected by the window $W^*(f - f_i)$, except the phase factor $\exp(-2\pi f_i t)$.

Introducing the function $h_{t,f}(\tau) = w(\tau - t)e^{j2\pi f\tau}$, the STFT can be seen as a decomposition of the signal $x(t)$ using the basis functions $h_{t,f}(\tau)$:

$$STFT_{x,w}(t,f) = \int_{-\infty}^{\infty} x(\tau) h^*_{t,f}(\tau) d\tau \qquad (9.6)$$

Then the STFT can be interpreted as a projection of the signal on a family of functions that are all derived from $w(t)$ through time and frequency shifting.

Figure 9.5 shows the function $h_{t,f}(\tau)$ and the module of its transform for four different frequencies. Note that with varying frequency, the duration of the time window is constant, the number of oscillations it includes increases, and the amplitude spectrum shifts.

The signal $x(t)$ can be reconstructed with the following formula:

$$x(t) = c \int_{-\infty}^{\infty} \int_{-\infty}^{\infty} STFT_{x,w}(\tau, v) g(t - \tau) e^{j2\pi v t} d\tau dv \qquad (9.7)$$

where there c is a normalization constant and the synthesis window $g(t)$ is any function that satisfies the condition $\int_{-\infty}^{\infty} h(t) g^*(t) dt$.

In particular, we can choose the synthesis window equal to that of the analysis $g(t) = h(t)$. Other relevant choices for the synthesis window are $g(t) = 1$ or $g(t) = \delta(t)$, as they lead to a simple synthesis formula (Equation 9.7).

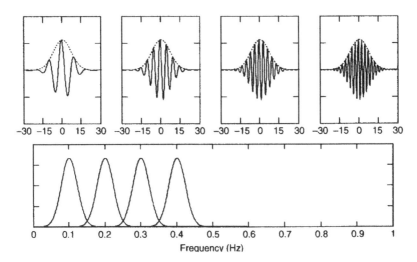

Figure 9.5. Basis functions of the STFT computed for four different frequency values (f = 0.1, 0.2, 0.3, and 0.4 Hz). On the top, the waveforms in the time domain; on the bottom, the respective amplitude spectra. In both the time and the frequency domains, the wave width does not change with the frequency.

The continuous STFT is extremely redundant. The discrete version of STFT can be obtained by discretizing the time-frequency plane with a grid of equally spaced points (nT, k/NT), where $1/T$ is the sampling frequency, N is the number of samples, and n and k are integers (Crochiere and Rabiner, 1983).

9.3 TIME-FREQUENCY RESOLUTION

The STFT is the local spectrum of the signal around the analysis time t. To get a good resolution in time, analysis windows of short duration should be used, that is, the function $w(t)$ should be concentrated in time. However, to get a good resolution in frequency, it is necessary to have a filter with a narrow band, that is, $W(f)$ must be concentrated in frequency.

We can define different measures of concentration (or dispersion) for a given function. In what follows, we consider as a measure of extension in time Δt and in frequency Δf the root mean square values (rms) or effective duration and effective bandwidth (Rioul and Vetterli, 1992). Given a function $w(t)$ with spectrum $W(f)$, without losing generality, we execute a shifting of the axes so that the center of gravity of $|w(t)|^2$ and $|W(f)|^2$ is zero, that is,

$$\overline{t} = \frac{1}{E_w} \int_{-\infty}^{\infty} t |w(t)|^2 \, dt = 0$$

and

$$\overline{f} = \frac{1}{E_w} \int_{-\infty}^{\infty} f |W(f)|^2 \, df = 0$$

where E_w is the energy of $w(t)$, given by

$$E_w = \int_{-\infty}^{\infty} |w(t)|^2 \, dt = \int_{-\infty}^{\infty} |W(f)|^2 \, df$$

It should be noted that for real oscillating functions $w(t)$, the previous expression of center of gravity should not involve the mirror image to negative frequencies of the function $|W(f)|^2$ to get the appropriate shift in the center of gravity of the single $|W(f)|^2$.

The measures of width in time and in frequency of function are defined according to the moments of inertia of $|w(t)|^2$ and $|W(f)|^2$:

$$\Delta t^2 = \frac{1}{E_w} \int_{-\infty}^{\infty} t^2 |w(t)|^2 \, dt$$

$$\Delta f^2 = \frac{1}{E_w} \int_{-\infty}^{\infty} f^2 |W(f)|^2 \, df$$

It should be noted that the values of duration Δt and width Δf are calculated from the squared module of $w(t)$ and $W(f)$, and, thus, are also defined for complex functions or oscillating functions with zero mean value.

Starting from these two measures, a global index of time-frequency resolution is the product $\Delta t \Delta f$. It may prove true the inequality

$$\Delta t \, \Delta f \geq \frac{1}{4\pi} \tag{9.8}$$

The lower limit is reached only by $w(t)$ functions of Gaussian type. This inequality is often referred as the Heisenberg uncertainty principle and it highlights that the frequency resolution Δf can be improved only at the expense of time resolution Δt and vice versa.

It should be noted that in the STFT, the window does not depend on t and f, and the values of resolution Δt and Δf are constant over the whole time-frequency plane.

There are other definitions of window duration and bandwidth such as the equivalent width or the half-width; these values are larger than the previously defined Δt and Δf.

For example, the choice of equivalent width to define Δt_e and Δf_e leads to the following condition:

$$\Delta t_e \, \Delta f_e \geq 1 \qquad (9.9)$$

9.4 MULTIRESOLUTION ANALYSIS

The use of an observation window constant with the frequency is somewhat unnatural. The concept of frequency, in fact, is associated with the concept of oscillation, and to be able to affirm that an oscillation at a certain frequency exists with reasonable accuracy and certainty, it is necessary to observe a phenomenon for a time interval including one or more periods. For example, if we are interested in detecting a circadian rhythm, we must observe a signal for a time length of one or more days, but if we are interested in quick changes such as power line interference (60 Hz), it is sufficient to have an observation window length on the order of tenths of a second. Besides, the frequency resolution is, many times, more significant not as an absolute measure but as a measure that is related to the frequency of interest. Considering the previous example, we can accept a frequency resolution Δf on the order of 0.1 Hz in power line frequency measurement, but it is necessary to have a Δf of the order of 10^{-5} Hz to distinguish a circadian rhythm from one with a period of 2 days. Moreover, it is easier to find in nature systems, such as the peripheral auditory system, that behave like a set of filters with constant factor of quality $Q = f/\Delta f$ rather than a set with constant frequency resolution. Older analog spectrum analyzers were often based on resonant circuits that behaved like a bank of bandpass filters with a factor $Q = f/\Delta f$ about constant, partially due to the difficulty of realizing filters with high Q. In digital signal processing, the discovery and performance of the fast Fourier transform (FFT) algorithm have compelled us to use constant Δf analysis. When the signal is nonstationary and contains a wide range of frequencies, the different resolution needs are satisfied calculating various STFTs, each one using analysis windows of different extent.

Only in some sectors such as engineering (electronics and telecommunications) have constant-Q filter banks continued to be used by implementing them digitally. The most common analysis algorithm foresees that the whole range of frequencies will be divided into two bands using two filters, a low-pass one and a high-pass one; this procedure is iterated on the low-frequency band. At each step, the signal is divided into two and yields half of the samples. Given that filters are not ideal, if they are not carefully designed, there are phenomena that can produce aliasing artifacts in the process of reconstruction.

This problem was solved in the mid 1970s with the discovery of the quadrature mirror filter (QMF), in which perfect cancellation of aliasing occurs. It was later on, in 1983, that the conjugate quadrature filter (CQF), couples of conjugate filters that allow an exact signal reconstruction, was discovered. In the following years, new structures were defined to implement filtering simultaneously, avoiding the procedure previously described in which the low frequency band is repeatedly divided into two and filtered.

On the other hand, in 1968–1971 Gambardella (Gambardella, 1971) generalized the STFT, introducing a frequency dependence in the window function. In particular, this generalization uses window functions that are scaled versions of each other and leads to a constant Q transform, reproducing the behavior of analog analyzers. This multiresolution approach leads to spectrum invariance for time-scale changes. Realizing the importance of this approach, Gambardella wrote in the conclusions of his paper, "This class of analyzers seems to have quite interesting properties, from both the physical and the mathematical points of view (...), and perhaps they deserve greater attention than they have received, at least theoretically, up to now." However, it would another twenty years before constant Q analysis received that attention and became popular in the scientific community under the name of "wavelet transform."

In fact, it was only in 1984 that Grossman and Morlet (Grossmann and Morlet, 1984) reintroduced the basic idea of constant Q analysis decomposing a signal into basic functions (wavelets) obtained from a cosine, windowed with a Gaussian function, and compressed or expanded over time at different scales. A few years later, Meyer (Meyer, 1987, 1992) started building a mathematical framework around the wavelet idea, and Daubechies (Daubechies, 1988, 1992), who was looking for orthonormal bases for wavelet transforms, found them by iterating CQF banks. Vetterli and Herley (1992) integrated the wavelet and filter bank theory, and Mallat (1989) derived wavelets from the concept of multiresolution analysis used in computer vision.

The well-known inadequacy of the constant resolution analysis and the need to overcome these Fourier-based analysis drawbacks led to many different ideas across the disciplines (electrical engineering: filter banks, subband coding, and pyramidal algorithms; physics: Littlewood–Paley theory; mathematics: Calderon formula), setting the ground for the constant-percentage-resolution approach. Therefore, although wavelet theory is a combination of these techniques, it has the merit of having integrated and structured them, providing an instrument even more powerful than the original ones.

9.5 WAVELET TRANSFORM

In the Wavelet transform (WT), the operation of frequency shifting characteristic of STFT is replaced by a change of scale over time (Rioul and

Vetterli, 1991; Daubachies, 1992; Flandrin, 1999). Indeed, given the function $h(\tau)$ (wavelet mother), the basic elements of the transform are obtained through changes of scale:

$$h_{t,a}(\tau)=\frac{1}{\sqrt{|a|}}h\left(\frac{\tau-t}{a}\right) \qquad (9.10)$$

where $a \in \Re$ is a scale factor and the term $1/\sqrt{|a|}$ is introduced to normalize the energy of the different wavelets.

The (continuous) wavelet transform is therefore defined as

$$WT_x(t,a)=\int_{-\infty}^{\infty} x(\tau)h_{t,a}{}^{*}(\tau)d\tau \qquad (9.11)$$

The WT, at a given time instant t and at a certain scale a, is the inner product between the signal $x(\tau)$ and wavelet $h_{t,a}(\tau)$ with scale a centered at the time t. Therefore, the WT indicates how the close signal $x(\tau)$ is to the function $h_{t,a}(\tau)$ and, as the $h_{t,a}(\tau)$ will be more or less concentrated in time around t, this similarity criterion will take into account the behavior of the signal $x(\tau)$ in a neighborhood of the time t.

The reconstruction formula of $x(t)$ is the sum of all orthogonal projections of the signal on the wavelet functions and results in

$$x(t)=C_h \int_{0}^{\infty}\int_{-\infty}^{\infty} WT_x(\tau,a)h_{\tau,a}(t)\frac{1}{a^2}d\tau da \qquad (9.12)$$

where constant C_h depends only on function h.

Although continuous functions $h_{a,\tau}(t)$ are very redundant and, therefore, not orthogonal, the inverse transformation is valid if the function $h(t)$ has finite energy (integrated in the square) and realizes a band-pass filter (zero mean) (Rioul and Vetterli, 1991; Daubachies, 1992; Flandrin, 1999):

$$\int_{-\infty}^{\infty}|h(t)|^2\, dt < \infty$$

and

$$\int_{-\infty}^{\infty} h(t)dt = 0 \qquad (9.13)$$

The latter condition is more difficult to satisfy and implies that $h(t)$ will fluctuate over time as a small wave, from which comes the name "wavelet." As we have seen for STFT, the function of synthesis, used in the reconstruction, may be different from that of analysis.

It should be remarked that underlying WT there is the concept of scale, whereas for STFT there is that of frequency. When we change the scale to a function,

$$f(t) \rightarrow f(at) \quad \text{with } a > 0$$

the function shrinks if $a > 1$ and expands if $a < 1$.

The WT can be written as

$$WT_x(t,a) = \frac{1}{\sqrt{a}} \int_{-\infty}^{\infty} x(\tau) h * \left(\frac{\tau - t}{a} \right) d\tau \qquad (9.14)$$

Highlighting that with increasing scale (growing a), the function h becomes wider over time and WT takes account of the slow behavior of signal $x(t)$. Then the WT behaves as a filter bank with impulse response h whose extension increases with a.

On the other hand, the following expression of WT,

$$WT_x(t,a) = \sqrt{a} \int_{-\infty}^{\infty} x(a\tau) h * \left(\tau - \frac{t}{a} \right) d\tau \qquad (9.15)$$

emphasizes that, as the scale grows, the signal is compressed, and then a wider interval is covered by the function h, which now remains constant in width. This way, ever more compressed signals $x(t)$ are filtered by the same filter with fixed impulse response h. The scale factor can be interpreted as the scale of a geographical map: large scale provides an overview, whereas small scale allows highlighting of details.

As the characterization of physical phenomena is very important to the concept of frequency, wavelets are often chosen to be well characterized in frequency and, therefore, they are, indeed, amplitude-modulated sinusoids. In applications, the most frequently used is the generalized wavelet of Morlet, which is a complex exponential modulated by a Gaussian:

$$h(t) = e^{-\frac{t^2}{2}} e^{j2\pi f_0 t} \qquad (9.16)$$

Though the WT is defined in terms of scale, it can be seen in terms of frequency choosing for parameter a the expression

$$a = \frac{f_0}{f} \qquad (9.17)$$

where f_0 is the center frequency of the spectrum Fourier of $h(t)$ with $f > 0$.

Then we obtain the WT defined in terms of time and frequency (Hlawatsch and Boudreaux-Bartels, 1992):

$$WT_x^h(t,f) = \sqrt{\left|\frac{f}{f_0}\right|} \int_{-\infty}^{\infty} x(\tau) h^* \left(\frac{f}{f_0}(\tau - t)\right) d\tau \qquad (9.18)$$

Figures 9.5 and 9.6 show the difference between the basis functions of STFT and those of WT. Whereas the former have constant width and, therefore, their amplitude spectrum is only shifted in frequency, the latter are scaled versions of each other, so with increasing frequency, the time width decreases and the spectrum widens.

The WT is a linear operator that maintains shifts in time and changes of scale (but not the frequency shifting):

$$\tilde{x}(t) = x(t - t_0) \quad \Rightarrow \quad WT_{\tilde{x},h}(t,f) = WT_{x,h}(t - t_0, f) \qquad (9.19)$$

$$\tilde{x}(t) = \sqrt{|a_0|} x(a_0 t) \quad \Rightarrow \quad WT_{\tilde{x},h}(t,f) = WT_{x,h}(a_0 t, f/a_0) \qquad (9.20)$$

The WT is the convolution between the signal $x(t)$ and the wavelet $h_a(-t)$ with

$$h_a(f) = \frac{1}{\sqrt{|a|}} h\left(\frac{t}{a}\right)$$

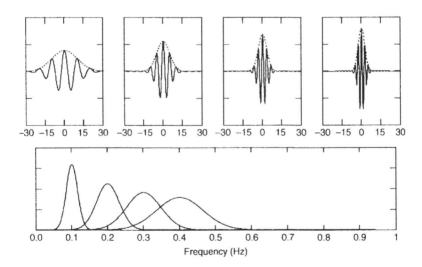

Figure 9.6. Morlet wavelet for four different values of scale ($f_0 = 1$ Hz; $f = 0.1$, 0.2, 0.3, and 0.4 Hz). On the top, the waveforms in the time domain; on the bottom, the respective amplitude spectra. With the increase of frequency, the wavelet becomes narrower in time but wider in frequency.

Then, as seen for the STFT, the WT can be written as the inverse transform of the product $X(f) \cdot \sqrt{|a|} H^*(af)$, where $H(f)$ is the transform of the wavelet $h(t)$:

$$WT_{x,h}(t,a) = \sqrt{|a|} \int_{-\infty}^{\infty} X(v) H^*(av) e^{j2\pi tv} dv \qquad (9.21)$$

This expression shows that the WT can be interpreted as signal filtering with a filter bank characterized by the scale parameter a.

The resolution of the WT depends on choice of the wavelet function and can be evaluated using measures Δt and Δf previously seen.

Let Δt_m and Δf_m be the RMS temporal extension and the RMS bandwidth of wavelet mother $h(t)$ ($a = 1$), for the generic function $h_a(t)$, then

$$\Delta t = a \, \Delta t_m$$

and

$$\Delta f = \frac{\Delta f_m}{a} \qquad (9.22)$$

Then product $\Delta t \Delta f$ is constant (changing the scale) and equal to $\Delta t_m \Delta f_m$ with the lower limit still represented by $\Delta t \Delta f \geq (1/4\pi)$.

The wavelets with Gaussian envelope, and then the Morlet, get the minimum value of product $\Delta t \Delta f$.

Figure 9.7 schematically shows the distribution of resolution in the time-frequency plane, on the left for the STFT and on the right for the WT; every point (t, f) is associated with an area that represents the concentration of the analysis function. The size of each cell is defined by the two intervals: $(t - \Delta t/2, t + \Delta t/2)$ and $(f - \Delta f/2, f + \Delta f/2)$. In the case of the STFT, these are constant throughout all the planes, whereas for the WT these dimensions vary with the frequency, which, in turn, is linked to the scale. The area is obviously in both cases constant and equal to $\Delta t \Delta f$.

Many scientific software packages contain routines for time-frequency signal representation. A complete Octave (and Matlab) package for time-frequency signal analysis was developed by F. Auger, P. Flandrin, P. Goncalves, and O. Lemoine. The software, with its manual and tutorial, can be freely downloaded at the link http://tftb.nongnu.org/.

Another large collection of routines (Octave, Matlab, and Scilab) for computing time-frequency representations can be downloaded at http://tfd.sourceforge.net/ (author, J. C. O'Neill).

The site http://www.wavelet.org/ is a good starting point to browse through the large research community involved in theory and applications of wavelets.

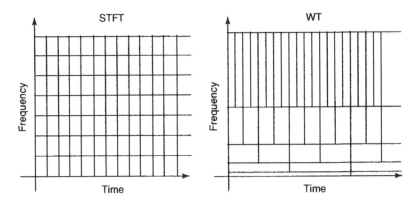

Figure 9.7. Tiling of the time-frequency plane: the plane coverage for the STFT (left) and for the WT (right). Each cell is drawn with width proportional to the time resolution Δt and height proportional to the frequency resolution Δf.

9.6   A GENERALIZATION OF THE SHORT-TIME FOURIER TRANSFORM

In many applications, the distribution of the resolution in the time-frequency plane obtained by constant Q analysis is not fully satisfactory; in fact, a Q factor that provides the temporal resolution required in the middle–low frequencies can produce a nonacceptable frequency resolution at high frequencies. In general, at different frequencies, it would be desirable to have a different compromise between resolution in time and frequency without the constraints of constant resolution of STFT or constant Q of wavelets. We can approximate our resolution requirements, dividing the frequency range into subbands and applying a STFT with a different analysis window in each subband, or using wavelet packets that allow a different Q factor in each subband. These approaches are computationally efficient; however, they do not provide a homogeneous spectrum at the subband boundaries. In what follows, we show a generalization of STFT (GSTFT) (Varanini et al., 1998), which allows one, for each frequency, to get a resolution in time (or frequency) that is believed to be most suitable according to the characteristics of signals or special purposes of analysis. Specific choices lead to standard STFT or WT while, in general, realizing a compromise between the constant resolution analysis and the constant Q analysis.

The generalization requested is obtained by allowing the analysis window $w(t)$ of STFT to change with frequency. GSTFT is therefore defined as

$$G(t,f) = \int_{-\infty}^{\infty} x(\tau)w*(\tau-t,f)e^{-j2\pi f\tau}d\tau \qquad (9.23)$$

Defining $h(t,f) = w(t,f)e^{j2\pi ft}$, we can obtain the equivalent expression,

$$G(t,f) = e^{-j2\pi ft} \int_{-\infty}^{\infty} x(\tau)h*(\tau-t,f)d\tau \qquad (9.24)$$

So, as was seen for STFT and WT, the spectral component at frequency f is obtained by filtering the signal $x(t)$ with a band-pass filter with impulsive response $h*(-t,f)$. The term $e^{-j2\pi ft}$ corrects the phase, changing the reference from local to the time axis origin $t = 0$. In the STFT, the band-pass filter is always the same but shifted in frequency; in the WT, the filter is scaled (constant Q); whereas for the GSTFT, the filter is a generic function of frequency.

Similarly to STFT and WT, the GSTFT can be written as the inverse transform of the product of $X(\upsilon)$ and $W(\upsilon,f)$, Fourier transforms, respectively, of the signal $x(t)$ and of the analysis window $w(t,f)$:

$$G(t,f) = e^{-j2\pi ft} \int_{-\infty}^{\infty} X(\upsilon)W*(\upsilon-f,f)e^{j2\pi \upsilon t}d\upsilon \qquad (9.25)$$

This expression allows us to view the GSTFT as a filter bank. Although the filters have constant bandwidth in STFT and constant relative bandwidth in WT, here these constraints are not present and, at each frequency, we can get the most appropriate resolution in frequency.

A simple formula for reconstruction of the signal $x(t)$ can be obtained as follows. Let $a_w(f) = \int_{-\infty}^{\infty} w(t,f)$. Then we have

$$\int_{-\infty}^{\infty} G(t,f)dt = \int_{-\infty}^{\infty} x(\tau)a_w(f)e^{-j2\pi f\tau}d\tau = a_w(f)X(f) \qquad (9.26)$$

Therefore, expressing $x(t)$ as inverse Fourier transform, we have the formula of reconstruction of the GSTFT:

$$x(t) = \int_{-\infty}^{\infty}\int_{-\infty}^{\infty} G(\tau,f)\frac{e^{j2\pi ft}}{a_w(f)}d\tau df \qquad \text{with } a_w(f)\neq 0 \;\; \forall f \qquad (9.27)$$

It should be noted that if $w(t,f) = w(t)$, we have as a special case the reconstruction formula of STFT with the synthesis window $g(t) = 1$.

We assume now that the generic $w(t,f)$ belongs to the class of functions so defined:

$$w(t, f) = \frac{\alpha(f)}{\sqrt{s(f)}} w_m \left(\frac{t}{s(f)} \right) \tag{9.28}$$

where $s(f)$ is a generic function, defined by the user, that expresses the desired resolution in time at each frequency f; $w_m(t)$ is a mother function chosen from the classic analysis windows (Gaussian, Von Hann, triangular, etc.) and normalized so that $\Delta t(w_m) = 1$ with $\Delta t(w_m)$ resolution in time defined as root mean square duration.

It is easy to prove, by using this normalization, that the resolution in time $\Delta t(f)$ of the generic window $w(t, f)$ results the user's requested resolution $s(f)$.

The factor $\alpha(f)$ allows us to normalize the functions $w(t, f)$ to the same energy in order to get the energy distribution in the time-frequency plane or to the same area in order to get the amplitude distribution. Energy distribution can be obtained by normalizing the mother function $w_m(t)$ to unitary energy and setting $\alpha(f) = 1$, whereas the amplitude distribution can be obtained by normalizing the mother function $w_m(t)$ to the unitary area and setting $\alpha(f) = 1/\sqrt{s(f)}$.

With appropriate choices and $s(f)$ we get the STFT and WT as special cases:

- Setting $(f) = k$, we obtain the STFT.
- Setting $s(f) = f_0/f$, we obtain the WT with a phase correction term:

$$G(t, f) = e^{-j2\pi ft} WT(t, f) \tag{9.29}$$

The GSTFT is a linear operator and it preserves the time shift. In its more general form, however, it does not preserve the frequency shift as STFT, or scale changes as WT:

$$\tilde{x}(t) = x(t - t_0) \quad \Rightarrow \quad G_{\tilde{x},h}(t, f) = G_{x,h}(t - t_0, f) \tag{9.30}$$

It should be remembered that, although there is neither constraint of constant resolution as in STFT nor of constant Q as in WT, in every point of the time-frequency plane, the resolutions in time and frequency must satisfy the principle of uncertainty given by Equation 9.8.

Two classical mother windows $w_m(t)$ normalized to satisfy the conditions of unitary energy and unitary time resolution are reported as examples.

Gaussian window:

$$w_g(t) = \frac{1}{\sqrt[4]{2\pi}} e^{-\frac{t^2}{4}}$$

Von Hann window:

$$w_h(t) = \frac{1}{3d}\left(1 + \cos(\pi\, t/d\,)\right) \text{ if } -d < t < d, \text{ otherwise } w_h(t) = 0$$

with $d = 3.535$

The dependence on the frequency of time resolution $\Delta t(f)$ or of frequency resolution $\Delta f(f)$ can be chosen according to the requirements of the specific application. It is often handy to use a sigmoidal function for $\Delta t(f)$ in order to have a constant Q in the central range of frequencies and a time-frequency resolution approximately constant near the extremities, as in the following expression:

$$\Delta t(f) = \frac{1}{2}(\Delta_{\max} - \Delta_{\min})\left\{1 - \tanh[2\, p\,(f - f_p)]\right\} + \Delta_{\min} \qquad (9.31)$$

The parameters $\Delta_{\min}$ and $\Delta_{\max}$ are the minimum and maximum resolution in time. The parameters p and f_p regulate, respectively, the slope and the position of the constant Q zone. Figure 9.8 shows $\Delta t(f)$ (left) and $\Delta f(f)$ (right) as functions of frequency.

A more general behavior of the time resolution $\Delta t(f)$, which may be desired for some applications, can be obtained by defining specific frequencies of interest and their required resolutions in time (or frequency). From these couples $[f_i, \Delta t(f_i)]$, we can get to the function $\Delta t(f)$ by interpolation.

In Figure 9.9 the STFT, WT, and GSTFT transforms are compared on processing of an artificial signal. The top panel shows the signal and its components: an oscillation at frequency $f_a = 0.03$ Hz with an interruption (dead zone) in the interval $\Im = 200–333$ s, two stationary oscillations close togeth-

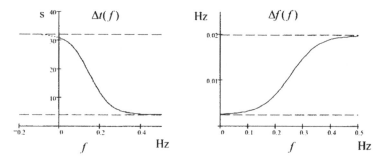

Figure 9.8. Sigmoidal function behavior: on the left, the time resolution $\Delta t(f)$; on the right, the respective frequency resolution $\Delta f(f)$ based on the hypothesis of a Gaussian window.

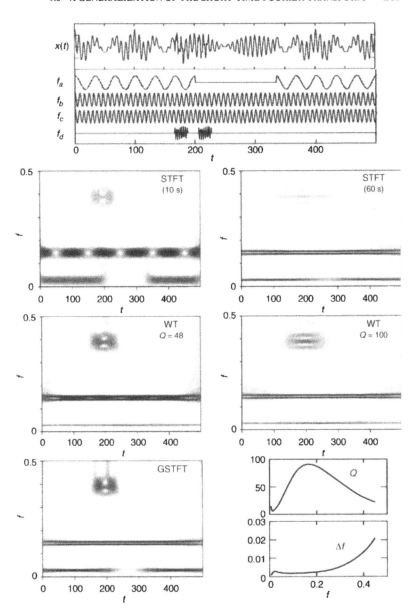

Figure 9.9. Application example of STFT, WT, and GSTFT on a synthetic signal. From top to bottom: the signal, its oscillatory components, the time-frequency representation of the amplitude spectra obtained by STFT with two different windows, that obtained by WT with two different Q, and, finally, that obtained by GSTFT with the plot of the Q factor, and Δf frequency resolution as function of frequency.

er in frequency (f_b = 0.13 Hz and f_c = 0.14 Hz), and two high-frequency bursts (f_d = 0.4 Hz) at the time 171 s and 211 s. The other panels show, in linear scale, the time-frequency representations of the amplitude of the transforms.

The STFT with a short window (10 s) separates the two bursts in time and identifies the interval 3, but does not separate the two oscillations close in frequency f_b and f_c. With a window 60 s in length, we are able to distinguish the oscillations but are unable to identify the two previous bursts. The WT with a factor Q = 48 allows us to distinguish the two bursts but not the f_b and f_c components and the interval 3. With a factor Q = 100, the WT separates f_b and f_c but does not identify events close in time. The latest figures on the bottom show the representation achieved with GSTFT and the plot of the relative Q factor and frequency resolution. These were obtained by starting from knowledge of the resolution needs at some specific frequencies, and then building the function $\Delta t(f)$ by interpolation. The resulting representation highlights that now we can both identify events close in time and separate different oscillations close in frequency.

The discrete version of GSTFT can be derived from Equation 9.24, sampling the time-frequency plane. The most natural definition follows from the resolution properties of the basic functions $w(t, f)$: the time and the frequency axes are sampled proportionally to their respective resolutions. Therefore, we build an array of frequencies $f[n]$, which are spaced proportionally to the frequency resolution $\Delta f(f)$, and a sparse matrix of time instants $t[n, m]$, whose elements, for each frequency index n, are spaced proportionally to the resolution in time $\Delta t(f)$. It yields the formula

$$G[m,n] = e^{-j2\pi f[n] t[n,m]} \sum_{k=-N/2}^{N/2-1} x[\,t[n,m]+k\,] w*(k, f[n]) e^{-j2\pi f[n]k} \quad (9.32)$$

where N denotes the width of the window and the array $x[m]$ is assumed to be extended with zeros at the beginning and end. The sum can be limited to the actual duration $2d\Delta t(f)$ of the window of frequency index n. The exponential term outside the sum can be omitted when the phase is not of interest or when using a local reference as in WT. Commonly, we are interested in relations with other uniform sampled signals, and then we use $t[n, m] = m$ that implements a uniform sampling over time. A uniform sampling in frequency can be obtained by placing $f[n] = n/N$.

When we use a high resolution in time, the frequency resolution can be very low and, hence, the wide band $W(t, f)$ may include the negative frequency of $X(f)$. In the case of real signals, this autoaliasing error can be avoided by using the analytical signal.

It is important to stress that for the discrete GSTFT there is no fast algorithm as those available for STFT and for DWT.

9.7   WAVELET TRANSFORM AND DISCRETE FILTER BANKS

The continuous versions of STFT, WT, and GSTFT are highly redundant and the formula for reconstruction is not uniquely defined. Dealing with the discrete versions, we can maintain a certain redundancy in the representation or achieve, in certain special cases, its total elimination, representing the signal on a basis of orthogonal functions. In order to ensure the reconstruction, we should nevertheless sample the time-frequency plane so that the basic function set is at least complete.

A natural way to discretize the time-frequency plane is to use a sampling step proportional to the resolution. Therefore, in STFT we can use a uniform sampling, and in WT use a sampling that at increasing frequency becomes more dense over time and sparse in frequency. Finally, in GSTFT, we can sample at instants of time and at frequencies that are chosen according to the distribution of TF resolution. However, uniform sampling over time remains the most appropriate when the signal needs to be processed or represented graphically involving other signals. In this latter case, the basis functions will be redundant and, in order to be complete, the sampling grid will have to be tight both in time and in frequency in order to be sufficiently dense in the parts of the plane of maximum resolution.

In signal analysis, a redundant basis is often used so that the resulting description can be truly invariant over time and that the TF representation provides higher detail, yielding a clearer separation between the various components forming the signal. On the other hand, nonredundant (or nearly nonredundant) bases of functions (orthogonal, semiorthogonal, or biorthogonal wavelets) are of great importance in the fields of compression and coding of the signals. Whereas in the STFT it is impossible to have orthonormal bases with well-localized functions both in time and frequency— Balian–Low theorem (Daubachies, 1992; Flandrin, 1999)—these bases exist for WT. Moreover, if the sampling in the TF plane is dense, the set of wavelet functions is more than complete and the reconstruction of the signal can be made with weak constraints on the mother wavelet $h(t)$. If, however, the sampling is minimal, we obtain an orthonormal basis and achieve a perfect reconstruction uniquely for specific choices of $h(t)$ (Daubachies, 1992).

We choose to discretize the parameters scale (a) and time (t) of WT in accordance with the respective resolutions:

$$a = a_0^j$$

$$t = k\, a_0^j\, t_0$$

We assume $a_0 > 1$ and $b_0 > 0$; j and k are integers. Then, from Equation 9.10, we get the family of wavelets:

$$h_{j,k}(\tau) = a_0^{-j/2} h(a_0^{-j}\tau - kt_0)$$

The wavelet transform becomes

$$WT_{j,k} = \int_{-\infty}^{+\infty} x(\tau) h_{j,k}^{*}(\tau) d\tau$$

The reconstruction formula is:

$$x(t) = c \sum_{j} \sum_{k} WT_{j,k} \, h_{j,k}(t)$$

where c is a constant.

For values of a_0 close to 1 and for small values of t_0, the basis of wavelet functions is redundant; therefore, the reconstruction is not unique and the restrictions on function $h(t)$ are few. On the other hand, with increasing a_0 the sampling becomes more and more sparse and for some of the values of a_0 it is possible to build an orthonormal wavelet basis, but only for specific choices of the function $h(t)$ (Daubachies, 1992).

Of particular interest is the discretization on a dyadic grid, which is obtained by placing $a_0 = 2$ and $t_0 = 1$. In this case, it is possible to build the functions $h_{j,k}(\tau)$ so that they form an orthonormal basis. These orthogonal continuous wavelets can be obtained with a procedure to the limit from the discrete wavelets of the WT for discrete time signals.

The wavelet transform is especially useful for discrete time signals and is practically identical to a decomposition based on a particular filter bank widely used in the field of signal subband coding (conjugate quadrature filter, CQF). The complexity of computing is on the order of $O(n)$ and is, therefore, lower than FFT $[O(n \log n)]$.

Figure 9.10 shows a bank of filters with two channels: on the left is the analysis part (subband decomposition) and on the right the synthesis part (reconstruction of the signal).

At each stage of analysis, a signal with lower resolution is derived by means of a low-pass filter G with cutoff frequency at half band. Then the signal is decimated by two, hence doubling the scale. Note that the change of resolution is obtained by low-pass filtering, whereas the change

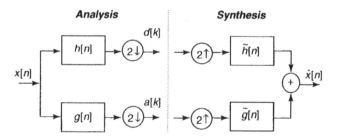

Figure 9.10. Two-channel filter bank: decomposition on the left, reconstruction on the right.

of scale is obtained by subsampling. A high-pass filter H, applied to the signal $x[n]$, selects the half band at high frequency, that is, the details lost in the previous filtering. Also in this case, it follows the decimation by two. In synthesis phase (on the right of Figure 9.10), the subband signals are interpolated and then added to reproduce the original signal. Interpolation is carried out by entering a zero between one sample and the next and by applying the reconstruction filter $\tilde{G}$ (or $\tilde{H}$).

It can be seen that although the filters are not ideal, the signal $x[n]$ can be perfectly reconstructed from the filtered and decimated sequences (Smith and Barnwell, 1984, 1986; Mintzer, 1985; Vaidyanathan, 1987). As filters G and H are not ideal and they must cover the whole range of frequencies, their frequency responses will overlap in the middle of band. The Nyquist theorem is not, therefore, verified and aliasing occurs. A requirement for perfect reconstruction is the cancellation of aliasing, which can be obtained with the condition

$$G(-z)\tilde{G}(z)+H(-z)\tilde{H}(z)=0 \qquad (9.33)$$

where Z transform notation has been used.

The global transfer function of decomposition–reconstruction is

$$T(z)=\frac{1}{2}\left[G(z)\tilde{G}(z)+H(z)\tilde{H}(z)\right] \qquad (9.34)$$

For a perfect reconstruction, that is, without amplitude and phase distortion, the $T(z)$ must introduce a simple delay:

$$T(z)=z^{-k} \qquad (9.35)$$

A choice for the two reconstruction filters $\tilde{G}$ and $\tilde{H}$ (quadrature mirror filter, QMF, Esteban and Galand, 1977) that meets the requirement of aliasing cancellation is

$$\tilde{G}(z)=H(-z) \quad\text{and}\quad \tilde{H}(z)=-G(-z) \qquad (9.36)$$

It is common to choose

$$H(z)=G(-z) \quad\Rightarrow\quad h(n)=(-1)^{n}g(n) \qquad (9.37)$$

The low-pass filter $g(n)$ is turned into a high-pass filter $h(n)$ through the modulation $(-1)^{n}$, with the transfer function $H(z)$ becoming symmetrical with respect to the center of the band.

From the condition on $T(z)$ to have no distortion, it is required that

$$G^{2}(z)-G^{2}(-z)=2z^{-k} \qquad (9.38)$$

With this choice, a linear phase filter FIR (symmetric impulse response) leads to a $T(z)$ with linear phase. The phase distortion is easily eliminated but it is not possible to completely eliminate the distortion in amplitude. More generally, there is no FIR filter simultaneously leading to the absence of distortion phase and amplitude thus satisfying the condition in Equation 9.35.

However, once we have eliminated one form of distortion we can design the filter with an optimization method and manage the minimization of the other.

Smith and Barnwell (1984, 1986) and Mintzer (1985) have demonstrated that it is possible to choose the analysis filters and synthesis filters in such a way as to have perfect reconstruction (CQF or Smith–Barnwell filters). The cancellation of aliasing (Equation 9.33) is satisfied with the choice

$$\tilde{G}(z) = -H(-z) \quad \text{and} \quad \tilde{H}(z) = G(-z) \tag{9.39}$$

and also

$$H(z) = z^{-(L-1)}G(-z^{-1}) \quad \Rightarrow \quad h(n) = (-1)^n g(L-1-n) \tag{9.40}$$

where the FIR filters of even length L are considered.

The condition on $T(z)$ to get no distortion is

$$G(z)G(z^{-1}) + G(-z)G(-z^{-1}) = 2 z^{-L-k} \tag{9.41}$$

We can check (Vetterli and Herley, 1992; Vetterli, 2001) that this condition is satisfied with a filter whose impulse response $g(n)$ is orthogonal for an even shift, that is $\langle g[n]g[n-2k]\rangle = \delta_k$, where the symbol $\langle\,\rangle$ indicates the inner product. Furthermore, the choice of the high-pass filter $h[n]$ as in Equation 9.40 causes it to be orthogonal for the even shift $\langle h[n]h[n-2k]\rangle = \delta_k$ and to be orthogonal to $g[n]$, that is $\langle g[n]h[n-2k]\rangle = 0$.

The orthonormal set $\{g[n-2k], h[n-2l]\}_{k,l\in Z}$ is also complete and, therefore, forms an orthonormal basis for $L_2(Z)$, the space of the square summable sequences.

The signal $x[n]$ is decomposed by the filter bank in two sequences, $a_1(k) = \langle x[n]g[n-2k]\rangle$ and $d_1(k) = \langle x[n]h[n-2k]\rangle$, which are obtained as orthogonal projections of $x[n]$ in the subspaces identified by the orthogonal sets $A_1 = \{g[n-2k]\}_{k\in Z}$ and $D_1 = \{h[n-2l]\}_{l\in Z}$.

The sequence $a_1[k]$ can be seen as an approximation of $x[n]$, whereas $d_1[k]$ represents the detail. This concept of analysis of a signal at different levels of resolution was formalized by Mallat (Mallat, 1989), who showed such decomposition carried out by subband filters from a mathematical point of view and formulated a theory for multiresolution analysis.

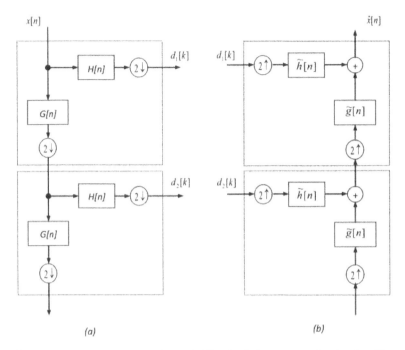

Figure 9.11. (a) A two-channel cascade filter bank decomposes the signal and implements the DWT. (b) The signal reconstruction with filter bank (IDWT).

The subband decomposition can be iterated by applying the decomposition to the sequence $a_j[n]$ on exit from a low-pass filter, as in Figure 9.11a. Therefore, given a signal $x[n]$ with $N = 2^J$ samples, at each step j with $1 \leq j \leq J$, we get a signal of detail $d_j[k]$ at the output from a high-pass filter and a signal of approximation $a_j[k]$ at the exit from low-pass filter. For both signals, a subsampling is made, so the number of samples of $d_j[k]$ and of $a_j[k]$ is half that of $a_{j-1}[n]$ in the previous step. The total number of samples is always N. At every step, the resolution is reduced to two by filtering, while the scale is increased by subsampling. In the end, we obtain the set $\{d_1, d_2, \ldots, d_J, a_J\}$ that is a multiresolution representation of the signal $x[n]$ and represents the desired decomposition, that is, the coefficients of DWT.

Each element d_j of the set has dimension 2^{J-j} and is the outcome of a band-pass filtering in the frequency interval $2^{-(j+1)} \leq f \leq 2^{-j}$ (normalized frequency).

The inverse of the discrete wavelet transform (IDWT) is equivalent to the scheme of reconstruction (filter bank) shown in Figure 9.11b. The procedure is similar to the previous one: the scale is doubled by inserting one zero between one sample and the following one (symbol $2 \uparrow$), and the reconstruction filters are applied.

For every iterative step, the wavelets of the DWT are given by the band-pass filters obtained by the cascade of low-pass filters relative to all the previous iterations followed by the high-pass filter $h[n]$.

Continuous orthogonal wavelets can be obtained by associating a stepwise linear function to discrete wavelets and then passing to the limit with the iteration of the previous procedure (Daubachies, 1992; Vetterli, 2001). Not all the banks of orthogonal filters generate continuous wavelets; the frequency response of the low-pass filter must have a sufficient number of zeros for $f = 0.5$.

So far, we have considered a particular structure of filter banks obtained by iterating a bank with two channels only on the low-pass branch. However, we can get orthogonal filter banks with more general structures when we decompose some high-pass branches also. In such cases, we speak of wavelet packets. Moreover, the filter banks can generally have a structure with $m \geq 2$ channels.

Lastly, the following observations should be made (Rioul and Vetterli, 1991; Daubachies, 1992; Vetterli and Herley, 1992; Vetterli, 2001):

1. Although every orthogonal wavelet basis is always associated with a bank of perfect reconstruction filters, the reverse is not true.
2. There is only one bank of orthogonal real filters with linear phase and it has only two coefficients different from zero.
3. Perfect reconstruction filter banks with reconstruction filters different from the analysis filters lead to the definition of bases of biorthogonal wavelets whose waveforms can be symmetrical (i.e., linear phase).

9.8   MATCHING PURSUIT

A basis of functions may not be suitable for the decomposition of a signal and, if the signal is not stationary, a basis appropriate for one signal segment may not be appropriate for the other segments. For example, for some time intervals an analysis with constant Q might be suitable, whereas other intervals require one with constant resolution.

The method of segmenting the signal into blocks and of choosing the appropriate basis for each block (best basis or wavelet packets) often produces in the reconstruction of the signal (synthesis) artifacts near the boundaries between blocks.

A more synthetic and accurate representation of the signal can be obtained by considering a set of redundant functions (which are, therefore, not orthogonal). In general, given a redundant dictionary of functions (atoms), the signal can be decomposed in several ways using different subsets. The aim is, therefore, to determine which subset of atoms of a dictionary produces the best decomposition. The simple inner product provides a measure of similarity between the signal to be analyzed and each function; however, although the similarity criterion has been fixed (mean square error), it is unthinkable to determine the best decomposition through exhaustive search. A suboptimal search method is the "matching pursuit" (Mallat and Zhang, 1993), which consists of the following iterative algorithm: once the atom dictionary that correlates most closely with the signal is chosen, the residual signal is calculated by subtracting the contribution of the atom found; the algorithm is iterated on the residual signal until its energy is higher than a given threshold. The orthogonality between the contribution and the residue allows, at each step of the algorithm, conservation of energy. The convergence is guaranteed if the dictionary is complete.

Mallat and Zhang have introduced a dictionary of Gabor functions that specializes the matching pursuit as a time-frequency transform. Each atom of the dictionary is obtained from a window function $g(t)$ by means of shifting, modulation, and change of scale:

$$g_{s,t,f}(\tau) = \frac{1}{\sqrt{s}} g\left(\frac{\tau - t}{s}\right) e^{j2\pi f \tau} \qquad (9.42)$$

where $g(t)$ is a Gaussian function normalized to unitary energy $g(\tau) = 2^{1/4} \cdot e^{-\pi \tau^2}$.

The time-frequency energy distribution of the signal can be obtained from its decomposition by summing the Wigner distributions of every atom. In any case, this dictionary—the parameters of which vary continually—has a potentially infinite dimension.

Therefore, for the execution of the MP algorithm we need to select a subset of atoms by sampling the three-dimensional space of the parameters (s, t, f). Searching in this subset produces approximate values of scale, time, and frequency, which are subsequently refined with a search in a dense interval around the found values. Each fixed a priori sampling introduces a polarization of the results that can be avoided with a randomization of parameters before the decomposition (Durka et al., 2001). The software of Mallat for the MP is available on the site ftp://cs.nyu.edu /pub/wave/software; moreover, Durka provides the software for the implementation of the MP a with stochastic dictionary for its application to the the EEG signal at http://brain.fuw.edu.pl/~durka/.

It must be observed that the form of the functions of decomposition proposal by Mallat is the same normally used in STFT, WT, or GSTFT; however, in this case, parameters t, f, and s are independent, whereas, even in more general GSTFT, s depends on frequency with a prefixed law. The matching pursuit method is seen by Qian and Chen (1994) from the perspective of adaptive approximation of signals and is classified as a parametric method since the functions of the dictionary can be selected according to a model of the signal to analyze.

9.9   APPLICATIONS TO BIOMEDICAL SIGNALS

Most biomedical signals include both impulse-like events (time-concen-trated information) and nonstationary oscillations (frequency-localized in-formation). The identification and quantification of such changes can be of great importance from the physiological–pathological–clinical point of view. In this context, TFR is a valuable tool for exploring biomedical sig-nal contents. Therefore, the scope of application of linear TFR in biomed-ical signal processing is quite large, especially for WT, ranging from the traditional ECG and EEG to such recent "signals" as genomic sequences (Haimovich et al., 2006).

The interested reader can refer to the book edited by Akay (Akay, 1998), which collects applications of TFR to biomedical signal processing. Specific WT applications can be found in a review article by Unser and Aldroubi (1996) and in a book by the same authors (Aldroubi and Unser, 1996). There is also a special issue of the *IEEE Engineering in Medicine and Biology Magazine* (March/April 1995) on time-frequency and wavelet analysis of biomedical signals. In the following section, we will focus on two specific, examples of application of GSTFT.

9.9.1   Analysis of Spectral Variability of Heart Rate

An example of application of the methods of time-frequency representation is shown in Figure 9.12. The series of time intervals between two successive heartbeats (RR), represented on the bottom of the figure, is relative to a tilt test and consists of two periods. In the first, the subject is in clinostatism; the RR duration is about one second and shows an oscillatory component of res-piratory origin. In the second, the subject is under orthostatism; the RR in-terval is much shorter and the respiratory component is absent. The TF rep-resentation at the top has been achieved with GSTFT using a Von Hann analysis window and a time resolution $\Delta t(f)$ that passes from 60 s at the low frequencies to 8 s at the high ones, following a sigmoidal trend as shown in Figure 9.8. We, therefore, obtain a high resolution in time in the high-fre-quency band (0.15–0.4 Hz) containing the respiratory component, which al-lows us to track the changes occurring during the tilt. At the same time, we get a high-frequency resolution at low frequencies (0.04–0.15 Hz).

The middle panel in Figure 9.12 has been achieved with STFT, us-ing a Von Hann window with resolution in time $\Delta t = 36$ s. Although this choice allows a discrete frequency resolution in the low-frequency band, it provides an inadequate temporal localization of the changes in power in the high-frequency band related to the tilt maneuver.

The example in Figure 9.13 shows the time-frequency representation relative to a series of RR intervals with high variability of respiratory com-ponent. The three-dimensional view allows us to grasp the small details of

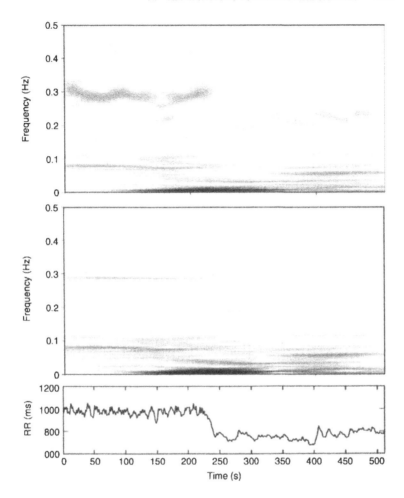

Figure 9.12. Time-frequency representation of RR interval (heart cycle period) series of a subject submitted to a tilt test. The time-frequency representation obtained with GSTFT (top panel), and that obtained with STFT (middle panel); the RR interval series (bottom panel).

nonstationary oscillatory phenomena. The series in this case has been analyzed with the STFT using a relatively narrow window. The good temporal resolution obtained allows us to assess the power of the respiratory component of origin (0.3–0.4 Hz) and its evolution over time.

9.9.2 Analysis of a Signal from a Laser Doppler Flowmeter

The laser Doppler signal measures the blood velocity and contains fluctuations related to activity of various physiological systems (heart, respirato-

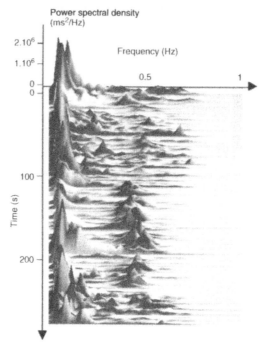

Figure 9.13. Time-frequency representation of an RR interval series obtained by STFT.

ry, autonomic, local myogenic, and endothelial). Because these oscillatory components cover a wide range of frequencies, the spectral analysis must be multiresolution. Figure 9.14 shows, on the left, the Q factor trend, obtained by interpolating some couples (frequency, Q) chosen according to the specific requirements of time-frequency resolution in each band. On the right, the spectrum obtained with the GSTFT, calculated at frequencies spaced proportionally to the frequency resolution, is shown.

9.10 CONCLUSIONS

Oscillatory behavior characterizes many physical, chemical, and physiological phenomena. Biomedical observable signals contain nonstationary oscillatory components coming from different unobservable systems. The time-frequency representation is a powerful tool to identify these nonstationary components and obtain information on the state of the generating systems. The extreme variability of the features that characterize each component of biomedical signals makes the robust nonparametric methods described herein more suitable than model-based approaches. The latter

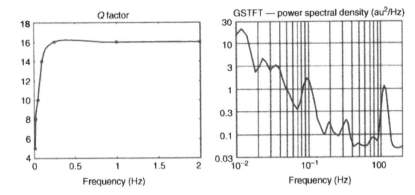

Figure 9.14. Spectral analysis of a laser Doppler perfusion signal. On the left, the Q factor as function of frequency, obtained by interpolation of user defined Q values (*). On the right, the spectrum obtained integrating in time the time-frequency distribution achieved by GSTFT.

can be optimal under specific assumptions but can perform very poorly otherwise. However, we should remark that time-frequency representation is an explorative technique and must be tuned to the specific signal according to the purpose of the analysis.

Although this review on linear time-frequency representation covers different methods, from standard STFT to matching pursuit, it does, however, leave out some other interesting approaches such as empirical mode decomposition and other instantaneous frequency-estimation methods.

REFERENCES

Akay, M. (Ed.), *Time Frequency and Wavelets in Biomedical Signal Processing,* Wiley/IEEE Press, 1998.

Aldroubi, A., and Unser, M., *Wavelets in Medicine and Biology,* CRC Press, 1996.

Allen, J. B., and Rabiner, L. R., A Unified Theory of Short-Time Spectrum Analysis and Synthesis, *Proceedings of the IEEE,* Vol. 65, No. 11, 1158–1564, 1977.

Crochiere, R. E., and Rabiner L. R., *Multi-Rate Digital Signal Processing,* Prentice-Hall, 1983.

Daubechies, I., Orthogonal Bases of Compactly Supported Wavelets, *Commun. Pure Appl. Math.,* Vol. 41, Nov., 909–996, 1988.

Daubachies, I., *Ten Lectures on Wavelets,* CBMS, SIAM Publications, 1992.

Durka, P. J., Ircha, D., and Blinowska, K. J., Stochastic Time-Frequency Dictionaries for Matching Pursuit, *IEEE Trans. on Signal Processing,* Vol. 49, No. 3, 507–510, 2001.

Esteban, D., and Galand, C., Application of Quadrature Mirror Filters to Split

Band Voice Coding Schemes, in *IEEE International Conference on Acoustics, Speech, and Signal Processing,* Vol. 2, pp. 191–195, 1977.

Flandrin, P., *Time-Frequency/Time-Scale Analysis,* Academic Press, 1999.

Gambardella, G., A Contribution to the Theory of Short-Time Spectral Analysis with Nonuniform Bandwidth Filters, *IEEE Trans. Circuit Theory,* Vol. CT-18, July, 455–460, 1971.

Grossmann, A., and Morlet, J., Decomposition of Hardy Functions into Square Integrable Wavelets of Constant Shape, *SIAM J. Math. Anal.,* Vol. 15, July, 723–736, 1984.

Haimovich, A. D., Byrne, B., Ramaswamy, R., and Welsh, W. J., Wavelet Analysis of DNA Walks, *Journal of Computational Biology,* Vol. 13, No. 7, 1289–1298, 2006.

Hlawatsch, F., and Boudreaux-Bartels, G. F., Linear and Quadratic Time-Frequency Signal Representations, *IEEE SP Magazine,* Vol. 2, 21–67, 1992.

Mallat, S., A Theory for Multiresolution Signal Decomposition: The Wavelet Representation, *IEEE Trans. Pattern Anal. Machine Intell.,* Vol. 11, No. 7, 674–693, 1989.

Mallat, S., and Zhang, Z., Matching Pursuits with Time-Frequency Dictionaries, *IEEE Trans. on Signal Processing,* Vol. 41, No. 12, 3397–3415, 1993.

Meyer, Y., *Wavelets with Compact Support,* Zygmund Lectures, University of Chicago, 1987.

Meyer, Y., *Wavelets and Operators,* Cambridge University Press, 1992.

Mintzer, F., Filters for Distortion-Free Two-band Multi-Rate Filter Banks, *IEEE Trans.on Acoust., Speech, and Signal Processing,* Vol. 32, 626–630, 1985

Portnoff, M. R., Time Frequency Representations of Digital Signals and Systems Based on Short Time Fourier Analysis, *IEEE Trans. on Acoustics, Speech, and Signal Processing,* Vol. 28, No. 1, 55–69, 1980.

Qian, S., and Chen, D., Signal Representation Using Adaptive Normalized Gaussian Functions, *Signal Processing,* Vol. 36, No. 3, 1–11, 1994.

Rioul, O., and Vetterli, M., Wavelets and Signal Processing, *IEEE Signal Processing Magazine,* Vol. 4, 14–38, 1991.

Smith, M., and Barnwell, T., A Procedure for Designing Exact Reconstruction Filter Banks for Free Structured Subband Coders, in *IEEE International Conference on Acoustics, Speech, and Signal Processing,* Vol. 9, pp. 421–424, 1984.

Smith, M., and Barnwell, T., The Design of Digital Filters for Exact Reconstruction in Subband Coding, *IEEE Trans. on Acoustics, Speech, and Signal Processing,* Vol. 34, No. 3, 434–441, 1986.

Unser, M., and Aldroubi, A., A Review of Wavelets in Biomedical Applications, *Proceedings of the IEEE,* Vol. 84, No. 4, 626–638, 1996.

Vaidyanathan, P. P., Quadrature Mirror Filter Banks, m-Band Extensions and Perfect Reconstruction Techniques, *IEEE Acoustics, Speech, and Signal Processing Magazine,* Vol. 4, 4–20, 1987.

Varanini, M., De Paolis, G., Emdin, M., Macerata, A., Pola, S., Cipriani, M., and Marchesi, C., A Multiresolution Transform for the Analysis of Cardiovascular Time Series, *Computers in Cardiology,* IEEE Computer Society Press, 1998.

Vetterli, M., and Herley, C., Wavelets and Filter Banks: Theory and Design, *IEEE Trans. on Signal Proc.,* Vol. 40, No. 9, 2207–2232, 1992.

Vetterli, M., Wavelets, Approximation, and Compression, *IEEE Signal Processing Magazine,* Vol. 4, 59–73, 2001.

CHAPTER 10

QUADRATIC TIME-FREQUENCY REPRESENTATION

Luca Mainardi

10.1 INTRODUCTION

In a previous chapter, we learned how to decompose a signal using elementary blocks of different shapes and dimensions: sinusoids, mother functions, or time-frequency distributions. These blocks are efficient tools for describing, in a synthetic way, morphological features of signals, such as waves, trends, or spikes. In a dual way, the same signal can be investigated in the frequency domain by using the Fourier transforms of these elementary functions. However, time and frequency domains are treated as separate worlds, often in competition because the need to locate a feature in time is usually paid for in terms of frequency resolution. A conceptually different approach aims to jointly look at the two domains and to derive a joint representation of a signal $x(t)$ in the combined time and frequency domain.

This chapter deals with a description of quadratic time-frequency distributions and their applications to the analysis of biomedical signals. A quadratic time-frequency distribution is designed to represent the signal energy simultaneously in the time and frequency domains and, thus, it provides temporal information and spectral information simultaneously.

The chapter is organized as follows. First, a possible route toward the definition of time-frequency representations (TFRs) is considered, leading to the introduction of Wigner–Ville distribution and the related Cohen's class distribution. Theoretical issues and practical ones will then be discussed with respect to the peculiar application to the processing of biomedical signals. The second part of the chapter describes the analysis of biological signals through time-frequency representations, with applications to electroencephalogram (EEG), electrocardiogram (ECG), and heart rate variability (HRV) signals.

10.2   A ROUTE TO TIME-FREQUENCY REPRESENTATIONS

A link between time and frequency domains may be obtained through the signal energy E_x. The following relation holds:

$$E_x = \int |x(t)|^2 dt = \int |X(\omega)|^2 d\omega \qquad (10.1)$$

where $X(\omega)$ is the Fourier transform of the signal and $|X(\omega)|^2$ is its power spectrum. It is therefore intuitive to derive a *joint* time-frequency representation, $TFR(t, \omega)$, able to describe the energy distribution in the t–f plane and to combine the concept of instantaneous power $|x(t)|^2$ with that of the power spectrum $|X_t(\omega)|^2$. Such a distribution, to be eligible as an *energetic* distribution, should satisfy the marginals

$$\int TFR_x(t,\omega)d\omega = |x(t)|^2 \qquad \text{and} \qquad \int TFR_x(t,\omega)dt = |X(\omega)|^2 \quad (10.2)$$

Thus, for every instant t, the integral of the distribution over all the frequency should be equal to the instantaneous power, whereas, for every angular frequency ω, the integral over time should equal the power spectral density of the signal. As a consequence of Equations 10.1 and 10.2, the total energy is obtained by integration of the TFR over the whole t–f plane:

$$E_x = \iint TFR_x(t,\omega)d\omega dt \qquad (10.3)$$

As the energy is a quadratic function of the signal, the $TFR(t, \omega)$ is expected to be quadratic.

From the pioneer works by Page (1952) and Gabor (1946), the problem of deriving the time-frequency representation of a signal has been investigated from different points of view. Cohen (1989) provided a detailed and exhaustive description on the topic. An interesting way to define energetic TFR starts from the definition of a time-varying spectrum (Page, 1952). Using the relationship that links power spectral density and TFR imposed by marginals (Equation 10.2), we derive a simple definition of a TFR:

$$TFR(t,\omega) = \frac{\partial}{\partial t}|X_t(\omega)|^2 \qquad (10.4)$$

The subscript t indicates that the quantity is a function of time and, thus, $|X_t(\omega)|^2$ is a time-varying spectrum. The latter can be derived by generalization of the relationship between the power spectrum of a signal and its autocorrelation function $R_t(\tau)$:

$$|X_t(\omega)|^2 = \frac{1}{2\pi} \int R_t(\tau)e^{-j\omega\tau} d\tau \qquad (10.5)$$

where

$$R_t(\tau) = \int x(t)x*(t-\tau)dt = \int x\left(t+\frac{\tau}{2}\right)x*\left(t-\frac{\tau}{2}\right)dt \qquad (10.6)$$

is a function of time. By substituting Equation 10.5 in Equation 10.4, a new definition of TFR is obtained:

$$TFR(t,\omega) = \frac{1}{2\pi}\int \frac{\partial}{\partial t} R_t(\tau)e^{-j\omega\tau}d\tau = \frac{1}{2\pi}\int K_t(\tau)e^{-j\omega\tau}d\tau \qquad (10.7)$$

where $K_t(\tau)$ is known as a *local autocorrelation function*. The above relation shows that a TFR can be obtained as the Fourier transform of a time-dependent autocorrelation function. We may observe that due to the derivative operation, the integral that characterizes the $R_t(\tau)$ disappears in $K_t(\tau)$, which de facto describes local properties of the signal. Among all the possible choices of $K_t(\tau)$, the most simple (Mark, 1970) is to select

$$K_t(\tau) = x\left(t+\frac{\tau}{2}\right)x*\left(t-\frac{\tau}{2}\right) \qquad (10.8)$$

The derived time-frequency distribution,

$$E_x = \frac{1}{2\pi}\int x\left(t+\frac{\tau}{2}\right)x\left(t-\frac{\tau}{2}\right)\exp\left\{-j\omega\tau\right\}d\tau \qquad (10.9)$$

is known as the Wigner–Ville (WV) distribution (Wigner, 1932; Ville, 1948).

10.3 WIGNER–VILLE TIME-FREQUENCY REPRESENTATION

The Wigner–Ville (WV) representation was originally introduced by Wigner (1932) in the field of quantum mechanics and successively applied to signal analysis by Ville (1948). It plays a fundamental role among the quadratic time-frequency distributions and it is a fundamental part of the

Cohen class. It satisfies several desirable mathematical properties for a quadratic time-frequency distribution (Table 10.1): it is real valued and preserves the time and frequency shifts of the signal, it satisfies the marginal properties (as evident from the way we derive it), and the instantaneous frequency can be estimated from the first moment of W_{xx}. In addition, for a linear chirp [a signal whose instantaneous frequency varies linearly with time according to $f(t) = f_0 + \alpha t$] it can be shown that

Table 10.1. Desirable properties (P) for time-frequency representations and kernel requirements (R) that guarantee the properties

Property of TFR	Kernel requirement
P1 Real value $T_x(t,\omega) \in R$	R1 $\varphi(\theta,\tau) = \varphi*(-\theta,-\tau)$
P2 Time-shift invariance $g(t) = x(t-t_0) \Rightarrow T_g(t,\omega)$ $\quad = T_x(t-t_0,\omega)$	R2 $\varphi(\theta,\tau)$ not dependent on t
P3 Frequency-shift invariance $g(t) = x(t)e^{j\omega_0 t} \Rightarrow T_g(t,\omega)$ $\quad = T_x(t,\omega-\omega_0)$	R3 $\varphi(\theta,\tau)$ not dependent on ω
P4 Marginals (time) $\int T_x(t,\omega)d\omega = x(t)x*(t)$	R4 $\varphi(\theta,0) = 1 \quad \forall \theta$
P5 Marginals (frequency) $\int T_x(t,\omega)dt = X(\omega)X*(\omega)$	R5 $\varphi(0,\tau) = 1 \quad \forall \tau$
P6 Instantaneous frequency $\dfrac{\int \omega T_x(t,\omega)d\omega}{\int T_x(t,\omega)d\omega} = \omega_i(t)$	R6 R4 and $\left.\dfrac{\partial\phi(\theta,\tau)}{\partial\tau}\right|_{\tau=0} = 0 \quad \forall\theta$
P7 Group delay $\dfrac{\int t T_x(t,\omega)dt}{\int T_x(t,\omega)dt} = t_g(\omega)$	R7 R5 and $\left.\dfrac{\partial\varphi(\theta,\tau)}{\partial\theta}\right|_{\theta=0} = 0 \quad \forall\tau$
P8 Interference reduction	R8 $\varphi(\theta,\tau)$ is a two-dimensional low-pass filter

$$W_{xx}(t, f) = \delta[t, f - f_x(t)] \qquad (10.10)$$

and the WV is a line in the t–f plane, concentrated at any instant around the instantaneous frequency of the signal (Figure 10.1a). From a practical point of view, this property shows that the representation is able to correctly localize (jointly in time and frequency) a sinusoidal component whose properties are varying with time. Equation 10.10 can be demonstrated by observing that $K_x(\tau) = x_t(\tau/2)x_t^*(-\tau/2)$ is characterized by the instantaneous frequency $f_K = \{f_x(+\tau/2) + f_x(-\tau/2)\}$. As $f_x(t)$ is linearly varying, we have $f_K = \{f_x(-t/2) + f_x(+t/2)\} = f_x(t)$ for every t; thus, Equation 10.10 holds.

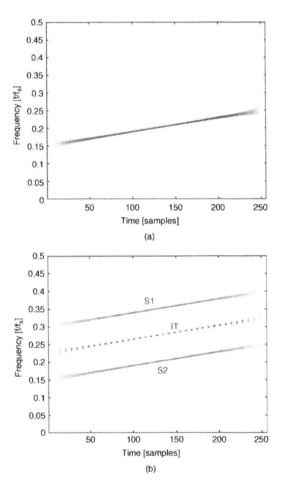

Figure 10.1. (a) WV of a monocomponent signal (linear chirp). The representation localizes correctly the signal in the t–f plane. (b) WV of a two-component signal. Interference terms (IT) appear in addition to the signal contributions (S1, S2).

This is a peculiar feature of WV and one of its key properties. A detailed description of this and other properties of WV can be found in Classen and Mecklenbrauker (1980a).

10.4   INTERFERENCE TERMS

Even if the WV representation is attractive for representing single-component, nonstationary signals, it becomes of poor utility when multicomponent signals are considered. In these cases, the distribution may assume negative values (and this is in contrast with the interpretation of energetic distribution) and interference terms (or cross terms) appear. The cross terms disturb the interpretation of the TFR as they are redundant information that may mask the true characteristics of the signal. An example of interference terms is shown in Figure 10.1b, where a signal composed by two chirps is considered. In addition to the signal components (S1,S2), some unwanted interference terms appear. These terms appear because the WV is a bilinear transform for which the quadratic superimposition principle holds (Flandrin, 1984). In the case of an N-component signal

$$x(t) = \sum_{k=1}^{N} c_k x_k(t)$$

the WV becomes

$$W_{xx}(t,f) = \sum_{k=1}^{N} |c_k|^2 W_{x_k x_k}(t,f) + \sum_{j=1}^{N} \sum_{i=j+1}^{N} (c_j c_i^* W_{x_j x_i} + c_i c_j^* W_{x_i x_j}) \quad (10.11)$$

where $W_{xk.xk}$ represents the auto-WV representation of each signal term, whereas

$$W_{x_i x_j}(t,\omega) = \int x_i \left(t + \frac{\tau}{2} \right) x_j^* \left(t - \frac{\tau}{2} \right) \exp \{ -j\omega\tau \} d\tau \quad (10.12)$$

is the cross-WV representation. Therefore, the resulting WV representation is the sum of the auto-WV plus the cross-WF, which generates interferences. Globally, the representation will be characterized by N signal terms and

$$\binom{N}{2} = \frac{N(N-1)}{2}$$

interference terms. The latter grows quadratically in respect to the number of components and may overwhelm the signal contributes quite rapidly.

Taking into account the relevant role of ITs, it is useful to analyze their characteristics in more detail. Let us consider a signal composed of two complex sinusoids:

$$x(t) = x_1(t) + x_2(t) = exp\{j\omega_1 t\} + exp\{j\omega_2 t\} \qquad (10.13)$$

The autocorrelation function becomes

$$K(\tau) = \left(exp\{j\omega_1(t + \tau/2)\} + exp\{j\omega_2(t + \tau/2)\}\right)\left(exp\{-j\omega_1(t - \tau/2)\}\right.$$

$$\left. + exp\{-j\omega_2(t - \tau/2)\}\right) \qquad (10.14)$$

and, after a few rearrangements, we may write

$$K_x(\tau) = K_{x1}(\tau) + K_{x2}(\tau) + \cos\left(\frac{\omega_1 - \omega_2}{2}t\right)exp\left\{j\frac{\omega_1 + \omega_2}{2}\tau\right\} \qquad (10.15)$$

It is worth noting the characteristics of the interference terms. They are located at in the middle of the frequency of the two components, $(\omega_1 + \omega_2)/2$ and their amplitudes oscillate in time at a frequency related to the distance of these components $(\omega_1 - \omega_2)/2$. It is the oscillatory nature of cross terms that generates negative values in the WV.

The above observations are general and it is possible to determine the position of the IT when the locations of the signal terms are known (Hlawatsch and Baudreaux-Bartels, 1992; Flandrin, 1984). An example is shown in Figure 10.2a where the two signal terms are centered in (t_1, f_1) and (t_2, f_2). It is possible to observe that interference terms are located around the central point $[t_{12} = (t_1 + t_2)/2, f_{12} = (f_1 + f_2)/2]$ and their amplitude oscillates in time with a period of $1/|f_1 - f_2|$ and in frequency with a period of $1/|t_1 - t_2|$. Therefore, the oscillation frequency grows with the distance between signal terms and the direction of oscillation is perpendicular to the line connecting the signal points (t_1, f_1) and (t_2, f_2). It is worth noting that the interference terms may be located in time intervals where no signal is present, for example between t_1, and t_2 in Figure 10.2, showing signal contributions in an area where no activity is expected (like a mirage in the desert).

These effects make the WV hardly readable, especially when a wideband noise is superimposed, and many authors have labeled the WV as a "noisy" representation (Cohen, 1989). For the sake of curiosity, we notice that IT may appear also in the case of a single component, as shown in Figure 10.2b. Interferences are located in the concavity of the distribution and are related to the interaction between past and future signal frequencies. Finally it is worth noting that any real signals, such as a sinusoid, $x = \cos(\omega t)$ $= exp\{jwt\} + exp\{-jwt\}$, generate interference between positive and negative frequencies of their spectrum (Classen, 1980a). To avoid this effect in

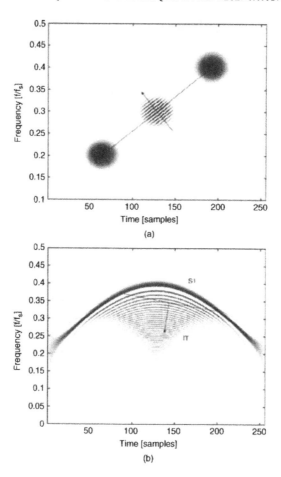

Figure 10.2. Localization of the interference terms in the t–f plane. (a) The interferences are generated for every signal pair (S1, S2). They are located in between the signal terms and oscillate in the direction perpendicular to the line connecting the signal terms. Oscillation frequency is related to the distance between the two terms (see text for details). (b) Example of interference generated by a monocomponent signal.

practical applications, the Hilbert transform is applied to the real signal to generate the analytic signal in which the negative frequencies are canceled.

10.5 COHEN'S CLASS

The characteristics of cross terms (oscillating) suggest the strategy for their suppression: the idea is to perform a two-dimensional low-pass filtering of the TFR, in order to suppress the higher frequency oscillations. If the properties of the selected filter do not depend on their position in the t–f plane (i.e., the filter characteristics are invariant to shifts in the t–f plane), we derive the class of shift-invariant, quadratic TFRs, known as Cohen's class (Cohen, 1989):

$$C_{xx}(t,f) = \int\int \Psi(u-t, \upsilon-f) W_{xx}(u,\upsilon) du d\upsilon \qquad (10.16)$$

As evident from the above relation, every member of the class can be obtained as the convolution between the W_{xx} and a function Ψ, the *kernel*. Every TFR of this class can be interpreted as a filtered version of W_{xx} (Hlawatsch and Boudreaux-Bartels, 1992). It is unequivocally defined by Ψ, and a new TFR of the class can be generated by projecting a new kernel. The possibility of generating TFR from a single class has practical advantages: the ability to prove general results and to study the common aspects or the peculiarities of each TFR. In addition, by imposing constraints on the kernel one obtains a subclass of TFR with a particular property (Cohen, 1966). A few examples of TFRs obtained using different kernels are shown in Figure 10.3. It is worth noting that the introduction of the kernel reduces the cross terms, but also the capability to localize the signal component. In fact, the useful property (Equation 10.10) is lost in C_{xx} due to the low-pass filtering effect of Ψ. Therefore, we are facing a compromise between the entity of the cross term and the preservation of joint time-frequency resolution in the t–f plane. Whereas in the linear time-frequency representations the compromise is between time or frequency resolution, in the quadratic TFR the compromise is between the maximization of joint t–f resolution and the minimization of cross terms.

The question is: which tools should be used to project the TFR with desired properties? An important tool is the *ambiguity function* (AF)

$$A_{xx}(\theta,\tau) = \int x\left(t+\frac{\tau}{2}\right) x^*\left(t-\frac{\tau}{2}\right) \exp\{j\theta t\} dt \qquad (10.17)$$

It is worth noting the structural analogy with the WV, with the difference that integration is performed over time. The AF has been largely applied in radar, sonar studies, and astronomy as the maximum of A_{xy} computed between an emitted signal $x(t)$ and the received one $y(t)$ provides information on the distance (lag ϕ) and speed (θ, doppler-shift) of a moving object. The relation linking AF and W_{xx} is

$$A_{xx}(\theta,\tau) = \int\int W_{xx}(t,\omega) e^{-(j\omega\tau + j\theta t)} dt d\omega \qquad (10.18)$$

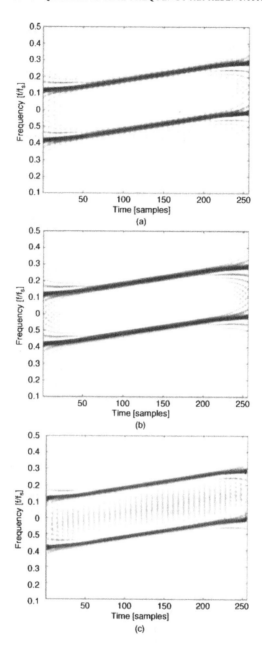

Figure 10.3. Reduction of IT obtained by time-frequency representation of the Cohen's class: (a) ED, (b) SPWV, and (c) RID. It is worth noting that the lines corresponding to the chirps are larger than in Figure 10.1; thus, the kernels reduce time-frequency localization.

which can be easily obtained from the combination of Equations 10.9 and 10.17. It is worth noting the AF and WV are linked by a two-dimensional Fourier transform; the two transformations can be considered as dual. In general, W_{xx} is real valued but A_{xx} may not be. The AF is the projection of W_{xx} in the plane θ–τ (known as the correlative domain). In this plane, signal and cross terms tend to separate. The former are mainly located close to the origin; the latter are located far from it. The effect is evident in Figure 10.4.

It is worth noting that in the correlative domain Cohen's class is simply described by a product:

$$C_{xx}(\theta,\tau) = \phi(\theta,\tau)A(\theta,\tau) \qquad (10.19)$$

where $\phi(\theta, \tau)$ is the two-dimensional Fourier transform of Ψ. In Equation 10.19, the effect of the kernel can be immediately appreciated; it weights the points of the θ–τ-plane. Therefore, in order to perform an efficient reduction of cross terms, the function $\phi(\theta, \tau)$ should have higher values close to the origin than far from it. Thus $\phi(\theta, \tau)$ should be the transfer function of a two-dimensional low-pass filter. The effect of various kernels in the correlative planes is shown in Figure 10.5.

The properties of $\phi(\theta, \tau)$ define the properties of the derived TFR. The relationship between the kernel and characteristics of TFR are listed in Table 10.1. Using these relationships, it is straightforward to verify, for example, the WV properties (already reported at the beginning of the chapter): W_{xx} is real valued (the kernel satisfies R1 in Table 1), is invariant to time and frequency shifts (constraints R2 and R3 are verified), and it satisfies marginals (R4 and R5). The reader can do the same exercise with other TFRs of the Cohen's class.

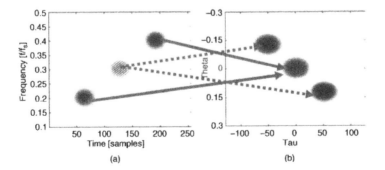

Figure 10.4. Relationship between WV (a) and AF (b). Signal components are mapped around the origin of the correlative plane, whereas the interference terms are located far from it.

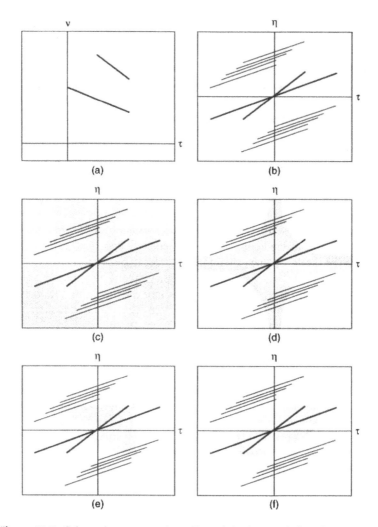

Figure 10.5. Schematic representation of kernels in the correlative plane. (a) TFR of the signal and (b) its projection in the θ–τ plane. Signal terms are the two lines passing from the origin; the others are the IT. Different kernels are superimposed on the AF. (c) WV kernel, (d) BJD, (e) SPWV, and (f) generic time-frequency filter. Freely modified from Flandrin (1984).

In the following, the TFRs that have been largely applied to the analysis and synthesis of biological signals will be briefly reviewed. They include the Choi–Williams or exponential distribution (ED) (Choi and Williams, 1989), the reduced interference distribution (RID) (Jeong and Williams, 1992), and the smoothed pseudo Wigner–Ville (SPWV) (Martin and Flandrin, 1985). See Table 10.2 for a complete list.

Table 10.2. Common time-frequency representations belonging to Cohen's class and their kernels

Representation	Kernel $\phi(\theta, \tau)$
Born and Jordan (BJD)	$\dfrac{\sin(\pi\tau\vartheta)}{\pi\tau\vartheta}$
Choi–Williams (CWD) or exponential (ED)	$\exp\left[-\dfrac{(2\pi\tau\vartheta)^2}{\sigma}\right]$
Cone kernel (CD)	$g(\tau)\lvert\tau\rvert\dfrac{\sin(\pi\tau\vartheta)}{\pi\tau\vartheta}$
Generalized exponential (GED)	$\exp\left[-\left(\dfrac{\tau}{\tau_0}\right)^{2M}\left(\dfrac{\vartheta}{\vartheta_0}\right)^{2N}\right]$
Generilized Wigner (GWD)	$\exp\left[j2\pi\alpha\vartheta\tau\right]$
Levin (LD)	$\exp\left[j\pi\lvert\tau\rvert\vartheta\right]$
Page (PD)	$\exp\left[-j\pi\lvert\tau\rvert\vartheta\right]$
Pseudo Wigner–Ville (PWV)	$\eta\left(\dfrac{\tau}{2}\right)\eta^{\cdot}\left(-\dfrac{\tau}{2}\right)$
Reduced interference (RID)	$S(\tau\vartheta)$
Rihaczek (RD)	$\exp\left[j\pi\tau\vartheta\right]$
Smoothed Pseudo Wigner–Ville (SPWV)	$\eta\left(\dfrac{\tau}{2}\right)\eta^{\cdot}\left(-\dfrac{\tau}{2}\right)G(\vartheta)$
Wigner–Ville (WV)	1

10.5.1 Exponential Distribution (ED)

The ED is an attempt to improve the WV. The kernel is a Gaussian function, centered in the origin, whose width depends on the parameter σ. This parameter can be tuned in order to find the best compromise between time-frequency resolution and cross-term suppression. The larger σ is, the more similar $\phi(\theta, \tau)$ will be to the kernel of WV [for $\sigma \to \infty$, $\phi(\theta, \tau) \to 1$]. When observed in the θ–τ plane, the kernel preserves the central part of the AF and excludes the most external parts where cross terms are supposed to be located. The effect of selection of σ on the characteristics of TFR obtained from the analysis of cortical EEG is shown in Figure 10.6.

10.5.2   Reduced Interference Distribution (RID)

The RID is described by

$$RID(t,\omega) = \iint \frac{1}{|\tau|} h\left(\frac{u-t}{\tau}\right) x(u+\tau/2)x*(u-\tau/2)e^{-j\omega\tau}dud\tau \quad (10.20)$$

where $h(t)$ is known as the *primitive function*. The function $h(t)$ is real valued, with unit area, symmetric with respect to t, $h(t) = h(-t)$, limited in the interval $[-1/2, 1/2]$ and going to zero on the extremes. These requirements guarantee the series of properties of Table 10.1, with a few requirements imposed on the kernels. It is worth noting the similarity between the char-

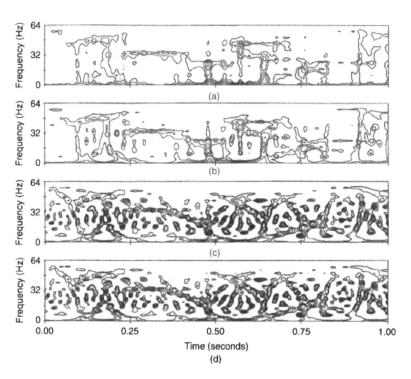

Figure 10.6. Analysis of cortical EEG by ED. Effect of σ selection on the reduction of interference terms. For lower σ, the IT energy is low, but when σ is increased, the IT contribution is augmented. (a) $\sigma = 0.1$, (b) $\sigma = 10$, (c) $\sigma = 1000$, (d) WV. For larger values of σ, the ED is similar to the WV (Zaveri et al., 1992).

acteristic of $h(t)$ and the characteristics of the window used in the design of digital filters. The primitive can be either the window or the impulse response of a filter and, therefore, it is possible to transfer the wide spectrum of knowledge used in filter design to the design of the RID.

10.5.3 Smoothed Pseudo Wigner–Ville (SPWV)

The peculiar property of this distribution is in the separable kernel $\Psi(t, f)$ = $g(t)H(f)$, where $g(t)$ and $H(f)$ are two windows that control *independently* the filtering in the time and frequency domains, respectively. The kernel structure provides a great flexibility in the selection of time/frequency resolution, ease of application, and straightforward implementation.

10.6 PARAMETER QUANTIFICATION

The TFR provides a synthetic and immediate image of the energy distribution in the t–f plane. However, some postprocessing methods are needed to derive quantitative parameters of interest. The latter are often obtained by computing amplitude and frequency of the spectral components in different bands. In this regard, properties such as P6 of Table 10.1 are used. The instantaneous frequency can be obtained as a weighted average of TFR in a certain range of frequency. The computation can be performed in different frequency bands to identify the various signal components.

It has been observed (Mainardi et al., 2004) that, in the case of SPWV, the $K_t(\tau)$ can be described by a sum of complex sinusoids by extension of Equation 10.15 to the N-component case:

$$K_x(\tau) = \sum_{k=1}^{M} A_k \exp\{j\omega_k \tau - \beta_k \tau\} \qquad (10.21)$$

where M takes into account both signal and cross terms. The problem of quantifying the signal components becomes the problem of identifying the amplitude and phase of complex damped sinusoids using well-known signal-processing algorithms (Kumaresan and Tufts, 1982).

10.7 APPLICATIONS

In this section, the analysis of biological signals through time-frequency representations will be presented, showing possible application fields ranging from analysis of electroencephalogram (EEG), electrocardiogram (ECG), and heart rate variability (HRV) signals, up to the analysis of electromyograms (EMGs). The selected applications, far from being exhaustive, are aimed at pointing out the usefulness of these methodologies for the analysis of biomedical signals. A more detailed and wide discussion can be found in review articles (Mainardi et al., 2002; Lin and Chen, 1996).

10.7.1 EEG Signal Analysis

Quadratic $t\!-\!f$ distributions have been applied to the analysis of evoked or event-related potentials (Morgan and Gevens, 1986; Williams et al., 1987) and to the characterization of surface or cortical EEG activity (Zaveri et al., 1992). De Weerd and Kap (1981) compared the performances of different TFRs in the analysis of evoked potentials (EPs) using approaches based on both filter banks and the Rihaczek TFR (Rihaczek, 1968). They preferred the approach based on filter banks due to the difficulty in interpreting the TFR results due to interference. Lately, Moragn and Gevins (1986) demonstrated the applicability of WV for the $t\!-\!f$ representation of EP and for the analysis of the EP waves, stressing the excellent capability of separating components in time and frequency domains. In an interesting work (Williams et al., 1987), the parameters extracted by RID are used for classification/differentiation of the event-related evoked responses (ERP). The authors hypothesized that the representation of the signal in the $t\!-\!f$ plane can reveal, in rich detail, the ERP complexity. In this work, the RID was applied to the analysis of the evoked response obtained from different classes of stimuli composed by words (a) pleasant, (b) unpleasant, (c) linked to phobia and anxiety of the patient, or (d) to his internal conflicts. The aim was to evaluate if the characteristics of ERP can reveal the presence of unconscious responses related to a patient's problems or personality. As the readability of RID is difficult, the authors approximated the TFR by Gabor functions and demonstrated that only five approximating functions are needed to capture the most significant features of the $t\!-\!f$ of the signal (Figure 10.7). The parameters computed from the approximating functions are able to discriminate between the different responses as a function of patient and personality.

Concerning the EEG signal, the analysis of dynamic changes in the spectrum is usually approached by STFT or AR adaptive methods. Nevertheless, there are a few relevant applications of quadratic time-frequency representations. In (Zaveri et al., 1992), the TFRs are used to characterize the cortical activity during epilepsy of the temporal lobe. Spectrogram, PWD, and RID were compared for the identification of short bursts in the signal. An example is shown in Figure 10.8, where a cortical signal is composed of two high-frequency bursts separated by a short stationary period characterized by slow rhythms. Every TFR is represented by two contour-level plots in which positive and negative components of the transform (presence of negative values are a hallmark of cross terms) are

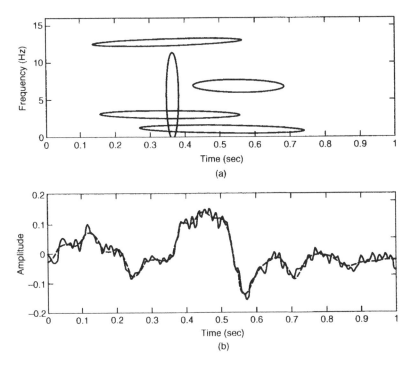

Figure 10.7. (a) Approximation by Gabor functions of the RID calculated for an ERP. The first five approximations are shown and the contours represent half of the component amplitude. (b) ERP (continuous line) and its reconstruction by approximants (dotted line). Taken from Williams (1996).

presented. We may observe that the spectrogram spread the activity in the t–f plane; thus, a precise localization and identification of the burst is not feasible. A better resolution is obtained with the PWD at the expense of the interference terms. In this case, the RID provides the best result with a significant suppression of cross terms which does not prevent the possibility to resolve signal features.

Some authors (Nayak et al., 1994) evidenced situations in which the ED of EEG signal provides better performance than traditional spectral analysis. Thanks to the better frequency localization of short-time activity, the authors evidenced small changes in the TFR related to the level of consciousness during anesthesia that were blind to traditional spectral analysis.

10.7.2 ECG Signal Analysis

Quadratic TFRs have been widely used for the dynamic analysis of the frequency contents of the ECG and its characteristic waveforms (P-QRS-T).

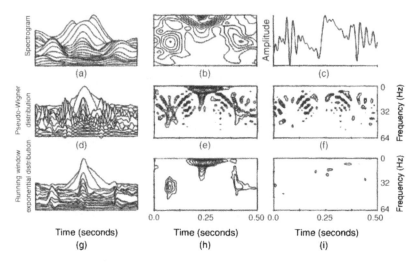

Figure 10.8. Comparison among spectrogram (a), WV (d), and RID (g) in the analysis of a cortical EEG signal. (c), (b), (d), and (e) represent the positive values of the transform, whereas (f) and (i) represent the negative ones (Zaveri et al., 1992).

Applications have been proposed for the detection of P-waves (Abeysekera and Boashash, 1989), for the analysis of the QRS complex (Novak et al., 1994), for the quantification of ventricular late potentials (Chouvarda et al., 1999), and for the evaluation of fibrillatory atrial waves during atrial fibrillation (Stridh et al., 2001).

The characteristics of QRS complex were investigated in normal subjects and postinfarcted patients using PWV (Novak et al., 1994). High-frequency components (> 90 Hz) were present in all the examined cases and were significantly more pronounced in the postinfarct patients. In addition, the authors observed these high-frequency oscillations during the whole QRS duration, not only in the terminal part where ventricular late potentials are usually observed.

A description of the t–f characteristics of both the QRS complex and the ST-T segment in patients with acute myocardial infarction after thrombolitic treatment can be found in (Chouvarda et al., 1999). The authors used the WV for the computation of TFR of ECG signals during the cardiac cycle and they quantified the characteristics in various portions of t–f plane (during the QRS complex, in the ST segment, and on the T wave, and in three frequency ranges of DC: 25 Hz, 25–50 Hz, and 50–100 Hz). The parameters extracted were used for the definition of a predictive index on the efficiency of trombolitic treatment. The index was able to differentiate between patients who had successful trombolitic treatment from those who had not.

The CD was used, and compared with the SPVW and spectrogram, to study the capability of these transformations to detect ventricular fibrillation events (Zhao et al., 1990). CD and SPWV, thanks to their time-frequency resolution, were identified as ideal candidates for the detection and classification of those cardiac arrhythmias that modify the spectral characteristics of ECG signals. The cross WV has been used to track the dynamic variations of the frequency of fibrillatory waves, which characterize the residual ECG signal during AF (Stridh et al., 2001). The residual EECG signal is obtained from surface ECG by subtracting waves related to ventricular activity. This signal is characterized by a dominant spectral component, relevant from a clinical point of view as it is correlated to the level of atrial organization. The authors observed dynamic changes that depend on the recording site and are induced by pharmacological infusions, and they concluded that the atrial activity can be variegated and more or less organized both spatially and temporarily.

10.7.3 Heart Rate Variability Signal

The first detailed study on the application of quadratic TFR to the analysis of HRV was by Novak and Novak (1993). The authors suggested the use of a SPWV, taking into account the characteristic of the HRV signal and the different requirements for time and frequency resolution. In fact, the spectrum of these signals is usually composed of a few components, usually two: a high-frequency component (HF) modulated by the parasympathetic system and a low-frequency component (LF) related to sympathetic activation. These components may evolve quickly in time and their changes must be tracked as fast as possible. In this context, the possibility of using a separable kernel appears advantageous. For these signals, the authors suggest using a rectangular window for time smoothing (with a variable length between 5 and 25 samples) and a Gaussian window [$h(k) = e^{-1/2[\alpha k/N/2]^2}$, $\alpha = 2.5$, and $N = 128$] for frequency smoothing. The study was conducted during provocative tests able to enhance sympathetic (rest-to-tilt maneuver) or parasympathetic (Valsalva maneuver or controlled breathing) activations. An example of an extracted TFR is shown in Figure 10.9, where the dynamic changes in the HRV spectra are evident. Analogous results were obtained by others (Pola et al., 1996), who concluded that the SPWV is the most appropriate method for the investigation of HRV signals. These methods allowed the investigation of dynamic of the autonomic nervous system (ANS) in the first phases of tilt. After tilt, vagal and sympathetic changes were nearly simultaneous, but the early, almost instantaneous disappearance of the HF component was accompanied by a slower, progressive increase of LF components (see Figure 10.9). When the tilt was prolonged, oscillations and nonstationarity characterize the spectral power, especially in the LF components and before syncope (Novak et al., 1995; Furlan et al., 1988). In these

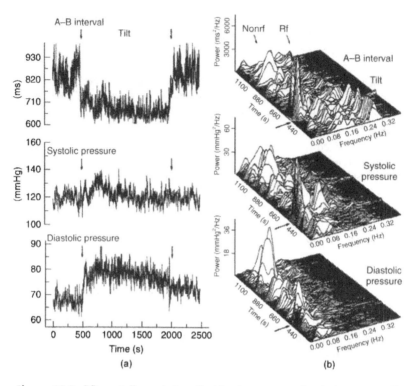

Figure 10.9. RR, systolic, and diastolic blood pressure series during rest-to-tilt maneuver (a) and derived TFR obtained by SPWV (b). The reduction of respiratory components during tilt is evident, as well as the appearance of a dominant LF peak during tilt. The amplitude of this peak is varying in time. The arrow marks the start and the end of the tilt procedure (Novak and Novak, 1993).

cases, the activation of the LF component is maximum before the onset of a syncopal event. Lepikovska and coworkers (1992) also observed a different response in the early phases of tilt, manifesting as an augmented activation of LF components of heart-rate and blood-pressure variability in patients who develop syncope. Dynamic response to tilt has been assessed using tilt at different angles; both the the early and late responses depend on the amplitude of the stimulus (i.e., the tilt angle). In particular, if the time course of the LF power component is described by fitting an exponential function (as depicted in Figure 10.10), both the time constant and amplitude of the fitting curve are functions of the tilt angle. Higher tilt angles evoked faster and higher sympathetic responses, thus suggesting that the ANS may grade the response to the stimulus (Mainardi et al., 2004).

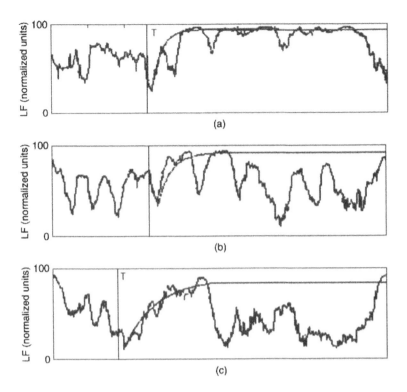

Figure 10.10. Trend of LF component (in normalized units) obtained by tilt at different angles: (a) 90°, (b) 60°, and (c) 45°. Analysis was performed by SPWV. The early phase of tilt (<60 s) is approximated by an exponential to evidence the different dynamics in the ANS responses (Mainardi et al., 1999).

10.7.4 Other Applications

Other applications of quadratic TFR can be found in the analysis of electromyographic signals (EMG) and the electrogastrogram (EGG). The WV has been applied to the analysis of EMG signals, to monitor muscle fatique during slow dynamic contractions (Knalfitz and Bonato, 1999), and to evaluate uterine contractions (Duchène et al., 1995). In the latter study, a certain number of TFRs were compared to verify the applicability of t–f representation to uterine contractions. The general conclusion was that an adaptive AR method allows correct tracking of the main spectral peaks of EMG signals. However, the TFR, computed using different kernels, provides better selectivity in the localization of harmonic components, even if the quantification of amplitude and frequency parameters require a complex postprocessing of the TFR. Finally, in the work by Zheng and coworkers (1989) the time-frequency analysis of a single-motor unit potential is described.

The ED has been used for the analysis of the surface EGG recorded at the abdominal level (Lin and Chen, 1994) in order to identify the gastric contractions and an index of stomach mobility. The EGG is characterized by a sequence of spikes and slow waves that can be properly analyzed by ED. This analysis allowed the authors to characterize the frequency content of the signal and revealed an augmentation of signal energy in the bands of interest (0–2 cpm and 2–4 cpm) during contractions with respect to the periods of gastric quiescence. The authors concluded that ED was a reliable and accurate way to describe the rapid frequency and amplitude changes of EGG signals.

10.8 CONCLUSIONS

This chapter explored the properties of quadratic time-frequency representations and their applications to biomedical signal analysis. Some theoretical concepts about t–f representations were introduced in order to provide the reader with the basic elements to appreciate the potential of these transformations and to judge their correct application. As a general comment, the optimal TFR does not exist; every TFR must be a compromise between time-frequency resolution and cross-term attenuation. However, the way this compromise is managed may vary from representation to representation with advantages/shortcomings that depend on the considered signal and on the measurement to be done. It is a common experience (or frustration) to observe that the same TFR may provide excellent performance in one application and poor results in others. Therefore, it is extremely important to be able to project, or to select, the most appropriate TFR for a certain signal or context. In this regard, the analysis of kernels in the correlative plane may be an immediate and easy-to-understand tool. The presented examples have been selected to give an idea of the flexibility of these transforms and their adaptability to a large variety of biological signals.

REFERENCES

Abeysekera, R. M. S. S., and Boashash, B., Time-Frequency Domain Features of ECG Signals: Their Application in P Wave Detection Using The Cross Wigner–Ville Distribution, in *ICASSP-89*. 23–26 May, 3, pp. 1524–1527, 1989.

Bonato, P., Roy, S. H., Knaflitz, M., and De Luca, C. J., Time-Frequency Parameters of The Surface Myoelectric Signal for Assessing Muscle Fatigue During Cyclic Dynamic Contractions, *IEEE Trans. Biomed Eng.*, Vol. 48, No. 7, 745–53, 2001.

Choi, H. I., and Williams, W. J., Improved Time-Frequency Representation of Multicomponent Signals Using Exponential Kernels, *IEEE Transactions on ASSP,* Vol. 37, 862–871, 1989.

Chouvarda, I., Maglaveras, N., Boufidou, A., Mohlas, S., and Louridas, G., Wigner–Ville Analysis and Classification of Electrocardiograms During Thrombolysis, *Med. Biol. Eng. Comput.,* Vol. 41, No. 6, 609–617, 2003.

Classen, T. A. C. M., and Mecklenbrauker, W., The Wigner Distribution—A Tool for Time-Frequency Signal Analysis—Part I: Continuous-Time Signals, *Philips J. Res.,* Vol. 35, 217–250, 1980a.

Classen, T. A. C. M., and Mecklenbrauker, W., The Wigner Distribution—A Tool for Time-Frequency Signal Analysis—Part II: Discrete-Time Signals, *Philips J. Res.,* Vol. 35, 276–300, 1980b.

Cohen, L., Generalized Phase-Space Distribution Functions, *J. Math. Phys.,* Vol. 7, 781–786, 1966.

Cohen, L., Time-Frequency Distributions—A Review, *Proceedings of the IEEE,* Vol. 77, 941–981, 1989.

De Weerd, J. P. V., and KAP, I., Sepctro-Temporal Representations and Time-Varying Spectra of Evoked Potentials: A Methodological Investigation, *Bio. Cybern.,* Vol. 41, 101–117, 1981.

Duchène, J., Devedeux, D., Mansour, S., and Marque, C., Analizing Uterine EMG: Tracking Instantaneous Burst Frequency, *IEEE Eng. and Med. Biol. Magazine,* Vol. 41, 1867–1880, 1995.

Flandrin, P., Some Features of Time-Frequency Representations of Multicomponent Signals, Acoustics, Speech, and Signal Processing, in *IEEE International Conference on ICASSP '84,* 9, pp. 266–269, 1984.

Furlan, R., Piazza, S., Dell'Orto, S., Barbic, F., Bianchi, A., Mainardi, L., Cerutti, S., Pagani, M., and Malliani, A., Cardiac Autonomic Patterns Preceding Occasional Vasovagal Reactions in Healthy Humans, *Circulation,* Vol. 98, 1756–1761, 1998.

Gabor, D., Theory of Communication, *J. of the IEE (London),* Vol. 93, 429–457, 1946.

Hlawatsch, F., and Boudreaux-Bartels, G. F., Linear and Quadratic Time-Frequency Signal Representations, *IEEE SP Magazine,* April, 21–67, 1992.

Jasson, S., Medigue, C., Maison-Blanche, P., Montano, N., Meyer, L., Vermeiren, C., Mansier, P., Coumel, P., Malliani, A., and Swynghedauw, B., Instant Power Spectrum Analysis of Heart Rate Variability During Orthostatic Tilt Using a Time-Frequency-Domain Method, *Circulation,* Vol. 96, No. 10, 3521–3526, 1997.

Jeong, J., Williams, W. J., Kernel Design for Reduced Interference Distributions, *IEEE Transactions on Signal Processing,* Vol. 40, 402–412, 1992.

Karlsson, S., Yu, J., and Akay, M., Time-Frequency Analysis of Myoelectric Signals During Dynamic Contractions: A Comparative Study, *IEEE Trans. Biomed. Eng.,* Vol. 47, No. 2, 228–238, 2000.

Knalfitz, M., and Bonato, P., Time-Frequency Method Applied to Muscle Fatigue Assessment During Dynamic Contractions, *J. of Electromyography and Kinesiology.* Vol. 9, 337–350, 1999.

Kumaresan, R., and Tufts, D., Estimating the Parameters of Exponentially Damped Sinusoids and Pole-Zero Modeling in Noise, *IEEE Trans. Acoust. Speech Signal Proc.,* Vol. 30, 833–840, 1982.

Lepicovska, V., Novak, P., and Nadeau, R., Time-Frequency Dynamics in Neural-ly Mediated Syncope, *Clin. Auton. Res.*, Vol. 2, 317–326, 1992.

Lin, Z., and Chen, J. D., Time-Frequency Representation of the Electrograstro-gram—Application of the Exponential Distribution, *IEEE Trans. on Biomedical Engineering*, Vol. 41, 267–275, 1994.

Lin, Z., and Chen, J. D., Advances in Time-Frequency Analysis of Biomedical Signals, *Crit. Rev. Biomed. Eng.*, Vol. 24, 1–72, 1996.

Mainardi, L. T., Montano, N., Bianchi, A., Malliani, A., and Cerutti, S., Assessment of the Dynamics of Neurovegetative Response Using Wigner–Ville Transform, in *BMES/EMBS Conference, Proceedings of the First Joint Conference*, 1130–130, 1999.

Mainardi, L. T., Bianchi, A. M., and Cerutti, S., Time-Frequency and Time-Varying Analysis for Assessing the Dynamic Responses of Cardiovascular Control, *Crit. Rev. Biomed. Eng.*, Vol. 30, No. 1–3, 175–217, 2002.

Mainardi, L. T., Montano, N., and Cerutti, S., Automatic Decomposition of Wigner Distribution and its Application to Heart Rate Variability, *Methods Int. Med.*, Vol. 1, 17–21, 2004.

Mark, W. D., Spectral Analysis of the Convolution and Filtering of Non-Stationary Stochastic Process, *J. Sound Vib.*, Vol. 11, 19–63, 1970.

Martin, W., and Flandrin, P., Wigner–Ville Spectral Analysis of Nonstationary Processes, *IEEE Transactions on ASSP*, Vol. 33, 1461–1470.

Morgan, N. H., and Gevins, A. S., Wigner Distributions of Human Event-Related Brain Potentials, *IEEE Trans. Biomed. Eng. Jan.*, Vol. 33, No. 1, 66–70, 1986.

Nayak, A., Roy, R. J., and Sharma, A., Time-Frequency Spectral Representation of the EEG as an Aid in The Detection of Depth of Anesthesia, *Ann. Biomed. Eng.*, Vol. 22, No. 5, 501–513, 1994.

Novak, P., and Novak, V., Time/Frequency Mapping of the Heart Rate, Blood Pressure and Respiratory Signals, *Med. Biol. Eng. Comput.*, Vol. 31, No. 2, 103–110, 1993.

Novak, P., Li, Z., Novak, V., and Hatala, R. Time-Frequency Mapping of the QRS Complex in Normal Subjects and in Postmyocardial Infarction Patients, *J. Electrocardiol.*, Vol. 27, 49–60, 1994.

Novak, V., Novak, P., Kus, T., and Nadeau, R., Slow Cardiovascular Rhythms in Tilt and Syncope, *J. Clin. Neurophysiol.*, Vol. 12, 64–71, 1995.

Page, C. H., Instantaneous Power Spectra, *J. Appl. Phys.*, Vol. 23, pp. 103–106, 1952.

Pola, S., Macerata, A., Emdin, M., and Marchesi, C., Estimation of the Power Spectral Density in Nonstationary Cardiovascular Time Series: Assessing the Role of the Time-Frequency Representations (TFR), *IEEE Trans. Biomed. Eng.*, Vol. 43, No. 1, 46–59, 1996.

Rihaczek, A., Signal Energy Distribution in Time and Frequency, *IEEE Transactions on Information Theory*, Vol. 14, 369–374, 1968.

Stridh, M., Sornmo, L., Meurling, C. J., and Olsson, S. B., Characterization of Atrial Fibrillation Using the Surface ECG: Time-Dependent Spectral Properties, *IEEE Transactions on Biomediacl Engineering*, Vol. 48, 19–27, 2001.

Ville, J., Thérorie et Applications de la Notation de Signal Analytique, *Cables et Trasmission*, Vol. 2A, 61–74, 1948.

Wigner, E. P., On the Quantum Correction for Thermo-Dynamic Equilibrium, *Physics Rev.*, Vol. 40, 1932.

Williams, W. J., Shevrin, H., and Marshall, R. E., Information Modeling and Analysis of Event Related Potentials, *IEEE Trans Biomed Eng.*, Vol. 51, No. 5, 737–743, 1987.

Williams, W. J., Reduced Interference Distribution: Biological Application and Interpretations, *Proceeding of IEEE*, Vol. 84, 1264–1280, 1996.

Zaveri, H. P., Williams, W. J., Iasemidis, L. D., and Sackellares, J. C., Time-Frequency Representation of Electrocorticograms in Temporal Lobe Epilepsy, *IEEE Trans Biomed Eng.*, Vol. 39, No. 5, 502–509, 1992.

Zhao, Y, Atlas, L. E., and Marks, R. J., II, The Use of Cone-Shaped Kernels for Generalized Time-Frequency Representations of Nonstationary Signals, *IEEE Transactions on ASSP*, Vol. 38, 1084–1091, 1990.

Zheng, C., Widmalm, S. E., and Williams, W. J., New Time-Frequency Analyses of EMG and TMJ Sound Signals, in *Proceedings of IEEE International Conference on Engineering in Medical Biology*, Vol. 6, 741–742, 1989.

Thispageintentionallyleftblank

CHAPTER 11

TIME-VARIANT
SPECTRAL ESTIMATION

Anna M. Bianchi

11.1 INTRODUCTION

The parametric approach to the estimation of power spectral density assumes that the time series under analysis is the output of a given process whose parameters are, however, unknown. Sometimes, some a priori information about the process is available, or it is possible to take into account some hypothesis on the generation mechanism of the series, and this can lead to a more targeted selection of the model structure to be used. More frequently, however, the model is independent from the physiology or the anatomy of the biological process generating the data and its formulation is based on input–output relationships, according to the so-called black-box approach. In order to obtain a reliable spectral estimation, a further a posteriori validation of the identified model is needed.

The parametric spectral approach is a procedure that can be summarized in three steps:

1. Choice of the correct model for the description of the data
2. Estimation of the model parameters based on the recorded data
3. Calculation of the power spectral density (PSD) through proper equations (according to the selected model) into which the parameters of the estimated model are inserted

The literature reports various models that can be used for this purpose; see for example (Kay and Marple, 1981). In practice, however, linear models with rational transfer functions are most frequently used; in fact, they can reliably describe a wide range of different signals. Among them, the autoregressive (AR) models are preferred for their all-pole transfer function; in fact, their identification is reduced to the solution of a linear equation system. This is not, however a limitation, as the Wald theorem assures that an AR model, if its order is sufficiently high, is equivalent to an ARMA (autoregressive, moving average) or MA (moving average) model. A proper validation of the selected model is, however, always needed (Box and Jenkins, 1970). In Appendix 1, the procedure for the spectral estimation though parametric linear models is described, whereas Figure 11.1 shows the schematic representation of the parametric spectral estimation performed through an autoregressive model.

The identification of the AR parameters is usually performed by solving a linear prediction problem, as described in Appendix 2. The described method provides an estimation based on a known sequence of data,

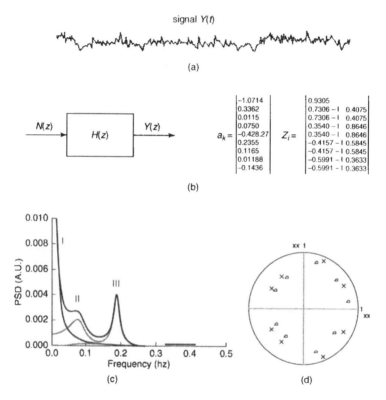

Figure 11.1. (a) The signal $y(t)$ is represented as the output of an AR model whose transfer function $H(z)$ is described as a function of its coefficients a_k or as a function of its poles z_i (b). The corresponding power spectral density is shown in panel (c). The spectral peaks are strictly connected to the poles shown in (d). (Mainardi et al., 1995.)

and when a new value is made available (for example, because a new sample of the signal has been acquired), the whole identification procedure should be restarted. This could lead to considerable problems, for example, in real-time applications. It could be useful in such cases to maintain the already obtained information and evaluate only the innovation that the new sample provides to the model, using recursive methodologies. In the literature, different methods for recursive parametric identification do exist. They allow one to update the set of autoregressive parameters each time a new sample is made available, and find application in real-time processing systems. As better explained in the following, the use of proper forgetting factors makes the updating dependent mainly on the more recent data, allowing the model to track changes in the signal each time the hypothesis of stationarity is not verified. We can then obtain time-variant AR models from which we have spectral estimations that vary in time according to the dynamic changes of the signal.

Adaptive spectral estimation algorithms belong to two main categories: approaches based on the approximation of a gradient (these include the well-known least-mean squares or LMS algorithm) and recursive estimation of least squares algorithms (recursive least squares, RLS). The literature also presents algorithms for the recursive updating of lattice filters that do not update directly the AR coefficients, but rather the reflection coefficients, from which the AR parameters are subsequently evaluated through the Levinson algorithm (Levinson, 1947; Durbin, 1960).

In the present description, we will focus only on LMS and RLS methods, as they are the most used in the literature. As regards the lattice filters, they are particularly useful when the adaptive estimation of the optimal order of the model is also required. For an extensive and exhaustive description, see (Friedlander, 1982a).

11.2 LMS METHODS

As pointed out above, it is possible to obtain the exact least squares solution of the problem by solving Equation 11.A.2.8 in Appendix 2. This implies the inversion of the data autocorrelation matrix S. Alternatively, it is possible to reach the solution in successive steps, through a recursive algorithm able to properly update the model coefficients. Taking into consideration that the optimal solution is on the minimum of the paraboloid that represents the mean square error in the p-dimensional space of the parameters (where p is the order of the autoregressive model), one can reach this minimum in successive steps, moving along the steepest descending path of the paraboloid. At each iteration step, the vector of the parameters is modified proportionally and in the opposite direction of the gradient, according to the relation

$$\hat{a}(t+1) = \hat{a}(t) - \mu \nabla(t) \qquad (11.1)$$

where μ is a scalar value that controls the stability and the convergence speed of the algorithm.

In the stationary case, the minimization corresponds to the adjustment of the AR parameters moving along the paraboloid surface until the minimum is reached.

The choice of the optimal μ value needs analyses that sometimes can be rather complex. However, it is possible to show that, in the stationary case, the algorithm converges for $0 < \mu < \lambda_{max}^{-1}$, where λ_{max} is the maximum eigenvalue of the data autocorrelation matrix S. On the other hand, the convergence speed depends on the minimum eigenvalue and can be approximated as: $T = 1/4\mu\lambda_{min}$ (Widrow and Stearns, 1985). In conclusion, the performance of the algorithm is strongly dependent on the features of the input data. Further, it is worth noting that this is true only when the gradient is exactly evaluated according to the definition $\nabla(t) = -2Q + 2Sa(t)$, whereas, in the real practice, we can only have estimations of the S matrix and the Q vector, which are obtained from the available data. Thus, errors are introduced in the calculation of the gradient, which makes less predictable the behavior of the algorithm.

In the nonstationary case, the minimum to be reached is continuously moving and the algorithm needs to track it. This is possible when the input data are slowly varying in respect to the convergence speed of the algorithm. In such a case, the estimation of S and Q also needs to be updated for each new sample added to the known sequence. There is, however, the possibility of updating these quantities recursively, according to these relations:

$$
\begin{aligned}
Q(t) &= Q(t-1) + \varphi(t)\varphi(t) \\
S(t) &= S(t-1) + \varphi(t)\varphi(t)^T
\end{aligned}
\qquad (11.2)
$$

In such a way, the LMS algorithm is able to update the AR model by keeping in consideration each new sample in the signal and then providing estimations of the power spectral density that are functions of the time, according to Equation A.1.4 (Appendix 1).

11.3   RLS ALGORITHM

The solution of the least squares method, given by Equation 11.A.2.8 in Appendix 2, can be easily implemented in recursive form by simply taking into consideration Equation 11.2. It is then possible to obtain the following formulation (Soderstom and Stoica, 1989):

$$\begin{cases} \hat{a}(t) = \hat{a}(t-1) + K(t)\varepsilon(t) \\ K(t) = S(t)^{-1}\varphi(t) \\ \varepsilon(t) = y(t) - \varphi(t)^T \hat{a}(t-1) \\ S(t) = S(t-1) + \varphi(t)\varphi(t)^T \end{cases} \tag{11.3}$$

In such a case, the parameter vector $\hat{a}(t)$ is given by the sum of the same parameters obtained at the previous time instant $(t-1)$ and of a correction term that is proportional to the estimation error $\varepsilon(t)$ weighed according to a gain vector $K(t)$. Further, thanks to the matrix inversion lemma, the algorithm is made more efficient, as it is possible to directly update the matrix $P(t) = S(t)^{-1}$ without inversions at each iteration:

$$\begin{cases} \hat{a}(t) = \hat{a}(t-1) + K(t)\varepsilon(t) \\ K(t) = \dfrac{P(t-1)\varphi(t)}{1 + \varphi(t)^T P(t-1)\varphi(t)} \\ \varepsilon(t) = y(t) - \varphi(t)^T \hat{a}(t-1) \\ P(t) = \left[P(t-1) - \dfrac{P(t-1)\varphi(t)\varphi(t)^T P(t-1)}{1 + \varphi(t)^T P(t-1)\varphi(t)} \right] \end{cases} \tag{11.4}$$

If the samples of the signal come from a nonstationary process, we can introduce into the recursive formulation, a forgetting factor, λ, that modifies the figure of merit J (Equation 11.A.2.8 of Appendix 2) according to the following relation

$$J = \frac{1}{t}\sum_{i=1}^{t} \lambda^{t-i}\varepsilon(i)^2 \tag{11.5}$$

The forgetting factor (which assumes values $\lambda \leqslant 1$), exponentially weights the samples of the prediction error in the calculation of J, then gives importance to the more recent values in the definition of the updating while the oldest ones are progressively forgotten with a time constant, $T = 1/(1 - \lambda)$, that can be interpreted as the "memory length" of the algorithm.

In conclusion, by the combination of the recursive formulation and the forgetting factor, we can obtain the following updating formulation of the autoregressive model, which is recursive and, in addition, allows the adaptation of the model according to the dynamical variations of the signal (Bittanti and Campi, 1994a):

$$
\begin{cases}
\hat{a}(t) = \hat{a}(t-1) + K(t)\varepsilon(t) \\[2mm]
K(t) = \dfrac{P(t-1)\varphi(t)}{\lambda + \varphi(t)^T P(t-1)\varphi(t)} \\[2mm]
\varepsilon(t) = y(t) - \varphi(t)^T \hat{a}(t-1) \\[2mm]
P(t) = \dfrac{1}{\lambda}\left[P(t-1) - \dfrac{P(t-1)\varphi(t)\varphi(t)^T P(t-1)}{\lambda + \varphi(t)^T P(t-1)\varphi(t)} \right]
\end{cases}
\qquad (11.6)
$$

Now, for each new time instant t we can obtain a new parameter vector $a(t)$ that in Equation 11.A.1.4 allows us to obtain a $P_y(f, t)$ and gives a description of the frequency content of the signal at the same time in the frequency and time domains.

11.4   COMPARISON BETWEEN LMS AND RLS METHODS

The criteria for the evaluation of the performances of a recursive algorithm are quite different and are often linked to the specific application, but are, in general, based on the following factors:

- *Speed of Convergence.* This is defined as the number of iterations needed for the algorithm to converge to the optimal least squares solution when the input is stationary. In general, a high convergence speed allows the algorithm to quickly adapt in a stationary situation, or to track the changes of the model in nonstationary cases. On the other hand, this could lead to an excessive sensitivity to the casual noise superimposed on the signal and thus to too variable and noisy estimations.

  Figure 11.2 shows the typical convergence curves for LMS and RLS algorithms (Marple, 1987). In general, the RLS algorithm converges

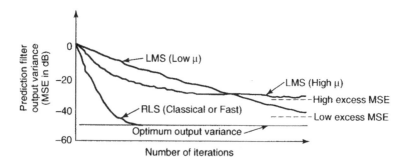

Figure 11.2. Convergence times for the algorithms LMS and RLS (Marple, 1987).

faster than the LMS. In noisy situations, or when the difference among the eigenvalues of the autocorrelation matrix is small, the two algorithms have similar convergence speeds. However, the LMS algorithm autogenerates noise as a consequence of using an estimation of the gradient and not its true value. This leads to bias in the estimation, the amplitude of which can be decreased by using lower values for μ and then increasing the convergence time.

- *Computational load.* This is related to the number of operations needed to perform a full iteration of the algorithm, the memory allocation needed for recording the data, and the software. All these considerations may assume relevant importance in real-time and portable applications. From this point of view, the LMS algorithm has some advantages as it is simpler and requires an operation number proportional to N (number of input samples), whereas the RLS algorithm is more complex and the computational load is proportional to N^2.

- *Numerical properties.* These deal with the way an error introduced in a certain point of the algorithm is propagated to the future estimations, and also with the effects of the truncation error on the output.

Both the algorithms are affected by quantization and rounding errors due to the use of a finite number of bits in the memory words. The rounding error can increase as the number of the iterations increases, and this could be an untrivial problem when the algorithms are implemented on hardware systems for the spectral estimation in real time. It has been experimentally proved that the RLS algorithm needs al least 10 bits, whereas for LMS seven are sufficient. The two methods provide comparable results when al least 12 bits are used (Ling and Proakis, 1984).

11.5   DIFFERENT FORMULATIONS OF THE FORGETTING FACTOR

Of the two presented algorithms, RLS is the most used in literature. It is worth remembering that its performance is strongly dependent on the choice of the forgetting factor λ. Of course, the choice of the optimal forgetting factor is a critical point in the use of the time-varying models. In fact, high values of λ may lead to inability to reliably track the fast dynamics of the signal, whereas too low values may make the algorithm too sensitive to the casual variations due to the noise.

For these reasons, in the literature different formulations of the forgetting factor have been proposed that attempt to finding an optimal balance between the convergence speed and noise rejection. Some of these are really interesting, at least from a theoretical point of view (Fortescue and Ydstie, 1981; Kulhavy and Karrny, 1984; Lorito, 1993), even if at expense of a increased complexity of the algorithm and, sometimes, of the need for a priori information about the process under examination. In the following, we are presenting some of the proposed solutions that have relevant advantages at a practical level.

11.5.1 Varying Forgetting Factor

The prediction error contains relevant information about the goodness of the estimation. In fact, if its variance is small, the model is properly fitted to the data and the dynamic of the signal variation is slower than the adaptation of the algorithm. Thus, we can think of using a higher forgetting factor for making the estimation more reliable from a statistical point of view. If, on the contrary, the noise variance is high, the model is still converging, or the dynamics of the signal changes are faster than the adaptation capability of the algorithm. In such conditions, it could be useful to decrease the value of the forgetting factor in order to allow a faster convergence. Based on these considerations, Fortescue and Ydstie (1981) proposed the use of a varying forgetting factor able to self-adapt to the signal characteristics, increasing when the signal is slowly varying, and decreasing when transitions are fast. For doing that, it is necessary to define a parameter that is a measure of the *information content* of the algorithm. For this purpose, Fortescue and Ydstie (1981) proposed the parameter

$$\Gamma_0(t_0) = \sigma_0^2 N_0 \tag{11.7}$$

where σ_0^2 is an estimation of the variance of the prediction error and N_0 is the nominal length of the adaptation window, or the "memory" of the algorithm, that defines the convergence speed. $\Gamma(t)$ can be recursively updated:

$$\Gamma(t) = \lambda(t)\Gamma(t-1) + [1 - \varphi^T(t)K(t)]\varepsilon(t)^2 \tag{11.8}$$

If the information content $\Gamma(t)$ is constant, that is,

$$\Gamma(t) = \Gamma(t-1) = \ldots = \Gamma(t_0) \tag{11.9}$$

It is possible to obtain from Equation 11.8 the value of the forgetting factor for each time instant t:

$$\lambda(t) = \frac{1 - \phi(t)^T K(t)\varepsilon(t)^2}{\Gamma_0(t_0)} \tag{11.10}$$

This approach guarantees constant estimation performance during the whole signal recording, in spite of possible variations in the dynamics. However, it is worth remembering that the algorithm performance is strongly dependent on the $\Gamma_0(t_0)$ value assigned at the beginning and, in particular, on the σ_0^2 value that needs to be known or a priori estimated with good precision.

11.5.2 Whale Forgetting Factor

From the approximate analysis of the estimation error (Lorito, 1993), it is possible to calculate how casual noise in the input data can affect the estimation error of the parameters. This relation is described by the transfer function that in case of the exponential forgetting factor (EF) has the following expression:

$$G^{EF}(z) = \frac{1-\lambda}{1-\lambda z^{-1}} \tag{11.11}$$

This is a low-pass filter with only one pole in $z = \lambda$, on which the properties of speed, adaptation, and noise rejection depend.

The compromise between noise sensitivity and adaptation speed can be made less restrictive if we increase the degrees of freedom of the filter, for example, by increasing the number of the coefficients of its transfer function. With a higher number of poles, in fact, it is possible to modulate the shape of the impulse response and then the sensitivity to the noise and the adaptation speed. A solution adopted in literature uses a second-order transfer function:

$$G^{WF}(z) = \frac{1-a_1-a_2}{1-a_1 z^{-1}-a_2 z^{-2}} \tag{11.12}$$

where the coefficients are chosen in order to guarantee the filter stability (poles inside the unitary circle). Figure 11.3 shows the comparison between the impulse responses (a) and the frequency responses (b) of the exponential forgetting coefficient (EF) and the two-pole coefficient, that, because of its impulse-response characteristic shape, is called *whale forgetting factor* (WF). It is possible to see how the $G^{WF}(z)$ algorithm is able to give a weight that rapidly decays with time [and is comparable with $G^{EF}(z)$], but better averages the contribution of the most recent data and then provides a low-pass filtering of the high-frequency noise that is superimposed on the signal. The updating equations for $Q(t)$ and $S(t)$ then become:

$$Q(t) = a_1 Q(t-1) + a_2 Q(t-2) + (1-a_1-a_2)\varphi(t)y(t) \tag{11.13}$$
$$S(t) = a_1 S(t-1) + a_2 S(t-2) + (1-a_1-a_2)\varphi(t)\varphi(t)$$

It is worth noting that in such a case the matrix inversion lemma cannot be applied; then, at each iteration, it is necessary to invert the matrix

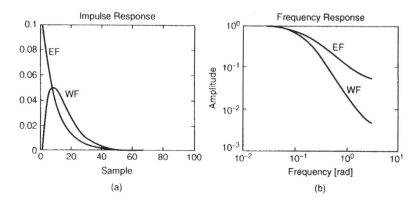

Figure 11.3. (a) Weight functions of the errors for the exponential forgetting factor (EF) and the whale forgetting factor (WF). (b) Corresponding frequency responses. (Bianchi et al., 1997.)

$S(t)$. Appendix 3 shows some simulations for clarifying the improvements due to the varying forgetting factor and to the whale forgetting factor in comparison with the time-constant exponential forgetting factor.

11.6 EXAMPLES AND APPLICATIONS

The methods of time-variant autoregressive spectral estimation have remarkable advantages that make them suitable to many different applications. Among them the most diffused are in the field of the studies of heart rate variability and, from a more general point of view, beat-to-beat variability signals related to the cardiovascular and cardiorespiratory systems. Many studies that can be found in the literature mainly deal with the autonomic nervous system during myocardial ischemia episodes, both spontaneous and drug induced (Bianchi et al., 1993; Mainardi et al., 1995; Petrucci et al., 1996), with the response to autonomic tests (Barbieri et al., 1997), with monitoring of patients in intensive care (Cesarelli et al., 1997), and with the autonomic response to drug infusion, stress tests, and transition among different sleep stages (Mendez et al., 2010).

In neurology, these methods are mainly applied to the dynamic variations of the EEG signal, for example, during anaesthesia induction (Witte et al., 2008), cases of brain damage, (Muthuswamy and Thakor, 1998), transitions toward epileptic seizures (Gath et al., 1992; Sun et al., 2001), study of desynchronization and synchronization of the different EEG rhythms during the execution of motor tasks (event-related potentials) (Tarvainen et al., 2004), for the study of the brain connectivity (Wilke et al., 2008), and also in the study of fMRI BOLD signals (Hemmelmann et al., 2009).

Other applications are related to the electrogastrogram (Chen, 1992), ventricular late potentials (Haberl et al., 1994), cardiac sounds and the electromyogram (Duchene et al., 1995), and, in general, all those signals with a frequency content that varies in time. Exhaustive reviews of the different applications are in Lin and Chen (1996), Cerutti et al. (2001), and Mainardi et al. (2002). We now provide a brief description of the most relevant ones from a clinical and physiological point of view.

11.6.1 Myocardial Ischemia

Myocardial ischemia is a pathology widely diffused in the industrialized countries. It can be defined as an imbalance between oxygen need and oxygen supply in the myocardial tissues. In general, a reduction of the coronary cross section, due to the presence of arterial plaques or to a vasospasm, is the cause of reduced blood flow that is insufficient for cardiac activity in some situations such as physical stress. At the electrophysiological level, the ischemic episodes are evidenced by displacements of the ST segment in the ECG signal. Several studies, performed both on animals and on human subjects, documented the simultaneous involvement of the autonomic nervous system. Autonomic alterations have been proven to be a consequence of ischemic events (Rimoldi et al., 1990; Joho et al., 1999). On the other hand, sympathetic activation seems also to be one of the possible causes (Lanza et al., 1996; Chierchia et al., 1990; Goseki et al., 1994). The sympatho–vagal balance is studied mainly in the frequency domain, but the main problem is, in the fact, that the analyzed phenomena have marked transient characteristics. A complete ischemic episode may vary in time from a few tens of seconds to a few minutes, during which the heart rate variability signal shows high levels of nonstationarity. Further, the involved phenomena are quite complex and are the consequence of many different factors, such as pathogenesis and ischemic site. In the study of Bianchi et al. (1997), two different populations of cardiac patients have been analyzed: one was affected by chronic stable ischemia (with characteristic plaques at the coronary level), and the other was affected by variant or Printzmetal ischemia (due to vasospasm). Figure 11.4 shows an example related to both the examined groups. We can observe the tachograms corresponding to the ischemic event (B and E mark the beginning and the end, respectively, as identified from the ST displacement on the ECG), the sequences of the power spectral densities calculated on a beat-to-beat basis, and the temporal trend of the LF/HF ratio, whose increases can quantify the sympathetic activation at the autonomic level (Task Force, 1996). In both the subjects, an increase in the sympathetic activation is well evident, corresponding to the ischemic event; however, it is possible to distinguish different pathogenetic mechanisms. In the former case, in fact, the sympathetic activation is clearly a reflex response, as it appears with a delay after

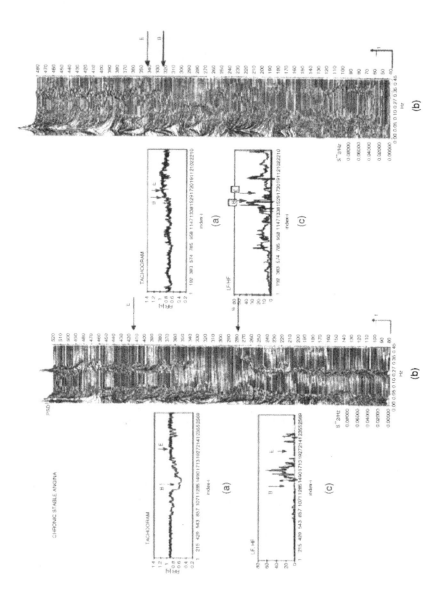

the beginning of the ST segment displacement. On the contrary, in the latter case the sympathetic activation reaches its maximum before the ischemia, thus provoking the event. In such a case, a more severe ischemia may lead to a temporary denervation of the sympathetic fibers and to a following vagal response (Lombardi, 1984). Similar results have great importance for a better comprehension of the mechanisms at the genesis of the ischemic attacks. It is then possible to gain an improvement in the therapy, which can be better focused. It is, however, worth noting that in analyzing spontaneous events coming from Holter recordings, several subjective factor related to daily life may lead to different interpretations: rest or stress conditions, day or night events, attack with or without pain, and so on. In order to operate in more controlled situations, and to obtain more comparable results, some clinical tests are based on the pharmacological induction of ischemia through infusion of dipyridamole or dobutamine. A study performed on subjects who underwent the dypiridamole test (Petrucci et al., 1996), showed that patients with a positive test showed a marked increase in sympathetic activation in respect to the basal conditions; further, such an increase was directly correlated with the ischemic indices obtained on an echocardiographic basis. In conclusion, the sympathetic activation was not only a typical response to the ischemic status, but it was also correlated to the severity of the pathology.

11.6.2 Monitoring EEG Signals during Surgery

During carotid endoarterectomy surgery, the carotid artery is temporarily clamped in order to allow the removal of the arterosclerotic plaques. This implies a dramatic reduction in the blood flow toward the brain that, in some cases, is not replaced by the collateral circulation and may cause permanent cerebral damage. A clinical indication of brain distress is the decrease in spectral power in the high-frequency range (7–21 Hz) of the EEG signal (Minicucci et al., 2000). Monitoring is usually performed by an expert neurophysiologist who visually inspects the traces scrolling on the monitor. The frequency domain analysis can provide a quantitative and more objective evaluation of the traces (Bianchi et al., 2000). In particular, methods of time-variant autoregressive estimation are suited for this purpose for two main reasons: the possibility of real-time application for the

Figure 11.4. (a) Heart rate variability signals from Holter recordings during ischemic events. (b) Seqeunce of the power spectral densities obtained on a beat-to-beat basis. (c) Trend of the LF/HF ratio. B and E mark, respectively, the beginning and the end of the ischemic event as detected on the basis of the ST segment displacement on the ECG signal. On the left, chronic stable angina; on the right, variant angina. (Bianchi et al., 1997.)

evaluation of the frequency content of the signal sample by sample, and the possibility of obtaining directly the spectral parameters of clinical interest through the automatic spectral decomposition (Zetterberg, 1969; Baselli et al., 1987).

11.6.3 Study of Desynchronization and Synchronization of the EEG Rhythms during Motor Tasks

It is well known that during the execution of voluntary movements or even only through movement imagery, the electroencephalographic rhythms recorded from the motor cortex present some modifications (Pfurtscheller and Lopes da Silva, 1999). In particular there is power decreasing in the alpha range (alpha desynchronization or alpha ERD—event-related desynchronization) related to the beginning of the movement, but preceding its execution. In fact, the alpha rhythm is characteristic of the resting phase and comes back to its basal value at the end of the movement. The beta rhythm is also desynchronized during the movement, even to a lesser degree, and there is a marked power increase (beta ERS) at the end of the movement. Graimann and coworkers (2003) analyzed EcoG (electrocorticograms obtained directly in the cortex) from patients who underwent this examination for presurgical evaluation in the treatment of epilepsy. Figure 11.5 shows as an example some traces recorded during the execution of the forefinger voluntary movements. It is possible to appreciate the fast dynamics in the signal. Usually, such dynamics are quantified through time-frequency techniques already described in this book (short-time Fourier transform, wavelet transform, quadratic distributions). However, the peculiar characteristic of these ERD/ERS patterns, which can be consciously generated and modified by the subject, make them the ideal input for brain–computer interface (BCI) applications, for which real-time processing is mandatory.

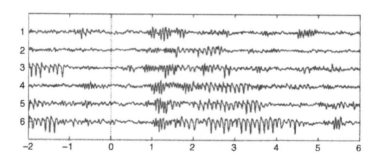

Figure 11.5. Example of EcoG during finger movements. The dotted line represents the starting time of the movement. The rhythm desynchronization is well visible around the start of the movement. (Graimann et al., 2003.)

The adaptive parametric models are particularly suitable for this application. Figure 11.6 shows the ERD/ERS trend (evaluated as power percentage variation in respect to the basal value, before the beginning of the movement) in the frequency range 14–16 Hz during the movement of the forefinger. Similar trends can be obtained in different cortical sites according to the movements of different body parts (right or left forefinger, tongue, lips, etc.). Proper classification techniques can detect and localize

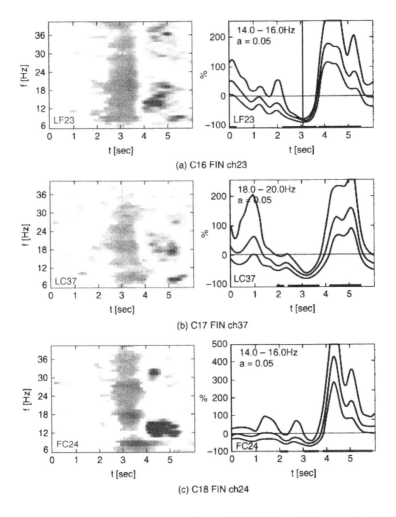

Figure 11.6. Time-frequency maps of ERD/ERS (left) and time trend of the ERD/ERS in the range 14–16 Hz (right). The dotted vertical line marks the movement starting point. ERD/ERS values are represented with a confidence interval of 95%. (Graimann et al., 2003.)

the temporal trends of ERD/ERS associated with different movements. It is thus possible to detect in the EEG signal different activation states that can be the needed inputs for communicating with a PC.

11.7   EXTENSION TO MULTIVARIATE MODELS

The described methodologies of time-variant spectral estimation can be easily extended to multivariate models, allowing the quantification of interactions among different signals, in particular through the calculation of the phase and coherence relationships among different rhythms, and of their causal dependencies. Their temporal evolutions can be evaluated when physiological conditions change or in the presence of a pathology.

The structure of the AR model is extended to a linear multichannel system S with M variables:

$$S = \begin{cases} y_1(t) = \sum_{k=1}^{p} a_{11}(k)y_1(t-k) + \sum_{k=1}^{p} a_{12}(k)y_2(t-k) + \ldots + \sum_{k=1}^{p} a_{1M}(k)y_M(t-k) + e_1(t) \\ \\ y_2(t) = \sum_{k=1}^{p} a_{21}(k)y_1(t-k) + \sum_{k=1}^{p} a_{22}(k)y_2(t-k) + \ldots + \sum_{k=1}^{p} a_{2M}(k)y_M(t-k) + e_2(t) \\ \vdots \\ y_M(t) = \sum_{k=1}^{p} a_{M1}(k)y_1(t-k) + \sum_{k=1}^{p} a_{M2}(k)y_2(t-k) + \ldots + \sum_{k=1}^{p} a_{MM}(k)y_M(t-k) + e_M(t) \end{cases} \tag{11.14}$$

where $y_i(t)$ $\{i = 1, 2, \ldots, M\}$ are the signals, p is the model order, and $e_i(t)$ are the white noises input to the system. The matrix formulation of the system is

$$Y(t) = \sum_{k=1}^{p} A(k)Y(t-k) + W(t) \tag{11.15}$$

Also, in this case it is possible to apply the recursive estimation, simply extending to the multivariate case the expressions in Equation 11.7, as described in detail in Mainardi et al. (1997).

Once the coefficients of the multivariate model are known for each sample in the signal, these can be used for the calculation of the multichannel-power spectral density according to the equation

$$S_t(f) = \left| [I - \Theta_t(z)]^{-1} E_t [I - \Theta_t(z)]^{-H} \right|_{z=e^{i2\pi f}} \tag{11.16}$$

where $-H$ indicates the hermitian matrix (traspose of the inverse matrix), $\Theta_n(z)$ is the z transform of the parameter matrix $\theta(t)$, and E_t is the covariance matrix of the input multichannel noise at time t. The diagonal elements in the matrix S_t are the power spectral densities of the signals, whereas the extradiagonal elements are the cross spectra between signal pairs. From them, we can calculate coherence and phase relationships.

This structure is quite general and can be adapted to any set of signals, such as beat-to-beat variability signals obtained from the cardiovascular system or multichannel EEG recordings.

In the analysis of the cardiovascular variability series, bivariate models ($M = 2$) are particularly interesting, because they are well suited for the description of the reciprocal interactions between the RR (r) series of variability and the blood pressure systolic (s) series of variability. In such a case, the model is reduced to the following structure expressed in the domain of the z transform:

$$S = \begin{cases} s(z) = A_s(z) \cdot s(z) + B_s(z) \cdot r(z) + w_1(z) \\ r(z) = B_r(z) \cdot s(z) + A_r(z) \cdot r(z) + w_2(z) \end{cases} \quad (11.17)$$

in which r and s act on each other through the transfer functions H_{sr} and H_{rs}. In the model that is graphically represented in Figure 11.7, H_{sr} describes the effects of r on s (prevalent mechanical effects), while H_{rs} describes the action of s on r that is mediated by neural reflexes, mainly of baroreceptive origin, but reflecting also other mechanisms. Thus, the gain of H_{rs} can be a measure of the α baroreceptive gain. The beat-to-beat identification of the model coefficients thus allows one to obtain the transfer functions, spectra, coherences and phases, baroreceptive gain, and the spectral power in the frequency range of interest during transient phenomena, and provides the interpretation of dynamic phenomena such as vaso–vagal syncope. In this event, the hemodynamical situation is characterized by hypotension and bradicardia, as shown in Figure 11.8a, in which the variability series related to the cardiac cycle, pressure values, and respiration are shown during a vaso–vagal syncope induced by tilting. The T marks the tilting time, whereas S marks the beginning of the syncope. From the variability signals, we can observe synchronous phenomena of bradicardia and hypotension that are commonly associated with inhibitory reflexes, mediated by the autonomic nervous system, that act on the cardiovascular system. The time-variant spectra related to the tachogram and to the systogram are shown in Figure 8b. After the tilting maneuver, the LF component increases in power both in the tachogram and in the systogram, and is high until the beginning of syncope. From a detailed analysis of the temporal trend of the spectral and cross-spectral parameters (not shown in this chapter), it was possible to see that the systogram LF power reaches its maximum before the beginning of hypotension then gradually decreases, whereas the LF tachogram power reaches its maximum later and abruptly decreases. At the same time, in the LF range the coherence

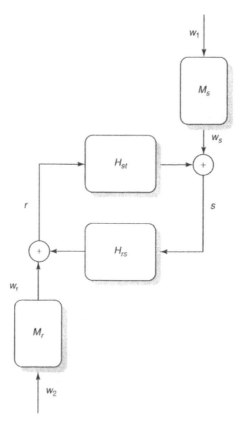

Figure 11.7. Closed-loop model describing the interactions between the tachogram t and the systogram s. The block H_{st} describes the short-time mechanical actions through which t affects s, whereas H_{ts} accounts for the baroreflex action neurally mediated from s to t.

between the two signals decreases as well as the baroreceptive gain (Mainardi et al., 1997). These results suggest the idea of a different neural control action acting on the heart and on the vessels, and confirm the hypothesis of an inhibition of the baroreflex control during syncope. Thus, a reduced link between cardiac and vascular regulation is identified as a likely cause at the origin of the syncope (Furlan et al., 1998).

Other authors (Barbieri et al., 1997) extended the model to three inputs, in order to include also the action of the respiratory activity on the cardiovascular variables. Sometimes, a similar approach can be even too generic, whereas the knowledge of the underlying anatomy and physiology may enable a simpler but surely best-fitted formulation of the model for the description of the phenomena. In fact, we should take into consid-

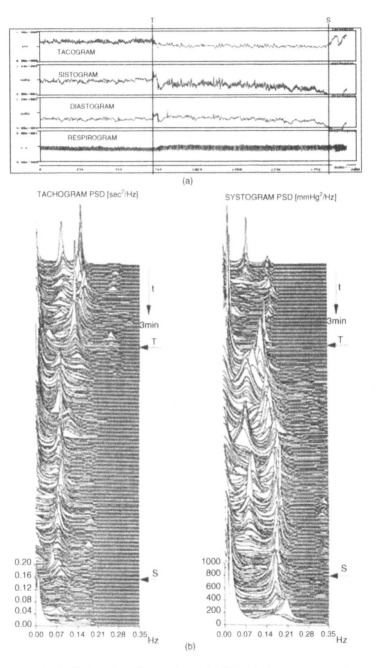

(a)

(b)

Figure 11.8. (a) Series of cardiovascular variabilities during a vaso–vagal syncope episode. (b) Time-variant power spectral densities obtained from the tachogram (on the left) and the systogram (on the right). T marks the tilting time; S is the syncope.

eration that when analyzing interrelations between variability of the heart rate and the arterial pressure, a closed loop must be considered, and while modeling the effects of respiration on the heart frequency, we should be aware that feedbacks from heart frequency to respiratory activity are considered unreliable or, at least, can be neglected (Baselli et al., 1992; Korhonen et al., 1996a,b; Barbieri et al., 1997). Thus, the model assumes the structure shown in Figure 11.9 and can be interpreted, in the time-variant formulation, as an adaptive filter in which the respiration is the reference input (Bianchi et al., 1994). Its usefulness lies in its capability to separate, in the HRV signal, the respiratory sinus arrhythmia (RSA) from the variability coming from other sources (nonrespiratory sinus arrhythmia, or NRSA).

In fact, it is well known that the respiratory arrhythmia acts on the heart-rate variability signal, mainly through neural mechanisms of vagal origin; thus, it is crucial to correctly separate such a contribution from the whole variability in order to correctly evaluate the sympatho–vagal balance acting in the regulation of heart frequency. In controlled situations (i.e., during laboratory experiments), the separation between LF and HF components is rather simple. In other cases, it would be more useful to separate the whole variability in power coherent or not coherent with the breath frequency (Bianchi et al., 1990). When the respiratory activity is nonstationary, it is necessary to use the model in its time-variant implementation. The model was used in the study of the heart-rate variability during sleep. Sleep, in fact, is not a stationary condition but actually constituted by different alternating sleep stages, classified according the different sleep depths. Among them, REM (rapid eye movement) sleep seems related to oneiric activity. The different stages are classified on the basis of the EEG signal, but they also affect the activity of the autonomic nervous

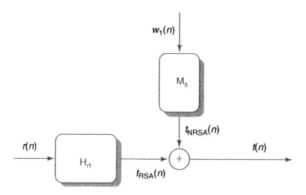

Figure 11.9. Model describing the tachogram as superimposition of activity coherent with respiration (RSA) and activity not coherent with respiration (NRSA).

system with rapid variations in the sympatho–vagal balance (Scholz et al., 1997). The batch frequency analysis is then improper. Figure 11.10(a) shows a tachogram obtained during transition from sleep stage 2 to REM stage. Panel (b) shows the corresponding respirogram, whereas panels (c) and (d) show, respectively, RSA and NRSA. The sequences of the power spectral densities are plotted in Figure 10.11(a) (tachogram), and (b) (respirogram), whereas panels (c) and (d) show, respectively, the power of the tachogram that is coherent and not coherent with respiration. It is worth noting that during REM sleep, respiration also affects the low-frequency band of the tachogram (due to its irregularity); however, it is always possible to see that the most of the power at low frequency is not dependent on respiration and can be related to orthosympathetic activations (Bianchi et al., 1994).

11.8 CONCLUSION

Time-variant parametric models are not as diffused in the literature as widely as linear decompositions or quadratic distributions are. However their use as well as the number of applications is increasing. From an application point of view, they can seem more complex than other presented

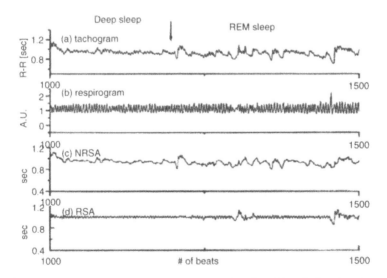

Figure 11.10. Tachogram (a) and respirogram (b) of a subject during sleep in correspondence with a transition from deep sleep to REM sleep. In (c) and (d), the components of the tachogram NRSA (nonrespiratory sinus arrhythmia) and RSA (respiratory sinus arrhythmia) are represented.

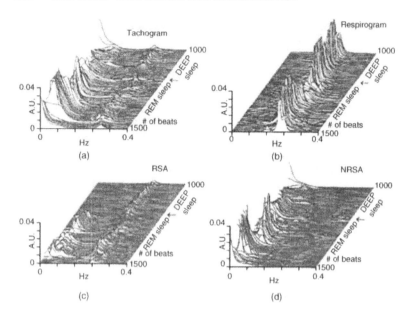

Figure 11.11. Time variant power spectral densities obtained trough the model shown in Figure 11.9. (a) Tachogram, (b) respirogram, (c) power of the tachogram coherent with respiration, (d) power of the tachogram not coherent with respiration.

methods. In fact, they require that the parameters of the algorithm (i.e., the forgetting factor λ and the model order) be properly chosen in order to guarantee reliable results. However, they present remarkable advantages that make them particularly useful for the analysis of biological signals. Among them are the possibility of choosing, independently, time resolution and frequency resolution; the former depending on the value of the forgetting factor and the latter on the model order. That guarantees a high flexibility of the algorithms that can be tuned according to the application. Further, autoregressive models can be easily extended to the description of multivariate systems with any number M of input variables, which enables the description of complex interrelationships among many signals in complex physiological systems such as cardiovascular or multichannel EEG recordings. Another important advantage is that only time-variant models can be implemented in real time; in fact, they are updated sample by sample. Finally, autoregressive models, thanks to the automatic spectral decomposition, directly provide the spectral parameters of interest such as power and frequency values related to the different physiological rhythms. On the other hand, it is worth noting that none of the presented methods can be viewed as optimal, but each of them is able to highlight specific characteristics of the analyzed process. Sometimes, a proper combination of more time-frequency techniques may result in a more complete tool for the analysis of a biological signal or phenomenon.

APPENDIX 1. LINEAR PARAMETRIC MODELS

In a linar system (with a rational transfer function), the input $\{w(t)\}$ and output $\{y(t)\}$ sequences (which represent the data under analysis) are linked by the following difference equation:

$$y(t) = \sum_{j=1}^{q} b_j w(t-j) - \sum_{i=1}^{p} a_i y(t-i) + w(t) \qquad (11.A.1.1)$$

where $w(t)$ is a white noise with zero mean and variance σ^2 and accounts for the nonpredictable part of the signal. This is an autoregressive moving average model (ARMA) and represents the more general situation for the models with a rational transfer function. In the z-transform domain, we can obtain the transfer function of the system:

$$H(z) = \frac{B(z)}{A(z)} = \frac{\sum_{m=0}^{q} b_m \cdot z^{-m}}{\sum_{k=0}^{p} a_k \cdot z^{-k}} \qquad (11.A.1.2)$$

The power spectral density of the output signals, $P_y(z)$, is evaluated according to the expression

$$P_y(f) = P_{ARMA}(f) = T\sigma^2 |H(z)|^2_{z=\exp(2\pi fT)} \qquad (11.A.1.3)$$

where T is the sampling frequency of the signal.

If all the parameters a_k (except for $a_0 = 1$) are set to 0, the process is a q-order moving average. On the contrary, if all the b_k are zero except for $b_0 = 1$, the process is autoregressive average of order p and the corresponding power spectral density (PSD) or $P_y(f)$ is calculated as

$$P_y(f) = P_{AR}(f) = \sigma^2 T \left| \frac{1}{A(z)} \right|^2_{z=e^{j\cdot}} \qquad (11.A.1.4)$$

APPENDIX 2. LEAST SQUARES IDENTIFICATION

The following expression represents an order p autoregressive (AR) linear model:

$$y(t) = a_1 y(t-1) + a_2 y(t-2) + \ldots + a_p y(t-p) + w(t) \quad \text{(11.A.2.1)}$$

where $w(t)$ is a white noise with zero mean and variance σ^2 and the parameter vector is defined as

$$a = [a_1, a_2, \ldots a_p]^T \qquad \text{(11.A.2.2)}$$

and the observation vector is

$$\phi(t) = [y(t-1), y(t-2), \ldots, y(t-p)]^T \qquad \text{(11.A.2.3)}$$

Then a more synthetic expression is given by

$$y(t) = \phi^T(t) a + w(t) \qquad \text{(11.A.2.4)}$$

The corresponding model in prediction form is:

$$\hat{y}(t) = \phi(t)^T a \qquad \text{(11.A.2.5)}$$

where $\hat{y}(t) = \hat{y}(t/t-1)$, and $y(t)$ is the acquired sample. Thus, the prediction error can be defined as

$$\varepsilon(t) = y(t) - \phi(t)^T a \qquad \text{(11.A.2.6)}$$

The minimization of the quadtratic cost function,

$$J_N = (1/N) \sum_{t=1}^{N} \varepsilon_\theta^2(t) \qquad \text{(11.A.2.7)}$$

that is, of the mean square error, leads to the unique solution

$$\hat{a} = \left[\sum_{t=1}^{N} \phi(t)\phi(t)^T \right]^{-1} \sum_{t=1}^{N} \phi(t)y(t) = S(N)^{-1}Q(N) \quad \text{(11.A.2.8)}$$

that is the least square (LS) estimator of a linear regression problem.

APPENDIX 3. COMPARISON OF DIFFERENT FORGETTING FACTORS

For a direct comparison among the different formulations of the described forgetting factors, some simulations have been carried out. Ten realizations of an autoregressive process have been generated according to the model described by the equations

$$y(t) = 1.6\,y(t-1) - 0.95\,y(t-2) + e(t) \quad t < 300$$
$$y(t) = -1.6\,y(t-1) - 0.95\,y(t-2) + e(t) \quad t \geq 300$$

(11.A.3.1)

Thus, it was possible to compare the different algorithms in stationary conditions ($t < 300$) and in response to a sudden change. The parameters for the evaluation are the mean square error (MSE) in the estimation of the coefficients a_1 and a_2 in stationary conditions, and the adaptation time T, expressed as the number of iterations after the sudden change before convergence to the new condition. Figure 11.12(a) shows the comparison between the Fortesque and the exponential forgetting factors. The graphs represent the MSE of the two algorithms (expressed as the mean value ±

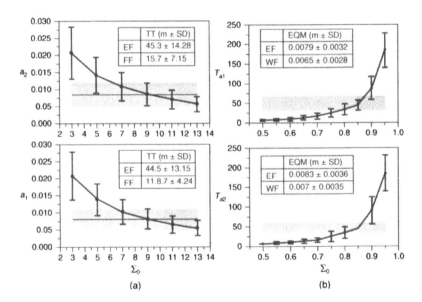

Figure 11.12. Comparison among the different forgetting factors. In panel (a) there are the MSE values for the estimations of the parameters a_1 and a_2 (espressed as mean value ± the standard deviation) obtained through the Fortesque method for different values of $\Gamma_0(t_0)$. When $\Gamma_0(t_0) = 9$, we obtain values comparable to the ones obtained with the exponential forgetting factor $\lambda = 0.9$; however, the transition times are decreased (see the inset tables). In panel (b), the transition times τ of the parameters a_1 and a_2 (expressed as mean value ± the standard deviation) are obtained through the whale forgetting factor for different values of the poles. For poles = 0.85, values of transition times comparable with the exponential forgetting factor are obtained; however, the MSEs are decreased (see the inset tables).

the standard deviation evaluated on ten realizations) as a function of the different values of the $\Gamma_0(t_0)$ for the Fortesque algorithm. The gray interval represents the MSE value (mean value ± the standard deviation) obtained through the constant forgetting factor $\lambda = 0.9$. It is possible to note that the estimations are equivalent when $\Gamma_0(t_0) = 9$. However, it is possible to observe that the variant forgetting factor makes the algorithm more reactive; in fact, the adaptation time is considerably shortened. Figure 11.12(b) describes the comparison between the EF and the WF forgetting factors. In that case, the transition times for the coefficients a_1 and a_2, respectively τ_1 and τ_2, are represented as a funcion of the poles of WF (both with the same value according to theoretical considerations reported by Lorito (1993), whereas the gray interval represents the T values (mean value ± the standard deviation) obtained when the exponential forgetting factor is $\lambda = 0.9$. It is possible to note that, for the same values of the reaction times (obtained when the poles are both = 0.85), the whale forgetting factor has a decreased sensitivity to the casual noise that is reflected in a decreased MSE.

In conclusion, both the methods are able to improve on the traditional method; however, it is worth remembering that this is obtained at expense of an increased computational load, which could considerably affect a real-time application.

REFERENCES

Barbieri, R., Bianchi, A. M., Triedman, J. K., Mainardi, L. T., Cerutti, S., and Saul, J. P., Model Dependency of Multivariate Autoregressive Spectral Analysis, *IEEE Eng. Med. Biol. Mag.*, Vol. 16, No. 5, 74–85, 1997.

Barbieri, R., Di Virgilio, V., Triedman, J. K., Bianchi, A. M., Cerutti, S., and Saul, J. P., Continuous Quantification of Respiratory and Baroreflex Control of Heart Rate: Use of Time-Variant Bivariate Spectral Analysis, in *IEEE Proceedings of Computers in Cardiology*, pp. 765–768, 1995.

Barbieri, R., Parati, G., and Saul, J. P., Closed- Versus Open-Loop Assessment of Heart Rate Baroreflex, *IEEE Eng. Med. Biol. Mag.*, Vol. 20, No. 2, 33–42, 2001.

Baselli, G., Biancardi, L., Porta, A., et al., Multichannel Parametric Analysis of the Coupling Between Respiration and Cardiovascular Variabilities, *J. Ambul. Monit.*, Vol. 5, No. 2–3, 153–165, 1992.

Baselli, G., Cerutti, S., Civardi, S., Lombardi, F., Malliani, A., Merri, M., Pagani, M., and Rizzo, G., Heart Rate Varianibility Signal Processing: A Quantitative Approach as an Aid to Diagnosis in Cardiovascular Pathologies, *Intern. J. of Bio-Medical Comp.*, Vol. 20, 51–70, 1987.

Bianchi, A. M., Mainardi, L. T., Signorini, M. G., Mainardi, M., and Cerutti, S., Time-Variant Power Spectrum Analysis for the Detection of Transient Episodes in HRV Signal, *IEEE Trans. on Biomed. Eng.*, Vol. 40, 136–144, 1993.

Bianchi, A. M., Mainardi, L. T., Meloni, C., Chierchia, S., and Cerutti, S., Continuous Monitoring of the Sympatho-Vagal Balance through Spectral Analysis: Recursive Autoregressive Techniques for Tracking Transient Events in Heart-Rate Signals, *IEEE Eng. Med. Biol. Mag.*, 64–73, 1997.

Bianchi, A. M., Meloni, C., Mainardi, L. T., Chierchia, S., and Cerutti, S., Sympatho-Vagal Evaluation During Thrombolytic Therapy in Patients with Acute Myocardial Infarction, in *IEEE Proceedings of Computers in Cardiology Conference*, 245–248, 1999.

Bianchi, A. M., Scholz, U. J., Mainardi, L. T., Orlandini, P., Pozza, G., and Cerutti, S., Extraction of the Respiration Influence from the Heart Rate Variability Signal by Means of Lattice Adaptive Filter, in *Proceedings of IEEE International Conference on BME*, 121–122, 1994.

Bianchi, A., Bontempi, B., Cerutti, S., Gianoglio, P., Comi, G., and Natali Sora, M. G., Spectral Analysis of Heart Rate Variability Signal and Respiration in Diabetic Subjects, *Med. Biol. Eng. Comput.*, Vol. 28, No. 3, 205–211, 1990.

Bianchi, M. T., Locatelli, L., Mainardi, T., Cursi, M., Comi, G., and Cerutti, S., Event-Related Brain Potentials: Laplacian Transformation for Multichanel Time-Frequency Analysis, *Methods of Information in Medicine*, Vol. 39, pp. 160–163, 2000.

Bianchi, A., Mainardi, T., and Cerutti, S., Time-Frequency Analysis of Biomedical Signals, *Trans. of the Institute of Measurement and Controls*, Vol. 22, pp. 321–336, 2000.

Bittanti, S., and Campi, M., Bounded Error Identification of Time-Varying Parameters by RLS Techniques, *IEEE Transactions on Automatic Control*, Vol. 39, No. 5, 1106–1110, 1994a.

Bittanti, S., and Campi, M., Least Squares Identification of Autoregressive Models with Time-Varying Parameters, in *Decision and Control, Proceedings of the 33rd IEEE Conference*, 4, 3610–3611, 1994b.

Box, G. E. P., and Jenkins, G. M., *Time Series Analysis Forecasting and Control*, Holden Day, 1970.

Cerutti, S., Bianchi, A. M., and Mainardi, L. T., Advanced Spectral Methods for Detecting Dynamic Behaviour, *Autonom. Neurosc.: Basic & Clin.*, Vol. 90, 3–12, 2001.

Cesarelli, M., Bifulco, P., and Bracale, M., Evaluating Time-Varying Heart Rate Variability Power Spectral Density, *IEEE Eng. in Med. and Biol.*, Vol. 6, 76–79, 1997.

Chen J., A Computerized Data Analysis System for Electrogastrogram, *Comput., Biol. Med.*, Vol. 22, 45–58, 1992

Chierchia, S., Muiesan, L., Davies, A., Balasubramian, V., Gerosa, S., and Raftery, E. B., Role of the Sympathetic Nervous System in the Pathogenesis of Chronic Stable Angina: Implications for the Mechanism of Action of Betablockers, *Circulation, Supplement II*, 82II-71–II-81, 1990.

di, Virgilio, V, Barbieri, R., Mainardi, L., Strano, S., and Cerutti, S., A Multivariate Time-Variant AR Method for the Analysis of Heart Rate and Arterial Blood Pressure, *Med. Eng Phys.*, Vol. 19, No. 2, 109–124, 1997.

Duchene, J., Devedeux, D., Mansour, S., and Marque, C., Analysing Uterine EMG: Tracking Instantaneous Burst Frequency, *IEEE Eng. Med. Biol. Mag.*, Vol. 14, 125–132, 1995.

Durbin J., The Fitting of Time Series Models, *Rev. Inst. Int. Stat.*, Vol. 28, 233–244, 1960.

Fortescue, T. R., and Ydstie, B. E., Implementation of Self-Tuning Regulators with Variable Forgetting Factors, *Automatica*, Vol. 17, No. 6, 617–26, 1981.

Friedlander, B., Lattice Filters for Adaptive Processing, *Proc. IEEE*, Vol. 70, 829–867, 1982a.

Friedlander, B., Recursive Lattice Forms for Spectral Estimation, *IEEE Trans. Acoust. Speech Signal Process.*, ASSP-30, 920–930, 1982b.

Furlan, R., Piazza, S., Dell'Orto, S., Barbic, F., Bianchi, A., Mainardi, L., Cerutti, S., Pagani, M., and Malliani, A., Cardiac Autonomic Patterns Preceding Occasional Vasovagal Reactions in Healthy Humans, Circulation, Vol. 98, No. 17, 1756–1761, 1998.

Gath, I., Feuerstein, C., Pham, T. D., and Rondouin, G., On the Tracking of Rapid Dynamic Changes in Seizure EEG, *IEEE Trans. Biomed. Eng.*, Vol. 39, 952–958, 1992.

Goseki, Y., Matsubara, T., Takahashi, N., Takeuchi, T., and Ibukiyama, C., Heart Rate Variability before the Occurrence of Silent Myocardial Ischemia during Ambulatory Monitoring, *Am. J. Cardiol.*, Vol. 73, No. 12, 845–849, 1994.

Graimann, B., Huggins, J. E., Schlogl, A., Levine, S. P., and Pfurtscheller, G., Detection of Movement-Related Desynchronization Patterns in Ongoing Single Channel Electrocorticogram, *IEEE Trans. on Neur. Sysyt. & Rehab Eng.*, Vol. 11, No. 3, 276–281, 2003.

Haberl, R., Steinbilerr, P., and Jilge, G., Spectral Analysis of the High Resolution ECG: Current Concepts and Future Directions, *Pace*, Vol. 17, 446–450, 1994.

Hemmelmann, D., Ungureanu, M., Hesse, W., Wüstenberg, T., Reichenbach, J. R., Witte, O. W., Witte, H., and Leistritz, L., Modelling and Analysis of Time-Variant Directed Interrelations Between Brain Regions Based on BOLD-Signals, *Neuroimage*, Vol. 45, No. 3, 722–737, 2009.

Joho, S., Asanoi, H., Remah, H. A., Igawa, A., Kameyama, T., Nozawa, T., Umeno, K., and Inoue, H., Time-Varying Spectral Analysis of Heart Rate and Left Ventricular Pressure Variability During Balloon Coronary Occlusion in Humans: A Sympathoexicitatory Response to Myocardial Ischemia, *J. Am. Coll. Cardiol.*, Vol. 34, No. 7, 1924–1931, 1999.

Kay, S. M., and Marple, S. L., Spectrum Analysis—A Modern Perspective, *Proc. IEEE*, Vol. 69, No. 11, 1981.

Korhonen, I. L., Mainardi, T., Carrault, G., Baselli, A., Bianchi, M., and Loula, P., Linear Multivariate Models for Physiological Signal Analysis: Applications, *Computer Methods & Programs in Biomedicine*, Vol. 51, 121–130, 1996.

Korhonen, I. L., Mainardi, T., Loula, P., Carrault, G., Baselli, G., and Bianchi, A. M., Linear Multivariate Models for Physiological Signal Analysis: Theory, *Computer Methods & Programs in Biomedicine*, Vol. 51, 85–94, 1996.

Kulhavi, R., and Karny, D., Tracking of Slow Varying Parameters by Directional Forgetting. In *Proceedings of 9th IFAC World Congress*, pp. 79–84, 1984.

Lanza, G. A., Pedrotti, P., Pasceri, V., Lucente, M., Crea, F., and Maseri, A., Autonomic Changes Associated with Spontaneous Coronary Spasm in Patients with Variant Angina, *J. Am. Coll. Cardiol.*, Vol. 28, No. 5, 1249–1256, 1996.

Levinson, N., The Wiener (Root Mean Square) Error Criterion in Filter Design And Prediction, *J. Math. Phys.*, Vol. 25, 261–278, 1947

Lin, Z., and Chen, J. D., Advances in Time-Frequency Analysis of Biomedical Signals, *Crit. Rev. Biomed. Eng.*, Vol. 24, No. 1, 1–72, 1996.

Ling F., and Proakis J., G., Numerical Accuracy and Stability: Two Problems of Adaptive Estimation Algorithms Caused by Round Off Error, in *IEEE Proceedings of 1984 International Conference on Acoustic, Speech and Signal Processing*, 1984.

Lombardi, F., Casalone, C., Della Bella, P., Malfatto, G., Pagani, M., Malliani, A., Global Versus Regional Myocardial Ischaemia: Differences in Cardiovascular and Sympathetic Responses in Cates, *Cardiovas. Res.*, Vol. 18, No. 1, 14–23, 1984.

Lorito, F., RLS Techniques with Generalized Forgetting Devices for the Identification of Time-Varying Systems, Internal report 92.069, Dept. of Electronics, Politecnico di Milano, 1993.

Mainardi, L. T., Bianchi, A. M., Baselli, G., and Cerutti, S., Pole-Tracking Algorithms for the Extraction of Time-Variant Heart Rate Variability Spectral Parameters, *IEEE Trans. Biomed. Eng*, Vol. 42, No. 3, 250–259, 1995.

Mainardi, L. T., Bianchi, A. M., Furlan, R., Piazza, S., Barbieri, R., di, Virgilio, V, Malliani, A., and Cerutti, S., Multivariate Time-Variant Identification of Cardiovascular Variability Signals: A Beat-to-Beat Spectral Parameter Estimation in Vasovagal Syncope, *IEEE Trans. Biomed. Eng.*, Vol. 44, No. 10, 978–989, 1997.

Mainardi, L. T., Bianchi, A. M., and Cerutti, S., Time-Varying Analysis for Assessing the Dynamic Responses of Cardiovascular Control, *Critical Reviews in Biomedical Engineering*, Vol. 30, No. 1–2, 181–223, 2002.

Marple, S. L., *Digital Spectral Analysis with Applications*, Prentice-Hall, 1987.

Mendez, M. O., Matteucci, M., Castronovo, V., Ferini-Strambi, L., Cerutti, S., and Bianchi, A. M., Sleep Staging from Heart Rate Variability: Time-Varying Spectral Features and Hidden Markov Models, *Int. J. of Biomed. Eng. And Techn.* (in press).

Minicucci, F., Cursi, M., Fornara, C., Rizzo, C., Chiesa, R., Tirelli, A., Fanelli, G., Meraviglia, M. V., Giacomotti, L., and Comi, G., Computer-Assisted EEG Monitoring During Carotid Endarterectomy. *J. Clin. Neurophysiol.*, Vol. 17, No. 1, 101–107, 2000.

Muthuswamy, J., and Thakor, N. V., Spectral Analysis Methods for Neurological Signals, *J. Neurosci. Methods*, Vol. 83, No. 1, 1–14, 1998.

Petrucci, E., Mainardi, L. T., Balian, V., Ghiringhelli, S., Bianchi, A. M., Bertinelli, M., Mainardi, M., and Cerutti, S., Assessment of Heart Rate Variability Changes During Dipyridamole Infusion and Dipyridamole-Induced Myocardial Ischemia: A Time Variant Spectral Approach, *J. Am. Coll. Cardiol.*, Vol. 28, No. 4, 924–934, 1996.

Pfurtscheller, G., and Lopes da Silva, F. H., Event-Related EEG/MEG Synchronization and Desynchronization: Basic Principles, *J. Clin. Neurophysiol.*, Vol. 110, 1842–1857, 1999.

Rimoldi, O., Pierini, S., Ferrari, A., Cerutti, S., Pagani, M., and Malliani, A., Analysis of Short-Term Oscillations of R-R and Arterial Pressure in Conscious Dogs, *Am. J. Physiol.*, 258(4 Pt 2), H967–H976, 1990.

Sarno, A. J., Pearson, M. A., Nabors-Oberg, R., Sollers, J. J., III, and Thayer, J. F., Autonomic Changes During Orthostasis: A Time-Frequency Analysis, *Biomed. Sci. Instrum.*, Vol. 36, 251–256, 2000.

Scholz, U. J., Bianchi, A. M., Kubicki, S., and Cerutti, S., Time-Variant Spectral Analysis of the Heart Rate Variability During Sleep: A Multiple Model Approach, in *Medicon—Conference on Medical & Biological Engineering,* VI, pp. 1189–1192, 1992.

Soderstrom, T., and Stoica, P., *SystemI,* Prentice-Hall, 1989.

Sroka, K., Peimann, C. J., and Seevers, H., Heart Rate Variability in Myocardial Ischemia During Daily Life, *J. Electrocardiol.,* Vol. 30, No. 1, 45–56, 1997.

Sun, M., Achheuer, M. L., and Sclabassi, R. J., Extraction and Analysis of Early Ictal Activity in Subdural Electroencephalogram, *Ann. Biomed. Eng.,* Vol. 29, No. 10, 878–886, 2001.

Tarvainen, M. P., Hiltunen, J. K., Ranta-aho, P. O., and Karjalainen, P. A., Estiamtion of Nonstationary EEG with Kalman Smoother Approach: An Application to Event-Related Synchronization (ERS), *IEEE Trans. Biomed. Eng.,* Vol. 51, No. 3, 516–524, 2004.

Task Force of the European Society of Cardiology and the North American Society of Pacing and Electrophysiology, Heart Rate Variability: Standards of Measurements, Physiological Interpretation, And Clinical Use, *Circulation,* Vol. 93, 1043–1065, 1996.

Widrow, B., and Stearn, S. D., *Adaptive Signal Processing,* Prentice-Hall, 1985.

Wiggins, R. A., and Robinson, E. A., Recursive Solution to Multi-Channel Filtering Problem, *J. Geophys. Res.,* Vol. 70, 1885–1891, 1965.

Wilke, C., Ding, L., and He, B., Estimation of Time-Varying Connectivity Patterns Through the Use of an Adaptive Directed Transfer Function, *IEEE Trans. Biomed. Eng.,* Vol. 55, No. 11, 2557–2564, 2008.

Witte, H., Putsche, P., Hemmelmann, C., Schelenz, C., and Leistritz, L., Analysis and Modeling of Time-Variant Amplitude-Frequency Couplings of and Between Oscillations of EEG Bursts, *Biol. Cybern,* 2008

Zetterberg, L. H., Estimation Parameters for a Linear Difference Equation with Application to EEG Analysis, *Math. Biosc.,* Vol. 5, 227–275, 1969.

PART V

COMPLEXITY ANALYSIS AND NONLINEAR METHODS

Thispageintentionallyleftblank

CHAPTER 12

DYNAMICAL SYSTEMS AND THEIR BIFURCATIONS

Fabio Dercole and Sergio Rinaldi

IN THIS CHAPTER, we summarize the basic definitions and tools of analysis of dynamical systems, with particular emphasis on the asymptotic behavior of continuous-time autonomous systems. In particular, the possible structural changes of the asymptotic behavior of the system under parameter variation, called bifurcations, are presented together with their analytical characterization and hints on their numerical analysis. The literature on dynamical systems is huge and we do not attempt to survey it here. Most of the results on bifurcations of continuous-time systems are due to Andronov and Leontovich (see Andronov et al., 1973). More recent expositions can be found in Guckenheimer and Holmes (1997) and Kuznetsov (2004), whereas less formal but didactically very effective treatments, rich in interesting examples and applications, are given in Strogatz (1994) and Alligood et al. (1996). Numerical aspects are well described in Allgower and Georg (1990) and in the fundamental papers by Keller (1977) and Doedel et al. (1991a,b), but see also Beyn et al. (2002) and Kuznetsov (2004). This chapter mainly combines material from two previous contributions of the authors, the first part of the book *Biosystems and Complexity* (Rinaldi, 1993, in Italian) and Appendix A of a recent book on evolutionary dynamics (Dercole and Rinaldi, 2008).

12.1 DYNAMICAL SYSTEMS AND STATE PORTRAITS

The dynamical systems considered in this chapter are continuous-time, finite-dimensional dynamical systems described by n autonomous (i.e., time-independent) ordinary differential equations (ODEs) called state equations:

$$\dot{x}_1(t) = f_1(x_1(t), x_2(t), \ldots, x_n(t))$$
$$\dot{x}_2(t) = f_2(x_1(t), x_2(t), \ldots, x_n(t))$$
$$\vdots$$
$$\dot{x}_n(t) = f_n(x_1(t), x_2(t), \ldots, x_n(t))$$

where $x_i(t) \in \mathbf{R}$, $i = 1, 2, \ldots, n$, is the ith state variable at time $t \in \mathbf{R}$, $\dot{x}_i(t)$ is its time derivative, and functions $f_1, \ldots, f_n$ are assumed to be smooth.

In vector form, the state equations are

$$\dot{x}(t) = f(x(t)) \tag{12.1}$$

where x and $\dot{x}$ are n-dimensional vectors (the state vector and its time derivative) and $f = (f_1, \ldots, f_n)^T$ (the T superscript denotes transposition).

Given the initial state $x(0)$, the state equations uniquely define a trajectory of the system, that is, the state vector $x(t)$ for all $t \geq 0$. A trajectory is represented in state space by a curve starting from point $x(0)$, and vector $\dot{x}(t)$ is tangent to the curve at point $x(t)$. Trajectories can be easily obtained numerically through simulation (numerical integration) and the set of all trajectories [one for any $x(0)$] is called the state portrait. If $n = 2$ (second-order or planar systems), the state portrait is often represented by drawing a sort of qualitative skeleton, that is, strategic trajectories (or finite segments of them), from which all other trajectories can be intuitively inferred. For example, in Figure 12.1a the skeleton is composed of 13 trajectories; three of them (A, B, C) are just points (corresponding to con-

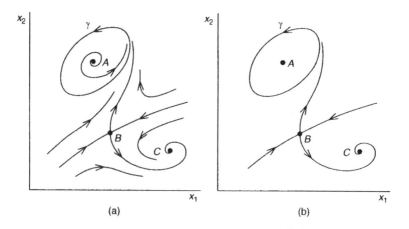

(a)   (b)

Figure 12.1. Skeleton of the state portrait of a second-order system. (a) Skeleton with 13 trajectories; (b) reduced skeleton (characteristic frame) with eight trajectories (attractors, repellors, and saddles with stable and unstable manifolds).

stant solutions of Equation 12.1 and are called equilibria, whereas one (γ) is a closed trajectory corresponding to a periodic solution of Equation 12.1 called a limit cycle. The other trajectories allow one to conclude that A is a repellor (no trajectory starting close to A tends or remains close to A), B is a saddle (almost all trajectories starting close to B go away from B, but two trajectories tend to B and compose the so-called stable manifold; the two trajectories emanating from B compose the unstable manifold and both manifolds are also called saddle separatrices), whereas C and γ are attractors (all trajectories starting close to C [γ] tend to C [γ]). Attractors are said to be (asymptotically) stable if all nearby trajectories remain close to them, globally stable if they attract all initial conditions (technically with the exclusion of sets with no measure in state space), whereas saddles and repellors are unstable. Notice, however, that attractors can also be unstable, as shown in Figure 12.2, where the equilibrium A attracts all nearby initial conditions, part of which lie along trajectories going away from it.

The skeleton of Figure 12.1a also identifies the basin of attraction of each attractor; in fact, all trajectories starting above (or below) the stable manifold of the saddle tend toward the limit cycle γ (or the equilibrium C). Notice that the basins of attraction are open sets since their boundaries are the saddle and its stable manifold. Often, the full state portrait can be more easily imagined when the skeleton is reduced, as in Figure 12.1b, to its basic elements, namely, attractors, repellors, and saddles with their stable and unstable manifolds. From now on, the reduced skeleton is called the characteristic frame.

The asymptotic behaviors of continuous-time, second-order systems are quite simple because in the case $n = 2$ attractors can be equilibria (stationary regimes) or limit cycles (cyclic or periodic regimes). But in higher

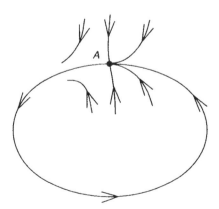

Figure 12.2. An example of unstable attractor, the equilibrium A.

dimensional systems, that is, for $n \geq 3$, more complex behaviors are possible since attractors can also be tori (quasiperiodic regimes) or strange attractors (chaotic regimes).

A torus attracting nearby trajectories is sketched in Figure 12.3a. A trajectory starting from a point of the torus remains forever on it (i.e., the torus is invariant for the dynamics of the system) but, in general, never passes again through the starting point. For example, two frequencies characterize a three-dimensional torus, namely two positive real numbers, $1/T_1$ and $1/T_2$, measuring the number of rotations around the cross section of the torus and the number of revolutions along it, per unit of time. Generically, the ratio T_1/T_2 is irrational, so that there is no period T such that

$$T = T_1 r_1 = T_2 r_2 \tag{12.2}$$

where r_1 and r_2 are positive integers. In words, there is no time T in which a trajectory on the torus carries out an integer number of cross-section rotations and an integer, possibly different, number of torus revolutions, that is, no time T after which the trajectory revisits the starting point. As a consequence, a single trajectory on the torus covers it densely in the long run, and the corresponding regime is called quasiperiodic, being the result of two (or more in higher dimensions) frequencies.

In special cases, however, the ratio T_1/T_2 can be rational, that is, trajectories on the torus can be periodic ($(r_1:r_2)$ cycles on the torus, for the minimum r_1 and r_2 satisfying Equation 12.2). A cycle on the torus can be stable (i.e., attracting nearby trajectories on the torus) or unstable. For obvious topological reasons, the existence of a stable $(r_1:r_2)$ cycle on the torus requires the existence of an unstable $(r_1:r_2)$ cycle on the same torus and rules out cycles characterized by different pairs.

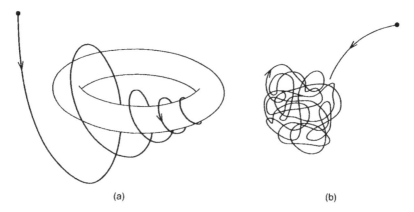

(a)   (b)

Figure 12.3. Sketch of an attracting torus (a) and of a strange attractor (b).

A strange attractor (a sort of "tangle" in state space) is shown in Figure 12.3b. Trajectories starting in the vicinity of the tangle tend to it and then remain in it forever. The most striking difference among attractors is that equilibria, cycles, and tori have integer dimension (0, 1, and 2, respectively), whereas strange attractors are fractal sets and, therefore, have noninteger dimension (see next chapter). Another important difference is that two trajectories starting from very close points in an attractor remain very close forever if the attractor is an equilibrium, a cycle, or a torus, whereas they alternatively diverge (stretching) and converge (folding) forever if the attractor is a tangle. The mean rate of divergence of nearby trajectories is measured by the so-called Lyapunov exponents, and turns out to be the most important indicator in the study of deterministic chaos (see Chapter 14).

In the simple but very important case of linear systems,

$$\dot{x}(t) = Ax(t)$$

the state portrait can be immediately obtained from the eigenvalues and eigenvectors of the $n \times n$ matrix A (we recall that the eigenvalues of an $n \times n$ matrix A are the zeros $\lambda_1, \lambda_2, \ldots, \lambda_n$ of its characteristic polynomial $\det(\lambda I - A)$, where det denotes matrix determinant, and that the eigenvectors associated with an eigenvalue λ_i are nontrivial vectors $x^{(i)}$ satisfying the relationship $Ax^{(i)} = \lambda_i x^{(i)}$). There are five generic state portraits of second-order continuous-time linear systems: three of them are shown in Figure 12.4 (the other two are obtained from cases a and b by reversing the

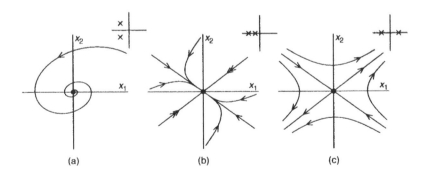

Figure 12.4. Three state portraits of generic second-order, continuous-time linear systems ($\lambda_1 \neq \lambda_2$, both with a nonzero real part; see the complex plane associated with each panel): (a) (stable focus) and (b) (stable node) are attractors; the unstable focus (positive real part complex conjugate eigenvalues) and the unstable node (positive real eigenvalues—repellors) are obtained by reversing all arrows in the state portraits (a) and (b), respectively; (c) is a saddle. Straight trajectories correspond to eigenvectors associated with real eigenvalues. Double arrows indicate the straight trajectories along which the state varies more rapidly.

sign of the eigenvalues and all arrows in the state portraits). When the two eigenvalues are complex (case a), the trajectories spiral around the origin and tend to (or diverge from) it if the real part of the eigenvalues is negative (or positive). By contrast, when the two eigenvalues are real (cases b and c), the trajectories do not spiral and there are actually special straight trajectories (corresponding to the eigenvectors) converging to (or diverging from) the origin if the corresponding eigenvalue is negative (or positive). Along the straight trajectories, both state variables vary in time as $\exp(\lambda_i t)$, whereas along all other trajectories they follow a more complex law of the kind $c_1\exp(\lambda_1 t) + c_2\exp(\lambda_2 t)$. Since in generic cases $\lambda_1 \neq \lambda_2$, one of the two exponential functions dominates the other for $t \to \pm\infty$ and all curved trajectories tend to align with one of the two straight trajectories. In particular, in the case of a stable node (characterized by $\lambda_2 < \lambda_1 < 0$, see Figure 12.4b), both exponential functions tend to zero for $t \to +\infty$, but in the long run $\exp(\lambda_1 t) \gg \exp(\lambda_2 t)$ so that all trajectories, except the two straight trajectories corresponding to the second eigenvector $x^{(2)}$, tend to zero tangentially to the first eigenvector $x^{(1)}$.

Very similar definitions can be given for discrete-time systems described by n difference state equations of the form

$$x(t + 1) = f(x(t)) \tag{12.3}$$

where the time t is an integer. In this case, trajectories are sequences of points in state space and, again, asymptotic regimes can be stationary, cyclic, quasiperiodic, and chaotic. The major difference between continuous-time and discrete-time dynamical systems is that the former are always reversible, since under very general conditions the system of Equation 12.1 has a unique solution for $t < 0$, whereas the latter can be irreversible. This implies that discrete-time systems can have quasiperiodic and chaotic regimes even if $n = 1$.

The equilibria of the system of Equation 12.1 can be found by determining all solutions $\bar{x}$ of Equation 12.1 with $\dot{x} = 0$. In second-order systems, the equilibria are often determined graphically through the so-called isoclines, which are nothing but the lines in state space on which $f_1(x_1, x_2) = 0$ (x_1 isoclines) and $f_2(x_1, x_2) = 0$ (x_2 isoclines). Obviously, the equilibria are at the intersections of x_1 and x_2 isoclines. Moreover, all trajectories cross x_1 (or x_2) isoclines vertically (or horizontally) because $\dot{x}_1$ (or $\dot{x}_2$) is zero on x_1 (or x_2) isoclines. This property is often useful for devising qualitative geometric features of the state portrait.

The stability of an equilibrium $\bar{x}$ is not as easy to ascertain. However, it can very often be discussed through linearization, that is, by approximating the behavior of the system in the vicinity of the equilibrium through a linear system. This can be done in the following way. Let

$$\delta x(t) = x(t) - \bar{x}$$

so that

$$\dot{\delta x}(t) = f(\bar{x} + \delta x(t))$$

Under very general conditions, we can expand the function f in Taylor series, thus obtaining

$$\dot{\delta x}(t) = f(\bar{x}) + \left.\frac{\partial f}{\partial x}\right|_{x=\bar{x}} \delta x(t) + O(\|\delta x(t)\|^2)$$

where $\|\cdot\|$ is the standard norm in R^n and $O(\|\delta x(t)\|^2)$ stand for a term that vanishes as $\|\delta x(t)\|^2$ when $\delta x(t) \to 0$. Noticing that $f(\bar{x}) = 0$, since $\bar{x}$ is a constant solution of Equation 12.1, we have

$$\dot{\delta x}(t) = \left.\frac{\partial f}{\partial x}\right|_{x=\bar{x}} \delta x(t) + O(\|\delta x(t)\|^2) \qquad (12.4)$$

where the $n \times n$ constant matrix

$$J = \left.\frac{\partial f}{\partial x}\right|_{x=\bar{x}} = \begin{bmatrix} \dfrac{\partial f_1}{\partial x_1} & \cdots & \dfrac{\partial f_1}{\partial x_n} \\ \vdots & & \vdots \\ \dfrac{\partial f_n}{\partial x_1} & \cdots & \dfrac{\partial f_n}{\partial x_n} \end{bmatrix}_{x=\bar{x}} \qquad (12.5)$$

is called the Jacobian matrix (or, more simply, Jacobian). One can easily imagine that, under suitable conditions, the behavior of the system of Equation 12.4 (which is still the system of Equation 12.1) can be well approximated in the vicinity of $\bar{x}$ by the so-called linearized system, which, by definition, is

$$\dot{\delta x}(t) = \left.\frac{\partial f}{\partial x}\right|_{x=\bar{x}} \delta x(t) \qquad (12.6)$$

This is, indeed, the case. In particular, it can be shown that if the solution $\delta x(t)$ of Equation 12.6 tends to 0 for all $\delta x(0) \neq 0$ (as in Figures 12.4a and b), then the same is true for the system of Equation 12.4 provided $\|\delta x(0)\|$ is sufficiently small. In other words, the stability of the linearized system implies the (local) stability of the equilibrium $\bar{x}$. This result is quite interesting because the stability of the linearized system can be numerically ascertained by checking if all eigenvalues λ_i, $i = 1, \ldots, n$ of the

Jacobian matrix (Equation 12.5) have negative real parts. A similar result holds also for the case of unstable equilibria. More precisely, if at least one eigenvalue λ_i of the Jacobian matrix has positive real part (as in Figure 12.4c), then the equilibrium $\bar{x}$ is locally unstable (i.e., the solution of Equation 12.4 diverges at least temporarily from zero for suitable $\delta x(0)$, no matter how small $\|\delta x(0)\|$ is). Similarly, the local stability of an equilibrium of a discrete-time system of the form of Equation 12.3 can be studied by simply looking at the module $|\lambda_i|$ of the n eigenvalues λ_i. In fact, if all $|\lambda_i| < 1$, that is, if all eigenvalues are inside the unit circle in the complex plane, the equilibrium is stable, whereas if at least one eigenvalue is outside the unit circle ($|\lambda_i| > 1$), the equilibrium is unstable.

The study of the stability of limit cycles can also be carried out through linearization, following a very simple idea suggested by Poincaré (see Figure 12.5). In the case of second-order systems (see Figure 12.5a), the Poincaré method consists in cutting locally and transversally the limit cycle with a manifold P, called the Poincaré section, and looking at the sequence $z(0), z(1), z(2), \ldots$ of points of return of the trajectory to P. Since P is one-dimensional, $z(t)$ is a scalar coordinate on P and the state equation (Equation 12.1) implicitly defines a first-order discrete-time system called the Poincaré map:

$$z(t + 1) = P(z(t)) \qquad (12.7)$$

The intersection $\bar{z}$ of the limit cycle γ with P is an equilibrium of the Poincaré map (since $\bar{z} = P(\bar{z})$) and γ is stable if and only if the equilibrium $\bar{z}$ of Equation 12.7 is stable. One can, therefore, use the linearization technique by taking into account that the eigenvalue of the linearized Poincaré map,

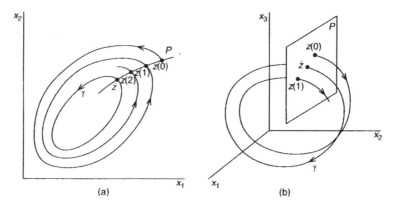

Figure 12.5. A stable limit cycle γ, the Poincaré section P, and the sequence $z(0)$, $z(1), z(2), \ldots$ of return points.

$dP/dz|_{z=\bar{z}}$ (called the Floquet multiplier, or simply multiplier, of the cycle), cannot be negative, since trajectories cannot cross each other. Thus, a sufficient condition for the (local) stability of the limit cycle γ is

$$\left.\frac{dP}{dz}\right|_{z=\bar{z}} < 1 \tag{12.8}$$

whereas the reverse inequality implies the instability of γ.

Similarly, in the case of third-order systems (see Figure 12.5b) the Poincaré section is a two-dimensional manifold P and the points of return $z(0)$, $z(1)$, $z(2)$, ... are generated by a two-dimensional Poincaré map (Eqaution 12.7). Again, the cycle is stable if and only if the equilibrium $\bar{z}$ of the discrete-time system (Equation 12.7) is stable. Thus, if the two multipliers of the cycle, that is, the two eigenvalues of the Jacobian matrix $\partial P/\partial z|_{z=\bar{z}}$, are smaller than 1 in the module, the cycle γ is stable, whereas if the module of at least one multiplier is greater than 1 the cycle is unstable. These sufficient conditions for the stability and instability of a cycle can obviously be extended to the n-dimensional case, where $\partial P/\partial z|_{z=\bar{z}}$ is an $(n-1) \times (n-1)$ matrix. It must be noticed, however, that they can only be verified numerically, since the cycle γ is, in general, not known analytically.

The Poincaré section is also very useful for distinguishing quasiperiodic from chaotic regimes in third-order systems. In fact, a torus appears on a Poincaré section as a regular closed curve, whereas strange attractors appear as clouds of points (with fractal geometry), as shown in Figure 12.6.

Figure 12.6. The image of a strange attractor on a Poincaré section.

12.2 STRUCTURAL STABILITY

Structural stability is a key notion in the theory of dynamical systems, since it is needed to understand interesting phenomena like catastrophic transitions, bistability, hysteresis, frequency locking, synchronization, subharmonics, deterministic chaos, and many others. The final target of structural stability is the study of the asymptotic behavior of parameterized families of dynamical systems of the form

$$\dot{x}(t) = f(x(t), p) \tag{12.9}$$

for continuous-time systems, and

$$x(t + 1) = f(x(t), p) \tag{12.10}$$

for discrete-time systems, where p is a vector of constant parameters. Given the parameter vector p, all the definitions that we have seen in the previous section apply to the particular dynamical system of the family identified by p. Thus, all geometric and analytical properties of systems of Equations 12.9 and 12.10, such as trajectories, state portrait, equilibria, limit cycles, their stability and associated Jacobian matrices and Poincaré maps, the basins of attraction, and, consequently, the asymptotic behavior of the system, now depend upon p.

Structural stability allows one to rigorously explain why a small change in a parameter value can give rise to a radical change in the system behavior. More precisely, the aim is to find regions P_i in parameter space characterized by the same qualitative behavior of the system of Equation 12.9, in the sense that all state portraits corresponding to values $p \in P_i$ are topologically equivalent (i.e., they can be obtained one from the other through a simple deformation of the trajectories). Thus, by varying $p \in P_i$ the system conserves all the characteristic elements of the state portrait, namely, its attractors, repellors, and saddles. In other words, when p is varied in P_i, the characteristic frame varies but conserves its structure. Figure 12.7 shows the typical result of a study of structural stability in the space (p_1, p_2) of two parameters of a second-order system. The parameter space is subdivided into three regions, P_1, P_2, and P_3, and for all interior points of each one of these regions the state portrait is topologically equivalent to that sketched in the figure. In P_1, the system is an oscillator, since it has a single attractor, which is a limit cycle. Also, in P_2 there is a single attractor, which is, however, an equilibrium. Finally, in P_3 we have bistability since the system has two alternative attractors (two equilibria), each with its own basin of attraction delimited by the stable manifold of the saddle equilibrium.

If p is an interior point of a region P_i, Equation 12.9 is said to be structurally stable at p since its state portrait is qualitatively the same as

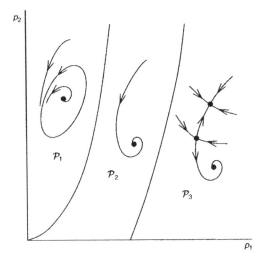

Figure 12.7. Bifurcation diagram of a second-order system. The curves separating regions P_1, P_2, and P_3 are bifurcation curves.

those of the systems obtained by slightly perturbing the parameters in all possible ways. By contrast, if p is on the boundary of a region P_i, the system is not structurally stable since small perturbations can give rise to qualitatively different state portraits. The points of the boundaries of the regions P_i are called bifurcation points, and, in the case of two parameters, the boundaries are called bifurcation curves. Bifurcation points are, therefore, points of degeneracy. If they lie on a curve separating two distinct regions P_i and P_j, $i \neq j$, they are called codimension-1 bifurcation points, whereas if they lie on the boundaries of three distinct regions they are called codimension-2 bifurcation points, and so on.

Notice that the simplest dynamical system, namely the first-order linear system $\dot{x}(t) = px(t)$, has a bifurcation at $p = p^* = 0$, that is, when its eigenvalue p is equal to zero. In fact, such a system is stable for $p < 0$ and unstable for $p > 0$, whereas it is neutrally stable (i.e., the equilibrium $x = 0$ is not unstable but does not attract all nearby trajectories) for $p = p^* = 0$.

In the following, we mainly focus on second-order, continuous-time systems and limit the discussion to codimension-1 bifurcations.

12.3 BIFURCATIONS AS COLLISIONS

A generic element of the parameterized family of dynamical systems (Equation 12.9) must be imagined to be structurally stable because if p is selected randomly it will be an interior point of a region P_i with probability 1. Under generic conditions, attractors, repellors, saddles, and their stable and unstable manifolds are separated one from each other. Moreover, the eigenvalues of the Jacobian matrices associated with equilibria have nonzero real parts, whereas the eigenvalues of linearized Poincaré maps associated with cycles have modules different from 1. By continuity, small parametric variations will induce small variations of all attractors, repellors, saddles, and their stable and unstable manifolds which, however, will remain separated if the parametric variations are sufficiently small. The same holds for the eigenvalues of Jacobian matrices and linearized Poincaré maps, which, for sufficiently small parametric variations, will continue to be noncritical. Thus, in conclusion, starting from a generic condition, it is necessary to vary the parameters of a finite amount to obtain a bifurcation, which is generated by the collision of two or more elements of the characteristic frame, which then changes its structure at the bifurcation, thus involving a change of the state portrait of the system.

A bifurcation is called local when it involves the degeneracy of some eigenvalue of the Jacobians associated with equilibria or cycles. For example, the bifurcation described in Figure 12.8, called saddle-node bifurcation, is a local bifurcation. Indeed, the bifurcation can be viewed as the collision, at $p = p^*$, of two equilibria: for $p < p^*$ the two equilibria (elements of the characteristic frame) are distinct and one is stable (the node N), whereas the other is unstable (the saddle S). Then, as p increases, the two equilibria approach each other and finally collide when $p = p^*$ (and then disappear). Notice that the characteristic frame is degenerate at $p = p^*$ because it is composed of one element (an equilibrium), whereas there are two equilibria for $p < p^*$ and none for $p > p^*$. But the bifurcation can also be interpreted in terms of eigenvalue degeneracy. In fact, the eigenvalues of the Jacobian evaluated at the saddle are one positive and one negative, whereas the eigenvalues of the Jacobian evaluated at the node are both negative, so that when the two equilibria coincide, one of the two eigenvalues of the unique Jacobian matrix must be equal to zero.

By contrast, global bifurcations cannot be revealed by eigenvalue degeneracies. One example, known as heteroclinic bifurcation, is shown in Fig-

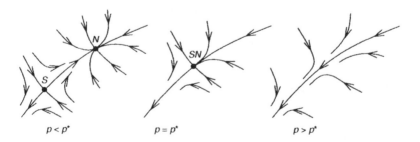

$p < p^*$ $p = p^*$ $p > p^*$

Figure 12.8. Example of local bifurcation: saddle-node bifurcation.

ure 12.9, which presents the characteristic frames (two saddles and their stable and unstable manifolds) of a system for $p = p^*$ (bifurcation value) and for $p \neq p^*$. The characteristic frame for $p = p^*$ is structurally different from the others because it corresponds to the collision of the unstable manifold X_1^+ of the first saddle with the stable manifold X_2^- of the second saddle. However, the two Jacobian matrices associated with the two saddles do not degenerate at p^*, since their eigenvalues remain different from zero. In other words, the bifurcation cannot be revealed by the behavior of the system in the vicinity of an equilibrium, but is the result of the global behavior of the system.

When there is only one parameter p and there are various bifurcations at different values of the parameter, it is often advantageous to represent the dependence of the system behavior upon the parameter by drawing in the three-dimensional space (p, x_1, x_2), often called control space, the characteristic frame for all values of p. This is done, for example, in Figure 12.10 for the same system described in Figure 12.7, with $p = p_1$ and constant p_2. Figure 12.10 shows that for increasing values of p a so-called Hopf bifurcation occurs, as the stable limit cycle shrinks to a point, thus colliding with the unstable equilibrium that exists inside the cycle. This is a local bifurcation, because the equilibrium is stable for higher values of p, so that the bifurcation can be revealed by an eigenvalue degeneracy. The figure also shows that a saddle-node bifurcation occurs at a higher value of the parameter, as two equilibria, namely a stable node and a saddle, become closer and closer until they collide and disappear. The Hopf and the saddle-node bifurcations are perhaps the most popular local bifurcations of second-order systems and are discussed in some detail in the next section.

12.4 LOCAL BIFURCATIONS

In this section, we discuss the seven most important local bifurcations of continuous-time systems. Three of them, called transcritical, saddle-node

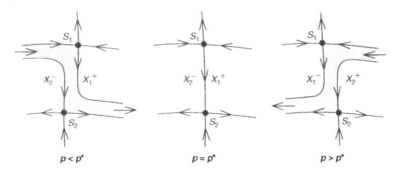

$p < p^*$ $p = p^*$ $p > p^*$

Figure 12.9. Example of global bifurcation: heteroclinic bifurcation.

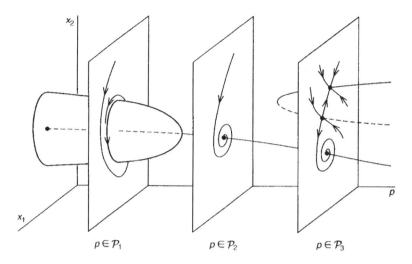

Figure 12.10. Characteristic frame in the control space of a system with a Hopf and a saddle-node bifurcation. Continuous lines represent trajectories in the three illustrated state portraits and stable equilibria, or limit cycles otherwise; dashed lines represent unstable equilibria. The symbols P_1, P_2, and P_3 refer to Figure 12.7.

(already encountered above), and pitchfork, can be viewed as collisions of equilibria. Since they can occur in first-order systems, we present them in that context. The other bifurcations involve limit cycles. Two of them can occur in second-order systems, namely the Hopf bifurcation (already seen), that is, the collision of an equilibrium with a vanishing cycle, and the tangent of limit cycles, which is the collision of two cycles. The last two bifurcations, the flip (or period-doubling) and the Neimark–Sacker (or torus), are more complex because they can occur only in three- (or higher) dimensional systems. The first is a particular collision of two limit cycles, one with period double the other, whereas the second is the collision between a cycle and a vanishing torus.

12.4.1 Transcritical, Saddle-Node, and Pitchfork Bifurcations

Figure 12.11 shows three different types of collisions of equilibria in first-order systems of the form of Equation 12.9. The state x and the parameter p have been normalized in such a way that the bifurcation occurs at $p^* = 0$ and the corresponding equilibrium is zero. Continuous lines in the figure represent stable equilibria, whereas dashed lines indicate unstable equilibria. In Figure 12.11a, the collision is visible in both directions, whereas in Figures 12.11b and 12.11c the collision is visible only from the left or from the right. The three bifurcations are called, respectively, transcritical,

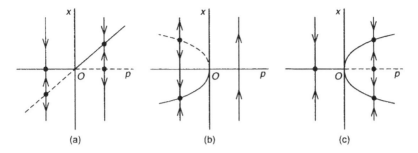

Figure 12.11. Three local bifurcations viewed as collisions of equilibria: (a) transcritical; (b) saddle-node; (c) pitchfork.

saddle-node, and pitchfork, and the three most simple state equations (called normal forms) giving rise to Figure 12.11 are

$$\dot{x}(t) = px(t) - x^2(t) \qquad \text{(transcritical)} \qquad (12.11\text{a})$$

$$\dot{x}(t) = p + x^2(t) \qquad \text{(saddle-node)} \qquad (12.11\text{b})$$

$$\dot{x}(t) = px(t) - x^3(t) \qquad \text{(pitchfork)} \qquad (12.11\text{c})$$

The first of these bifurcations is also called exchange of stability since the two equilibria exchange their stability at the bifurcation. The second is called saddle-node bifurcation because in second-order systems it corresponds to the collision of a saddle with a node, as shown in Figure 12.8, but it is also known as fold, in view of the form of the graph of its equilibria. Due to the symmetry of the normal form, the pitchfork has three colliding equilibria, two stable and one unstable in the middle.

It is worth noticing that in changing the sign of the quadratic and cubic terms in the normal forms (Equation 12.11), three new normal forms are obtained, namely

$$\dot{x}(t) = px(t) + x^2(t) \qquad \text{(transcritical)} \qquad (12.12\text{a})$$

$$\dot{x}(t) = p - x^2(t) \qquad \text{(saddle-node)} \qquad (12.12\text{b})$$

$$\dot{x}(t) = px(t) + x^3(t) \qquad \text{(pitchfork)} \qquad (12.12\text{c})$$

which have the bifurcation diagrams shown in Figure 12.12. Comparing Figures 12.11 and 12.12, it is easy to verify that nothing changes from a phenomenological point of view in the first two cases. However, for the pitchfork bifurcation this is not true, since in the case of Equation 12.11c

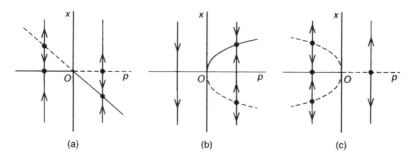

Figure 12.12. Bifurcation diagrams corresponding to the normal forms (Equation 12.12).

there is at least one attractor for each value of the parameter, whereas in case of Equation 12.12c, for $p > 0$, there is only a repellor. To distinguish the two possibilities, the pitchfork (Equation 12.11c) is called supercritical, whereas the other is called subcritical.

12.4.2 Hopf Bifurcation

The Hopf bifurcation (actually discovered by A. A. Andronov for second-order systems; see Andronov et al., 1973, and Marsden and McCracken, 1976 for the English translation of Andronov and Hopf's original works) explains how a stationary regime can become cyclic as a consequence of a small variation of a parameter, a rather common phenomenon not only in physics but also in biology, economics, and the social sciences. In terms of collisions, this bifurcation involves an equilibrium and a cycle that, however, shrinks to a point when the collision occurs. Figure 12.13 shows the

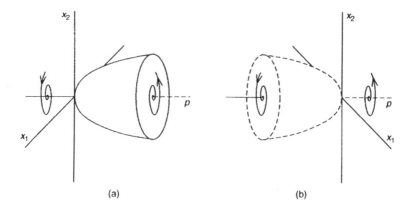

Figure 12.13. Hopf bifurcation: (a) supercritical; (b) subcritical.

two possible cases, known as supercritical and subcritical Hopf bifurcations, respectively. In the supercritical case, a stable cycle has in its interior an unstable focus. When the parameter is varied, the cycle shrinks until it collides with the equilibrium and after the collision only a stable equilibrium remains. By contrast, in the subcritical case the cycle is unstable and is the boundary of the basin of attraction of the stable equilibrium inside the cycle. Thus, after the collision there is only a repellor.

The normal form of the Hopf bifurcation is

$$\dot{x}_1(t) = px_1(t) - \omega x_2(t) + cx_1(t)(x_1^2(t) + x_2^2(t))$$

$$\dot{x}_2(t) = \omega x_1(t) + px_2(t) + cx_2(t)(x_1^2(t) + x_2^2(t))$$

which, in polar coordinates, becomes

$$\dot{\rho}(t) = p\rho(t) + c\rho^3(t)$$

$$\dot{\theta}(t) = \omega$$

This last form shows that the trajectory spirals around the origin at constant angular velocity ω, whereas the distance from the origin varies in accordance with the first ODE, which is the normal form of the pitchfork. Thus, the stability of the cycle depends upon the sign of c, called the Lyapunov coefficient.

Taking into account Figures 12.11c and 12.12c, it is easy to check that the Hopf bifurcation is supercritical (or subcritical) if $c < 0$ ($c > 0$). In the case $c = 0$, the system is linear and for $p = p^* = 0$ the origin is neutrally stable and surrounded by an infinity of cycles. For $p = p^*$, the origin of the state space is stable in the supercritical case and unstable in the opposite case. The Jacobian of the normal form, evaluated at the origin, is

$$J = \begin{bmatrix} p & \omega \\ -\omega & p \end{bmatrix}$$

and its two eigenvalues $\lambda_{1,2} = p \pm i\omega$ cross the imaginary axis of the complex plane when $p = 0$. This is the property commonly used to detect Hopf bifurcations in second-order systems. In fact, denoting by $\bar{x}(p)$ an equilibrium of the system, the Jacobian evaluated at $\bar{x}(p)$ is

$$J = \begin{bmatrix} \dfrac{\partial f_1}{\partial x_1} & \dfrac{\partial f_1}{\partial x_2} \\[2ex] \dfrac{\partial f_2}{\partial x_1} & \dfrac{\partial f_2}{\partial x_2} \end{bmatrix}_{x=\bar{x}(p)}$$

and such a matrix has a pair of nontrivial and purely imaginary eigenvalues if and only if

$$\text{trace}(J) = \left.\frac{\partial f_1}{\partial x_1}\right|_{x=\bar{x}(p)} + \left.\frac{\partial f_2}{\partial x_2}\right|_{x=\bar{x}(p)} = 0$$

$$\det(J) = \left.\frac{\partial f_1}{\partial x_1}\right|_{x=\bar{x}(p)} \left.\frac{\partial f_2}{\partial x_2}\right|_{x=\bar{x}(p)} - \left.\frac{\partial f_1}{\partial x_2}\right|_{x=\bar{x}(p)} \left.\frac{\partial f_2}{\partial x_1}\right|_{x=\bar{x}(p)} > 0$$

In practice, one annihilates the trace of the Jacobian evaluated at the equilibrium and finds in this way the parameter values that are candidate Hopf bifurcations. Then, the test on the positivity of the determinant of J is used to select the true Hopf bifurcations among the candidates. Under suitable nondegeneracy conditions, the emerging cycle is unique and its frequency is $\omega = \sqrt{\det(J)}$, because $\sqrt{\det(J)} = \lambda_1 \lambda_2$, whereas its amplitude increases as $\sqrt{-c(p - p^*)}$.

Determining if a Hopf bifurcation is supercritical or subcritical is not easy. One can try to find out if the equilibrium is stable or unstable but this is quite difficult since linearization is unreliable at a bifurcation. Alternatively (but equivalently), one can determine the sign of the Lyapunov coefficient c by following a procedure that is often quite cumbersome (see, e.g., Guckenheimer & Holmes, 1997; Kuznetsov, 2004) and is, therefore, not reported here.

12.4.3 Tangent Bifurcation of Limit Cycles

Other local bifurcations in second-order systems involve limit cycles and are somehow similar to transcritical, saddle-node, and pitchfork bifurcations of equilibria. In fact, the collision of two limit cycles can be studied as the collision of the two corresponding equilibria of the Poincaré map defined on a Poincaré section cutting both cycles. Thus, the transcritical, saddle-node, and pitchfork bifurcations of such equilibria correspond to analogous bifurcations of the colliding limit cycles.

The most common case is the saddle-node bifurcation of limit cycles, more often called fold or tangent bifurcation of limit cycles, where two cycles collide for $p = p^*$ and then disappear, as shown in Figure 12.14. On the Poincaré section P, the bifurcation is revealed by the collision of two equilibria of the Poincaré map, S unstable and N stable, which then disappear. In terms of eigenvalue degeneracy, the eigenvalue of the linearized Poincaré map evaluated at S (or N) is larger (or smaller) than 1, so that when the two equilibria coincide, the eigenvalue of the unique linearized Poincaré map must be equal to 1.

Varying the parameter in the opposite direction, this bifurcation explains the sudden birth of a pair of cycles, one of which is stable. While in

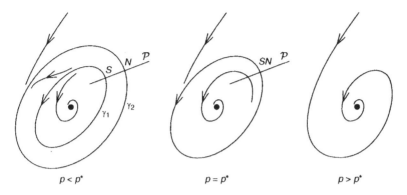

Figure 12.14. Tangent bifurcation of limit cycles: two cycles γ_1 and γ_2 collide for $p = p^*$ and then disappear.

the case of the Hopf bifurcation the emerging cycle is degenerate (it has zero amplitude), in this case the emerging cycles are not degenerate.

12.4.4 Flip (Period-Doubling) Bifurcation

The flip bifurcation is the collision of two particular limit cycles, one tracing twice the other and, therefore, having a double period, in a three- (or higher) dimensional state space. In the supercritical (or subcritical) case, it corresponds to a bifurcation of a stable (or unstable) limit cycle of period T into a stable (or unstable) limit cycle of period $2T$ and an unstable (or stable) limit cycle of period T, as sketched in Figure 12.15 (for the supercritical case) just before and after the bifurcation.

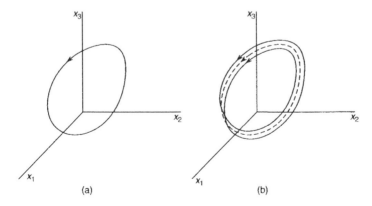

Figure 12.15. Flip bifurcation. (a) Stable limit cycle of period T; (b) unstable limit cycle of period T and stable limit cycle of period $2T$.

Physically speaking, the stable limit cycle becomes only slightly different, but the key feature is that the period of the limit cycle doubles through the bifurcation. In other words, if before the bifurcation the graph of one of the state variables, say x_1, has a single peak in each period T, after the bifurcation the graph has two slightly different peaks in each period $2T$. On a Poincaré section, looking only at second return points, the flip bifurcation resembles the pitchfork bifurcation, where two stable equilibria, $\bar{z}'$ and $\bar{z}''$ (corresponding to the two intersections of the period-$2T$ cycle with the Poincaré section), collide with a third unstable equilibrium $\bar{z}$ (the intersection of the period-T cycle) and disappear, while $\bar{z}$ becomes stable.

Mathematically speaking, the flip bifurcation is characterized by a multiplier of the period-T cycle equal to -1. In fact, when the cycle is unstable, the divergence from it, seen on a Poincaré section, is characterized by (first) return points that tend to alternate between points $\bar{z}'$ and $\bar{z}''$. This is due to a negative multiplier < -1. Just after the bifurcation (from right to left in Figure 12.15), the cycle is stable but the multiplier is still negative, between -1 and 0, that is, the multiplier is equal to -1 at the bifurcation.

12.4.5 Neimark–Sacker (Torus) Bifurcation

This bifurcation, when supercritical, explains how a stable limit cycle can become a stable torus by slightly varying a parameter. Figure 12.16 clearly represents this bifurcation and shows that it can be interpreted (from right to left) as the collision of a stable vanishing torus with an unstable limit cycle inside the torus. On a Poincaré section, one would see a stable equilibrium (intersection of the cycle of Figure 12.16a with the Poincaré section) bifurcating into an unstable equilibrium and a small regular closed

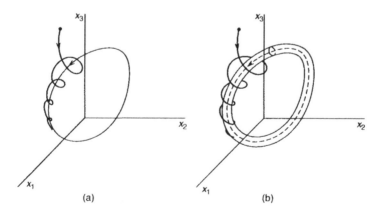

Figure 12.16. Neimark–Sacker bifurcation. (a) Stable limit cycle; (b) unstable limit cycle and small stable torus.

curve (the intersection of the torus of Figure 12.16b with the Poincaré section). In a sense, on the Poincaré section, one would observe invariant sets with the same geometry observed in the case of the supercritical Hopf bifurcation (see Figure 12.16a). For this reason, the Neimark–Sacker bifurcation is sometimes confused with the Hopf bifurcation. Similarly, the subcritical Neimark–Sacker bifurcation resembles the subcritical Hopf bifurcation (see Figure 12.16b).

In terms of cycle multipliers, the Neimark–Sacker bifurcation corresponds to a pair of complex conjugate multipliers crossing the unit circle in the complex plane. When the cycle is stable, nearby trajectories converge to the cycle by spiraling around it, whereas, when unstable, trajectories diverge from the cycle and spiral toward the torus.

In a two-parameter space, the Neimark–Sacker bifurcation curve separates the region in which the system has periodic regimes from that in which the asymptotic regime is quasiperiodic. However, as shown in Figure 12.17, in the region where the attractor is a torus, there are very narrow subregions, each delimited by two curves merging on the Neimark–Sacker curve. In these subregions, called Arnold's tongues, the attractor is a cycle on the torus, and the two curves delimiting each tongue are tangent bifurcations of limit cycles. The points on the Neimark–Sacker curve from which the Arnold's tongues emanate are, therefore, codimension-2 bifurcation points. Although the Arnold's tongues are infinitely many but

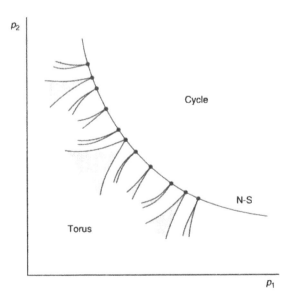

Figure 12.17. Neimark–Sacker bifurcation curve and Arnold's tongues emanating from it.

countable (generically, there is a tongue for each possible $(r_1:r_2)$ pair characterizing a cycle on the torus), only a few of them can be numerically or experimentally detected, since the others are too thin. Nevertheless, Arnold's tongues are quite important because they explain the subtle and intriguing phenomenon known as frequency locking.

12.5 GLOBAL BIFURCATIONS

As discussed in Section 12.3, global bifurcations cannot be detected through the analysis of the Jacobians associated with equilibria or cycles. However, they can still be viewed as structural changes of the characteristic frame.

12.5.1 Heteroclinic Bifurcation

In Figure 12.9, we reported the bifurcation corresponding to the collision of a stable manifold of a saddle with the unstable manifold of another saddle. This bifurcation is called heteroclinic bifurcation since the trajectory connecting the two saddles is called heteroclinic trajectory (or connection).

12.5.2 Homoclinic Bifurcation

A special but important global bifurcation is the so-called homoclinic bifurcation, characterized by the presence of a trajectory connecting an equilibrium with itself, called homoclinic trajectory (or connection).

There are two collisions that give rise to a homoclinic trajectory. The first and most common collision is that between the stable and unstable manifolds of the same saddle, as depicted in Figure 12.18. The second collision, shown in Figure 12.19, is that between a node and a saddle whose unstable manifold is connected to the node. The corresponding bifurcations are called homoclinic bifurcations to a standard saddle, or simply homoclinic bifurcation, and homoclinic bifurcations to a saddle node.

Figure 12.18 shows that the homoclinic bifurcation to a standard saddle can also be viewed as the collision of a cycle $\gamma(p)$ with a saddle $S(p)$. When p approaches p^*, the cycle $\gamma(p)$ gets closer and closer to the saddle $S(p)$ so that the period $T(p)$ of the cycle becomes longer and longer, since the state of the system moves very slowly when it is very close to the saddle. By contrast, Figure 12.19 shows that the homoclinic bifurcation to a saddle node can be viewed as a saddle-node bifurcation on a cycle $\gamma(p)$, which therefore disappears. When p approaches p^*, the system "feels" the forthcoming appearance of the two equilibria and, therefore, the state slows down close to the point at which they are going to appear. Thus, in

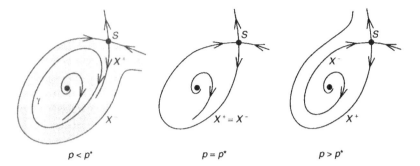

Figure 12.18. Homoclinic bifurcation to standard saddle. For $p = p^*$, the stable manifold X^- of the saddle S collides with the unstable manifold X^+ of the same saddle. The bifurcation can also be viewed as the collision of the cycle γ with the saddle S.

both cases, $T(p) \to \infty$ as $p \to p^*$, and this property is often used to detect homoclinic bifurcations through simulation. Another property used to detect homoclinic bifurcations to standard saddles through simulation is related to the form of the limit cycle, which becomes "pinched" close to the bifurcation, the angle of the pinch being the angle between the stable and unstable manifolds of the saddle.

Looking at Figures 12.18 and 12.19 from the right to the left, we can recognize that the homoclinic bifurcation explains the birth of a limit cycle. As in the case of Hopf bifurcations, the emerging limit cycle is degenerate, but this time the degeneracy is not in the amplitude of the cycle but in its period, which is infinitely long. The emerging limit cycles are stable in the figures (the gray region in Figure 12.18 is the basin of attraction),

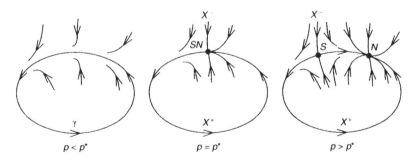

Figure 12.19. Homoclinic bifurcation to a saddle node. For $p = p^*$, the unstable manifold X^+ of the saddle node SN comes back to SN transversally to the stable manifold X^-. The bifurcation can also be viewed as a saddle-node bifurcation on the cycle γ.

but by reversing the arrows of all trajectories the same figures could be used to illustrate the cases of unstable emerging cycles. In other words, homoclinic bifurcations in second-order systems are generically associated with a cycle emerging from the homoclinic trajectory existing at $p = p^*$ by suitably perturbing the parameter. It is interesting to note that the stability of the emerging cycle can be easily predicted by looking at the sign of the so-called saddle quantity χ, which is the sum of the two eigenvalues of the Jacobian matrix associated with the saddle, that is, the trace of the Jacobian (notice that one eigenvalue is equal to zero in the case of homoclinic bifurcation to the saddle node). More precisely, if $\chi < 0$ the cycle is stable, whereas if $\chi > 0$ the cycle is unstable. As proved by Andronov and Leontovich (see Andronov et al., 1973), this result holds under a series of assumptions that essentially rule out a number of critical cases. A very important and absolutely not simple extension of the Andronov and Leontovich theory is Shil'nikov's theorem (Shil'nikov, 1968) concerning homoclinic bifurcations in three-dimensional systems.

12.6   CATASTROPHES, HYSTERESIS, AND CUSP

We can now present a simple but comprehensive treatment of a delicate problem, that of catastrophic transitions in dynamical systems. A lot has been said on this topic in the last decades and the so-called catastrophe theory (Thom, 1972) has often been invoked improperly, thus generating expectations that will never be satisfied. Reduced to its minimal terms, the problem of catastrophic transitions is the following: assuming that a system is functioning in one of its asymptotic regimes, is it possible that a microscopic variation of a parameter triggers a transient toward a macroscopically different asymptotic regime? When this happens, we say that a catastrophic transition occurs.

To be more specific, assume that an instantaneous small perturbation from p to $p + \Delta p$ occurs at time $t = 0$ when the system is on one of its attractors, say $A(p)$, or at a point $x(0)$ very close to $A(p)$ in the basin of attraction $B(A(p))$. A first possibility is that p and $p + \Delta p$ are not separated by any bifurcation. This implies that the state portrait of the perturbed system $\dot{x} = f(x, p + \Delta p)$ can be obtained by slightly deforming the state portrait of the original system $\dot{x} = f(x, p)$. In particular, if Δp is small, by continuity, the attractors $A(p)$ and $A(p + \Delta p)$, as well as their basins of attraction $B[A(p)]$ and $B(A(p + \Delta p))$, are almost coincident, so that $x(0) \in B(A(p + \Delta p))$. This means that after the perturbation, a transition will occur from $A(p)$, or $x(0)$ close to $A(p)$, to $A(p + \Delta p)$. In conclusion, a microscopic variation of a parameter has generated a microscopic variation in system behavior.

The opposite possibility is that p and $p + \Delta p$ are separated by a bifurcation. In such a case, it can happen that the small parameter variation triggers a transient, bringing the system toward a macroscopically different attractor. When this happens for all initial states $x(0)$ close to $A(p)$, the bifurcation is called catastrophic. By contrast, if the catastrophic transition is not possible, the bifurcation is called noncatastrophic, whereas in all other cases the bifurcation is said to be undetermined.

We can now revisit all bifurcations we have discussed in the previous sections. Let us start with Figure 12.11 and assume that p is small and negative, that is, $p = -\varepsilon$; that $x(0)$ is different from zero but very small, that is, close to the stable equilibrium; and that $\Delta p = 2\varepsilon$ so that, after the perturbation, $p = \varepsilon$. In case (a) (transcritical bifurcation), $x(t) \rightarrow \varepsilon$ if $x(0) > 0$ and $x(t) \rightarrow -\infty$ if $x(0) < 0$. Thus, this bifurcation is undetermined because it can, but does not always, give rise to a catastrophic transition. In a case like this, the noise acting on the system has a fundamental role since it determines the sign of $x(0)$, which is crucial for the behavior of the system after the parametric perturbation. We must notice, however, that in many cases the sign of $x(0)$ is a-priori fixed. For example, if the system is positive because x represents the density of a population, then for physical reasons $x(0) > 0$ and the bifurcation is, therefore, noncatastrophic. However, under the same conditions, the transcritical bifurcation of Figure 12.12a is catastrophic. Similarly, we can conclude that the saddle-node bifurcation of Figure 12.11b is catastrophic, as well as that of Figure 12.12b, and that the pitchfork bifurcation can be noncatastrophic (as in Figure 12.11c) or catastrophic (as in Figure 12.12c).

From Figure 12.13, we can immediately conclude that the supercritical Hopf bifurcation is noncatastrophic, whereas the subcritical one is catastrophic. This is why the two Hopf bifurcations are sometimes called catastrophic and noncatastrophic. Finally, Figures 12.14, 12.18, and 12.19 show that tangent and homoclinic bifurcations are catastrophic.

When a small parametric variation triggers a catastrophic transition from an attractor A' to an attractor A'', it is interesting to determine if it is possible to drive the system back to the attractor A' by suitably varying the parameter. When this is possible, the catastrophe is called reversible. The most simple case of reversible catastrophes is the hysteresis, two examples of which (concerning first-order systems) are shown in Figure 12.20. In case (a), the system has two saddle-node bifurcations, whereas in case (b) there is a transcritical bifurcation at p_1^* and a saddle-node bifurcation at p_2^*. All bifurcations are catastrophic (because the transitions $A \rightarrow B$ and $C \rightarrow D$ are macroscopic) and if p is varied back and forth between $p_{\min} < p_1^*$ and $p_{\max} > p_2^*$ through a sequence of small steps with long time intervals between successive steps, the state of the system follows closely the cycle $A \rightarrow B \rightarrow C \rightarrow D$ indicated in the figure and called hysteretic cycle (or, briefly, hysteresis). The catastrophes are, therefore, reversible, but after a transition from A' to A'' it is necessary to pass through a second catastro-

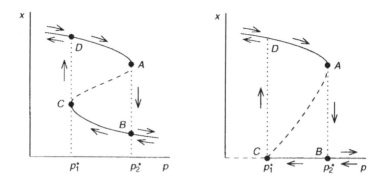

Figure 12.20. Two systems with hysteresis generated by two saddle-node bifurcations (a), and a saddle-node and a transcritical bifurcation (b).

phe to come back to the attractor A'. This simple type of hysteresis explains many phenomena in physics, chemistry, and electromechanics, but also in biology and social sciences. For example, the hysteresis of Figure 12.20b was used by Noy-Meir (1975) to explain the possible collapse (saddle-node bifurcation) of an exploited population described by the equation

$$\dot{x} = rx\left(1 - \frac{x}{K}\right) - \frac{ax}{1 + a\tau x}p$$

where x is resource density (e.g., density of grass) and p is the number of exploiters (e.g., number of cows). If p is increased step by step (e.g., by adding one extra cow every year), the resource declines smoothly until it collapses to zero when a threshold p_2^* is passed. To regenerate the resource, one is obliged to radically reduce the number of exploiters to $p < p_1^*$.

Hysteresis can be more complex than in Figure 12.20, not only because the attractors involved in the hysteretic cycle can be more than two, but also because some of them can be cycles. To show the latter possibility, we consider the so-called Rosenzweig–MacArthur model (Rosenzweig and Mac Arthur, 1963) that describes the dynamics of a tritrophic food chain (x_1: prey; x_2: predator; p: superpredator) in which, however, the top population is (or is kept) constant. Without entering into the details of the analysis of this second-order model (see Kaznetsov et al., 1995), we show in Figure 12.21 the equilibria and the cycles of the system in the control space (p, x_1, x_2) for a specified value of all other parameters. The figure points out five bifurcations: a transcritical (*TR*), two homoclinic (h_1 and h_2), a supercritical Hopf (*H*), and a saddle-node (*SN*). Two of these bifurcations, namely the second homoclinic h_2 and the saddle-node *SN*, are catastrophic and irreversible. In fact, catastrophic transitions from h_2 and *SN* bring the system toward the trivial equilibrium (K, 0) (extinction of the

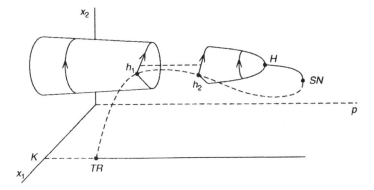

Figure 12.21. Equilibria and limit cycles (characteristic frame) of the Rosen-zweig–MacArthur tritrophic food chain model with constant superpredator population p.

predator population), and from this state it is not possible to return to h_2 or SN by varying p step by step. By contrast, the two other catastrophic bifurcations, namely the first homoclinic h_1 and the transcritical TR, are reversible and identify a hysteretic cycle obtained by varying back and forth the parameter p in an interval slightly larger than $[p_{TR}, p_{h1}]$. On one extreme of the hysteresis, we have a catastrophic transition from the equilibrium $(K, 0)$ to a prey–predator limit cycle. Then, by increasing p, the period of the limit cycle increases (and tends to infinity as $p \to p_{h1}$), and on the other extreme of the hysteresis we have a catastrophic transition from a homoclinic cycle (in practice, a cycle of very long period) to the equilibrium $(K, 0)$. Thus, if p is varied smoothly, slowly, and periodically from just below p_{TR} to just above p_{h1}, one can expect that the predator population will

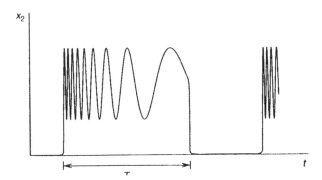

Figure 12.22. Periodic variation of a predator population induced by a periodic variation of the superpredator.

vary periodically in time, as shown in Figure 12.22. In conclusion, the predator population remains very scarce for a long time and then suddenly regenerates, giving rise to high-frequency prey–predator oscillations, which, however, slow down before a crash of the predator population occurs. Of course, tritrophic food chains do not always have such wild dynamics. In fact, many food chains are characterized by a unique attractor and, therefore, cannot experience catastrophic transitions and hysteresis.

An interesting variant of the hysteresis it the so-called cusp, described by the normal form

$$\dot{x} = p_1 + p_2 x - x^3$$

which is still a first-order system, but with two parameters. For $p_1 = 0$, the equation degenerates into the pitchfork normal form, whereas for $p_2 > 0$ the equation points out a hysteresis with respect to p_1 with two saddle nodes. The graph of the equilibria $\bar{x}(p_1, p_2)$ is reported in Figure 12.23, which shows that for the parameters (p_1, p_2) belonging to the cusp region in parameter space, the system has three equilibria: two stable and one unstable (in the middle). In contrast with the hysteresis shown in Figure

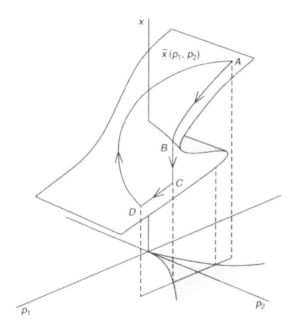

Figure 12.23. Equilibria of the cusp normal form. The unstable equilibria are on the gray part of the surface, which corresponds to the gray cusp region in the parameter space.

12.20, this time after a catastrophic transition from an attractor A' to an attractor A'' (transition $B \to C$ in the figure), one can find the way to come back to A' without suffering a second catastrophic transition (path $C \to D \to A \to B$ in the figure).

12.7   ROUTES TO CHAOS

The bifurcations we have seen in the previous sections deal with the most common transitions from stationary to cyclic regimes and from cyclic to quasiperiodic regimes. Only one of them, namely the homoclinic bifurcation in third-order systems, can mark, under suitable conditions specified by the Shil'nikov theorem, the transition from a cyclic regime to a chaotic one. In an abstract sense, the Shil'nikov bifurcation is responsible for one of the best-known "routes to chaos," called the torus explosion, characterized by the collision in a three-dimensional state space of a saddle cycle with a stable torus. Observed on a Poincaré section, the bifurcation is revealed by a gradual change in shape of the intersection of the torus with the Poincaré section, which becomes more and more pinched while approaching the collision with the saddle cycle. After the collision, the torus breaks into a complex fractal set, which, however, retains the geometry of a pinched, closed curve, as shown in Figure 12.24.

Another, and perhaps best known, route to chaos is the Feigenbaum cascade, which is an infinite sequence $\{p_i\}$ of flip bifurcations in which the p_is accumulate at a critical value p_∞, after which the attractor is a genuine strange attractor. Very often, this route to chaos is depicted by plotting the local peaks of a state variable, say x_1, as a function of a parameter p, as shown in Figure 12.25. Physically speaking, the attractor remains a cycle until $p = p_\infty$, but the period of the cycle doubles at each bifurcation p_i, whereas unstable, actually saddle, cycles of longer and longer periods ac-

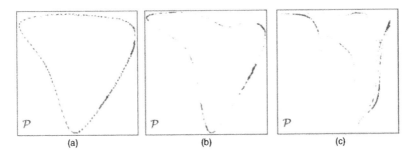

Figure 12.24. Torus explosion route to chaos viewed on a Poincaré section. (a) Regular torus; (b) pinched torus; (c) strange attractor.

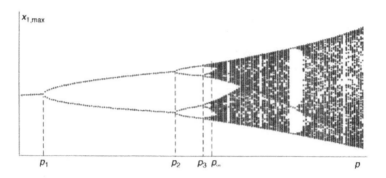

Figure 12.25. The Feigenbaum route to chaos. For $p \leq p_1$ the attractor is a cycle; for $p_1 < p < p_\infty$ the attractor is a longer and longer cycle; for $p \leq p_\infty$ the attractor is a strange attractor.

cumulate in state space. This route to chaos points out a general property of strange attractors, namely, the fact that they are basically composed of an aperiodic trajectory visiting a bounded region of the state space densely filled with saddle cycles, repelling along some directions (stretching) and attracting along others (folding).

12.8   NUMERICAL METHODS AND SOFTWARE PACKAGES

All effective software packages for numerical bifurcation analysis are based on continuation (see, e.g., Keller, 1977; Allgower and Georg, 1990; Doedel et al., 1991a,b; Beyn et al., 2002; Kuznetsov, 2004, Chapter 10), which is a general method for producing in $\mathbf{R}^q$ a curve defined by $(q - 1)$ equations:

$$F_1(w_1, w_2, \ldots, w_q) = 0$$
$$F_2(w_1, w_2, \ldots, w_q) = 0$$
$$\vdots$$
$$F_{q-1}(w_1, w_2, \ldots, w_q) = 0$$

or, in compact form,

$$F(w) = 0, \qquad w \in \mathbf{R}^q, \qquad F: \mathbf{R}^q \to \mathbf{R}^{q-1} \qquad (12.13)$$

Given a point $w^{(0)}$ that is approximately on the curve, that is, $F(w^{(0)}) \neq 0$, the curve is produced by generating a sequence of points $w^{(i)}$, $i = 1, 2, \ldots$ that are approximately on the curve (i.e., $F(w^{(i)}) \neq 0$), as shown in Figure

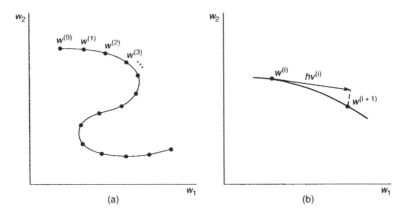

Figure 12.26. Generation of the curve defined by Equation 12.13 through continuation.

12.26a. The ith iteration step, from $w^{(i)}$ to $w^{(i+1)}$, is a so-called prediction–correction procedure with adaptive step size and is illustrated in Figure 12.26b. The prediction $hv^{(i)}$ is taken along the direction tangent to the curve at $w^{(i)}$, where $v^{(i)}$ is computed as the vector of length 1 such that $F/w|_{w=w(i)}v^{(i)} = 0$, the absolute value of h, called the step size, is the prediction length, and the sign of h controls the direction of the continuation. Then suitable corrections try to bring the predicted point back to the curve with the desired accuracy, thus determining $w^{(i+1)}$. If they fail, the step size is reduced and the corrections are tried again until they succeed or the step size goes below a minimum threshold, at which the continuation halts with failure. By contrast, if corrections succeed at the first trial, the step size is typically increased.

   Given a second-order system $\dot{x} = f(x, p)$, where p is a single parameter, assume that an equilibrium $\bar{x}^{(0)}$ is known for $p = p^{(0)}$. Thus, starting from point $(\bar{x}^{(0)}, p^{(0)})$ in $\mathbf{R}^3$, the equilibria $\bar{x}(p)$ can be easily produced, as shown in Figure 12.27, through continuation by considering Equation 12.13 with

$$F(w) = f(x, p), \qquad w = \begin{bmatrix} x \\ p \end{bmatrix}$$

Moreover, at each step of the continuation, the Jacobian $J(\bar{x}(p), p)$ and its eigenvalues $\lambda_1(p)$ and $\lambda_2(p)$ are numerically estimated and a few indicators $\phi(\bar{x}(p), p)$, called bifurcation functions, are computed. These indicators annihilate at specific bifurcations, as shown in Figure 12.27. For example, ϕ' = det(J) is a bifurcation function of transcritical, saddle-node, and pitchfork bifurcations, since at these bifurcations one of the eigenvalues of the

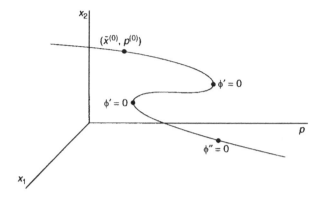

Figure 12.27. The curve $\bar{x}(p)$ produced from $(\bar{x}^{(0)}, p^{(0)})$ through continuation in the three-dimensional control space (p, x_1, x_2) and three bifurcation points, detected through the annihilation of the bifurcation functions ϕ' and ϕ''.

Jacobian matrix is zero and $\det(J) = \lambda_1 \lambda_2$. Similarly, $\phi'' = \text{trace}(J)$ is a Hopf bifurcation function (see Section 12.4). Once a parameter value annihilating a bifurcation function has been found, a few simple tests are performed to check if the bifurcation is really present or to detect which is the true bifurcation within a set of potential ones. For example, as clearly pointed out by Figure 12.11b, at a saddle-node bifurcation the p-component of the vector tangent to the curve $\bar{x}(p)$ annihilates. By contrast, at transcritical and pitchfork bifurcations (see Figures 12.11a and c) two equilibrium curves, one of which is $\bar{x}(p)$, transversally cross each other, so that there are two tangent vectors at $p = p^*$, one with a vanishing p-component in the pitchfork case. Analogously, if $\phi''(\bar{x}(p^*), p^*) = 0$ one must first check that $\phi'(\bar{x}(p^*), p^*)$ is positive before concluding that $p = p^*$ is a Hopf bifurcation (see Section 12.4).

Once a particular bifurcation has been detected through the annihilation of its bifurcation function ϕ, it can be continued by activating a second parameter. For this, Equation 12.13 is written with

$$F(w) = \begin{bmatrix} f(x, p) \\ \phi(x, p) \end{bmatrix}, \qquad w = \begin{bmatrix} x \\ p \end{bmatrix}$$

where w is now four dimensional since p is a vector of two parameters. If the curve obtained through continuation in $\mathbf{R}^4$ is projected on the two-dimensional parameter space, the desired bifurcation curve is obtained.

In the case of local bifurcations of limit cycles and global bifurcations, the functions ϕ are quite complex and their evaluation requires the solution of the ODEs $\dot{x} = f(x, p)$. Actually, a rigorous treatment of the

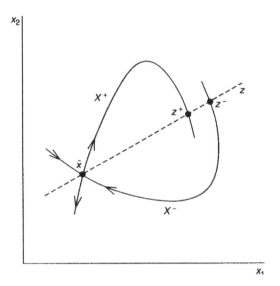

Figure 12.28. The bifurcation function $\phi = z^+ - z^-$ is zero when there is a homoclinic bifurcation, that is, when the stable and unstable manifolds X^- and X^+ of the saddle collide.

problem brings one naturally to the formulation of two-boundary-value problems (Doedel et al., 1991b; Beyn et al., 2002). For example, as shown in Figure 12.28, homoclinic bifurcations can be detected by the function $\phi = z^+ - z^-$, where z^+ and z^- are the intersections of the unstable and stable manifolds of the saddle with an arbitrary axis z passing through the saddle. Thus, ϕ is zero if and only if the saddle has a homoclinic connection.

There are many available software packages for bifurcation analysis, but the most interesting ones are AUTO (Doedel, 1981; Doedel et al., 1997, 2007), LOCBIF (Khibnik et al., 1993), CONTENT (Kuznetsov and Levitin, 1997), and MATCONT (Dhooge et al., 2002). They can all be used to study systems with more than two state variables and they can detect and continue all bifurcations mentioned in this chapter. AUTO is the most popular software for bifurcation analysis and is particularly suited for the analysis of global bifurcations. LOCBIF is more effective than AUTO for local bifurcations, since it can also continue codimension-2 bifurcations. However, LOCBIF runs only on MS-DOS and has, therefore, been reimplemented and improved in CONTENT, which runs on several software platforms. MATCONT, continuously updated, is aimed at encapsulating the best features of all previously mentioned software packages in a MATLAB environment.

REFERENCES

Allgower, E. L., and Georg, K., *Numerical Continuation Methods: An Introduction,* Springer-Verlag, 1990.

Alligood, K. T., Sauer, T. D., and Yorke, J. A., *Chaos: An Introduction to Dynamical Systems,* Springer-Verlag, 1996.

Andronov, A. A., Leontovich, E. A., Gordon, I. J., and Maier, A. G., *Theory of Bifurcations of Dynamical Systems on a Plane,* Israel Program for Scientific Translations, Jerusalem, 1973.

Beyn, W.-J., Champneys, A. R., Doedel, E. J., Govaerts, W., Kuznetsov, Yu, A., and Sandstede, B., Numerical Continuation, and Computation of Normal Forms, in *Handbook of Dynamical Systems,* Fiedler, B. (Ed.), vol. 2, pp. 149–219, Elsevier Science, Burlington, 2002.

Dercole, F., and Rinaldi, S., *Analysis of Evolutionary Processes: The Adaptive Dynamics Approach and its Applications,* Princeton University Press, 2008.

Dhooge, A., Govaerts, W., and Kuznetsov, Yu, A., MATCONT: A MATLAB Package for Numerical Bifurcation Analysis of ODEs, *ACM T. Math. Software,* Vol. 29, 141–164, 2002.

Doedel, E. J., AUTO: A Program for the Automatic Bifurcation Analysis of Autonomous Systems, *Cong. Numer.,* Vol. 30, 265–384, 1981.

Doedel, E. J., Champneys, A. R., Dercole, F., Fairgrieve, T. F., Kuznetsov, Yu, A., Oldeman, B., Paffenroth, R. C., Sandstede, B., Wang, X. J., and Zhang, C. H., AUTO-07p: Continuation and Bifurcation Software for Ordinary Differential Equations, Department of Computer Science, Concordia University, Montreal, 2007.

Doedel, E. J., Champneys, A. R., Fairgrieve, T. F., Kuznetsov, Yu, A., Sandstede, B., and Wang, X. J., AUTO97: Continuation and Bifurcation Software for Ordinary Differential Equations, Department of Computer Science, Concordia University, Montreal, 1997.

Doedel, E. J., Keller, H. B., and Kernévez, J.-P., Numerical Analysis And Control Of Bifurcation Problems (I): Bifurcation In Finite Dimensions, *Int. J. Bifurcat. Chaos,* Vol. 1, 493–520, 1991a.

Doedel, E. J., Keller, H. B., and Kernévez, J.-P., Numerical Analysis and Control of Bifurcation Problems (II): Bifurcation in Infinite Dimensions, *Int. J. Bifurcat. Chaos,* Vol. 1, 745–772, 1991b.

Guckenheimer, J., and Holmes, P., *Nonlinear Oscillations, Dynamical Systems and Bifurcations of Vector Fields,* 5th ed., Springer-Verlag, 1997.

Keller, H. B., Numerical Solution of Bifurcation and Nonlinear Eigenvalue Problems, in *Applications of Bifurcation Theory,* pp. 359–384, Rabinowitz, P. H. (Ed.), Academic Press, 1977.

Khibnik, A. I., Kuznetsov, Yu, A., Levitin, V. V., and Nikolaev, E. V., Continuation Techniques and Interactive Software for Bifurcation Analysis of ODES And Iterated Maps. *Physica,* Vol. D, 62, 360–370, 1993.

Kuznetsov, Yu, A., *Elements of Applied Bifurcation Theory,* 3rd ed., Springer-Verlag, 2004.

Kuznetsov, Yu, A. and Levitin, V. V., CONTENT: A Multiplatform Environment for Analyzing Dynamical Systems, Dynamical Systems Laboratory, Centrum

voor Wiskunde en Informatica, Amsterdam, The Netherlands, ftp.cwi.nl/pub/
CONTENT, 1997.

Kuznetsov, Yu, A., Muratori, S., and Rinaldi, S., Homoclinic Bifurcations in Slow-
Fast Second-Order Systems, *Nonlinear Anal.,* Vol. 25, 747–762, 1995.

Marsden, J., and McCracken, M., *Hopf Bifurcation and its Applications,* Springer-
Verlag, 1976.

Noy-Meir, I., Stability of Grazing Systems: An Application of Predator-Prey
Graphs, *J. Ecol.,* Vol. 63, 459–483, 1975.

Rinaldi, S., The Theory of Complex Systems, in *Biosystems and Complexity,* Be-
lardinelli, E. and Cerutti, S. (Eds.), pp. 15–64, Patron, Bologna, (in Italian),
1993.

Rosenzweig, M. L., and MacArthur, R. H., Graphical Representation and Stability
Conditions of Predator–Prey Interactions, *Am. Nat.,* Vol. 97, 209–223, 1963.

Shil'nikov, L. P., On the Generation of Periodic Motion from Trajectories Doubly
Asymptotic to an Equilibrium State of Saddle Type, *Math. USSR-Sb+,* Vol. 6,
427–437, 1968.

Strogatz, S. H., *Nonlinear Dynamics and Chaos,* Addison-Wesley, 1994.

Thom, R., *Structural Stability and Morphogenesis,* Benjamin-Cummings (English
translation 1975), 1972.

Thispageintentionallyleftblank

CHAPTER 13

FRACTAL DIMENSION
From Geometry to Physiology

Rita Balocchi

T HE NOTION OF dimension plays an important role in mathematics because it supplies a precise parameterization of the conceptual complexity of any geometrical object and of the space it is contained in. Aristotle made reference to it in the Book Delta (the book of definitions) of his *Metaphysics*:

> But everywhere the one is indivisible either in quantity or in kind. Now that which is indivisible in quantity is called a unit if it is not divisible in any dimension and is without position, a point if it is not divisible in any dimension and has position, a line if it is divisible in one dimension, a plane if in two, a body if divisible in quantity in all—that is, in three—dimensions. (Ross, 2006)

Many centuries later, Benoit Mandelbrot opened his *The Fractal Geometry of Nature* with

> Why is geometry often described as "cold" and "dry"? One reason lies in its inability to describe the shape of a cloud, a mountain, a coastline, or a tree. Clouds are not spheres, mountains are not cones, coastlines are not circles, and bark is not smooth, nor does lightning travel in a straight line. More generally, I claim that many patterns of nature are so irregular and fragmented, that, compared with Euclid—a term used in this

Advanced Methods of Biomedical Signal Processing. Edited by S. Cerutti and C. Marchesi
Copyright © 2011 the Institute of Electrical and Electronics Engineers, Inc.

work to denote all of standard geometry—nature exhibits not simply a higher degree but an altogether different level of complexity. . . . The existence of these patterns challenges us to study these forms that Euclid leaves aside as being "form-less," to investigate the morphology of the "amorphous." Mathematicians have disdained this challenge, however, and have increasingly chosen to flee from nature by divising theories unrelated to anything we can see or feel.

Responding to this challenge, I conceived and developed a new geometry of nature and implemented its use in a number of diverse fields. It describes many of the irregular and frag-mented patterns around us, and leads to full-fledged theories, by identifying a family of shapes I call fractals.* (Mandelbrot, 1983)

If on the one hand the innovative content of Mandelbrot's statement is undeniable, on the other hand the tone surprises, being decidedly assertive in criticizing geometry, defining it as arid and separated from nature. Even though I do not wish to go into an argument that would be marginal with re-spect to the aim of this chapter, I do, however, wish to submit a short com-ment. Geometry—kept alive for centuries despite many adversities—was, at its beginning, conscious of the reductive nature of its sphere of applica-tion.[†] However, this is exactly why it has great cultural value, summarizable by the twofold aim of educating to abstraction and unification. This con-sciousness became lost starting from the seventeenth century (and this somehow justifies the assertion of Mandelbrot), particularly for the cultural hegemony exerted by Galileo and Descartes.[‡]

This chapter builds on the novel approach proposed by Mandelbrot with the aim of introducing the cornerstone concepts of fractal geometry, passing from the description of fractal properties in space (objects) to that in time (time series), to conclude with some practical applications in the biomedical field.

---

*The term *fractal* comes from the Latin verb *frangere* (to break, to create irregular fragments) and from its past participle *fractus*, and expresses the concept of irregu-larity.

[†]As is known, Euclid came from the Platonic Academy where the study of Geome-try was cultivated at the highest levels, but this discipline was thought to be de-rived from and subordinated to a much higher and more complex knowledge that defined its aims and limits.

[‡]Galileo Galilei wrote, "The book of nature is written in the language of mathemat-ics and its characters are triangles, circles and other geometrical figures; if we can-not understand that language, we will be doomed to wander about as if in a dark labyrinth."

The chapter is organized as follows. Section 13.1 contains, in the first part, references to topology and is intended for those readers interested in formal definitions. This part can be used at the beginning of the reading or can be referred to if necessary during reading. The second part contains some definitions of the notion of dimension necessary to define the fractal object. Section 13.2 deals with fractal geometry in space. Section 13.3 uses the concepts previously defined and modifies them to be applied to time series. Section 13.4 is a review of characteristics and computation methods and discusses some biomedical applications. Finally, Section 13.5 provides some suggestions for a correct practical use of the techniques presented.

13.1 GEOMETRY

Many concepts underlying this chapter are surely well known by readers, at least in an intuitive form. In any case, for the reader keen on formal aspects, this section provides all the definitions directly or indirectly referred to in what follows.

It is not necessary to read this section in a sequential way so that the concepts can be consulted when encountered in what follows. On the contrary, it would be convenient to pay attention to the Section 13.2 before starting to read further.

13.1.1 Topology

Topology of the Set X. Given a set X, the family $\tau = \{A_i\}$, $A_i \subseteq X$ is a topology on X if the following conditions hold:

$X \in \tau$ (X belongs to τ)
$\varnothing \in \tau$ (the empty set belongs to τ)
Every union (finite or infinite) of A_i is $\cup A_i \in \tau$
Every finite intersection of A_i is $\cap A_i \in \tau$

Topologic Set. The couple (X, τ) is a topological space and A_i are the openings of the topological space. A classical example of a topological space is $\Re$ with the topology $\tau = \{]a,b[, a$ and $b \in \Re\}$.

Distance. Let X be a set and $X \times X$ its Cartesian product. A distance (metric) on X is a function $d: X \times X \to \Re$ such that any choice of x, y, z in X is:

1. $d(x, y) > 0$
2. $d(x, y) = d(y, x)$
3. $d(x, z) \leq d(x, y) + d(y, z)$
4. $d(x, y) = 0$ if and only if $x = y$

Metric Space. A metric space is a set X where the distance (called a metric) between elements of the set is defined.

Neighborhood. Let X be a topological space and M a subset of X. A subset U is a neighborhood of M if an opening A exists such that $M \subseteq A \subseteq U$. In particular, each opening containing M is a neighborhood of M.

Adherent Point. Let X be a topological space and M a subset of X. A point $x \in X$ is an adherent point (also called a closure point) for M if $U \cap M \neq \emptyset$ for every neighborhood U of x.

Frontier. In a topological space X, a point x is a frontier point of a set M if x is adherent to M and to its complement. The set of frontier points of M is the frontier (or boundary) of M.

Cover. A family $\{X_i\}$ of subsets of a set X is a cover of X if $\cup X_i = X$, that is, if every $x \in X$ belongs to at least one X_i.

13.1.2 Euclidean, Topologic, and Hausdorff–Besicovitch Dimension

We shall see in what follows the role and meaning of dimension for fractal objects. In the meantime, to pave the way, it is necessary to introduce the definition of at least three different dimensions, two of them most commonly used and relatively simple, the third a little less familiar and fairly more complicated.

Euclidean Dimension. DE indicates the number of coordinates necessary to specify an object in a given space:

1. $DE = 1$, point on a line
2. $DE = 2$, point on a plane
3. $DE = 3$, point on a space

Topological Dimension. Let (X, τ) be a topological space and $F \in (X, \tau)$. F has topological dimension Dt if F has an open neighboorhood with frontier of dimension $Dt - 1$. A point or a set of points totally disconnected have dimension $Dt = 0$.

Hausdorff–Besicovitch Dimension (Hausdorff, 1919). We start with the definition of a d-dimensional measure. Let X be a subset of a metric space and $d \in \mathfrak{R}$, $d > 0$. Let $\{S_i\}$ be the set of all the finite cover of X with diameter (diam $S_i) \leq \rho$, with $\rho > 0$:

$$m_d(X, \rho) = \inf \{\Sigma_i (\text{diam } S_i)^d\} \tag{13.1}$$

We define an external d-dimensional measure as the quantity $m_d(X)$:

$$m_d(X) = \lim_{\rho \to 0} m_d(X, \rho) \qquad (13.2)$$

The limit of Equation 13.2 can be either zero, finite and positive, or infinite.

Besicovitch showed that for every set X, a real value D exists (Hausdorff–Besicovitch dimension) such that the d-measure of X is infinite for $d < D$, and zero for $d > D$:

$$D = \sup\{d \in R^+ : m_d(X) = \infty\} \qquad (13.3)$$

Capacity Dimension. This definition applies to the particular condition in which all the covers have the same diameter ρ: (diam S_i) $= \rho$. Generally, this definition is identified with the D Hausdorff–Besicovitch (H-B) dimension.

13.2   FRACTAL OBJECTS

Although fractal geometry can be considered as a very young discipline, the objects involved are so fascinatingly shaped that they have been granted the honor to come out of their specific sphere of pertinence to appear in different contexts for merely ornamental purposes. Considering the wide fame of fractal images and of their descriptive basic concepts, it is always somehow embarrassing to start a treatment of this geometry using the well-known canonical figures; on the other hand, those figures are well suited for didactic purposes, so I hope that the readers will understand if I introduce this basic level as well. Let us start with the construction of some objects for which we will define fractals on the basis of some specific properties.

13.2.1   Koch Curve, Cantor Set, and Sierpinski Triangle

The Koch curve is part of a family of curves built by addition and generated according to the procedure shown in Figure 13.1. A segment of length l_0 is divided into three parts, each of length $l_0/3$. The middle third is removed and replaced with two edges of an equilateral triangle, each of length $l_0/3$. This new object consists of four segments instead of one, each of length $l_0/3$. Each of the four segments are operated using the same procedure of removal and substitution, obtaining a polygonal of 16 segments, each of length $l_0/9$. The length of the polygonal increases at each step; in particular, the first few values can be seen in the table inserted in Figure 13.1.

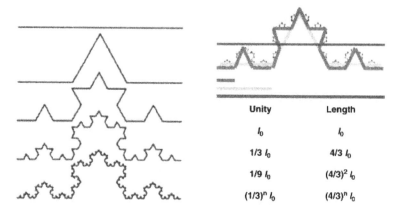

Unity	Length
l_0	l_0
$1/3\, l_0$	$4/3\, l_0$
$1/9\, l_0$	$(4/3)^2\, l_0$
$(1/3)^n\, l_0$	$(4/3)^n\, l_0$

Figure 13.1. Koch curve.

What are the salient characteristics of this object? From a computational point of view, the length depends on the choice of the ruler: the finer the ruler, the more details will be detected (after an infinite number of iterations, however, the length becomes infinite). So this object seems not to have a specific intrinsic measure, besides its obvious topologic dimension $Dt = 1$, which does not help to characterize it.

Let us analyze a second classic object, the Cantor set. This object belongs to the family of objects built by subtraction according to the scheme of Figure 13.2 (left panel). The segment of length l_0 is divided into three parts, and the open middle third is removed, leaving two line segments. The same procedure is applied to the remaining two line segments, and so on, ad infinitum. At each iteration, every segment is replicated in two parts and scaled by one-third of is length. In the limit to an infinite number of iterations, the set powders into points. How can we characterize this powder, which has a topologic dimension $Dt = 0$?

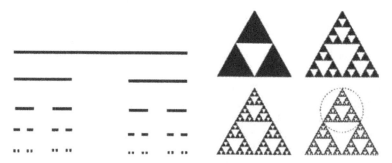

Figure 13.2. Left: Cantor set. Right: Sierpinski triangle.

The third object we are going to analyze is the Sierpinski triangle, whose generation is shown in Figure 13.2 (right panel). The procedure consists in removing from an equilateral triangle the internal triangle whose vertices are the middle points of the three edges. The removal is repeated for each of the newly generated triangles. Each of these triangles contains three equal copies of triangles, each scaled by one-third. The object of interest is the profile obtained following the perimeter of all the triangles. What kind of measure can be given to such an object, whose topologic dimension is, again, $Dt = 1$?

To answer the previous questions, let us approach the calculation of the d-dimension of each object mentioned above.

For the Koch curve. Let N be the number of spheres with radius $< \rho$ of the cover defined in Equation 13.1. Therefore,

$$m_d(X) = \lim_{\rho \to 0} N(\rho) \cdot \rho^d = \lim n_{\to \infty} 4^n/3^{nd} = 3^{n(\log 4/\log 3)}/3^{nd} \quad (13.4)$$

It is easy to see that

$$0 < m_d(X) < \infty \qquad \text{if and only if } d = \log 4/\log 3 \quad (13.5)$$

We could associate with the Koch curve the value $d = \log 4/\log 3 \approx 1.2618$. Likewise, for the Cantor set the result is $d = \log(2)/\log(3) \approx 0.6309$. Last, for the Sierpinski triangle, we have $d = \log(3)/\log(2) \approx 1.5849$.

It can be noted that the topologic dimension of each object does not agree with D. Furthermore, the topologic dimension is always strictly lower than D.

At this point, we can introduce the Mandelbrot definition of fractal object.

Definition: An object is said to be fractal if its H-B dimension strictly exceeds its topologic dimension, $Dt < D$.

13.2.2 Properties of Fractals

We have just defined a fractal as an object whose H-B dimension is strictly higher than the topological dimension. In addition, each object with a noninteger H-B dimension is fractal. This last statement does not lead to the deduction that an object with an integer dimension is not fractal; in fact, for an object to be fractal, a noninteger H-B dimension is a sufficient and nonnecessary condition.

An example of a fractal object with integer dimension D is the well-known Hilbert curve, whose construction is more complicated than the ones previously analyzed. Figure 13.3 shows one of the possible schemes of its generation. The basic unit is replicated four times with dimensions

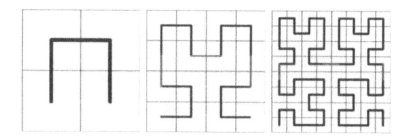

Figure 13.3. Hilbert curve. (Reproduced with permission from "Tutorial: Hilbert Curve Coloring" by Kerry Mitchell, http://www.fractalus.com/kerry/tutorials/hilbert-tutorial.html.)

reduced by one-half. While the elements on the top maintain their position, the lower elements are rotated left and right by 90°. The individual elements are then connected and the procedure is repeated with replications, rotations, and connections. The H-B dimension is $D = \log(4)/\log(2) = 2$, whereas the topological dimension is $Dt = 1$; therefore, by definition, the curve is fractal.

The noninteger dimension was interpreted by Mandelbrot as the fractal ability to protrude into higher dimensional space without completely filling it. Thus, a dimension between 0 and 1 is the ability of a set of points to partially fill a line, a dimension between 1 and 2 is the ability to partially fill a plane, and so on. It is interesting to note that the Hilbert curve has dimension 2 and actually touches all the points of a square.

This is the commonly given meaning of the fractal dimension. Recently, however, a new, interesting interpretation has been proposed by Ricourdeau (Ricourdeau, 1997), who believes that considering the fractal dimension in terms of coordinates is a way to restrain it and to deprive it of a richness that has nothing to do with the position of an object in space; fractal dimensions refer to a different system of relations he calls deformations. A fractal dimension is composed of both an integer and a decimal part. The hypothesis is the following:

Case a. The integer part is 0. This implies that we are in the presence of a phenomenon that does not involve a displacement in space. Therefore, a dimension with the integer part equal to 0 indicates mutations, that is, what is existing disappears and what does not exist appears.

Case b. The integer part is 1. In this case, the deformation also includes a spatial displacement and refers to a body movement (trajectory).

Case c. The integer part is 2. We are in presence of a deformation of a body into itself.

The decimal part of a fractal dimension indicates how the intensity of deformation changes. A dimension of 1.9999 with recurring decimal 9 is not very close to dimension 2. The two values are very far apart, inasmuch as the first expresses an extremely winding course, which can also happen in a three-dimensional space, whereas the second is related to deformation of a surface or of a curve. But there is more: dimension 2 can be seen as the limit to infinite of dimension 1.999, that is, of a course undergoing the maximal bearable random windings. This is the case of Brownian motion, whose trajectory has fractal dimension 2.

Besides these controversial interpretations, the other major fractal property is the that no matter the magnification level of observation, the detail is always similar to the whole object. This property is called self-similarity* or self-affinity.

13.3   FRACTALS IN PHYSIOLOGY

Self-similarity is an exact characteristic of fractals in that it is maintained for infinite scale lengths. Obviously, nothing similar can exist in nature; many objects can be found exhibiting scale invariance, but only for a limited number of magnifications.

Many structures with this characteristic can be found in physiology: the neural net, the bronchial tree, the pulmonary arterial tree, the retinal vessels, and many others.

A tree-shaped connection is a highly efficient mechanism for the passage from large to small or from complex to simple. To understand the significance of this efficiency, we have only to consider, for instance, that the exchange of respiratory gases between the external environment and the alveoli takes place through respiratory airways whose fractal ramification passes from a transversal section of 3 cm^2 in the trachea to a complex alveolar surface of 70 m^2 (Glenny et al., 1991).

Of particular interest is the fractal structure of the heart: the coronary tree, which through all its ramifications accomplishes the vital task of distributing and collecting blood in the cardiac muscle; the tendineous chords, which connect the tricuspid and mitral valves to the muscle; and the Purkinje fibers, the anatomical pathways through which the electrical stimuli pass from atria to ventricles to induce cardiac contraction.

It is obviously clear that the same mathematical formalism used to compute the dimension D is not applicable to those structures. It is, however, possible to characterize their fractality using a statistical approach. Toward this aim, many interesting applications can be found, for instance,

---

*Often in the literature, this concept is also referred to as self-affinity. Actually, the two concepts are not identical: selfsimilar objects are isotropic, that is, they scale similarly along each direction, whereas self-affine objects are still fractals but anisotropic, that is, they scale differently in different directions.

the analysis of regional blood flow distribution in the myocardium and lung (Balocchi et al., 1998).

13.3.1   Self-Similarity of Dynamic Processes

So far, we have introduced autosimilarity in space. Is this property also common to processes that develop in time? They are very frequent in physiology; for instance, ionic currents, blood flows and pressures, ventilation, heart rate, and many others.

Let X_i, $i = 1, \ldots, N$ be a discrete time series representing a dynamical process $X(t)$ sampled at intervals Δt. Figure 13.4 shows the RR time series, that is the series of times between successive heartbeats. It is interesting to note how the first series (from top), lasting about 2 hours, exhibits an apparently erratic time course and how this characteristic is maintained in the two subepochs of 30 and 15 minutes. This behavior suggests a way to extend the concept of autosimilarity to the temporal dimension: autosimilarity occurs when subepochs of a time series exhibit a time course

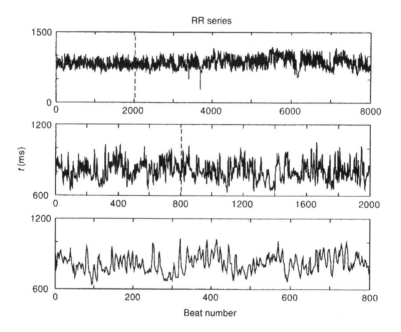

Figure 13.4. The RR series (time intervals between successive beats). All the series show the erratic behavior of the heartbeat for different temporal windows. From the first series on top (of a 2-hour length), the second is extracted (of about 30 min, up to the dashed line) and, from this, the third series is extracted (of about 15 min, up to the dashed line).

similar to that of the time series as a whole, at least for a certain number of magnification levels. In formal terms, a stochastic process $\{X(t), t \geq 0\}$ is self-similar if for each $a > 0$ there exists a $b > 0$ such that

$$\{X(at)\} \stackrel{d}{=} \{bX(t)\} \tag{13.6}$$

where the symbol $\stackrel{d}{=}$ means equality in statistical distribution.

For completeness, we add the following: $\{X(t), t \leq 0\}$ is said to be stochastically continuous in t if for each $\varepsilon > 0$ $\lim_{h \to 0} P\{|X(t + h) - X(t)| > \varepsilon\} = 0$. It can be demonstrated that if $\{X(t), t \geq 0\}$ is stochastically continuous and self-similar, then a unique value $H \geq 0$ exists, such that $b = a^H$ in Equation 13.6. Furthermore, $H > 0$ if and only if $X(0) = 0$. Some authors give a different definition: A stochastic process is self-similar if there is $H > 0$ such that for each $a > 0$ $\{X(at)\} \stackrel{d}{=} \{a^H X(t)\}$. It follows that $X(0) = 0$. Yet, the above proposition is incorrect because it does not imply the uniqueness of H. This value H is called the Hurst exponent (Hurst et al., 1965) or similarity exponent.

13.3.2   Properties of Self-Similar Processes

Mathematically, self-similarity manifests itself in a variety of equivalent modes. In the power spectrum, power P and frequency f are linked by the relation $P = f^{-\beta}$, $\beta > 0$. In other words, the power spectrum follows a "power law" and β is the scaling exponent (Noujaim, 2007). Another interesting aspect is given by the aggregate process to $X(t)$. A new time series $X_k^{(m)}$ is obtained by averaging the original series over nonoverlapping blocks of length m:

$$X_k^{(m)} = 1/m(X_{(k-1)m+1} \ldots + X_{km}), \qquad k = 1, 2, \ldots, [N/m] \tag{13.7}$$

For increasing values of m, the autocorrelation function of $X_k^{(m)}$ does not tend to 0, whereas the variance does, even though more slowly than a stationary stochastic process. In particular, $\mathrm{var}[X_k^{(m)}] \sim m^{-\gamma}$, $0 < \gamma < 1$ ($\gamma = 1$ in the stationary case), that is, the aggregate process shows persistency of the statisctical properties along different temporal scales.

In less informal terms, we can say that a self-similar process is characterized by the intrinsic and pervasive presence of irregularities that constitute its complex structure. One of the most specific instances of such a process is surely fractional Brownian motion (fBm), often indicated as $B^H(t)$ (Mandelbrot, 1983), where H is the Hurst exponent.

The series fBm is nonstationary, with stationary increments called fractional Gaussian noise (fGn) and $0 < H < 1$. Values of H close to 1 characterize a smooth signal, whereas values close to 0 are typical of rough signals (Figure 13.5). As we have just said, the increments of a signal fBm

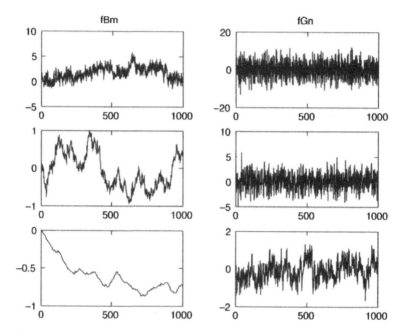

Figure 13.5. Left: three fBm series with different Hurst exponents, $H = 0.1$, $H = 0.5$, $H = 0.9$ (from top). Right: three fGn series with the same exponent of the corresponding fBm series. The fBm series become smoother and smoother as H increases.

give rise to a stationary fGn signal so, vice versa, cumulative sums of a signal fGn give rise to a fBm signal. Both series have the same exponent H, but not the same scaling exponent on the power spectrum (Figure 13.6).

13.4   HURST EXPONENT

Across centuries, the annual floods of the Nile river were the basis of the flourishing Egyptian agriculture, affording this people a very high level of civilization. However, the amount of overflowed water could either allow a good harvest, with the possibility of storage of wheat and other products, or a poor harvest, with consequent food shortages. The flood of the Nile started around April in Sudan and arrived at the town of Aswan in southern Egypt around July, reaching the north and its maximum peak in September–October. Starting from November–December, the level rapidly decreased, reaching its minimum around March–May.

Ancient Egyptians used to keep an accurate documentation of the annual maximum and minimum levels of the Nile, using a measuring unit

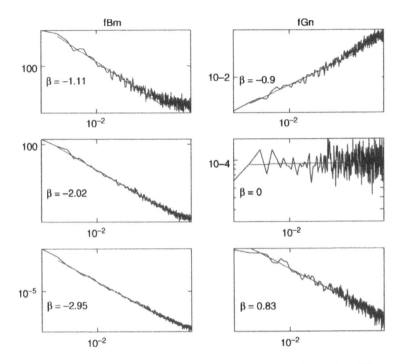

Figure 13.6. Power spectra in log–log scale of the fBm and fGn series of Figure 13.5. For each series, the value β indicates the opposite value of the slope of the regression line. As mentioned in the text, the relationship between β and H changes according to the typology (fFm or fGn) of the series.

called the nilometer, of which some traces still exist. In particular, the values of maximum and minimum levels annotated without interruption from year 641 to 1450 are still available. The British occupation of 1882 marks the beginning of a series of projects aimed at controlling waters in the entire basin of the Nile to guarantee a constant flow of water through the use of an intricate system of channels and dams.

To estimate how tall the Aswan dam should be in order to keep an appropriate amount of water always available, sir Harold Edwin Hurst (1880–1978), a British physicist and hydrologist who was fond of Egypt, carried out various simulations and empirical computations, including the analysis of 800 years of Nile river level values (Hurst et al., 1965). Surprisingly, he discovered an interesting phenomenon: instead of a random alternation between good and bad years, a good year was followed by other years of good levels of water and, similarly, the beginning of scarcity of water persisted in the following years. The approach used by Hurst was based on empirical considerations synthesized in the computation of the ratio R/S, or rescaled range analysis, a way to quantify correlation among values of a time series.

13.4.1   Rescaled Range Analysis

The Hurst analysis assumes a uniform outflow of water, computed as the average of the inflows. The time series of water volume increments is the cumulative difference of inflow and outflow:

$$X_j = \Sigma_{i=1}^{j} (x_i - \bar{x}_i) \tag{13.8}$$

The range of X_j in the period of observation, $R = \max(X_j) - \min(X_j)$, determines how tall the dam ought to be. By dividing R by the inflow standard deviation S, for different lengths N of the observation interval, the ratio R/S follows the law

$$R/S = pN^H \tag{13.9}$$

with p independent of time. From Equation 13.9, it is possible to obtain H as the slope of the line

$$\log(R/S) = \log(P) + H \log(N) \tag{13.10}$$

If inflows and outflows had been independent, the value of the exponent (Hurst exponent*) should have been $H = 1/2$; in the case of Nile, it resulted in $H = 0.72$.

Actually, natural phenomena are very rarely random independent processes and the Hurst exponent can be very useful to characterize a series under examination both from the point of view of morphology and correlation.

We have already mentioned that $0 \le H \le 1$ and the meaning of values of H close to 0 and 1. When dealing with an uncorrelated series X_i, its cumulative values Y_i constitute Brownian motion and $H = 1/2$ (independent increments and lack of correlation). All other values $0 < H < 1$ characterize fBm processes.

In particular, when $1/2 < H < 1$ we are facing processes with positive correlation, so-called persistent processes or those with anomalous diffusion. The case $0 < H < 1/2$, on the contrary, indicates negative correlation with characteristics of antipersistency and consequent diffusion suppression.

The value $H = 0.72$ for the fluctuations of the Nile water level indicates a phenomenon characterized by persistency and, indeed, the history

---

*In his work, Hurst used for the exponent the symbol K, but Mandelbrot changed it into H, and this remains the universally adopted notation.

of ancient Egypt lets us know of cycles of prosperity followed by famine, also narrated in the Bible in Pharaoh's dream:

> Then Pharaoh said to Joseph, "In my dream I was standing on the bank of the Nile, when out of the river there came up seven cows, fat and sleek, and they grazed among the reeds. After them, seven other cows came up, scrawny and very ugly and lean. . . . The lean, ugly cows ate up the seven fat cows that came up first. . . . I told this to the magicians, but none could explain it to me." Then Joseph said to Pharaoh, "The seven good cows are seven years. . . . The seven lean, ugly cows . . . they are seven years of famine. Seven years of great abundance are coming throughout the land of Egypt, but seven years of famine will follow them. Then all the abundance in Egypt will be forgotten, and the famine will ravage the land. . . ." (Genesis, 41:17)

13.4.2 Methods for Computing *H*

The specific procedure for computing the rescaled range can be found in many papers, and a huge number of methods exist to compute the exponent *H*, not necessarily related to the *R/S* analysis (Bassingthaighte and Raymond, 1994). Most of the methods are suitable for any kind of time series, not necessarily of a biological nature. Among the others, we can cite the Higuchi algorithm (Higuchi, 1988), the dispersional analysis (Bassingthaighte and Raymond, 1995), the Katz algorithm (Katz, 1988), and the signal summation conversion (Eke et al., 2000). Comparison and comments on the algorithms can be found in (Pilgram and Kaplan, 1998; Sevik, 1998; Esteller et al., 2001; Eke et al., 2002; Cerutti et al., 2007).

The technique we will describe here in more detail is detrended fluctuation analysis (DFA), introduced in 1994 by Peng [27] and widely used for analysis of long-term correlations of nonstationary time series, including heart rate variability (Peng et al., 1995; Balocchi et al., 1999; Goldberger et al., 2002; Schmitt and Ivanov, 2007; Lee et al., 2008).

13.4.2 Detrended Fluctuation Analysis. This technique was introduced to analyze strongly nonstationary time series such as 24-hour RR series recorded while the subject is attending to his/her daily activities (eating, walking, driving, sleeping, getting angry, etc.). The procedure is as follows:

1. Let $X = \{X_i\}$ be a series of length N. We build a new series $Y = \{Y_j\}$:

$$Y_j = \sum_{i=1}^{j} (X_i - \bar{X}_i) \qquad (13.11)$$

where $\overline{X}_j$ is the mean of the cumulative sum of the first j values of X.

2. The series Y is then divided into nonoverlapping windows of length n and the trend $Y_{j,n}$ is computed in each window.

3. The fluctuation function $F(n)$ defined as

$$F(n) = \sqrt{1/N \; \sum_{j=1}^{N} (Y_n - Y_{j,n})^2} \qquad (13.12)$$

It increases for increasing values of n. The existence of a linear relationship between $F(n)$ and n in log–log scale indicates the presence of scaling with exponent α (DFA exponent).

4. Characteristic values of α are:

$\alpha = 0.5$ for random series X_i (Gaussian noise)
$\alpha = 1.5$ for series X_i, such as Brownian noise
$\alpha = 1$ for series with $1/f$ power-law power spectrum
$0.5 < \alpha < = 1$ for persistent series
$0 < \alpha < 0.5$ for antipersistent series
$\alpha > 1$ for nonfractal series with correlation

Figure 13.7 shows the trend of $F(n)$ versus n (log–log scale) for a 24-hour RR series of a healthy subject and of one in a coma. It is interesting to note that the slope α shows a $1/f$ power-law behavior for the normal subject ($\alpha = 1$) and a Brownian noise trend for the subject in a coma ($\alpha = 1.5$).

From a macroscopic point of view, differences in α reflect intrinsic differences due to the main regulatory mechanisms of heart rate variability (lacking or highly injured neural connections). Sometimes a different slope is detected at very small scale levels, which is due to short-term variability related to respiratory activity. The two RR typologies of Figure 13.7 seem to differ also in this aspect, showing a brownian-like behavior

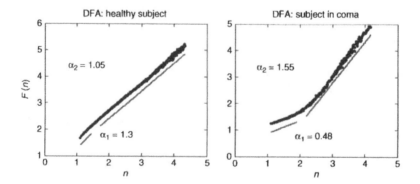

Figure 13.7. DFA of a healthy subject and of a subject in a coma. Axes are in base-10 logarithmic scale.

for the normal subject ($\alpha = 1.5$) and a white-noise behavior for the subject in a coma ($\alpha = 0.5$). Figure 13.8 shows the DFA behavior of the time between consecutive heelstrikes of the same foot (stride intervals) during free walking and under a metronomic pacing. Free walking is a $1/f$ power-law process, whereas, in the rhythmic case the slope is very close to zero for long-term observations and to $\alpha = 0.5$ in the short term. This behavior could suggest a predominance of uncertainty for rhythm adaptation in the short term, whereas gait regularity prevails in long term.

In the course of time, theoretical and, sometimes, corrective investigations have been performed for DFA computation, according to the specificity of the series analyzed (Heneghan and McDarby, 2000).

13.5   CONCLUDING REMARKS

We have seen that there are different methods to characterize fractality of a time series: dimension D, Hurst exponent H, β exponent of power-law spectrum, and α exponent of DFA. Given that all of them refer to the scaling property of a series, it is reasonable to guess that those indices are related to each other in some way. And this is actually so, but, as reported in Table 13.1, it is crucial to know the nature fGn or fBm of the series for a correct application of such correspondences.

To complete this lesson and to make it of practical use, I would like to call attention not only to the importance of a correct application of the relations among exponents, but also of always making a critical appraisal of the series under investigation before applying any method of fractal analysis.

I would like to remind that fractal analysis of physiological signals is aimed at identifying and characterizing the possible presence of self-simi-

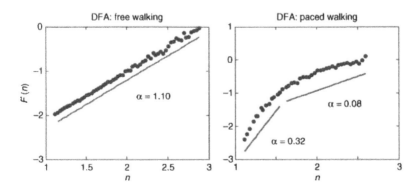

Figure 13.8. DFA of the stride intervals of 1 hour free walking (left), and paced walking following a metronome (right). Axes are in base-10 logarithmic scale.

Table 13.1. Correspondences among fractal measures

	α	β	H_{fBm}	H_{fGn}	D
DFA: exponent, α		$(\beta + 1)/2$	$H_{fBm} + 1$	H_{fGn}	$3 - D$
Spectrum: exponent, β	$2\alpha - 1$		$2H_{fBm} + 1$	$2H_{fGn} - 1$	$5 - 2D$
Hurst: fBm series, H_{fBm}	$\alpha - 1$	$(\beta - 1)/2$			$2 - D$
Hurst: fGn series, H_{fGn}	α	$(\beta + 1)/2$			
Fractal dimension, D	$3 - \alpha$	$(5 - \beta)/2$	$2 - H_{fBm}$		

larity along a certain number of orders of magnitude. Therefore, it is helpful to know the performances of the different techniques in function of the series length (Sevcik, 1998). Furthermore, as we have seen, the fGn or fBm nature of a process changes the relation between exponents; it is, therefore, important to check this aspect before any analysis. One possible approach can be, for example, to estimate the β exponent of the power spectrum: values of $\beta < 1$ indicate a fGn-type signal, whereas values $\beta > 1$ are typical of fBm signals.

For all the readers interested in undertaking practical applications on biomedical signals, I suggest the website www.physionet.org, where a very rich collection of data, software, and tutorials can be found.

Finally, from all discussed herein one could think that time series can be divided into two major categories: fractal and nonfractal. But that is not so. A third category exists—multifractal series. Their property is to be decomposable in short periods, each characterized by its own Hurst exponent h (also named the Lipschitz–Hölder or, simply, Hölder exponent). Those series are characterized not through a fractal dimension but through the function $f(h)$, called the multifractal spectrum or singularity spectrum. But this is another story!

REFERENCES

Aristotle, *Metaphysics,* Book V, Part 6, translated by W. D. Ross, http://classics. mit.edu/Aristotle/metaphysics.5.v.html (public domain), 2006.

Balocchi, R., Carpeggiani, C., Fronzoni, L., Peng, C.-K., Michelassi, C., Mietus, J., and Goldberger, A. L., Short and Long-Term Heart Rate Dynamics in Atrial Fibrillation, in *Methodology and Clinical Applications of Blood Pressure and Heart Rate Analysis,* M. Di Rienzo et al. (Eds.), IOS Press, pp. 91–96, 1999.

Balocchi, R., Michelassi, C., Carpeggiani, C., Castellari, M., and Trivella, M. G., Fractal Analysis of the Heart, in *Chaos and Noise in Biology and Medicine,* M. Barbi and S. Chillemi (Eds.), Vol. 7, pp. 86–96, World Scientific, 1998.

Bassingthwaighte, J. B., and Beyer, R. P., Fractal Correlation in Heterogeneous Systems, *Physica D,* Vol. 53, 71–84, 1991.

Bassingthwaighte, J. B., and Raymond, G. M., Evaluating Rescaled Range Analysis for Time Series, *Ann. Biomed. Eng.*, Vol. 22, 432–444, 1994.

Bassingthwaighte, J. B., and Raymond, G. M., Evaluation of the Dispersional Analysis Method for Fractal Time Series, *Ann. Biomed. Eng.*, Vol. 23, 491–505, 1995.

Bassingthwaighte, J. B., Physiological Heterogeneity: Fractals Link Determinism and Randomness in Structures and Functions, *News in Physiological Sciences*, Vol. 3, 5–10, 1988.

Bassingthwaighte, J., Liebovitch, L., and West, B., *Fractal Physiology*, Oxford University Press, 1994.

Cerutti, S., Esposti, F., Ferrario, M., Sassi, R., and Signorini, M. G., Long-Term Invariant Parameters Obtained from 24-h Holter Recordings: A Comparison Between Different Analysis Techniques, *Chaos*, Vol. 17, No. 15–108, 2007.

Eke, A., Hermán, P., and Hajnal, M., Fractal and Noisy CBV Dynamics in Humans: Influence of Age and Gender, *J. Cereb. Blood. Flow Metab.*, Vol. 26, No. 7, 891–898, 2006.

Eke, A., Herman, P., Bassingthwaighte, J. B., Raymond, G. M., Percival, D. B., Cannon, M., Balla, I., and Ikrenyi, C., Physiological Time Series: Distinguishing Fractal Noises from Motions, *Pflugers Arch—Eur. J. Physiol.*, Vol. 439, 403–415, http://ioz.seas.upenn.edu/Publications/BPubs_files/00778819IS-CAS.pdf, 2000.

Eke, A., Herman, P., Kocsis, L., and Kozak, L. R., Fractal Characterization of Complexity in Temporal Physiological Signals, Physiol. Meas., Vol. 23, R1–R38, 2002.

Esteller, R., Vachtsevanos, G., Echauz, J., and Litt, B., A Comparison of Fractal Dimension Algorithms, *IEEE Transactions on Circuits & Systems,* Vol. 48, No. 2, 177–183, 2001.

Glenny, R. W., Blood Flow Distribution in the Lung, *Chest,* Vol. 114, 8S–16S, 1998.

Glenny, R. W., Robertson, H. T., Yamashiro, S., and Bassingthwaighte, J. B.,Applications of Fractal Analysis to Physiology, *J. Appl. Physiol.,* Vol. 70, No. 6, 2351–2367, 1991.

Goldberger, A. L., Amaral, L. A. N., Hausdorff, J. M., Ivanov, P. C., Peng, C.-K., and Stanley, H. E., Fractal Dynamics in Physiology: Alterations with Disease and Aging. *Proceedings of the National Academy of Sciences,* Vol. 99, Suppl. 1, 2466–2472, 2002.

Harte, D., *Multifractals Theory and Applications, p. 220,* Chapman & Hall/CRC, 2001.

Hausdorff, F., Dimension und Äusseres Mass. *Mathematische Annalen,* Vol. 79, 157–179, 1919.

Heneghan, C., and McDarby, G., Establishing the Relation between Detrended Fluctuation Analysis and Power Spectral Density Analysis for Stochastic Processes, *Physical Review E,* Vol. 62, No. 5, 6103–6110, 2000.

Higuchi, T., Approach to an Irregular Time Series on the Basis of the Fractal Theory, *Physica D,* Vol. 31, p. 277–283, 1988.

Hurst, H. E., Black, R. P., and Simaika, Y. M., *Long-Term Storage: An Experimental Study,* Constable, 1965.

Kantelhardt, J. W., Koscielny-Bunde, E., Rego, H. H. A., Havlin, S., and Bunde,

A., Detecting Long-Range Correlations with Detrended Fluctuation Analysis, *Physica A*, Vol. 295, 441–454, 2001.

Katz, M. J., Fractals and the Analysis of Waveforms, *Comput. Biol. Med.*, Vol. 18, 145, 1988.

Lee, J.-M., Hu, J., Gao, J., Crosson, B., Peck, K. K., Wierenga, C. E., McGregor, K., Zhao, Q., White, and Keith, D., Discriminating Brain Activity from Task-Related Artifacts in Functional MRI: Fractal Scaling Analysis Simulation and Application, *J. Neuroimage*, Vol. 40, No. 1, 197–212, 2008.

Mandelbrot, B. B., *The Fractal Geometry of Nature*, W.H. Freeman and Company, 1983.

Noujaim, S. F., Berenfeld, O., Kalifa, J., Cerrone, M., Nanthakumar, K., Atienza, F., Moreno, J., Mironov, S., and Jalife, J., Universal Scaling Law of Electrical Turbulence in the Mammalian Heart, *Proc. Natl. Acad. Sci. USA*, Vol. 104, No. 52, 20985–20989, 2007.

Pallikari, F., and Boller, E., A Rescaled Range Analysis of Random Events, *Journal of Scientific Exploration*, Vol. 13, No. 1, 25–40, 1999.

Peng, C.-K., Buldyrev, S., Havlin, S., Simons, M., Stanley, H., and Goldberger, A., Mosaic Organization of DNA Nucleotides, *Phys. Rev. E*, Vol. 49, 1685–1689, 1994.

Peng, C.-K., Havlin, S., Stanley, H., and Goldberger, A., Quantification of Scaling Exponents and Crossover Phenomena in Nonstationary Heartbeat Time Series, *Chaos*, Vol. 5, 82–87, 1995.

Pilgram, B., and Kaplan, D. T., A Comparison of Estimators for 1/f noise, Physica D, Vol. 114, 108–122, 1998.

Ricordeau, C., *L'Adieu au Big-bang*, Aubin Eds., 1997.

Schmitt, D. T., and Ivanov, P. Ch., Fractal Scale-Invariant and Nonlinear Properties of Cardiac Dynamics Remain Stable with Advanced Age: A New Mechanistic Picture of Cardiac Control in Healthy Elderly, *Am. J. Physiol. Regul. Integr. Comp. Physiol.*, Vol. 293, No. 5, R1923–37, 2007.

Sevcik, C., A Procedure to Estimate the Fractal Dimension of Waveforms, Complexity International, 5, http://journal-ci.csse.monash.edu.au/ci/vol05/sevcik/sevcik.html, 1998.

Sprott, J. C., *Chaos and Time Series Analysis*, p. 305, Oxford University Press, 2003.

FURTHER READING

Mandelbrot B .B., *Gaussian Self-Affinity and Fractals*, Springer, 2001.
Peitgen H.-O., and Richter P. H., *The Beauty of Fractals*, Springer, 1986.
Schroeder M., *Fractals, Chaos, Power Laws*, W.H. Freeman and Company, 1991.

CHAPTER 14

NONLINEAR ANALYSIS OF EXPERIMENTAL TIME SERIES

Maria Gabriella Signorini and Manuela Ferrario

14.1 INTRODUCTION

The mathematical description of a physical system permits one to study the different types of dynamical behavior that can be generated: fixed points, limit cycles, tori up to chaos, and strange attractors. In this chapter, we take a different point of view: starting from a series of experimental measurements, that is, from a time series, we want to know how the observed data can help to understand the system that has generated them. We then proceed to try to find the equations of a system, which can explain the measured data, without knowing the actual physical system that has generated the time series. We will see that the study of nonlinear dynamical systems with chaotic behavior can be understood starting from a measured time series, even if the model is unknown or too complex.

The first researchers who explored the influence of the nonlinear dynamics on systems behavior based on the observation of real systems were Poincaré (1927) and Van der Pol (1954). It was a meteorologist, Edward Lorenz, in 1963, who, by applying a simplified model (three differential equations) of the convective motion of a fluid, demonstrated that simple systems do not necessarily exhibit simple behaviors. Lorenz was using a basic computer, a Royal McBee LGP-30, to run his weather simulation. He wanted to see a sequence of data again and to save time he started the simulation in the middle of its course. He was able to do this by entering a printout of the data corresponding to conditions in the middle of the simulation he had calculated previously. To his surprise, the weather that the machine began to predict was completely different from the weather calcu-

Advanced Methods of Biomedical Signal Processing. Edited by S. Cerutti and C. Marchesi
Copyright © 2011 the Institute of Electrical and Electronics Engineers, Inc.

lated before. Lorenz attributed this to the computer printout. The printout rounded variables off to three-digit numbers, but the computer worked with six-digit numbers. This difference is tiny and the consensus at the time would have been that it should have had practically no effect. However, Lorenz discovered that small changes in initial conditions produced large changes in the long-term outcome.

Even if the system one builds obeys deterministic laws, it is impossible to predict a state of the system because the imprecision of initial conditions produces an estimated orbit that diverges from the true one after a somewhat short time (Eckmann and Ruelle, 1985; Lorenz, 1963). The complexity is not caused by the interactions of many objects but is intrinsic to the dynamical system and reflected by the variables trends that can be observed from it. This behavior is called *deterministic chaos*.

The equilibrium that characterizes a chaotic dynamical system is the *strange attractor,* a portion of the state space with a noninteger (fractal) dimension that all the trajectories described by the state vector asymptotically converge to. As the attractor resides in a finite portion of the state space, the trajectories will develop inside the attractor (they will diverge following an exponential law), but once at the extremities of the attractor, the trajectories will have to compensate for the phase of stretching with a phase of folding. The phenomenon of stretching and folding is at the basis of the fractal structure of the attractor.

As we have previously anticipated, another property of the strange attractor is its sensible dependency on initial conditions: two points of the trajectories close to each other will tend to diverge and they will not cover the same path (Ruelle, 1979).

The study of the features of a system characterized by a deterministic dynamic and chaotic behavior requires the estimation of several parameters such as the fractal dimension of space filled by the points of the attractor and the Lyapunov exponents (generalization of the eigenvalues of a regular attractor), which measure the exponential divergence of the trajectory. In fact, a sufficient condition for a system to be chaotic is the divergence of the close trajectories, a condition that is identified by a positive Lyapunov exponent (Eckmann and Ruelle, 1985; Ott et al., 1994).

If the model of the system is known, that is, the number of state variables, it is possible to measure precisely the characteristics of the attractor. However, in many practical applications the knowledge of the system is often limited, both because of the complexity and the unfeasibility to know and measure all the variables, for example, the functions generating heartbeat or the electrical activity of the brain. Moreover, for these systems, the measurements of some variables is technically unfeasible.

Therefore, we need to reconstruct the strange attractor starting from a measured time series, even if the number of system variables is unknown. In fact, it is possible to obtain a reconstructed attractor with the same fractal dimension d and the same Lyapunov exponents of the original one, in a m-dimensional space, starting from a single measured variable as a result of the invariance property of the attractors. The measurement of the reconstructed attractor provides information about the true attractor. However, in contrast to the linear indexes, which are obtainable from raditional statistics, the nonlinear methods cannot be applied in a black-box manner. We will examine several indexes provided by different nonlinear methods; these indexes are values whose relevance is strongly related to a general knowledge of the features of the system generating the time series.

The objective of this chapter is to show how the application of some nonlinear methods permits one to assess parameters that are related to the system attractor, if it exists and has a finite, not integer, dimension. These parameters are computed starting only from the measured time series and are measures of some attractor invariants.

This approach has a great potential in the context of biological time series analysis, but the proof of its validity has not yet been concluded (Kantz et al., 1998). It must be said that the data available from biology and medicine are even less suitable for a simple investigation. Moreover, the linear statistical methods have provided limits in a diagnostic context. The nonlinear approach permits one, instead, to analyze the aperiodicity in a time series as an expression of a different class of systems, such as chaotic and deterministic systems.

For example, we can see as a paradigm of this approach the results of many works relating the cardiovascular system and the heart-rate variability signal. In fact, several attempts were made to find evidence of low-order deterministic chaos in the cardiovascular system (Poon and Merrill, 1997). The main approach was based on dimension calculation and time-delay reconstruction of experimental data. Over the years, though, the number of difficulties and contradictory results lessened the possibility of reaching definite conclusions. In fact, given the large number of interacting units, the cardiovascular system is likely highly dimensional. Although it is theoretically possible that under special conditions the observable part of the system might be limited to a small low-order subsystem, nobody has shown that this statement is actually valid in general.

Nevertheless, notwithstanding the lack of theoretical applicability, several independent research groups in the last 10 years verified that statistics originally meant to study properties of low-order dynamical systems were actually capable of capturing new facets buried in the HRV series better than traditional indexes (Signorini et al., 2000). "Better" here means that these nonlinear statistics proved often to be better or independent predictors of mortality and to better discriminate between populations in different physiological states (Cerutti et al., 2007).

In fact, these parameters, computed starting from experimental measurements, have revealed their usefulness as diagnostic markers of, for example, risk of sudden death after transplantation or in congestive heart failure patients (Bigger et al., 1996; Ho et al., 1997; Huikuri, 1998).

We can say that nonlinear analysis is a new approach to investigating the experimental data. Moreover, the mathematical instruments have became even more rigorous and refined so to be even more useful in research.

14.2   RECONSTRUCTION IN THE EMBEDDING SPACE

The analysis of a time series with deterministic features, measured when it is supposed to be at equilibrium on an attractor, has to face (and possibly solve) some problems. The first of them is the reconstruction of the state of the system that has generated the measured variable, that is, the data sample.

The second problem is the definition of the space dimension, in which the attractor can be reconstructed from the measured variable. A dynamical system $\dot{x}(t) = f(x(t))$ can be studied starting from the equations that describe it or by trying to find the principal properties of the system, for example, by analyzing the measurable outputs.

In this case, in order to know the dynamics of a system $\dot{\vec{x}}(t) = f(x(t))$ or $\vec{x}(t + 1) = f(\vec{x}(t))$, it should be very important to know and measure the time performance of all the variables that characterize the state: $\vec{x}(t) = [x_1(t), x_2(t), \ldots , x_n(t)]$. However, in laboratory experiments, in biological signal analysis, in medicine, or even in economics, that is, in most problems that deal with real-world data, the measure of all the variables is technically unfeasible. Moreover, in many cases the number and type of the state variables are unknown. For example, we do not have the dynamical model of many biological or physiological systems.

The most common situation is to have the measure of only one output variable $y(t) \in R$, which is a function, at each instant, of the state of the system $\vec{x}(t) \in R^n$, that is,

$$\begin{cases} \dot{\vec{x}}(t) = f(\vec{x}(t)) \\ y(t) = g(\vec{x}(t)) \end{cases} \tag{14.1}$$

A fixed sampling interval $\tau > 0$ (which we will name *delay*), will have a time series $y(n) = y(t_0 + n\tau_s)$ of finite length N and be affected by noise.

The reconstruction of the system attractor in the state space, starting from scalar measures, is possible only if the derivative values of the measured variables are known and if the relations between the derivatives and the state variables are identified, that is, the differential equations that generate the system.

However, in order to obtain $\dot{y}(n)$, it would be necessary to find an approximation of $[y(t_0 + (n + 1)\tau_s) - y(t_0 + n\tau_s)]/\tau_s$, as the time series has values defined only at τ_s intervals. Because τ_s is finite, the approximation should work as a high-pass filter and, thus, it should cause a loss of information. If we examine the formula for the derivatives, we see that at each step we are adding to the information already contained in the measurement $y(n)$; measurements at other times lagged by multiples of the observation time step τ_s.

Kalman showed in the 1960s how it is possible, in case of a linear system, to reconstruct the system state that has generated the observed variable, starting from a sufficiently large number N of delayed outputs (if the system is observable).

In the 1970s, almost simultaneously two research groups (Packard et al., 1980; Ruelle, 1979) developed this theory by extending it to the case of nonlinear systems. Thus was introduced the idea of using time-delay coordinates to reconstruct the phase space of an observed dynamical system (Mané, 1981; Takens, 1981).

The main idea of the embedding theorem is that we really do not need to know the true value of all the derivatives to form a coordinate system in which to capture the structure of orbits in phase space. In fact, the embedding theorem states that, given a dynamical system $\dot{\vec{x}}(t) = f[\vec{x}(t)]$ and the scalar measures $y(n)$ of a system variable, it is possible to reconstruct, by the method of delays, an attractor that has the same dynamical and geometrical features of the attractor of the original system. Whitney suggested the basic idea with his reconstruction theory (Whitney, 1936), which demonstrated that sets reconstructed by applying projections are diffeomorphic. Thus, they have the same properties as the original topological space (manifold) if the projection space has a dimension two times larger than the dimension of the system attractor A.

The embedding theorem consists, hence, in creating N vectors of m dimensions (embedding dimension or reconstruction dimension) starting from a time series $y(n)$ by using a time delay τ:

$$\vec{y}_n = [y(n), y(n + \tau), \ldots, y(n + (d - 1)\tau)] \qquad (14.2)$$

The reconstruction of an observed variable in an embedding space m by the method of delays is an approach to reconstruct all the information contained in the system and thus associated with the n variables constituting it, which are not directly measurable (Mané, 1981; Sauer et al., 1991; Takens, 1981).

If m is the embedding dimension, then R^m is the embedding space for the attractor A and the image of the attractor in this space is called the *reconstructed attractor*.

It is important to observe that the reconstruction dimension m must be big enough to provide a Euclidean space R^m sufficiently large to unfold the attractor of dimension d_A. The example in Figure 14.1 shows that, if

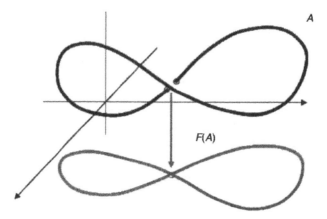

Figure 14.1. An example of limit cycle (A) and its representation in R^2. Notice that points far from each other on the attractor become self-crossing points.

the attractor is a cycle ($d = 1$) and is represented in $d = 2$, there may be a self-crossing of the orbit at isolated points, thus generating situations that do not permit a clear identification of points really close in the phase space. These ambiguities should be solved if the attractor was observed in $d = 3$. Naturally, we could add other coordinates, but usually the embedding dimension m is defined as the dimension that permits one to show the attractor in the phase space without ambiguities, that is, without the trajectory defining the reconstructed attractor A_m having a false self-crossing.

For this reason, it is necessary (but not sufficient) that $m > d$, where d is the first integer dimension larger than d_A. From the embedding theorem, by supposing that $A \subset R^n$ and A is an attractor of dimension d, m is an embedding dimension if $m > 2d$. It is a condition always sufficient but somewhat conservative. It is often useful to find a smaller embedding dimension $d < m \leq 2d$.

It is opportune here to remember that when we analyze a time series, d is unknown a priori. If d is assumed to be small, attempts should be made to search the embedding dimension.

Another crucial point of the reconstruction in the embedding space is the identification of the time delay τ. In theory, τ is generic and almost any value can be used for the reconstruction. However, as Figure 14.2 shows for the reconstruction of the Lorenz attractor, if we choose a too small value for τ, then the coordinates $y(n + k\tau)$ and $y(n + (k + 1)\tau)$ are so close that they are not distinguishable, and the effect is the alignment of the reconstructed points on the bisectrix. On the contrary, if τ is too large, the coordinates result independent of each other in a statistical sense and then tend to fill uniformly the phase space, as a stochastic noise should do.

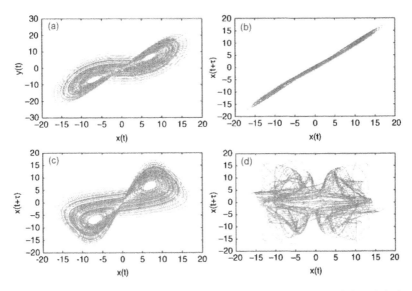

Figure 14.2. Lorenz attractor. (a) Representation of two variables of the original attractor, (b) reconstruction by adopting $\tau = 1$, (c) $\tau = 15$, (d) $\tau = 200$. The reconstruction procedure was performed by using the x coordinate of the Lorenz system. The parameters were $S = 10$, $R = 25$, $B = 8/3$, and initial conditions were $x_0 = 0.1$, $y_0 = 0.1$, and $z_0 = 0.1$. The Lorenz system is described by well-defined differential equations:

$$\begin{cases} \dot{x}(t) = S\big(y(t) - x(t)\big) \\ \dot{y}(t) = x(t) \cdot \big(R - z(t)\big) - y(t) \\ \dot{z}(t) = x(t) \cdot y(t) - Bz(t) \end{cases}$$

14.2.1 Choosing the Time Delay τ

The identification of the proper time delay τ can be made by using different empirical methods. Among these, the most well known are the first zero of the autocorrelation function of $y(t)$ (or other specified values instead of zero such as $1/e$, 0.5, and 0.1) and the first minimum of the mutual information (MI) function of $y(t)$.

However, the first criterion permits an identification of linear dependency among samples. The second approach is thus preferred because it takes into account an important feature of deterministic chaos, the nonlinear independency (Fraser, 1989a,b).

The function of mutual information arises from the idea that it is possible to quantify the information about a measurement at one time from a measurement taken at another time. For this reason, it is supposed that there should be a probability distribution associated with each system that regulates the observations in it (Galleger, 1968). Given a series of measurements y, the mutual information between the observations $y(n)$ and the delayed observations $y(n + \tau)$ is given by

$$I(\tau) = \Sigma_{n=1}^{N} P(y(n), y(n + \tau)) \cdot \log_2 \left[\frac{P(y(n), y(n + \tau))}{P(y(n)) \cdot P(y(n + \tau))} \right] \quad (14.3)$$

where $P(y(n))$ and $P(y(n + \tau))$ are the probabilities of finding a time series value in the nth interval and in the $n + \tau$th interval, respectively. $P(y(n), y(n + \tau))$ is the joint probability that an observation falls into the nth interval and later falls into the $n + \tau$th interval.

When there is an independence among the samples at a certain time delay τ, Equation 14.1 will be null, as $P(y(n), y(n + \tau)) = P(y(n)) \cdot P(y(n + \tau))$.

The time τ, which is associated with the first minimum of the mutual information $I(\tau)$, is indicated as the optimum time delay for the reconstruction of phase space. It is the value of the delay at which the observations are somewhat uncorrelated but not statistically independent (Fraser and Swinney, 1986).

14.2.2   Choosing the Embedding Dimension d_E: The False Neighbors Method

The theorem of Takens considers $d_E = 2d_A + 1$ as a sufficient condition for the choice of the embedding dimension. However, from a theoretical point of view, it should not be a problem if we consider an embedding space R^n larger than necessary, but from a practical point of view this is counterproductive. First of all, in order to extract interesting properties from the data, it is necessary to perform operations in R^d with a computational cost that is exponentially proportional to d. Second, in presence of noise or other high-dimensional contaminations, the additional dimension $d - d_E$ is not populated by the dynamics of the system, but entirely by the contaminating signals.

The method of the false nearest neighbors (FNN), which will be described in detail, permits one to identify the minimum embedding dimension (Kennel et al., 1992). However, there exist also other methods for the estimation of m, among these we can cite the singular value decomposition (SVD), the saturation of some invariants of the system, and the true vector fields (Broomhead and King, 1986; Kaplan and Glass, 1992).

The formulation of the FNN method results from asking, directly from the data, the basic question addressed in the embedding theorem: when has one eliminated false crossing of the orbit with itself, which arose by virtue of having projected the attractor into a too low dimensional space? The FNN method is based on the idea that by going from the d dimension to the $d + 1$ dimension (d is an integer value) we can discriminate the points on the orbit of $y(n)$ that are truly close from the points that are not. False neighbors are those points that are close to each other only because the orbit has been reconstructed in a R^d space that is too small. Only when a sufficiently large embedding dimension is reached are all the neighbors of the trajectory points really close (Kennel et al., 1992).

Let us consider in the reconstructed space R^d the point $y(k) = [y(k), y(k + \tau), \ldots, y(k + (d - 1)\tau)]$ and the closest point to it $y^{nn}(k)$, where τ is the optimum time delay and d is the chosen integer dimension for the reconstruction. Let $r_d(k)$ be the distance between the two points:

$$r_d(k) = \sqrt{(y_k - y_k^{nn})^2 + (y_{k+\tau} - y_{k+\tau}^{nn})^2 + \ldots + (y_{k+(d-1)\tau} - y_{k+(d-1)\tau}^{nn})^2} \quad (14.4)$$

By going from the d dimension to $d + 1$, the points previously considered will have one more coordinate and the square of the new distance will be $r_{d+1}^2(n) = r_d^2(n) + (y_{n+d\tau} - y_{n+d\tau}^{nn})^2$.

In general, we can state that if $r_{d+1}^2(n) \gg r_d^2(n)$ then the two points are close only because of the projection, that is, they are false neighbors. We can mathematically define a point as a false neighbor if

$$\frac{|y_{n+d\tau} - y_{n+d\tau}^{nn}|}{r_d(\tau)} > R_S \quad (14.5)$$

where R_S is an appropriate threshold value. In general, for $R_S \geq 10$, it can be observed that the number of false neighbors becomes stable.

However, the outlined criterion is not enough to determine the minimum embedding dimension when it is applied to randomly generated data or affected by noise, because the points tend to uniformly occupy the space. In fact, for high-dimensional signals, such as white noise, we will obtain a value of d rather small and, thus, wrong. This result can be explained by the fact that the distances of the close points for these signals are comparable with the dimension of the attractor R_A.

A second criterion was introduced: if the closest point $y^{nn}(k)$ has a distance $r_d(k) \approx R_A$ and it is a false neighbor, then the distance $r_{d+1}(k)$ will be about $2R_A$. This means that distant, but nearest, neighbors will be stretched to the extremities of the attractor when they are unfolded from each other if they are false nearest neighbors (Sauer et al., 1991). The second criterion is thus

$$\frac{r_{d+1}(n)}{R_A} \geq 2 \quad (14.6)$$

R_A is the average radius of the attractor, approximately estimated by

$$R_A^2 = \frac{1}{N} \Sigma_{n=1}^N [y_n - \bar{y}]^2 \quad (14.7)$$

where $\bar{y}$ is the sample mean.

As defined, false neighbors are those points that satisfy both of the described criteria. The result for the Lorenz system is reported in Figure 14.3c. The percentage of the FNN points decreases along with the increase of the reconstruction dimension and it becomes null for $m = n = 3$, the exact integer dimension for the Lorenz system. Thus, the value d_E is that value of d for which there are not false neighbors. It is important to emphasize that this method also provides some indications about the reconstruction reliability because it provides the percentage of false neighbors for each d.

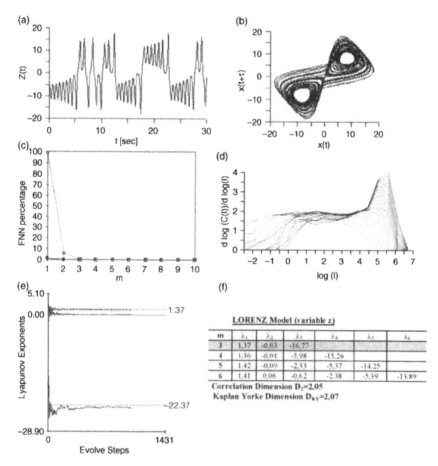

LORENZ Model (variable z)

m	λ_1	λ_2	λ_3	λ_4	λ_5	λ_6
3	1.37	-0.03	-16.77			
4	1.36	-0.01	-5.98	-15.26		
5	1.42	-0.09	-2.33	-5.37	-14.25	
6	1.41	0.06	-0.62	-2.38	-5.39	-13.89

Correlation Dimension D_2=2,05
Kaplan Yorke Dimension D_{ky}=2,07

Figure 14.3. Procedure of reconstruction starting from a time series obtained by the measure of the z coordinate of the Lorenz system with parameters $S = 10$, $R = 25$, and $B = 8/3$. (a) Time series, (b) reconstruction by applying the time-delay method, (c) estimation of the minimum embedding dimension m with the FNN method, (d) computation of the correlation dimension, (e) spectrum of Lyapunov exponents, (f) spectrum of the LE in dependence on the embedding dimension and computation of the Kaplan–Yorke dimension.

14.3 TESTING FOR NONLINEARITY WITH SURROGATE DATA

The choice of a nonlinear approach to analyze an experimental time series can be motivated by two factors. The first one is the nature of the signal itself. When the analyses with linear methods are performed, some dynamical structures are discovered that are not explainable by linear processes. The second motivation is the a priori knowledge of the existence of nonlinear components in the signal; a linear analysis would be unsatisfactory and incomplete.

However, the fact that a system contains nonlinear components may be not sufficient to assure the presence of nonlinearity in the measured signal. For this reason, the nonlinear methods can be performed only after the proof of the presence of nonlinearity in the time series of interest (Theiler, 1992).

In order to test for the presence of nonlinear structures, the hypothesis test is generally performed. In this case, the hypothesis to test against is that the data was generated by a stationary, Gaussian, linear stochastic process, or a more general null hypothesis including the possibility that the nonlinearity stays in the observational system and not in the dynamical system generating the data. In this latter case, a possible null hypothesis would be that there is a stationary, Gaussian, linear stochastic process that generates a sequence y_n, but the actual observations are $x_n = h(y_n)$, where h is a monotonic nonlinear function.

14.3.1 Surrogate Time Series

In dependence on the formulated null hypothesis, a proper surrogate time series should preserve the properties fixed by the null hypothesis (*constrained realizations*). The null hypothesis of a Gaussian stochastic process with arbitrary linear correlations can be reformulated in the following manner. All the structures in the time series can be exhaustively described by the first and second moments, in other words, mean, variance, or autocorrelation function. This means that a surrogate time series can be a sequence that is simply a random shuffling of the original time series, but which preserves such properties.

There exist several methods of surrogation that are compatible with this null hypothesis. One of the simplest methods is the phase-randomized method (Theiler, 1992). This methods consists in (1) determining the amplitudes of the discrete Fourier transform, that is, the computation of the periodogram of the discrete time series s_n:

$$S_k = \sum_{n=0}^{N-1} s_n e^{-j2\pi kn/N} \tag{14.8}$$

and (2) in multiplying each value of the Fourier transform by random phases $e^{j\alpha_k}$, where α_k is an independent random value uniformly distributed in $[0\ 2\pi]$, and, finally, transforming it back to the time domain:

$$\hat{s}_n = \sum_{n=0}^{N-1} S_k e^{j2\pi kn/N} \cdot e^{j\alpha_k} \tag{14.9}$$

In order to obtain a real surrogate time series, it is important to remember that α_k is required to be an odd function; for example, it can be constructed by generating the value for $k = 1, 2, \ldots, N/2$ and then by imposing $\alpha_k = -\alpha_{-k}$.

In the more general hypothesis (when the nonlinearity is supposed to reside in the observational system), it is necessary to generate surrogate data with a more refined approach, for example, by constraining the surrogates to have the same power spectrum as well as the same distribution of values as the data. A simple surrogation technique with this approach is the amplitude-adjusted Fourier transform (AAFT) (Theiler et al., 1992). Given the measured time series $\{x_n\}$, the method consists in generating a sequence $\{y_n\}$ of random numbers and then rank ordering them according to the original series; that is, if $x(k)$ is the smallest value of the series $\{x_n\}$, then $y(k)$ will be the smallest value of the sequence $\{y_n\}$. The resulting series $\{\tilde{y}_n\}$ is Gaussian but follows the measured time evolution of the original time series. Through the phase-randomized methods, we generate the surrogate of $\{\tilde{y}_n\}$, named $\{\hat{y}_n\}$. Finally, the series $\{x_n\}$ is rearranged according to the surrogate time series $\{\hat{y}_n\}$. The obtained time series $\{X_n\}$ has the property of having the same amplitude distribution of $\{x_n\}$.

However, the AAFT method produces a correct result only when a large number N of samples is used. Moreover, this method permits one to generate surrogate time series with the same amplitude distribution but not with the same spectrum of the original sequence. In particular, there is a bias effect that consists in getting the spectrum flatter.

The development of the iterative method of Schreiber has permitted us to obtain surrogate time series with the same amplitude distribution and the same spectrum of the original ones (Schreiber and Schmitz, 1996; Schreiber, 1998).

The algorithm consists of a simple iteration scheme. Let us consider a time series $\{x_n\}$ and its Fourier transform:

$$X_k = \sum_{n=1}^{N-1} x_n \cdot e^{-j\frac{2\pi kn}{N}} = |X_k| \cdot e^{j\theta_k} \tag{14.10}$$

The first iteration consists of a random shuffle (without replacement) of the data, thus obtaining the time series $\{x_n^{(0)}\}$ (first iteration). At each successive iteration, let us generate a new time series, starting from the one obtained at the preceding iteration by performing every time two consecutive steps:

1. Computation of the Fourier transform of $\{x_n^{(i)}\}$:

$$X_k^{(i)} = \sum_{n=0}^{N-1} x_n^{(i)} \cdot e^{-j\frac{2\pi kn}{N}} = \left| X_k^{(i)} \right| \cdot e^{j\theta_k^{(i)}} \qquad (14.11)$$

Replacing the squared amplitudes $|X_k^{(i)}|$ by the original ones $|X_k|$ and then transforming back creates a new time series $\{\bar{x}_n^{(i)}\}$. The phases of the complex Fourier components are kept. Thus, the first step enforces the correct spectrum but the distribution will usually be modified.
2. Rank-ordering of $\{x_n\}$ according to the resulting series $\{\bar{x}_n^{(i)}\}$. In this way we obtain a time series $\{x_n^{(i+1)}\}$, which will have exactly the same amplitudes distribution, but not the same power spectrum of $\{x_n\}$.

If the time series had infinite length, by continuing the iteration one should obtain a time series with the same power spectrum and the same amplitude distribution of the original one (Schreiber and Schmitz, 2000).

The iteration stops when the rearranged time series $\{x_n^{(i+1)}\}$ is identical to the one obtained at the preceding iteration (the transformation toward the correct spectrum will result in a modification that is too small to cause a reordering of the values) or when it has reached a certain level of accuracy or discrepancy.

The discrepancy is defined as the difference between the smoothed spectra of the original time series and the surrogate one. If

$$X_k^2 = \left| \sum_{n=1}^{N-1} x_n \cdot e^{-j\frac{2\pi kn}{N}} \right|^2 \qquad (14.12)$$

is the spectrum of the original time series, it is defined to smooth the spectrum integrated at intervals of fixed frequency f. In Schreiber and Schmitz (1996), $f = 21$ is proposed as a good compromise between the need to estimate the spectral differences and the necessity to obtain a surrogate time series with a spectrum as similar as possible to the spectrum of the original time series. The smoothed spectrum is thus defined as

$$\hat{X}_k^2 = \frac{\sum_{J=K-10}^{K+10} X_K^2}{21} \qquad (14.13)$$

The (relative) discrepancy is thus defined as

$$\frac{\sum_{k=0}^{N-1}(\hat{X}_k^{(i)} - \hat{X}_k)^2}{\sum_{k=0}^{N-1}\hat{X}_k^2} \qquad (14.14)$$

and it is considered also a measure of the accuracy of the surrogation. A too large value of discrepancy would indicate unreliability of the test if the adopted statistics were sensible to differences in the power spectrum. An example of a time series generated starting from a Gaussian linear process is shown in Figure 14.4; the time series was surrogated by adopting the iterative method of Schreiber.

14.3.2   Artifacts

The randomization schemes discussed so far all base the quantification of linear correlations on the Fourier amplitudes of the data. Unfortunately, the autocorrelation function given by

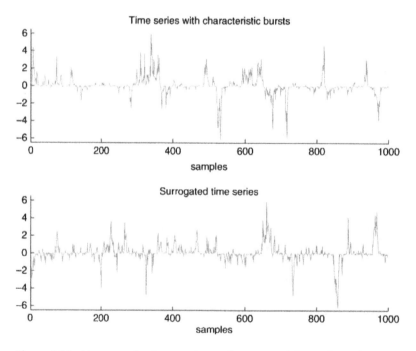

Figure 14.4. Upper panel refers to a time series generated from a linear Gaussian process ($x_n = 0.9x_{n-1} + \eta_n$, where $\{\eta_n\}$ is a white noise) observed through a nonlinear system $s_n = h(x_n) = \alpha x_n^3$. In the lower panel, the time series is surrogated with the recursive procedure of Schreiber (see text).

$$C(\tau) = \frac{1}{N-\tau} \sum_{n=\tau+1}^{N} s_n s_{n-\tau} \qquad (14.15)$$

corresponds to the Fourier amplitudes *only* if the time series is one period of a sequence that repeats itself every N time steps. This is compatible neither with the nature of the signal nor with the null hypothesis. This artifact can be estimated by the phase slip and the amplitude jump γ_{jump} between the first and the last points of the time series (s_1, s_2, s_{N-1}, s_N):

$$\gamma_{slip} = \frac{\left[(s_2 - s_1) - (s_N - s_{N-1})\right]^2}{\sum_{n=1}^{N} \left(s_n - \bar{s}\right)^2} \qquad (14.16)$$

$$\gamma_{jump} = \frac{(s_1 - s_N)^2}{\sum_{n=1}^{N} (s_n - \bar{s})^2} \qquad (14.17)$$

The fractions γ_{jump} and γ_{slip} give the contributions to the total power of the series of the mismatch of the end points and the first derivatives, respectively. To reduce the effect of these artifacts, it is necessary to select a subsequence of the time series so to minimize the quantities γ_{jump} and γ_{slip}.

14.3.3 A Particular Case: The Spike Train

A spike train is a sequence of N events (for example, neuronal spikes or heartbeats) occurring at times $\{t_n\}$. This very common kind of data is fundamentally different from the case of unevenly sampled time series in that the sampling instances $\{t_n\}$ are not independent of the measured process. Very often, the discrete sequence of interevent intervals $x_n = t_n - t_{n-1}$ is treated as if it were an ordinary time series. We must keep in mind, however, that the index n is not proportional to time anymore. It depends on the nature of the process if it is more reasonable to look for correlations in time or in event number. In particular, the literature on heart-rate variability (HRV) contains interesting material on the question of spectral estimation and linear modeling of spike trains, here usually interbeat (RR) interval series.

A very convenient and powerful approach that uses the real time t rather than the event number n is to write a spike train as a sum of Dirac delta functions placed at the spike instances, $s(t) = \sum_{n=0}^{N} \delta(t - t_n)$, considering a continuous time scale.

For the property $\int s(t) e^{i\omega t} dt = \sum_{n=1}^{N} e^{i\omega t_n}$, the periodogram estimation is then simply obtained by squaring the (continuous) Fourier transform of $s(t)$:

$$P(\omega) = \frac{1}{2\pi} \left| \sum_{n=1}^{N} e^{-i\omega t_n} \right| \qquad (14.18)$$

It is possible to generate surrogate spike trains that preserve the amplitude of the spectral estimator, previously outlined, but this is computationally very cumbersome. It is preferred to adopt the binned autocorrelation (used also to generate surrogate time series of unevenly sampled signals) (Schreiber and Schmitz, 2000).

Starting from the definition of correlation

$$C(\tau) = \alpha \int s(t)s(t-\tau)dt = \alpha \int \sum_{i,j=1}^{N} \delta(t-t_i)\delta(t-\tau-t_j)dt \quad (14.19)$$

where α is a normalization constant, the binned autocorrelation is defined as

$$C_\Delta(\tau) = \frac{1}{\Delta} \int_{\tau-\Delta}^{\tau} d\tau' C(\tau') \quad (14.20)$$

If we choose α such that $C(0) = 1$, we obtain

$$C_\Delta(\tau) = \frac{|B_{ij}(\tau-\Delta,\tau)|}{N\Delta} \quad (14.21)$$

where the numerator indicates the number of intervals $\Delta = t_i - t_j$ contained in a bin.

Even in this case, the choice of the amplitude Δ of the frequency intervals and the choice of the maximum value of τ are very delicate. In fact, if Δ is too small, the probability of having void bins increases, whereas the possibility to limit the maximum length τ permits one to have a reasonable computational time. The example in Figure 14.5 shows the time series of interbeat intervals and the surrogate time series obtained by imposing the same binned autocorrelation.

14.3.4 Test Statistics

Once the null hypothesis to be rejected is defined, one must define the discriminating statistics of the test, a number or a function that quantifies some properties of the time series, and the rejection criterion (i.e., the value of the discriminating statistics for which the null hypothesis has to be rejected).

Let us suppose that the observed time series x_n is obtained by the relation $x_n = h(y_n)$, where h is a nonlinear function and y_n is a realization of a linear Gaussian process (null hypothesis). Let us generate M surrogate time series that satisfy this hypothesis, for example, by adopting the iterative method of Schreiber. The number M of surrogate time series is determined by the significance level chosen for the test. In fact, if we consider α to be the residual probability of a false rejection (error of first type), the corresponding level of significance is $(1 - \alpha) \times 100\%$. In this case, we need $M = 1/\alpha - 1$ surrogate time series for a one-sided test and $M = 2/\alpha - 1$ for a two-sided test (Schreiber and Schmitz, 2000).

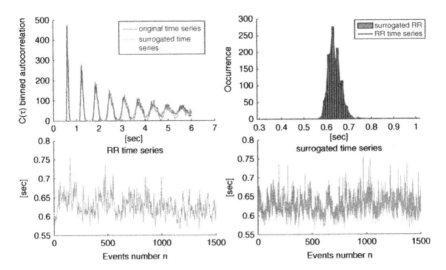

Figure 14.5. In the lower panels, the time series of interbeat RR and corresponding surrogated time series. In the upper panels, the binned autocorrelation and the value distribution of both the time series. Notice that the surrogate time series preserves the main features of the original ones (correlation and amplitudes distribution).

The discriminating statistics can be in this case the function of autocorrelation (discriminating for the surrogation) and the function of mutual information. It is necessary to compute the percentage variation, caused by the surrogation, of the autocorrelation and mutual information functions in the following manner:

$$\Delta r_{\%}(\tau) = \frac{r(\tau) - \bar{r}(\tau)}{r_m} = \frac{r_d(\tau)}{r_m} \qquad (14.22)$$

$$\Delta I_{\%}(\tau) = \frac{I(\tau) - \bar{I}(\tau)}{I_m} = \frac{I_d(\tau)}{I_m} \qquad (14.23)$$

where $r_d(\tau)$ and $I_d(\tau)$ are, respectively, the difference between the autocorrelation of the original time series and the mean of the surrogated ones, and the difference between the mutual information of the original time series and the mean of the surrogated ones. Here, r_m and I_m are the average values of the original autocorrelation function and the original mutual information function.

The function $\Delta r_{\%}(\tau)$ is the discriminating statistics that permit one to verify the "goodness" of the surrogation; it indicates if the linear properties of the original time series were preserved or not. As regards the null hypothesis and the surrogation method adopted, this difference should be null. A too large deviation of the function $\Delta r_{\%}(\tau)$ from the null value would mean that the surrogation was not correctly performed and it is impossible to state the nonlinearity of the signal; in other words, the test is unreliable for that signal.

The function $\Delta I_{\%}(\tau)$ is the discriminating statistics that identify the presence of nonlinear structure of the system generating the data. In fact, if one obtains a large deviation of the mutual information between the values computed for the original data and the surrogate time series, this means that the surrogation succeeds in destroying the nonlinear dependencies of the signal. In particular, it is possible to obtain a high peak in the function $\Delta I_{\%}(\tau)$, in correspondence to a certain value of $\bar{\tau}$. From experimental analysis, it was verified that such a value matches the optimum value of delay for the reconstruction of the trajectories in the phase space (Abarbanel et al., 1993). Figure 14.6 shows the trends of the function $\Delta I_{\%}(\tau)$ and $\Delta r_{\%}(\tau)$ for an experimental time series.

It is necessary to specify that the rejection of null hypothesis does not mean that the dynamical system generating the observed time series is deterministic, although this is possible.

14.4   ESTIMATION OF INVARIANTS: FRACTAL DIMENSION AND LYAPUNOV EXPONENTS

Once the attractor has been reconstructed, a classification of the physical system from which the time series was extracted can be performed by ana-

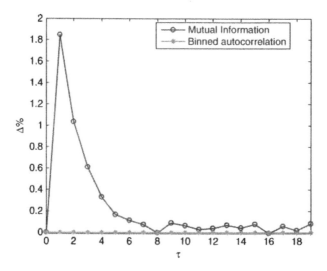

Figure 14.6. Test of nolinearity. The test was performed on the time series outlined in Figure 14.4. The autocorrelation function adopted in this case is thus the binned autocorrelation. Notice that MI < 5% and, therefore, the null hypothesis cannot be rejected, as expected.

lyzing some invariants such as the Lyapunov exponents or the fractal dimension of the attractor. The former dynamically characterizes the attractor by describing the expansion or contraction of the trajectories in the phase space; the latter describes instead the geometrical properties of the attractor.

14.4.1 Lyapunov Exponents

Lyapunov exponents entirely define the evolution of the trajectories in a dynamical system with dimension larger than one and permit the characterization of the structure of the time-ordered points making up a trajectory. To understand the meaning of the Lypunov exponents, let us consider the effect of the dynamics on a small spherical fiducial hypervolume in the phase space. Complex dynamics, like those associated with chaotic systems, can produce distortion of the ith element into extremely complicated shapes. However, if we consider sufficiently small orders of magnitude and very short time intervals, the initial effects due to the system dynamics could be only the contraction and the stretching of some directions and, thus, a global distortion of the hyperelement (e.g., hypersphere into an ellipsoid). The primary, longest axis of this ellipsoid will correspond to the most unstable direction of the flow, and the asymptotic rate of expansion of this axis is what is measured by the largest Lyapunov exponent (Abarbanel et al., 1993). If the infinitesimal radius of the initial hypersphere is $r(0)$ and the length of the ith principal axis at time t is $l_i(t)$ then the ith Lyapunov exponent can be defined as

$$\lambda_i = \lim_{t \to \infty} \frac{1}{t} \log \frac{l_i(t)}{r(0)} \qquad (14.24)$$

The set of all the Lyapunov exponents λ_i is called Lyapunov spectrum and, by convention, the Lyapunov exponents are always ordered so that $\lambda_1 > \lambda_2 \ldots > \lambda_m$, where $m = d$ is the dimension of the system (Eckmann et al., 1985). They characterize the dynamical system not only from a qualitative point of view, but also give important hints about quantitative properties of the chaotic dynamics. The sum of the positive Lyapunov exponents gives an indication about the average velocity of the divergence in the phase space.

Every dissipative dynamical system will have at least one negative exponent and the movement of the trajectories, after a transit time, will settle on a limit set, the attractor. However, the exponential expansion denoted by a positive exponent is incompatible with the limited dynamics caused by a finite attractor, unless there is a folding of the distant trajectories, characterized, thus, by negative exponents.

The main problem of the analysis of time series is that the physical phase space is unknown, whereas to reconstruct the spectrum of the exponents it is necessary to define an embedding space. It must be emphasized that the number of the exponents depends on the reconstruction dimension m and it can be much larger than the true physical dimension of the system. These additional exponents are called spurious and in the literature there are several techniques for identifying them and thus to avoid considering them in the computation of the spectrum (Brown et al., 1991; Bryant et al., 1990). For this reason, in the literature the estimation of the maximum Lyapunov exponent (LLE, largest Lyapunov exponent) is commonly used, which provides information about the global divergence of the trajectories in the reconstructed phase space.

Given a time series $\{x(t)\}$, let us construct the trajectories with the delay method $\bar{x}(t) = [x(t), (t + \tau), \ldots, x(t + (m - 1)\tau)]$. Let us consider now a point of the trajectory $\bar{x}(t)$ and a point close to it $\bar{x}(t')$ so that the initial distance is infinitesimal, $\delta_1 = \|\bar{x}(t) - \bar{x}(t')\|$. After T time intervals, with $T \in \{1, 2, 3, \ldots\}$, the distance between the two points will be $\delta_F = \|\bar{x}(t + T) - \bar{x}(t' + T)\|$. The rate of local separation of the trajectories per time unit is

$$\lambda_1^{\text{local}} = \frac{1}{T} \log\left(\frac{\delta_F}{\delta_1}\right) \qquad (14.25)$$

To obtain the rate of global separation of the trajectories, the largest Lyapunov exponent, it is necessary to average the rate of local separation $\lambda_1 = E[\lambda_1^{\text{local}}]$ for all the couples of points of the entire trajectory initially close to each other.

14.4.2 Fractal Dimension

The attractor dimension has been the most intensely studied invariant quantity for dynamical systems. This was motivated by the fact that attractors associated with chaotic dynamics have a fractional dimension, in contrast to regular or integrable systems which always have an integer dimension (see the previous chapter). Much effort was expended to search for evidence of fractional dimensions in experimental time series to demonstrate the existence of chaos in the real word.

Beyond the simplest concept of dimension as the number of coordinates needed to specify a state, for a generic hypervolume V the relationship with a characteristic length parameter L is $V \propto L^D$. Planar areas scale quadratically with the length of a side, and volumes scale as the cube of the side length. The dimension D can be simply defined as $\log V / \log L$.

As regards the generic attractor, let us consider its representation in the embedding space and a covering of the attractor constituted by elements of a certain geometry (spheres, cubes, squares, etc.). The parameter r defines the resolution of the partition and it corresponds to the characteristic length of the chosen geometrical element, for example, the radius of a sphere or the side of a cube. It is then defined as $N(r)$, the minimum number of elements needed to completely cover the attractor. The fractal dimension D_F is defined as the limit for $r \to 0$ of the ratio log $N(r)$/log$(1/r)$. When such limit does not exist, then the dimension cannot be defined. The upper limit of the estimation of D_F is d_E, the embedding dimension.

It might happen that, by adopting different coverings (cubes instead of spheres), we can obtain different values of D_F. In this case, it is preferred to talk about the Hausdorff dimension, a generalization of fractal dimension that is defined as the minimum value among those obtained by adopting different coverings (Abarbanel et al., 1993; Grassberger and Procaccia, 1991). In general, this procedure of calculus is performed through algorithms of box counting, which, unfortunately, cannot be used for systems of high dimensionality (Greenside et al., 1982).

There are, however, different approaches to formulate the concept of fractional dimension, which are in relation with each other, but whose details and relations can be very complex. All are particular forms of the Mandelbrot fractal dimension (Mandelbrot, 1983), which derives from consideration of the work of the mathematician Hausdorff (Hausdorff, 1919). Among different indexes of fractal dimension, we cite the most important: the capacity dimension (or Hausdorff dimension), the information dimension, the correlation dimension, and the Lyapunov dimension (Parker et al., 1987).

14.5   DIMENSION OF KAPLAN AND YORKE

As previously described, the computation of the entire spectrum of the Lyapunov exponents (LE) provides the local velocity of contraction or expansion in all the directions of the phase space. It consists of a number of exponents equal to the embedding dimension. Figure 14.3e shows an example for the Lorenz system that has a topological dimension $n = 3$.

When the space dimension is unknown, the number of the exponents depends on the reconstruction dimension m, which is often much larger than the true physical dimension of the system. Even if the experimental time series are not too noisy, the computation of the spectrum of LE is a difficult procedure, as witnessed by the variety of values obtained by applying the algorithm to the system. The calculus of the exponents requires the estimation of the Jacobian matrix (i.e., linear, local models) along with the reconstructed trajectory, a very delicate procedure from a numerical point of view. In particular, when the algorithm is blindly applied to data from a stochastic process, it is not easy to verify the consistency of the assumption of an underlying dynamical system.

In spite of all these difficulties, the computation of the LE spectrum permits one to establish interesting relations between different indicators of nonlinearity. Kaplan and Yorke (1978) have conjectured that the fractal dimension and the LE spectrum (the geometry and the dynamics of the system) are strongly related.

Let us consider a trajectory $x(t) = \Phi(t, x_0)$ whose exponents are $\lambda_1 \geq \lambda_2 \geq \cdots \geq \lambda_n$, where n is the topological dimension. For $m \leq n$, $S(m) = \lambda_1 + \lambda_2 + \cdots + \lambda_m$ represents the mean rate of expansion/contraction($>0/<0$) of the m-dimensional volume along with the trajectory. If, for some k, $S(k) > 0$, we will have k dimensions of the volume that expand themselves, but we will have also $S(k + 1) < 0$ in $k + 1$ dimensions that contract themselves so as to have a d-dimensional volume invariant (d not integer, $k < d < k + 1$), where d is the dimension of the attractor A.

The formula of Kaplan and Yorke estimates the fractal dimension d of the attractor as

$$d_{KY} = k + \frac{S(k)}{|\lambda_{k+1}|} \tag{14.26}$$

where k is the maximum integer such that the sum of the k largest exponents is still nonnegative. Even if a rigorous demonstration was performed only for a certain set of dynamical systems, the numerical simulations indicate that the relation generally holds. For example, for the Lorenz system ($n = 3$), the value of the exponents is $\lambda_1 = 0.905$, $\lambda_2 = 0$, and $\lambda_3 = -14.57$. For $k = 2$, we have $d_{KY} = 2 + 0.905/14.57 = 2.062$, which is the fractal dimension of the Lorenz attractor.

Figure 14.3f underlines how the value of the exponents, obtained by the numerical integration of the reconstructed trajectory, depends on the m value. In fact, for values $m > n$, not only do we introduce $m - n$ spurious exponents, but we modify the value of the "true" exponents.

14.6 ENTROPY

As we have previously outlined, a system is defined as chaotic if its attractor has a fractal dimension and if it exhibits sensitive dependence on initial conditions. It produces information as two initial conditions very close to each other evolve in completely different trajectories. To measure this property, the concept of mean rate of creation of information, also known as entropy or the Kolmogorov–Sinai invariant, was introduced. Let us partition the phase space into boxes of size ε and assume that there is an attractor and that the trajectory $\bar{x}(t)$ lies in the basin of attraction. Then the Kolmogorov entropy can be then defined as:

$$K = -\lim_{\substack{\tau \to 0 \\ \varepsilon \to 0 \\ d \to \infty}} 1/d\tau \sum_{i_1,\ldots,i_d} p(i_1,\ldots,i_d) \ln p(i_1,\ldots,i_d) \tag{14.27}$$

where $p(i_1, \ldots, i_d)$ is the joint probability that the state of the system $\bar{x}(t_0 + \tau)$ is in the box i_1 and $\bar{x}(t_0 + d\tau)$ is in the box i_d (d is the number of boxes covering the space state and τ is the interval time at which the state is measured).

A useful algorithm to estimate the Kolmogorov–Sinai (K-S) entropy is described in the work of Eckmann and Ruelle (1985). For an experimental time series $u(1), u(2), \ldots$, regularly spaced in time, it constructs a sequence of points $x_m(i)$ obtained by taking $x_m(i) = [u(i), \ldots, u(i + m - 1)]$. Therefore, this construction associates points $X(i)$ in the phase space of the system (which is, in general, infinite dimensional) with their projections $x_m(i) = \pi_m[X(i)]$ in $\Re^m$, obtained by the measures $\{u(i)\}$ as outlined above.

This is useful for the determination of the following quantity:

$$C_i^m(r) = N^{-1} \quad \{\text{number of } x_m(j) \text{ such that } d[x_m(i), x_m(j)] \leq r\} \quad (14.28)$$

where $d[x_m(i), x_m(j)]$ is the distance between the vectors, measured as $\max\{|x(i) - x(j)|, \ldots, |x(i + m - 1) - x(j + m - 1)|\}$.

We can interpret Equation 14.28 as the probability that $x_m(j)$ is close within r to $x_m(i)$; in other words, it is the probability that the signal $u(j)$ remains for m consecutive units of time in the sphere of radius r centered at $u(i)$.

If

$$\Phi^m(r) = \frac{1}{N} \sum_i \log C_i^m(r) \quad (14.29)$$

then the quantity $\Phi^{m+1}(r) - \Phi^m(r)$ is the average over i of the logarithm of the probability that $u(j + m)$ is close within r to $u(i + m)$, given that $u(j + k)$ is close within r to $u(i + k)$ for $k = 0, \ldots, m - 1$.

Entropy can finally be defined as

$$\text{K-S} = \lim_{r \to 0} \lim_{m \to \infty} \lim_{N \to \infty} [\Phi^{m+1}(r) - \Phi^m(r)] \quad (14.30)$$

The measure of entropy (Equation 14.30) guarantees that a deterministic system is chaotic when K-S assumes a finite nonzero value, and it converges to the attractor as N tends to infinite.

Later, Grassberger and Procaccia proposed an entropy estimation method that can be applied even if the signal length is limited [18]. Their estimated entropy K_2 is a lower bound of K ($K \geq K_2$).

Given an experimental time series $\{u(i)\}$ regularly spaced in time, we construct the quantity

$$K_{2,m}(\varepsilon) = 1/\tau \ln \frac{C_m(\varepsilon)}{C_{m+1}(\varepsilon)} \quad (14.31)$$

where $C_m(\varepsilon)$ is $\lim_{N \to \infty} 1/N^2$ {number of pair (i, j) such that $d[x_m(i), x_m(j)]$ $\leq \varepsilon$}. The K_2 entropy can be then estimated with

$$\lim_{m \to \infty} \lim_{\varepsilon \to 0} K_{2,m}(\varepsilon) \sim K_2 \qquad (14.32)$$

Pincus (1995) observed that the defined K-S entropy assumes an infinite value for all processes with superimposed noise and, thus, is unable to distinguish a class of processes that are different in complexity. In this case, the word complexity refers to the predictability of the system state location by knowing the initial conditions. The less predictable the states are, the more complex the system is. For instance, Gaussian noise spreads its states over all state space, so it has the highest level of complexity; in fact, it is unpredictable and all points in the phase space are probable states of the system. In contrast, a periodic dynamical system with period 2 is a system with a low level of complexity; in fact, all the trajectories meet two points with probability 1. The majority of biological dynamical systems lie between the two extremes of this complexity scale: systems that have a structured distribution of the trajectories, such as the strange attractor, or systems that occupy a linear region of the space, as in the quasiperiodic case.

Starting from these considerations, Pincus modified the K-S entropy algorithm, limiting its purpose to measure the signal "regularity," that is, the presence of similar patterns in a time series, thus allowing the analysis of different systems also corrupted by random noise. He called this family of statistics approximate entropy (ApEn) (Pincus et al., 1991), that is, not an approximation of the entropy previously outlined, though inspired by its definition (Equation 14.3).

Given N data points $\{u(i)\}$, the algorithm constructs sequences $x_m(i)$ and computes, for each $i \leq N - m + 1$, the quantity

$$C_i^m(r) = N^{-1} \quad \{\text{number of } j \leq N - m + 1 | d[x_m(i), x_m(j)] \leq r\} \qquad (14.33)$$

that measures, with a tolerance r, the regularity of patterns, comparing them to a given pattern of length m (m and r are fixed values; m is the detail level at which the signal is analyzed and r is a threshold, which filters out irregularities).

The parameter of regularity is defined as ApEn(m, r) = $\lim_{N \to \infty}[\Phi^m(r) - \Phi^{m+1}(r)]$, where $\Phi^m(r) = (N - m + 1)^{-1}\Sigma_{i=1}^{N-m+1} \ln C_i^m(r)$.

ApEn$(m, r, N) = [\Phi^m(r) - \Phi^{m+1}(r)]$ is the estimator of this parameter for an experimental time series of a fixed length N.

As Pincus noticed (Pincus, 1995; Pincus et al., 1991; Richman and Moorman, 2000), approximate entropy is affected by a bias effect. Let us consider A_i as the number of vectors $x_m(j)$ such that $d[x_m(i), x_m(j)] \leq r$ with $i \neq j$, and B_i as the number of vectors $x_{m+1}(j)$ such that $d[x_{m+1}(i),$ $x_{m+1}(j)] \leq r$ with $i \neq j$. The ApEn assigns to the template $x_m(i)$ a biased conditional probability, $(A_i + 1)/(B_i + 1)$. For a large number N of samples, the biased probability converges to the unbiased one A_i/B_i. Naturally, this discrepancy is more evident when a large number of templates have $A_i = B_i = 0$ in a time series. Furthermore, this statistic is strongly dependent on the record length. The elimination of bias is not, however, so easy; the straightforward removal of self-counting could produce a high sensitivity to the outliers. In fact, if there were a template that did not match other templates, ApEn could not be calculated, $\ln(0)$ being infinite.

As regards the consistency [the general property for which, if $\text{ApEn}(m_1, r_1)(S) \leq \text{ApEn}(m_1, r_1)(T)$, then it must be $\text{ApEn}(m_2, r_2)(S) \leq \text{ApEn}(m_2, r_2)(T)$], it is proved in the work of Richman and Moorman (2000) that ApEn lacks this attribute in some cases. The reason is that irregular time series are more affected overall by the bias, which overestimates the value of entropy when r values are very small. However, the ApEn statistic often works well with real data and it is a simple tool that gives general information about the regularity and the persistence of the signal. This is the reason why it has found a large application in clinical investigations, for example, for the analysis of the interbeat RR signal to study cardiovascular diseases (Fukuta et al., 2003; Ho et al., 1997; Mäkikallio et al., 1997).

Entropy methods exploit a symbolic representation of a time series. Severe reduction of information is used to enhance relevant features. ApEn and, more generally, coarse-grained entropies could be useful to track qualitative changes in time series patterns, without the need to precisely characterize the generating system.

Recently, Richman and Moorman (Lake et al., 2002; Richman and Moorman, 2000) developed a modification of this algorithm in order to remove what they considered the defects of ApEn. The name of this new statistic is sample entropy (SampEn).

The differences with respect to ApEn are: (i) self-matches are not counted, (ii) only the first $N - m$ vectors of length m are considered, and (iii) the conditional probabilities are not estimated in a template manner; they do not adopt as a probability measure the ratio of the logarithmic sums, but they compute directly the logarithm of conditional probability.

We define the following quantities for $i, j \leq N - m$:

$$A_i^m(r) = (N - m - 1)^{-1} \quad \{\text{number of } x_{m+1}(j) \text{ such that} \quad (14.34)$$
$$d[x_{m+1}(i), x_{m+1}(j)] \leq r, i \neq j\}$$

$$B_i^m(r) = (N - m - 1)^{-1} \quad \{\text{number of } x_m(j) \text{ such that} \quad (14.35)$$
$$d[x_m(i), x_m(j)] \leq r, i \neq j\}$$

$$A^m(r) = (N - m)^{-1} \sum_{i=1}^{N-m} A_i^m(r) \quad (14.36)$$

$$B^m(r) = (N - m)^{-1} \sum_{i=1}^{N-m} B_i^m(r) \quad (14.37)$$

The parameter SampEn(m, r) is then given by $\lim_{N \to \infty}\{-\ln[A^m(r)/B^m(r)]\}$, and the associated statistics SampEn(m, r, N) are defined by removing the limit. We can observe that this latest algorithm is very similar to the estimation of the Kolmogorov entropy by Grassberger and Procaccia (Equation 14.31).

In contrast to ApEn, sample entropy shows a relative consistency in cases where ApEn does not. Furthermore, the values of SampEn also agree with theoretical values for very short time series much more than the values of ApEn (Richman and Moorman, 2000).

It is clear that both ApEn and SampEn can supply only one index about a general behavior of the time series, but they cannot say anything about the underlying dynamics. Thus, if signal X has a lower value of entropy than signal Y, we can only say that X is more regular than Y. If the original purpose of entropy was to identify chaotic dynamics, the statistics ApEn and SampEn have changed the perspective as they give a figure related to the regularity or predictability of the time series at the original time scale.

14.7 NONLINEAR NOISE REDUCTION

The last point to be dealt with is the problem of noise reduction. This procedure is usually performed by the application of linear filters. In the case of nonlinear noise, the signal filtering requires the application of particular methods. The signals generated by nonlinear sources are often characterized by large-band spectra, which do not permit one to identify the noise component.

In these cases, the linear filters are not suitable because they can modify the nonlinear components by introducing linear correlations and they are only based on the frequency content of the signal and the noise.

Nonlinear noise reduction, instead, does not rely on frequency information in order to define the distinction between signal and noise. Instead, the structure in the reconstructed phase space is exploited. The nonlinear approach is based on the identification of a simple dynamical system, characterized by a reduced number of state variables, which is consistent with the data. The problem is not to separate a deterministic signal from random fluctuations because what we define noise can come from systems of high dimensionality. The problem is to identify dynamics of low dimension, which are dipped in a complex signal, on the basis of a model to be applied for a region of interest of the attractor (Cawley and Hsu, 1992; Kostelich and Schreiber, 1990).

The hypothesis is that the measured data are composed of the output of a low-dimensional dynamical system and by random or high-dimensional noise. This means that in an arbitrarily high-dimensional embedding space, the deterministic part of the data would lie on a low-dimensional manifold, whereas the effect of the noise would spread the data off this manifold. From a mathematical point of view, we can state that a separation in the singular values of the local or global sample covariance matrix of the data will occur between the signal, which is presumed to dominate the larger singular values, and the noise, which is presumed to dominate the smaller singular values. The idea is to form a sample covariance matrix in dimension d larger than the embedding dimension and then project the data onto the first singular values. This is then taken as new data, and a new time-delay embedding is made. The procedure is continued until the final version of the data is clean enough. The projection onto the singular directions is a linear filter of the data. A combination of local filters, each different, makes for a global nonlinear filter.

14.8 CONCLUSION

The development of chaos theory in the last decades has supplied the framework to study nonlinear dynamical systems through a new approach. Many efforts were made to explain the complex behavior of biological systems with the presence of nonlinear features instead of the usual stochastic models. The most direct link between chaos theory and the real world is the analysis of time series coming from real systems, in terms of nonlinear dynamics (Katz et al., 1997).

In some situations, when the a priori knowledge about the generating system is deep enough, it is possible to precisely characterize the structure of the system itself from the observed time series by applying the methods described in this chapter (embedding theory, Lyapunov exponents, attractor invariants, etc.). Nevertheless, it is often impossible to gain sufficient insight into biological systems from signals recorded by noninvasive techniques. The knowledge we have on their behavior is limited by their intrinsic complexity, which is the result of interacting mechanisms contributing to the physiological performance. For these reasons, measures such as the approximate entropy have been developed in order to characterize the time series so as to search for differences among different class, for example, among patient groups affected by different pathologies.

We want to conclude with useful suggestions to implement the described analyses. First of all, there is PhysioNet. PhysioNet was established in 1999 as the outreach component of the Research Resource for Complex Physiologic Signals, a cooperative project initiated by researchers at Boston's Beth Israel Deaconess Medical Center/Harvard Medical School, Boston University, McGill University, and MIT, originally established under the auspices of the National Center for Research Resources of the NIH. This organization has created a website that offers free access to large collections of recorded physiologic signals and related open-source software (http://www.physionet.org).

TISEAN (http://www.mpipks-dresden.mpg.de/~tisean/) is a software project for the analysis of time series with methods based on the theory of nonlinear deterministic dynamical systems, or chaos theory. It has grown out of the work of several groups during the last few years. Some of the routines are built around the programs given in the book by Kantz and Schreiber (1997). In this website, it is possible to find papers, tutorials, and software that implement many procedures outlined in this chapter (nonlinear noise reduction, search for embedding dimension, computation of Lyapunov spectrum, and so on).

APPENDIX

14.A1 Chaotic Dynamics

For a dynamical system to be classified as chaotic, it must have the following properties:

- It must be sensitive to initial conditions
- It must be topologically mixing
- Its periodic orbits must be dense

Sensitivity to initial conditions means that each point in such a system is arbitrarily closely approximated by other points with significantly different future trajectories. Thus, an arbitrarily small perturbation of the current trajectory may lead to significantly different future behavior.

Sensitivity to initial conditions is popularly known as the "butterfly effect," so called because of the title of a paper given by Edward Lorenz in 1972 to the American Association for the Advancement of Science in Washington, D.C., entitled "Predictability: Does the Flap of a Butterfly's Wings in Brazil Set off a Tornado in Texas?" The flapping wing represents a small change in the initial condition of the system, which causes a chain of events leading to large-scale phenomena. Had the butterfly not flapped its wings, the trajectory of the system might have been vastly different.

Topologically mixing means that the system will evolve over time so that any given region or open set of its phase space will eventually overlap with any other given region. Here, "mixing" is really meant to correspond to the standard intuition; the mixing of colored dyes or fluids is an example of a chaotic system.

14.A2 Attractors

Some dynamical systems are chaotic everywhere, but in many cases chaotic behavior is found only in a subset of phase space. The cases of greatest interest arise when the chaotic behavior takes place on an attractor, since then a large set of initial conditions will lead to orbits that converge to this chaotic region.

An easy way to visualize a chaotic attractor is to start with a point in the basin of attraction of the attractor, and then simply plot its subsequent orbit. Because of the topological transitivity condition, this is likely to produce a picture of the entire final attractor.

14.A3   Strange Attractors

Although most of the motion types mentioned above give rise to very simple attractors, such as points and circle-like curves called limit cycles, chaotic motion gives rise to what are known as strange attractors, attractors that can have great detail and complexity. For instance, a simple three-dimensional model of the Lorenz weather system gives rise to the famous Lorenz attractor. The Lorenz attractor is perhaps one of the best-known chaotic system diagrams, probably because not only was it one of the first, but it is one of the most complex, and, as such, gives rise to a very interesting pattern that looks like the wings of a butterfly. Another such attractor is the Rössler map, which experiences a period-two doubling route to chaos, like the logistic map.

Strange attractors occur in both continuous dynamical systems (such as the Lorenz system) and in some discrete systems (such as the Hénon map). The strange attractors typically have a fractal structure.

The Poincaré–Bendixson theorem shows that a strange attractor can only arise in a continuous dynamical system if it has three or more dimensions. However, no such restriction applies to discrete systems, which can exhibit strange attractors in two- or even one-dimensional systems.

REFERENCES

Abarbanel, H. D. I., Brown, R., Sidorowich, J. J., and Tsimring, L. S., The Analysis of Observed Chaotic Data in Physical Systems, *Review of Modern Physics,* Vol. 65, No. 4, 1331–1362, 1993.

Baselli, G., Cerutti, S., Porta, A., and Signorini, M. G., Short and Long Term Non-Linear Analysis of RR Variability Series, *Medical Engineering & Physics,* Vol. 24, No. 1, 21–32, 2002.

Bigger, T. J., Jr., Steinman, R. C., Rolnitzky, L. M., Fleiss, J. L., and Albrecht, P., et al., Power Law Behavior of RR Interval Variability in Healthy Middle-Aged Persons, Patients with Recent Acute Myocardial Infarction, and Patients with Heart Transplant, *Circulation,* Vol. 93, No. 12, 2142–2151, 1996.

Broomhead, D., and King, G., Extracting Qualitative Dynamics from Experimental Data, *Physica,* Vol. 20D, 217, 1986.

Brown, R., Bryant, P., and Abarbanel, H. D. I., Computing the Lyapunov Spectrum of a Dynamical System from an Observed Time Series, *Phys. Rev. A,* Vol. 43, 2787–2806, 1991.

Bryant, P., Brown, R., and Abarbanel, H. D. I., Lyapunov Exponents from Observed Time Series, *Physical Review Letters,* Vol. 65, 1523, 1990.

Cawley, R., and Hsu, G. H., Local-Geometric-Projection Method for Noise Reduction in Chaotic Maps and Flows, *Phys Rev A,* Vol. 46, 3057–3082, 1992.

Cerutti, S., Esposti, F., Ferrario, M., Sassi, R., Signorini, M. G., Long-Term Invariant Parameters Obtained from 24-h Holter Recordings: A Comparison between Different Analysis Techniques, *Chaos,* Vol. 17, No. 1, 1–9, 2007.

Eckmann, J.-P., and Ruelle, D., Ergodic Theory of Chaos and Strange Attractors, *Review of Modern Physics,* Vol. 57, 617, 1985.

Eckmann, J.-P., Oliffson Kamphorst, S., Ruelle, D., and Ciliberto, S., Lyapunov Exponents from a Time Series, *Phys. Rev. A.* Vol. 34, 4971, 1986.

Farmer, J. D., Ott, E., and Yorke, J. A., The Dimension of Chaotic Attractors, *Physica D.* Vol. 7, 153–180, 1983.

Fraser, A. M., Information and Entropy in Strange Attractors, *IEEE Trans. Information Theory,* Vol. 35, No. 2, 245–262, 1989.

Fraser, M., Reconstructing Attractors from Scalar Time Series: A Comparison of Singular System and Redundancy Criteria, *Physica D,* Vol. 34, 391–404, 1989.

Fraser, M., and Swinney, H. L., Independent Coordinates for Strange Attractors from Mutual Information, *Phys. Rev. A.* Vol. 33, 1134–1140, 1986.

Fukuta, H., et al., Prognostic value of nonlinear heart rate dynamics in hemodialysis patients with coronary artery disease, *Kidney Int.,* Vol. 64, No. 2, pp. 641–648, 2003.

Galleger, R. G., *Information Theory and Reliable Communication,* Wiley, 1968.

Grassberger, P., Generalized Dimensions of Strange Attractors, *Phys. Lett. A,* Vol. 97, 227, 1983.

Grassberger, P., and Procaccia, I., Measuring the Strangeness of Strange Attractors, *Physica,* Vol. 9D, 189, 1983.

Grassberger, P., Schreiber, T., and Schaffrath, C., Nonlinear Time Sequence Analysis, *International Journal of Bifurcation and Chaos,* Vol. 1, 521, 1991

Greenside, H. S., Wolf, A., Swift, J., and Pignataro, T., Impracticality of a Box-Counting Algorithm for Calculating Dimensionality of Strange Attractors, *Phys Rev A,* Vol. 25, No. 6, 3453–3456, 1982.

Ivanov, P. C., Amaral, L. A. N., and Goldberger, A. L., et al., Multifractality in Human Heartbeat Dynamics, *Nature,* Vol. 399, 461–465, 1999.

Hausdorff, F., Dimension und äusseres Mass, *Mathematische Annalen,* No. 79, 1919.

Hegger, R., Kantz, H., and Schreiber, T., Practical Implementation of Nonlinear Time Series Methods: The TISEAN Package, *Chaos.* Vol. 9, 413, 1999.

Ho, K. K. L., et al., Predicting Survival in Heart Failure Case and Control Subjects by Use of Fully Automated Methods for Deriving Nonlinear and Conventional Indices of Heart Rate Dynamics, *Circulation,* Vol. 96, No. 3, pp. 842–848, 1997.

Huikuri, H. V., Makikallio, T. H., Airaksinen, K. E., Seppanen, T., Puukka, P., Raiha, I. J., and Sourander, L. B., Power-Law Relationship of Heart Rate Variability as a Predictor of Mortality in the Elderly, *Circulation.* Vol. 97, No. 20, 2031–2036, 1998.

Kanters, J. K., Holsteinrathlou, N. H., and Agner, E., Lack of Evidence for Low-Dimensional Chaos in Heart-Rate Variability, *J. Cardiovasc. Electr.*, Vol. 5, No. 7, 591–601, 1994.

Kantz H., Kurths, J., and Mayer-Kress, G., (Eds), *Non-Linear Analysis of Physiological Data*, Springer-Verlag, 1998.

Kantz H., and Schreiber, T., *Nonlinear Time Series Analysis*, Cambridge University Press, 1997.

Kaplan, D. T., and Glass, L., A Direct Test for Determinism in a Time Series, *Physical Review Letters*, Vol. 68, 427–430, 1992.

Kaplan, D. T., and Glass, L., *Understanding Nonlinear Dynamics*, Springer-Verlag, 1995.

Kaplan, J. L., and Yorke, J. A., Chaotic Behaviour of Multidimensional Difference Equations, in *Lecture Notes in Mathemathics*, Vol. 730, Springer, 1978.

Kennel, M. B., Brown, R., and Abarbanel, H. D. I., Determining Embedding Dimension for Phase-Space Reconstruction Using a Geometrical Construction, *Phys. Rev. A*, Vol. 45, 3403–3411, 1992.

Kobayashi, M., and Musha, T., 1/f Fluctuation of Heartbeat Period, *IEEE Transactions on BME*, Vol. 29, No. 6, 456–457, 1982

Kostelich, E. J., and Schreiber, T., Noise Reduction in Chaotic Time Series: A Survey of Common Methods, *Phys Rev E*, Vol. 48, 1752–175, 1990.

Kostelich, E. J., and Yorke, J. A., Noise Reduction: Finding the Simplest Dynamical System Consistent with the Data, *Physica D*, Vol. 41, 183–196, 1990.

Lake, D. E., Richman, J. S., and Moorman, J. R., Sample Entropy Analysis of Neonatal Heart Rate Variability, *Am. J. Physiol. Regul. Integr. Comp. Physiol.*, Vol. 283, R789–R797, 2002.

Lorenz, E. N., Deterministic Nonperiodic Flow, *J. Atmos. Sci.*, Vol. 20, 130–141, 1963

Mandelbrot, B., *The Fractal Geometry of the Nature*, W.H. Freeman, 1983.

Mané, R., Dynamical Systems and Turbulence, in *Lecture Notes in Mathematics*, No. 898, p. 320, Springer, 1981.

Mäkikallio, T. H., et al., Dynamic Analysis of Heart Rate May Predict Subsequent Ventricular Tachycardia after Myocardial Infarction, *Am J Card*, Vol. 80, No. 6, 779–783, 1997.

Makikallio, T. H., Huikuri, H. V., Peng, C.-K., Goldberger, A. L., Hintze, U., and Moler, M., Fractal Correlation Properties of R-R Interval Dynamics and Mortality in Patients with Depressed Left Ventricular Function after an Acute Myocardial Infarction, *Circulation*, Vol. 101, 47–53, 2000.

Ott, E., Sauer, T. D., and Yorke, J. A., *Coping with Chaos—Analysis of Chaotic Data and the Exploitation of Chaotic Systems*, Wiley, 1994

Packard, N. H., Crutchfield, J. P., Farmer, J. D., and Shaw, R. S., Geometry from a Time Series, *Phys. Rev. Letters*, Vol. 45, 712–716, 1980.

Parker, T., and Chua, L., Chaos: A Tutorial for Engineers, *Proceedings of the IEEE*, Vol. 75, 982, 1987.

Pincus, S. M., Approximate Entropy (ApEn) as Complexity Measure, *Chaos*, Vol. 5, No. 1, 110–117, 1995.

Pincus, S. M., Gladstone, I. M., and Ehrenkranz, R. A., A Regularity Statistic for Medical Data Analysis, *J. Clin. Monit.*, Vol. 7, 335–345, Oct. 1991.

Peng, C.-K., Havlin, S., Hausdorff, J. M., Mietus, J. E., Stanley, H. E., and Goldberger, A. L., Fractal Mechanisms and Heart Rate Dynamics: Long-Range Correlation and their Breakdown with Diseases, *Journal of Electrocardiology,* Vol. 28, Suppl., 59–65, 1995.

Poon, C. S., and Merrill, C. K., Decrease of Cardiac Chaos in Congestive Heart Failure, *Biotechnology,* Vol. 389, 492–495, 1997.

Richman, J. S., and Moorman, J. R., Physiological Time-Series Analysis Using Approximate Entropy and Sample Entropy, *Am. J. Physiol. Heart Circ. Physiol.,* Vol. 278, H2039–2049, 2000.

Poincarè, H., *Ouevres,* Gauthier-Villar, 1954.

Ruelle, D., Ergodic Theory of Differentiable Dynamical Systems, *Publications Mathématiques of the Institut des Hautes Etudes Scientifique,* Vol. 50, 27, 1979.

Sauer, T. D., Yorke, J. A., and Casdagli, M., Embedology, *J. of Statistical Physics,* Vol. 65, Nos. 3/4, 579–616, 1991.

Schreiber, T., and Schmitz, A., Improved Surrogate Data for Non-Linearity Tests, *Phys. Rev. Lett.,* Vol. 77, 635–638, 1996.

Schreiber, T., Costrained Randomitation of Time Series, *Physical Review Letters,* Vol. 80, 2105–2109, 1998

Schreiber, T., and Schmitz, A., Surrogate Time Series, *Physica D,* Vol. 142, 346–382, 2000.

Signorini, M. G., Sassi R., and Cerutti, S., Nonlinear Biomedical Signal Processing, II: Dynamic Analysis and Modelling, in Metin Akay (Ed.), *Assessment of Nonlinear Dynamics in Heart Rate Variability Signals,* pp. 263–281, IEEE Press, 2000.

Takens, F., Dynamical Systems and Turbulence, in *Lecture Notes in Mathematics,* No. 898, p. 366, Springer, 1981.

Theiler, J., Eubank, S., Longtin, A., Galdrikian, B., and Farmer, J. D., Testing for Nonlinearity in Time Series: The Method of Surrogate Data, *Physica D,* Vol. 58, 77–94, 1992.

TISEAN (Nonlinear Time Series Analysis), http://www.mpipks-dresden.mpg. de/~tisean/TISEAN_2.1/index.html.

TSTOOL (Nonlinear Time Series Analysis), http://www.physik3.gwdg.de/ tstool/index.html.

Van der Pol, B., Forced Oscillations in a Circuit with Nonlinear Resistance, *Phil. Mag.,* Vol. 3, No. 65, 1927.

Whitney, H., Differentiable Manifolds, *Ann. Math.,* Vol. 37, 645–680, 1936.

CHAPTER *15*

BLIND SOURCE SEPARATION
Application to Biomedical Signals

Luca Mesin, Aleš Holobar, and Roberto Merletti

15.1 INTRODUCTION

Blind source separation (BSS) is a prominent problem in signal processing. In the past few decades, it was applied to many fields in which separation of compound signals, simultaneously observed by different sensors, is
of interest. The problem can be considered as built up of three physical elements: sources (also called transmitters), sensors (also called receivers),
and communication channels that reflect the properties of the physical
medium propagating the signals form the sources to the sensors. The signals detected by the sensors are commonly referred to as observations and
are assumed to be algebraic combinations of the unknown source signals.
The BSS approach assumes limited a priori information on the communication channels (linearity, memory properties, etc.) and tries to reconstruct
the source signals from the detected signals only. Analysis of the communication channels is important mainly for selection of a proper processing
technique, namely, communication channels weight and/or filter the signals coming from the sources and, together with them, determine the temporal and spectral characteristics of the detected mixtures.

An example of a physical medium propagating sound is air. Physical
properties of the air determine the weights of communication channels in
the speech separation (or cocktail-party separation) problem. This problem
deals with separation of different human voices or sounds from instruments, recorded by two or more microphones during simultaneous emission of two or more sources (Koutras et al., 2000; Anemüller and Gramss,
1999). The source separation problem for sonar is discrimination of the
echoes from different simultaneously present targets. Radar requires the
solution of problems equivalent to those of sonar. Communication systems
working underwater or in an orbit also face equivalent problems of source

Advanced Methods of Biomedical Signal Processing. Edited by S. Cerutti and C. Marchesi
Copyright © 2011 the Institute of Electrical and Electronics Engineers, Inc.

separation, but use signals with different spectral features. The problem of separating a mixture of echoes from different targets is also important in the earth sciences, for example in the study of different geological layers or in the search for water or oil reservoirs.

Source separation finds important applications also in the life sciences. Electroencephalographic (EEG), electrocardiographic (ECG), electromyographic (EMG), and mechanomyographic (MMG) signals are all compound biomedical signals, generated by several tens (EMG, MMG) or even millions (EEG) of biophysical sources. Separation of biomedical signals augments the power of human-body scanning techniques and plays an important role in understanding complex processes in biomedical phenomena (Vigàrio, 1997; Vigàrio et al., 1998, 1999; Makeig et al., 1996). This chapter is devoted to basic descriptions of frequently used source separation methods, with focus on the biomedical applications.

15.2   MATHEMATICAL MODELS OF MIXTURES

A mathematical model of the source–sensor communication, also called a mixing process, determines an analytical relation between the source signals and the observations. Mixing models can be classified as follows (Lacoume, 1999):

1. *Nonlinear model.* The most general model and very difficult to study as the source signals do not satisfy the superimposition principle (i.e., their contributions combine nonlinearly to form the observed mixtures).
2. *Post nonlinear model.* The process consists of linear mixing and an instantaneous nonlinear mapping of the source signals.
3. *Linear model.* The most widely studied model and the only one considered in this chapter. In each observation, contributions from different sources are linearly combined, that is, superimposed on each other.

Linear mixing model can further be divided in two subgroups:

- *Convolutive mixing model.* The mixing process is a causal multidimensional convolution,

$$\mathbf{x}(t) = \int \mathbf{A}(t - \tau)\, \mathbf{s}(\tau) d\tau \qquad (15.1)$$

where $\mathbf{s}(t)$ are source signals from N sources, $\mathbf{x}(t)$ are observations detected by M sensors, and $\mathbf{A}(t)$ is a mixing matrix comprising impulse responses of all the communication channels that relate the source signals $\mathbf{s}(t)$ to the observed signals $\mathbf{x}(t)$. The convolutive mixing model is typically assumed to be causal, with memory of the source signals received in the past.

- *Instantaneous mixing model.* The signals detected in a time instant are obtained as linear combinations of the source signals at the same instant:

$$\mathbf{x}(t) = \mathbf{As}(t) \tag{15.2}$$

An instantaneous mixing model has no knowledge of the source samples received in the past.

In numerical implementation, $\mathbf{s}(t)$ is a matrix of sampled source signals (with the T samples of the signal from the rth source in the rth row), and $\mathbf{x}(t)$ is a matrix of sampled observations, with observations from different sensors in different rows. Without loss of generality, the observations $\mathbf{x}(t)$ are also assumed to be zero-mean.

The linear BSS problem has two types of ambiguities. First, it is clear from Equations 15.1 and 15.2 that amplitude scaling of the sources can be compensated by an inverse scaling of the corresponding elements of matrix $\mathbf{A}$. Thus, with no a priori knowledge on the mixing matrix $\mathbf{A}$, the power of individual source signals cannot be determined and is, by convention, set equal to 1. A second ambiguity lies in the order in which the source signals are determined.

Now, let us assume that the signals $\mathbf{s}(t)$ are emitted from N different sources, while the observations $\mathbf{x}(t)$ are detected by the M different sensors, where $M \geq N$. Then, in order to reconstruct the source signals, we must first estimate the mixing matrix $\mathbf{A}$, invert it, and apply its inverse to the observed signals $\mathbf{x}(t)$. Thus, the unknowns of the BSS problem comprise both the elements of $\mathbf{A}$ and the source signals $\mathbf{s}(t)$. In the case of the discretized instantaneous model of Equation 15.2 with T samples of long source signals, we must estimate $M \times N$ entries of $\mathbf{A}$ and $N \times T$ samples of the source signals, given just $M \times T$ samples of observations $\mathbf{x}(t)$. The number of unknowns to be determined is usually greater than the number of equations imposed by Equations 15.1 or 15.2 (even when $M \geq N$) and further a priori conditions on the source signals or/and the mixing matrix $\mathbf{A}$ are required to solve the problem of source separation. Most of BSS methods do not use any information about the mixing matrix $\mathbf{A}$. Instead, they only rely on additional information about the sources. The latter are usually considered to be uncorrelated or statistically independent. Although somehow contraintuitive, these assumptions are often sufficient to estimate the source signals, except for the ambiguities on their amplitudes and order (as stated above).

In order to comply with practice, a random noise is usually added to Equations 15.1 and 15.2. Such a noise can be either additive or multiplicative. Typically, the noise is further assumed to be a zero-mean, temporarily and spatially white random process. Temporal whiteness implies the independence of noise samples belonging to the time series of each individual observation, whereas spatial whiteness refers to independence of samples of noise between different observations at the same time instant. Frequently, the noise is also assumed to be independent of the source signals.

15.3   PROCESSING TECHNIQUES

Practically all source separation techniques are based on maximization of the distance between the estimated source signals. The definition of the distance depends on the selected a priori assumptions on the sources and generates classification of different BSS approaches. In what follows, we will briefly describe only some of those BSS classes that found their way into the field of biomedical signal processing. The interested reader is referred to Hyvarinen and coworkers (2001) for a more thorough and complete overview of BSS approaches.

One of the best known signal decomposition techniques is principal component analysis (PCA), also known as the Karhunen–Loeve or Hotelling transform. Strictly speaking, PCA does not belong to the BSS family, as it does not truly reconstruct the original source signals. Nonetheless, it is a very popular decomposition technique and is used as a preprocessing step in numerous BSS approaches. PCA builds on the correlation of observed signals and decomposes the observations into uncorrelated signal components. If the source signals are Gaussian, uncorrelatedness also implies independence, and the signal components obtained by PCA are also statistically independent. A useful property of PCA is that it preserves the power of observations, removes any linear dependencies between the reconstructed signal components, and reconstructs the signal components with maximum possible energies (under the constraint of power preservation and uncorrelatedness of the signal components). Thus, PCA is frequently used for a lossy data compression (see Section 15.3.1 for details).

The second large class of signal decomposition techniques is independent component analysis (ICA). ICA belongs to the family of BSS and imposes statistical independence of sources, meaning that all the samples of the source signals are assumed to be independent, identically distributed (i.i.d.) random variables. ICA preserves the information contained in the observations and, at the same time, minimizes the mutual information of estimated source samples (mutual information is the information that the samples of the source signals have of each other). Thus, ICA is also useful in data compression, usually allowing higher compression rates than PCA.

Specific optimization techniques used to maximize the distance between the independent sources determines further classification of the ICA methods:

1. *Algebraic methods.* Matrix calculus is used to estimate the mixing matrix **A**.
2. *Neural-networks-based* methods. Neural networks perform recursive estimation of weights, which define linear combinations of the mixtures; these combinations are the estimates of the sources.

In the next subsection, PCA and ICA are discussed in more detail. In particular, examples of algebraic and neural-network-based source-separation methods are described, along with some indications of the most typical assumptions about the statistical independence of sources. References for further reading are also provided.

15.3.1   PCA and ICA: Possible Choices of Distance between Source Signals

Assume a simple mixing model with N source signals $s(t)$ and M observations $x(t)$:

$$x(t) = As(t) + n(t) \tag{15.3}$$

where $n(t)$ is a zero-mean additive Gaussian noise.

PCA is mathematical method that determines the amount of redundancy in the observations $x(t)$ and estimates a linear transformation P, which reduces this redundancy to a minimum. P is further assumed to have a unit norm, so that the total power of the observations $x(t)$ is preserved. Strictly speaking, PCA does not assume any mixing model. Redundancy of information in $x(t)$ is simply measured by the cross correlation between the different observations. Therefore, although PCA can be interpreted as a signal-decomposition technique, the estimated principal components $y(t) = Px(t)$ differ significantly from the original sources $s(t)$ (see Subsection 15.3.1.1). ICA, on the other hand, employs a much stronger assumption on statistical independence of sources, requires a-priori knowledge about the mixing model, and allows reconstruction of original sources $s(t)$.

The ICA problem was first proposed by Jutten (1987) and Hérault and Jutten (1991). The neural, iterative approach used by Hérault and Jutten underlines the similarities of ICA with PCA and is, for historical reasons, discussed in the next subsection. Independently from Hérault and Jutten, Bar-Ness proposed an equivalent method (Bar-Ness, 1982). Giannakis and coworkers (1989) addressed the issue of identifiability of ICA, using cumulants of the third order. Higher-order statistics were used by Ruiz and Laucoume (1989), and by Gaeta and Laucoume (1990), who introduced the maximum likelihood method for the estimation of the mixing matrix. The algebraic method introduced by Cardoso (1989) and Cardoso and Souloumiac (1991) is based on the properties of the fourth-order cumulants. Inouye and Matsui (1989) proposed an innovative solution to the problem of separation of two variables. At the same time, Comon (1994) proposed a method for separation of N sources, whereas Fety (1988) was the first to study source separation for a dynamic problem.

15.3.1.1 Principal Components Analysis (PCA). The decomposition into principal components provides the representation of a set of signals $x(t)$ as linear combinations of orthogonal components $y(t)$ (called principal components) to be determined. For consistency reasons, signals $x(t)$ are called observations here, even though PCA does not require $x(t)$ to be a mixture of any sources. Principal components $y(t)$ are directly related to the observations $x(t)$ and are chosen to minimize the mean square error (MSE):

$$\frac{1}{T} \sum_{k=1}^{M} \int_{0}^{T} \left| x_k(t) - \sum_{i=1}^{m} c_{ki} y_i(t) \right|^2 dt \qquad (15.4)$$

where T is the observation interval and $c_{ki} y_i(t)$ is the ith approximation of the kth observation by the ith principal component $y_i(t)$. An iterative method to obtain the principal components results directly from their definition (Equation 15.4) and is based on the following iterative steps.

1. Compute the first principal component, minimizing the sum of the M mean square errors in Equation 15.4.
2. Compute the second principal component under the constraint of being orthogonal to the previous one(s).
3. Repeat step 2 until M principal components are reconstructed.

The scalar weights c_{ki} form orthonormal directional vectors $\mathbf{c}_k = [c_{k1}, c_{k2}, c_{kM}]$, called principal directions in M-dimensional space of observations. Exploiting the orthonormal property of principal directions, it is possible to prove that the minimum mean square error of Equation 15.4 is equal to the sum of variances of remaining $M - m$ principal components:

$$P_k = \sum_{i=1}^{n} c_{ik}^2 \qquad (15.5)$$

Application of abovementioned PCA procedure to a pair of surface EMG (sEMG) signals is illustrated in Figure 15.1. sEMG signals were recorded at the surface of the skin, above the biceps brachii muscle. Pickup electrodes (sensors) were positioned close to each other in a linear array structure (interelectrode distance of 5 mm) and acquired electrical signals form approximately the same group of muscle fibers. As a result, both sEMG signals x_1 and x_2 are highly correlated, as demonstrated by joint vector space representation in panel c. PCA finds the directions of maximal vari-

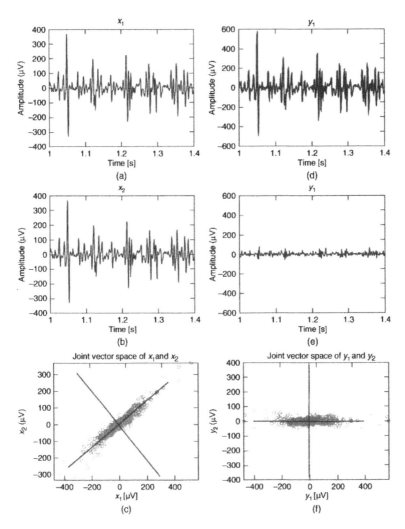

Figure 15.1. Application of PCA to the pair of surface EMG signals. Signals x_1 in panel (a) and x_2 in panel (b) were recorded by close-by sets of electrodes, placed over the skin above the biceps brachii muscle during a low-level contraction. Both electrode systems detected electrical activity of approximately the same group of muscle fibers. As a result, the signals x_1 and x_2 exhibit a high level of redundancy. As demonstrated by the joint vector space presentation in panel (c), more than 90% of variance is in the first principal direction (i.e., the direction of the first principal component). Each circle in (c), depicts a pair of values $x_1(t)$, $x_2(t)$ at a fixed time t. The first principal direction is denoted by a black dashed line, the second principal direction by a black dotted line. The two principal components $y_1(t)$ in panel (d) and $y_2(t)$ in panel (e) were reconstructed by projecting the observations $x_1(t)$ and $x_2(t)$ on the subspaces spanned by principal directions. The first principal component $y_1(t)$ resembles the main dynamics in the observations $x_1(t)$ and $x_2(t)$, whereas $y_2(t)$ can be interpreted as a low noise uncorrelated with $y_1(t)$. Panel (f) depicts the joint vector space representation of the principal components $y_1(t)$, $y_2(t)$ after rotation of the axes depicted in (c).

ance (so-called principal directions) and projects the observations x_1 and x_2 on these directions to reconstruct the principal components y_1 and y_2. In Figure 15.1c, the first principal direction is represented by a black dashed line, the second principal direction by a black dotted line. Reconstructed principal components (projections to the principal directions) are depicted in panels d and e. Note that, due to high level of redundancy in observations x_1 and x_2, more than 90% of total power is stored in the first principal component y_1.

According to Equation 15.4 and Figure 15.1, the first principal direction is a direction of maximum variance. This suggests a second PCA computation technique. Let $\mathbf{w}_1$ be the unit norm weight vector representing the first principal direction of observations $\mathbf{x}(t)$. By definition, the linear combination $\mathbf{w}_1^T \mathbf{x}$ is the first principal component with the maximum variance. The weight vector $\mathbf{w}_1$ can then be obtained as

$$\mathbf{w}_1 = \arg\max_{\|\mathbf{w}\|=1} E\left[\left(\mathbf{w}^T \mathbf{x}\right)^2\right] \tag{15.6}$$

Afterward, the projection of $\mathbf{x}$ on the subspace spanned by already reconstructed principle directions is calculated as $\mathbf{x} - \sum_{i=1}^{k-1}(\mathbf{w}_i^T \mathbf{x})\mathbf{w}_i$, and the kth ($k \geq 2$) principal direction is calculated as

$$\mathbf{w}_k = \arg\max_{\|\mathbf{w}\|=1} E\left[\left(\mathbf{w}^T\left(\mathbf{x} - \sum_{i=1}^{k-1}(\mathbf{w}_i^T \mathbf{x})\mathbf{w}_i\right)\right)^2\right] \tag{15.7}$$

This procedure is then repeated for all the remaining principal directions. Strictly speaking, principal directions reveal the directions of the maximum variance of M-dimensional random process. In the case of deterministic signals, we say that the principal directions reveal the directions of the maximum power in observations $\mathbf{x}(t)$.

Principal components, as introduced so far, reveal their usefulness in data compression, but their connection to the problem of source separation is weak. In Section 15.3.2, we show that principal components of the observations $\mathbf{x}(t)$ are associated with the sources $\mathbf{s}(t)$ by an unknown rotation matrix. The method for the estimation of this unknown rotation matrix is described in Section 15.4.2, where a biomedical application of a PCA-based BSS method is discussed.

15.3.1.2 Independent Component Analysis (ICA). Now, assume the source signals $\mathbf{s}(t)$ in Equation 15.3 are random processes. In ICA, source separation is achieved by additionally supposing that the source signals are statistically independent, instead of being just uncorrelated (PCA). Different measures of independence can be introduced, giving rise to different ICA methods.

When the number of observations M is greater than the number of sources N, the source signals can be estimated by applying the separation matrix $\mathbf{Q}$ to observations $\mathbf{x}(t)$:

$$\mathbf{y}(t) = \mathbf{Q}\mathbf{x}(t) \tag{15.8}$$

where $\mathbf{Q}$ is generally unknown. Neglecting the influence of noise, for $\mathbf{y}(t)$ to be equal to the original sources $\mathbf{s}(t)$, we should have $\mathbf{Q} = \mathbf{A}^{\#}$, where $\#$ indicates the matrix pseudoinverse (see the Appendix). As $\mathbf{A}$ is unknown, additional assumption of independence of the source signals is required. One of the most intuitive ways of realizing how the assumptions on statistical independence can be used to estimate the separation matrix $\mathbf{Q}$ is based on the central limit theorem, which guarantees that the linear combination of independent non-Gaussian random variables has a distribution that is closer to a Gaussian than the distribution of any individual variable. This implies that the samples of the vector of observations $\mathbf{x}(t)$ are more Gaussian than the samples of the vector of sources $\mathbf{s}(t)$. Thus, the source separation can be based on minimisation of Gaussianity of reconstructed sources $\mathbf{y}(t)$. All that we need is a measure of non-Gaussianity, which is used as an objective function by a given numerical optimization technique. Many different measures of Gaussianity have been proposed. Some of them are briefly summarized in the following.

1. *Kurtosis.* Kurtosis of a zero-mean random variable v is defined as

$$K(v) = E[v^4] - 3E[v^2]^2 \tag{15.9}$$

where $E[]$ stands for mathematical expectation. For a Gaussian variable v, $E[v^4] = 3E[v^2]^2$ and kurtosis of a Gaussian variable is 0. For most non-Gaussian distributions, kurtosis is nonzero (either positive or negative). Variables with positive kurtosis are called super-Gaussian (a typical example is the Laplace distribution). They have a more spiky distribution, with heavy tails and more pronounced peaks with respect to a Gaussian distribution. Variables with negative kurtosis are called sub-Gaussian, and have distribution that is flatter than Gaussian. A typical example of sub-Gaussian distribution is a uniform distribution. Being based on the fourth-order statistic, kurtosis is very simple to compute, but is highly sensitive to outliers. Its value might be significantly influenced by a single sample with a large value. Hence, it is not appropriate for separation of noisy measurements and measurements with severe signal artifacts.

2. *Negentropy.* Given the covariance matrix of a multidimensional random variable, negentropy is defined as the difference between the entropy of a Gaussian variable with the same covariance matrix and that of the considered random variable. It vanishes for Gaussian distributed variables and is

positive for all other distributions. From a theoretical point of view, negentropy is the best estimator of Gaussianity (in the sense of minimal mean square error of the estimators), but has a high computational cost as it is based on estimation of the probability density function of unknown random variables. For this reason, it is often approximated by kth order statistics, where k is the order of approximation (Hyvarinen, 1998; Jones and Sibson, 1987).

3. *Mutual Information.* Another method for source separation by ICA is associated with information theory. Mutual information between M random variables is defined as

$$I(y_1,...,y_m) = \sum_{i=1}^{m} H(y_i) - H(\mathbf{y}) \qquad (15.10)$$

where $\mathbf{y} = [y_1 \ldots y_m]$ is a M-dimensional random vector. Information entropy H of a discrete random vector $\mathbf{y}$ is defined as $H(\mathbf{y}) = \sum_i - P(\mathbf{y} = \mathbf{a}_i) \log P(\mathbf{y} = \mathbf{a}_i)$, where $\mathbf{a}_i$ are the possible values of $\mathbf{y}$. For a continuous random variable with probability density $f(\mathbf{y})$, entropy H is defined as $H(\mathbf{y}) = -\int_{-\infty}^{\infty} f(\mathbf{y}) \log [f(\mathbf{y})]d\mathbf{y}$. Mutual information is always nonnegative and equals zero only when variables $y_1 \ldots y_m$ are independent. It is possible to prove (Hyvarinen, 2000) that mutual information of variables with unitary variance is equivalent to negentropy except for the sign and a constant term, that is, maximization of negentropy is equivalent to minimization of mutual information.

Mutual information is also related to Kullback–Leibler divergence, defined as (Hyvarinen, 1999)

$$\delta(f,g) = -\int_{-\infty}^{\infty} f(\mathbf{y}) \log[f(\mathbf{y}) / g(\mathbf{y})]d\mathbf{y} \qquad (15.11)$$

which can be seen as a measure of a distance between probability density functions f and g. Kullback–Leibler divergence is always nonnegative and vanishes if and only if the probability densities f and g are equal. In ICA, Kullback–Leibler divergence measures the distance between the density $f(\mathbf{y})$ and the factorized density $g(\mathbf{y}) = f_1(y_1)f_2(y_2) \ldots f_n(y_n)$, where $f_i(y_i)$ is the marginal density of variable y_i. Mutual information and Kullback–Leibler divergence share the same practical drawbacks as negentropy. To use them in practice, we need to somehow approximate mutual entropy of unknown random variables. As a result, although theoretically different, all three measures of non-Gaussianity (i.e., mutual information, Kullback–Leibler, and negentropy) lead to essentially the same ICA algorithms.

4. *Maximum Likelihood Estimation.* Another well-known method to estimate the independent components is maximum likelihood (ML) estima-

tion. The ML approach is based on the log-likelihood function (i.e., logarithm of likelihood), defined as (Pham et al., 1992)

$$L = \sum_{t=1}^{T} \sum_{i=1}^{n} \log\{f_i[\mathbf{q}_i^T \mathbf{x}(t)]\} + T \log|\det(\mathbf{Q})| \qquad (15.12)$$

where time interval is discretized into T samples, $\mathbf{q}_i^T$ is the ith row of $\mathbf{Q}$, and f_i is the probability density of the ith source signal (f_i is assumed to be known). Likelihood can be represented as Kullback–Leibler divergence between the actual density of observations and the factorized density of source signals. Thus, the ML approach is essentially equivalent to minimization of mutual information.

There is an important limitation of the ICA method. As already indicated by the listed measures of non-Gaussianity, Gaussian variables are not separable by ICA (Comon, 1994). Indeed, an M-dimensional Gaussian distribution is invariant to any M-dimensional orthonormal transformation. Thus, two or more linearly combined Gaussian variables are not separable by ICA. The same applies to deterministic source signals with Gaussian distribution. ICA can separate them only if at most one source signal has a Gaussian distribution.

At the end of Section 15.3.1.1, we stated that PCA allows description of a set of statistical data (or a set of deterministic signals) using uncorrelated components (i.e., random variables or deterministic signals). Since PCA transformation is orthonormal, the variance (in the case of statistical data) or power (in the case of deterministic signals) of the observations is preserved by principal components (see Section 15.3.2). ICA is also useful to explore statistical data (deterministic signals). It provides independent random variables (independent deterministic signals) that preserve the information contained in the observations. In the following, two applications of PCA and ICA methods to the mixing model (Equation 15.3) are discussed.

15.3.2 Algebraic PCA Method: Application to an Instantaneous Mixing Model

The algebraic method for the computation of principal components is based on the correlation matrix of observations $\mathbf{x}(t)$:

$$\hat{\mathbf{R}}_{\mathbf{xx}} = \begin{bmatrix} r_{11} & \cdots & r_{1m} \\ \vdots & \ddots & \vdots \\ r_{m1} & \cdots & r_{mm} \end{bmatrix} \qquad (15.13)$$

where $r_{ij} = (1/T)\int_0^T x_i(t)x_j(t)dt$ (continuous signals) or $r_{ij} = (1/T)\Sigma_{t=1}^{T} x_i(t)x_j(t)$

(sampled signals) is the correlation between the ith and the jth observations. Note that $\hat{R}_{xx}$ is real, positive, and symmetric. Now, assume the observations $x(t)$ are deterministic and follow the mixing model (Equation 15.3). Consider the singular value decomposition (see the Appendix) of the $M \times N$ matrix A:

$$A = V\Lambda^{1/2}U^T \qquad (15.14)$$

where $U_{N \times N}$ and $V_{M \times M}$ (matrix) are unitary matrixes of sizes $N \times N$ and $M \times M$ (i.e., $UU^T = I$, $V^TV = I$) and Λ is a diagonal $M \times N$ matrix with the N nonzero eigenvalues λ_i of AA^T on the diagonal. Without loss of generality, we can assume that λ_i are arranged in decreasing order. The diagonal form of the correlation matrix $\hat{R}_{xx}$ (for the sampled signal) is given by

$$\hat{R}_{xx} = \frac{1}{T}\sum_{t=1}^{T} x(t)x^T(t) = \frac{1}{T}\sum_{t=1}^{T}[As(t)+n][s^T(t)A^T+n^T] = \qquad (15.15)$$

$$AA^T + I\sigma_n^2 = V\Lambda V^T + I\sigma_n^2$$

where $I\sigma_n^2$ is the covariance matrix of the noise (which, given the adopted assumptions on the white noise, is equal to the identity matrix multiplied by the noise variance). In Equation 15.15, the normalization of the source signals to the unit norm and the notion of uncorrelatedness of the source signals and noise were used. It is worth noticing that the eigenvalues of $\hat{R}_{xx}$ sum up to the total power of observations $x(t)$.

Now, neglect the influence of noise and consider the relation between the eigenvectors of $\hat{R}_{xx}$ and the principal components $y(t)$. Assume a signal $y_k(t)$ is a linear combination of the observations $x(t)$ [trace($\hat{R}_{xx}$) = trace(Λ)]. Then $y_k(t)$ can be expressed as a linear combination of eigenvectors of $\hat{R}_{xx}$ (completeness property of the eigenvectors of the correlation matrix) multiplied by a unit norm vector c:

$$y_k(t) = \sum_{i=1}^{m} c_i v_i x_i(t) = c^T V^T x(t) \qquad (15.16)$$

where v_i^T is the ith eigenvector of $\hat{R}_{xx}$ and $c = [c_1, \ldots, c_M]^T$. The power of the signal $y_k(t)$ can be expressed as

$$\sum_{t=1}^{T} y_k(t)y_k^T(t) = c^T V^T \sum_{t=1}^{T} x(t)x^T(t)Vc = c^T V^T V\Lambda V^T Vc = \sum_{i=1}^{m} c_i^2\lambda_i \qquad (15.17)$$

The right-most sum in Equation 15.17 is a convex combination (linear combination with unitary sum of weights) of the eigenvalues, which takes a maximum at $c_i = \delta(i - k)$ [$\delta(i - k)$ denoting the Kronecker delta]. Thus, the first eigenvector of $\hat{R}_{xx}$ indicates the direction of the maximum power (or variance) of the observations, which is, by definition, the first principal direction. The corresponding eigenvalue λ_i gives the power (variance) of the first principal component $\mathbf{V}_1^T\mathbf{x}(t)$. By repeating this procedure and limiting it to the subspace of eigenvectors $\mathbf{V}_2$ to $\mathbf{V}_M$, the second principal direction is found to be aligned with eigenvector $\mathbf{V}_2$, the third principal direction is aligned with $\mathbf{V}_3$, and so on. Therefore, the eigenvectors $\mathbf{V}$ of correlation matrix $\hat{\mathbf{R}}_{xx}$ reveal the principal directions of observations $\mathbf{x}(t)$.

It is worth noticing that a complete computation of the mixing matrix $\mathbf{A}$ requires not only the matrices $\mathbf{\Lambda}$ and $\mathbf{V}$, but also the unitary matrix $\mathbf{U}$ (known as the rotation matrix; see Section 15.4.2). As only $\mathbf{U}$ is estimated by PCA, principal components are not sufficient to reconstruct the original source signals.

Note also that the $M \times M$ matrix $\hat{\mathbf{R}}_{xx}$ has dimension larger or equal to that of $\mathbf{A}$ ($M \times N$) because $M > N$. According to Equation 15.15, the first N eigenvalues and eigenvectors of $\hat{\mathbf{R}}_{xx}$ provide information on the N source signals, whereas the additional $M - N$ eigenvalues provide information about the power (variance) of noise. This property will be used by the ICA approach in Section 15.4.2.

15.3.3 Neural ICA Method: Application to Instantaneous Mixing Model

Indicating with $\mathbf{y}(t) = \mathbf{Q}\mathbf{x}(t)$ an estimate of the source signals in Equation 15.3, the aim of ICA methods is to compute an estimate $\mathbf{A}^s$ of the mixing matrix $\mathbf{A}$ such that $\mathbf{x}(t) = \mathbf{A}^s\mathbf{y}(t)$. The algebraic PCA method, discussed in the previous section, relies on a well-known technique of eigenvalue decomposition (see Appendix) and requires a priori knowledge of all the samples of observations $\mathbf{x}(t)$. This knowledge is not assumed by neural methods, which are based on an intrinsically different approach that allows real-time implementations. Neural techniques utilize iterative updates of weights $\mathbf{A}^s$ in order to achieve convergence to the minimum of a predefined functional $F[\mathbf{A}^s]$. $F[\mathbf{A}^s]$ measures the aforementioned distance among the different estimates of the source signals $\mathbf{y}(t)$ and attains the minimum when $\mathbf{y}(t) = \mathbf{s}(t)$. Possible choices for $F[\mathbf{A}^s]$ are those listed in Section 15.3.1.

The weights $\mathbf{A}^s$ are updated iteratively (Karhunen et al., 1997). In each iteration step, the value of the functional $F[\mathbf{A}^s]$ is decreased [i.e., the distance between the $\mathbf{y}(t)$ estimates of the source signals is increased]. A widely used numerical minimization technique is the gradient descent algorithm (or stochastic gradient descent algorithm in the case of random processes), for which the weights are updated in the direction opposite to the gradient of $F[\mathbf{A}^s]$:

$$\mathbf{A}_{n+1}^s = \mathbf{A}_n^s - \mu_n \nabla F[\mathbf{A}^s] \tag{15.18}$$

where the notation ∇ is just shorthand for gradient. Performances and convergence of the minimization method depend (usually with opposite direction) on the parameter λ_n, called the learning rate, which determines the decrease of the weights in the opposite direction to the gradient. The learning rate is usually chosen to be adaptive, with smaller values close to the minimum of the functional $F[A^s]$.

As an example of application, we discuss the recursive neural network architecture introduced by Hérault and Jutten (1991) (Figure 15.2), with the aim to separate two sources from two observations:

$$\begin{cases} x_1(t) = a_{11}s_1(t) + a_{12}s_2(t) \\ x_2(t) = a_{21}s_1(t) + a_{22}s_2(t) \end{cases} \tag{15.19}$$

where both $x_i(t)$ and $s_i(t)$ are signals with T samples. Every neuron receives the sequence of samples of observation $x_i(t)$ as an input. The outputs of all neurons are connected to all the inputs of the neurons with $j \neq i$ weighted by a scalar weight c_{ji} (Figure 15.2). The estimate of the output $\mathbf{y}(t) = \mathbf{Q}\mathbf{x}(t)$ is obtained by the separation matrix $\mathbf{Q} = (\mathbf{I} + \mathbf{C})^{-1}$, where $\mathbf{C}$ is the matrix of weights c_{ji}. In the case of two sources considered in Equation 15.19 we have

$$\begin{cases} y_1(t) = x_1(t) - c_{12}y_2(t) \\ y_2(t) = x_2(t) - c_{21}y_1(t) \end{cases} \tag{15.20}$$

from which the following relation between source signals $\mathbf{s}(t)$ and their estimates $\mathbf{y}(t)$ arises:

$$\begin{cases} y_1(t) = \dfrac{1}{1 - c_{12}c_{21}} \left[\left(a_{11} - c_{12}a_{21} \right) s_1(t) + \left(a_{12} - c_{12}a_{22} \right) s_2(t) \right] \\ y_2(t) = \dfrac{1}{1 - c_{12}c_{21}} \left[\left(a_{21} - c_{21}a_{11} \right) s_1(t) + \left(a_{22} - c_{21}a_{12} \right) s_2(t) \right] \end{cases} \tag{15.21}$$

For the estimates to be proportional to the source signals, the following relations must hold:

$$a_{12} - c_{12}a_{22} = 0, \quad a_{21} - c_{21}a_{11} = 0 \tag{15.22}$$

In such a case, y_1, y_2 are proportional to s_1, s_2, respectively. Assuming instead

$$a_{11} - c_{12}a_{21} = 0, \quad a_{22} - c_{21}a_{12} = 0 \tag{15.23}$$

y_1, y_2 are proportional to s_2, s_1, respectively. The convergence of the method to either Equations 15.22 or 15.23 is associated with the ICA am-

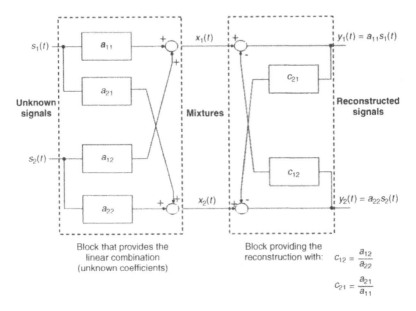

Figure 15.2. Iterative neural network architecture, introduced by Hérault and Jutten (1991), for separation of two sources $s_1(t)$ and $s_2(t)$ out of two observations $x_1(t)$ and $x_2(t)$. Two processing blocks are depicted: the mixing process (left) and separation algorithm (right). The separation block consist of two neurons (one per each source). Each neuron receives the samples of both observations, $x_1(t)$ and $x_2(t)$ as inputs. The output of each neuron is multiplied by a weight $c_{ij}, j \neq i$, and fed back to the input of the other neuron. The estimate of the output $y(t) = Qx(t)$ is obtained with a separation matrix $Q = (I + C)^{-1}$, where C is the matrix with the weights c_{ij} (see the text for details). This method is suitable for real-time implementation.

biguity on the order of the reconstructed sources. Convergence actually occurs only for one of the solutions Equations 15.22 and 15.23, as only one of the solutions is stable. Indeed, the condition that the loop gain $c_{21}c_{12}$ is less than 1 can be satisfied only by one of the solutions (Jutten and Hérault, 1991). For the gain $c_{21}c_{12} > 1$, the method diverges.

The theoretical study of M sources in Equation 15.3 is simply an extension of the previous example with two sources. In the case of M sources, the learning rule is based on the gradient method as discussed in the following. Assume that the first $M - 1$ sources are already determined (up to the multiplicative constant):

$$\begin{cases} y_1(t) = a_{11}s_1(t) \\ \quad\vdots \\ y_{M-1}(t) = a_{M-1,M-1}s_{M-1}(t) \end{cases} \tag{15.24}$$

Substituting Equation 15.24 into the expression for the estimation of the
Mth source $y_M(t) = x_M(t) - \sum_{i=1}^{M-1} c_{Mi} y_i(t)$ and considering that $x_M(t) = \sum_{i=1}^{M} a_{Mi} s_i(t)$ we get

$$y_M(t) = \sum_i \left(a_{Mi} - c_{Mi} a_{ii} \right) s_i(t) + a_{MM} s_M(t) \qquad (15.25)$$

In order to estimate the Mth source as a function $y_M(t)$ proportional to it,
the weights c_{Mi} must be chosen so that the first term on the right-hand side
of Equation 15.25 vanishes. By using the assumption of the uncorrelated
source signals we have

$$E[y_M(t)^2] = \sum_i \left(a_{Mi} - c_{Mi} a_{ii} \right)^2 E[s_i(t)^2] + a_{MM}^2 E[s_M(t)^2] \qquad (15.26)$$

Thus, $E[y_M(t)^2]$ can be considered as the functional $F[\mathbf{A}^s]$ because its minimum is attained when the estimated source signal $y_M(t)$ is proportional to
the source $s_M(t)$. Applying the gradient method to the functional $E[y_M(t)^2]$
we have

$$c_{ij}^k = c_{ij}^{k-1} + \mu_k E[y_i(t) y_j(t)] \qquad i \neq j \qquad (15.27)$$

where k is the iteration step and μ_k is the positive constant determining the
learning rate (i.e., the increment of the weights). In the case of the stochastic gradient method, the same equation (15.27) is obtained, but without the
expectation operator.

There are infinite solutions corresponding to noncorrelated sources
(Jutten and Hérault, 1991; Jutten, 1987), but only one for which the
sources are statistically independent. Thus, the rule must be modified so
that the method converges to the unique solution corresponding to statistically independent sources. A further problem in the learning rule (Equation 15.27) is related to its symmetry: coefficients c_{12} and c_{21} vary in the
same way; the solution to which the method converges is correct only if
the mixing matrix $\mathbf{A}$ is symmetric. To avoid these problems, the learning
rule (Equation 15.27) is substituted with

$$c_{ij}^k = c_{ij}^{k-1} + \mu_k E[f(y_i(t)) g(y_j(t))] \qquad i \neq j \qquad (15.28)$$

where f and g are two different nonlinear, even functions (in order to
break symmetry) with the same sign (in order for their product to have the
same sign as $E[y_i(t)y_j(t)]$), and the direction opposite to that of the gradient is taken. It is possible to prove that, in the case of source signals with
symmetrical probability densities, if the iterative rule (Equation 15.28)
converges, the obtained estimates $y_i(t)$, $y_j(t)$ are statistically independent.

15.4   APPLICATIONS

In this section, examples of BSS applications to the surface electromyographic (EMG) signals are discussed. First, a short overview of the electrical activity of human muscle is outlined. The main focus is on generation of electrical potentials in muscle fibers and on their asynchronous merging into the detectable EMG interference patterns. The descriptions provided should serve only as a coarse introduction to the field of electromyography. More advanced descriptions can be found in Merletti and Parker (2004).

15.4.1   Physiology of Human Muscles

Human muscles consist of 10 to 150 mm long and 5 to 90 μm diameter muscle fibers that are attached to the bones by tendons. Each muscle fiber is innervated by a single motoneuron that transmits the control commands from the central nervous system (CNS) in a form of firing pulse trains. Several muscle fibers are innervated by the same motoneuron, forming a basic functional unit of the muscles called the motor unit (MU). The number of fibers in each MU varies considerably within the same muscle and even more between different muscles. Typically, muscles are comprised of several tens to several hundreds of MUs.

Electrical signals sent by the CNS propagate along a nerve fiber and terminate in a neuromuscular junction (NMJ) where they excite membranes of all innervated muscle fibers. Every pulse in a motoneuron induces a local depolarization of the transmembrane potential of each muscle fiber, called the single-fiber action potential (SFAP). The depolarized zone (i.e., SFAP) propagates without attenuation along the muscle fiber from the NMJ to the tendon endings, causing the muscle fiber to contract. The sum of single-fiber action potentials corresponding to all the fibers of a single motor unit is called the motor unit action potential (MUAP).

Several tens of MUs are simultaneously active in the muscle tissue. Their MUAPs superimpose in time and space and form a highly complex interference pattern of EMG, which can be detected either within the muscle (with needle electrodes) or on the skin above the investigated muscle (with surface electrodes). The technical difficulties associated with interpretation of recorded EMG interference patterns limit the accuracy and diagnostic value of the EMG in practice and generate many source separation problems. In the following, only two representative examples of source separation are discussed. The first example deals with the problem of separation of EMG signals from different muscles (so-called muscle crosstalk). The second example addresses identification of single MU discharge patterns, that is, decomposition of the surface EMG into constituent MUAP trains.

15.4.2   Separation of Surface EMG Signals Generated by Muscles Close to Each Other (Muscle Crosstalk)

An important artifact of surface EMG signals is crosstalk from nearby muscles. Crosstalk is the signal detected over a muscle but generated by a nearby muscle (Figure 15.3). This complex phenomenon depends on the properties of the propagating medium (i.e., subcutaneous tissue interposed between the muscle fibers and the detection electrodes) and on sources (i.e., the firing patterns of active MUs). The exact physical properties of interposed subcutaneous tissue are not known; hence, as little as possible a priori information on the communication channel is assumed.

Crosstalk signals can be superimposed on the signal of interest both in the time and frequency domains and represents a serious problem for surface EMG. The distinction of signals generated by muscles close to each other is, hence, an example of a very important source-separation problem. By assuming that the EMG signals generated by different muscles are statistically independent, the problem can be addressed by ICA techniques.

One of the first applications of BSS to the problem of crosstalk was proposed by Farina and coworkers (2004). Their work is based on a separation algorithm called SOBI (second-order blind identification), introduced by Belouchrani and coworkers (1997). SOBI extends the PCA

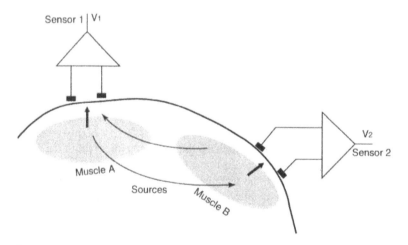

Figure 15.3. Representation of two muscles and two detection systems for surface EMG signals. Each detection system acquires both the EMG signal of the muscle over which the electrodes are placed and the EMG signal produced by the other, nearby muscle. This phenomenon is known as muscle crosstalk. Its linear instantaneous model is shown in Figure 15.2. A third detection system in an intermediate position could record a third mixture of the two sources, as shown in Figure 15.4.

method presented in Section 15.3.2 and consists of two steps: (1) whitening and (2) assessment of the unknown rotation matrix $\mathbf{U}$. The linear instantaneous mixing model (Equation 15.3) is assumed. Although not completely accurate, this model enables reasonably good approximation of surface EMG signals, especially when the investigated muscles are close to each other.

Step 1—Whitening. Spatial whitening of the observations $\mathbf{x}(t)$ (decorrelation in space) follows the procedure for estimation of the principal components in Section 15.3.2. The $N \times M$ matrix $\mathbf{W}$ is determined such that

$$\mathbf{WAA}^T\mathbf{W}^T = \mathbf{I} \tag{15.29}$$

By definition in Equation 15.29, matrix $\mathbf{WA} = \mathbf{U}$ is unitary. Application of $\mathbf{W}$ to the observations $\mathbf{x}(t)$ yields the so-called whitened observations $\mathbf{z}(t)$:

$$\mathbf{z}(t) = \mathbf{Wx}(t) = \mathbf{Us}(t) + \mathbf{Wn}(t) \tag{15.30}$$

By analogy with the procedure in Section 15.3.2, matrix $\mathbf{W}$ can be determined from the covariance matrix of observations $\mathbf{x}(t)$:

$$\hat{\mathbf{R}}_{\mathbf{xx}} = \frac{1}{T}\sum_{t=1}^{T}\mathbf{x}(t)\mathbf{x}^T(t) \tag{15.31}$$

which can be factorized as

$$\hat{\mathbf{R}}_{\mathbf{xx}} \approx \mathbf{A}\hat{\mathbf{R}}_{\mathbf{ss}}\mathbf{A}^T + \sigma^2\mathbf{I}_n \tag{15.32}$$

As the sources $\mathbf{s}(t)$ are uncorrelated, the covariance matrix $\hat{\mathbf{R}}_{\mathbf{ss}}$ is diagonal. Furthermore supposing all the sources of unitary power (ICA ambiguity on power of sources) $\hat{\mathbf{R}}_{\mathbf{ss}}$ can be made equal to identity. Under these assumptions, Equations 15.29 and 15.32 indicate that matrix $\mathbf{W}$ diagonalizes the matrix $\hat{\mathbf{R}}_{\mathbf{xx}}$. Thus, $\mathbf{W}$ and σ^2 can be computed from the eigenvalues and eigenvectors of $\hat{\mathbf{R}}_{\mathbf{xx}}$. First, an estimate $\hat{\sigma}^2$ of the variance of the noise is obtained from the average of the $M - N$ smallest eigenvalues of matrix $\hat{\mathbf{R}}_{\mathbf{xx}}$ (Section 15.3.2). Second, given the N greatest eigenvalues $\lambda_1, \ldots, \lambda_n$ and the correspondent eigenvectors $\mathbf{V}_1, \ldots, \mathbf{V}_M$ of $\hat{\mathbf{R}}_{\mathbf{xx}}$, $\mathbf{W}$ is given by

$$\mathbf{W} = [(\lambda_1 - \hat{\sigma}^2)^{-1/2}\,\mathbf{V}_1, \ldots, (\lambda_n - \hat{\sigma}^2)^{-1/2}\,\mathbf{V}_M]^T \tag{15.33}$$

Note that, although closely related, whitening by matrix $\mathbf{W}$ extends the PCA method described in Section 15.3.2 as it scales the whitened components $z_i(t)$ by a factor $(\lambda_i - \hat{\sigma}^2)^{-1/2}$ to make them of unit norm (a property not required by PCA). In order to estimate the matrix $\mathbf{A}$, the unitary matrix $\mathbf{U}$ must be estimated by a rotation operation in the second step.

Step 2—Rotation. From the matrix factorization $\mathbf{U} = \mathbf{WA}$, we have

$$\mathbf{A} = \mathbf{W}^{\#}\mathbf{U} \tag{15.34}$$

Thus, given the whitening matrix $\mathbf{W}$, the mixing matrix $\mathbf{A}$ can be determined by estimating the matrix $\mathbf{U}$. As $\mathbf{U}$ is unitary, it can be considered as an N-dimensional rotation matrix and estimated by the joint-diagonalization procedure (Belouchrani et al., 2001) of the correlation matrices of whitened observations $\mathbf{z}(t)$. From the definition of the covariance matrix,

$$\hat{\mathbf{R}}_{zz}(\tau) = \frac{1}{T} \sum_{t=1}^{T} \mathbf{z}(t)\mathbf{z}(t + \tau)^{T} \tag{15.35}$$

we have

$$\hat{\mathbf{R}}_{zz}(\tau) \approx \mathbf{U}\,\hat{\mathbf{R}}_{ss}(\tau)\mathbf{U}^{T} \qquad \tau \neq 0 \tag{15.36}$$

For nonzero lags τ the contribution of the temporarily white Gaussian noise $\mathbf{n}(t)$ vanishes. As a result, $\mathbf{U}$ can be determined from any matrix $\hat{\mathbf{R}}_{zz}(\tau)$ at nonzero lag τ. A more stable procedure consists in choosing a number of matrices $\hat{\mathbf{R}}_{zz}(\tau)$ for different values of $\tau \neq 0$ and determining the matrix $\mathbf{U}$ as a "best joint diagonalizer" of the set of selected matrices. By "best joint diagonaliser" we refer to a matrix that makes the matrices $\hat{\mathbf{R}}_{zz}(\tau)$ as close to diagonal as possible. Ideally, $\mathbf{U}$ diagonalizes all the matrices $\hat{\mathbf{R}}_{zz}(\tau)$, but this is seldom the case, mainly due the noise. Therefore, a criterion to measure the goodness of joint diagonalization is required. A criterion of the choice is the sum of squares of off-diagonal elements of matrices $\mathbf{U}^{T}\hat{\mathbf{R}}_{zz}(\tau)\mathbf{U}$ (Belouchrani et al., 1997, 2001; Belouchrani and Amin, 1998):

$$\text{off}(\hat{R}_{..}) = \sum_{t=1}^{\Upsilon} \sqrt{\sum_{i} \sum_{j \neq i} \left| r_{ij}(\tau) \right|^{2}} \tag{15.37}$$

where $r_{ij}(\tau)$ denotes the (i,j)-th element of selected matrix $\hat{\mathbf{R}}_{zz}(\tau)$ for $\tau = 1$, ..., Υ. Equation 15.37 leads to implementation of the so-called Jacobi joint-diagonalization method (Cardoso and Souloumiac, 1996), estimating the matrix $\mathbf{U}$ as

$$U = \min_{\|U\|=1} \arg \left(\text{off}(U^{T}\hat{R}_{..}U) \right) \tag{15.38}$$

Once the mixing matrix A is known, the sources can be estimated as $\mathbf{y}(t) = \mathbf{A}^{\#}\mathbf{x}(t)$. Exact technical description of joint diagonalization is beyond the scope of this chapter. The interested reader is referred to Cardoso and Souloumiac (1996), Belouchrani et al. (1997), and Holobar et al. (2006).

An example of application of the SOBI algorithm to experimental sEMG signals (Farina et al., 2004) is shown in Figure 15.4 (experimental setup) and Figure 15.5 (reduction of crosstalk). The algorithm was applied to two forearm muscles that allow rotation and flexion of the wrist. The two muscles are very close to each other and it is impossible to separate their EMG activity with classical methods. The BSS algorithm was applied to three mixtures of signals detected over the two muscles and in an intermediate region, respectively. As demonstrated in Figure 15.5, the selectivity of the detection is improved when either a rotation or flexion of the wrist is executed.

15.4.3 Separation of Single Motor Unit Action Potentials from Multichannel Surface EMG

The second BSS application includes the decomposition of surface EMG signals into constituent MUAP trains. As explained in Subsection 15.4.1, the surface EMG is a compound signal comprising the contributions of different MUs. Even at moderate muscle contraction levels, many MUs contract asynchronously. Their MUAPs superimpose both in space and time

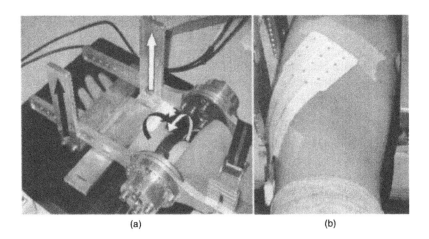

(a)   (b)

Figure 15.4. Experimental setup for the detection of surface EMG signals from two forearm muscles. The hand is fixed in an isometric brace measuring the force produced during rotation and flexion efforts. (a) The subject alternates wrist rotation and flexion efforts at regular time intervals. (b) EMG signal is detected with three electrode arrays placed over the pronator teres, the flexor carpi radialis, and between the two muscles. The signal detected over the pronator teres (which rotates the wrist) is not zero during flexion, even if this muscle is not active during this contraction (crosstalk signal from flexor carpi radialis muscle), and vice versa (see Figure 15.5). Reproduced with permission from Farina et al. (2004).

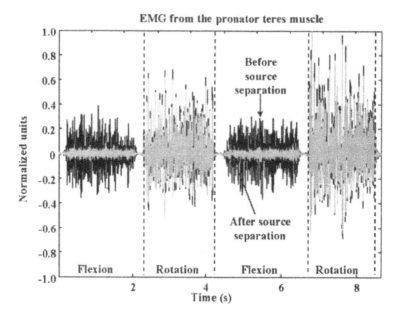

Figure 15.5. Application of source-separation technique to the three signals detected over the pronator teres, the flexor carpi radialis, and between the two muscles with the techniques described in Figure 15.4. These signals are the mixtures $x(t)$ of the source-separation problem. The SOBI algorithm provides the separation of the activity of the pronator teres, reducing the amplitude of the signal recorded over the pronator teres during flexion. The same holds for the flexor carpi radialis for the complementary time intervals (results not shown). Reproduced with permission from Farina et al. (2004).

and create complex interference patterns, which are very difficult to interpret. Nonetheless, surface EMG received remarkable attention over the past few decades and has become a mature measuring technique. The information extracted from the sEMG signals is currently being exploited in several different clinical studies mainly concerned with timing of muscle activation, EMG amplitude modulation, and electrical manifestations of fatigue.

The development of flexible high-density (HD) arrays of surface electrodes and multichannel amplifiers opened new possibilities of recording up to a few tens of sEMG signals over a single muscle. At the same time, source-separation techniques capable of processing and combining the information from such multichannel recordings emerged. De Luca and coworkers (2006) proposed the decomposition of four-channel sEMGs, whereas Kleine and coworkers (2007) demonstrated the importance of recording many sEMG signals over the skin surface for decomposition purposes. BSS methods have also been proposed. Garcia and coworkers (2005) and Nakamura and coworkers (2004) acquired sEMG signals with a linear array of surface electrodes, oriented transversally with respect to the muscle fibers, and demonstrated that, in this configuration and up to reasonable limitations, sEMG signals can be modeled as linear instantaneous mixtures (i.e., by Equation 15.3). On the other hand, Holobar and Zazula (2004) modeled sEMG signals as linear convolutive mixtures (i.e., by Equation 15.1) and proposed the convolution kernel compensation (CKC) decomposition technique. This technique proved to be highly accurate and robust, reconstructing MUAP trains of up to twenty MUs from multichannel sEMG recordings. In the following, the CKC decomposition technique is discussed in more detail.

15.4.3.1 *Convolution Kernel Compensation.* In the case of isometric muscle contractions, sampled sEMG signals $x(t)$ can be modeled as outputs of the convolutive linear mixing model (Equation 15.1):

$$x_i(t) = \sum_{j=1}^{N} \sum_{l=0}^{L-1} a_{ij}(l)s_j(t-l) + n_i(l) , i = 1, \ldots, M \qquad (15.39)$$

where $n_i(l)$ stands for zero-mean additive noise. Each model input $s_j(t)$ is modeled as binary pulse sequences, carrying the information about the MUAP activation times:

$$s_j(t) = \sum_{r=-\infty}^{\infty} \delta[t - T_j(k)], j = 1, \ldots, N \qquad (15.40)$$

where $\delta(\tau)$ denotes the Dirac impulse and $T_j(k)$ stands for the time instant in which the kth MUAP of the jth MU appeared. Activation times $T_j(k)$ are supposed to be random and statistically independent (though experimental observations show almost periodic discharge rates and correlated fluctuations of rate among different MUs). The channel response $a_{ij}(l)$, $l = 0$, $1, \ldots, L - 1$, corresponds to the L samples long the MUAP of the jth MU, as detected in the ith observation. The channel responses $a_{ij}(l)$ must be of limited time support, but can be of arbitrary shape. Hence, any physical property of the subcutaneous tissue can be taken into account.

Equation 15.39 can be rewritten in matrix form:

$$x(t) = A\bar{s}(t) + n(t) \qquad (15.41)$$

where $n(t) = [n_1(t), \ldots, n_M(t)]^T$ is a vector of white noise with a covariance matrix $\sigma_n^2 I$ and the mixing matrix A comprises all the MUAPs as detected by the different surface electrodes:

$$\mathbf{A} = \begin{bmatrix} a_{11}(0) & \cdots & a_{11}(L-1) & a_{12}(0) & \cdots & a_{12}(L-1) & \cdots \\ a_{21}(0) & \cdots & a_{21}(L-1) & a_{22}(0) & \cdots & a_{22}(L-1) & \cdots \\ \vdots & \cdots & \vdots & \vdots & \cdots & \vdots & \cdots \\ a_{M1}(0) & \cdots & a_{M1}(L-1) & a_{M2}(0) & \cdots & a_{M2}(L-1) & \cdots \end{bmatrix} \quad (15.42)$$

Vector $\bar{s}(t)$ stands for an extended form of sampled source signals:

$$\bar{\mathbf{s}}(t) = \left[s_1(t), s_1(t-1), \ldots, s_1(t-L+2), \ldots, s_N(t), \ldots, s_N(t-L+2) \right]^T \quad (15.43)$$

The CKC method (Holobar and Zazula, 2007) fully automates the identification of MU discharge sequences in Equation 15.41. In the first step, the cross-correlation vector $\mathbf{r}_{s_j x} = E[s_j(t)\mathbf{x}^T(t)]$ between the jth source signal and all the measurements is estimated (Holobar and Zazula, 2007). In the second step, the jth pulse train s_j is estimated by the linear minimum mean-square error (LMMSE) estimator:

$$\hat{s}_j(t) = \mathbf{r}_{s_j x} \hat{\mathbf{R}}_{xx}^{-1} \mathbf{x}(t) = \mathbf{r}_{s_j \bar{s}} \mathbf{A}^T \left[\mathbf{A}\hat{\mathbf{R}}_{\bar{s}\bar{s}} \mathbf{A}^T + \sigma_n^2 \mathbf{1} \right]^{-1} [\mathbf{A}\bar{s}(t) + \mathbf{n}(t)] \quad (15.44)$$

where $\hat{\mathbf{R}}_{xx} = E[\mathbf{x}(t)\mathbf{x}^T(t)]$ is the correlation matrix of measurements $\mathbf{x}(t)$, and $\mathbf{r}_{s_j \bar{s}} = E[s_j(t)\bar{s}^T(t)]$ is the vector of cross-correlation coefficients between the jth source and all the sources. By analogy with Subsection 15.4.2, $\hat{\mathbf{R}}_{\bar{s}\bar{s}} = \mathbf{I}$, whereas, in the case of statistically independent sources, $\mathbf{r}_{s_j \bar{s}} = [\delta(1-j), \delta(2-j), \ldots \delta(N-j)]$ equals the unit norm vector with the jth element equal to 1 and zeros elsewhere. When the influence of noise is neglected, the unknown mixing matrix $\mathbf{A}$ is compensated and Equation 15.44 simplifies to

$$\hat{s}_j(t) = \mathbf{r}_{s_j x} \hat{\mathbf{R}}_{xx}^{-1} \mathbf{x}(t) = \mathbf{r}_{s_j \bar{s}} \mathbf{A}^T \mathbf{A}^{-T} \hat{\mathbf{R}}_{\bar{s}\bar{s}} \mathbf{A}^{-1} \mathbf{A}\bar{s}(t) = \mathbf{r}_{s_j \bar{s}} \hat{\mathbf{R}}_{\bar{s}\bar{s}} \bar{s}(t) = s_j(t) \quad (15.45)$$

Equation 15.44 requires the cross-correlation vector $\mathbf{r}_{s_j x}$ to be known in advance. This is never the case and Holobar and Zazula (2007) proposed a probabilistic iterative procedure for its blind estimation. In the first iteration step, the unknown cross-correlation vector is approximated by vector of measurements $\mathbf{r}_{s_j x} = \mathbf{x}(t_1)$ where, without loss of generality, we assumed that the jth MU discharged at time instant t_1. Then, the first estimation of the jth source $s_j(t)$ is computed according to Equation 15.44. In the next step, the largest peak in $\hat{s}_j(t)$ is selected as the most probable candidate for the second discharge of the jth source, $t_2 = \max\arg[\hat{s}_j(t)]$, and the vector $\mathbf{r}_{s_j x}$ is updated as $\quad {}_{t_1 \neq t_2}$

$$\mathbf{r}_{s_j x} = \frac{\mathbf{r}_{s_j x} + \mathbf{x}(t_2)}{2} \quad (15.46)$$

This procedure is then repeated, until $\mathbf{r}_{s_j x}$ converges to a stable solution (Holobar and Zazula, 2007).

The CKC method inherently resolves MUAP superimpositions. Moreover, it implicitly combines all the available information provided by all the observations $x(t)$. By compensating for the shapes of the detected MUAPs (which are included in the mixing matrix A), it directly estimates the impulse sources without reconstructing the detected MUAP shapes. This significantly decreases the number of unknowns to be estimated in Equation 15.41 and reduces the computational time. When required, MUAP shapes can be estimated by spike-triggered averaging of sEMG signals, taking the MUAP activation times in $s_j(t)$ as triggers.

The problem with the CKC method is that the convolutive model (Equation 15.41) increases the number of sources $s(t)$ by the factor L. Thus, in order to decompose the sEMG signals, the number of observations must also be large (at least a few dozen). This calls for HD sEMG acquisition systems with at least several tens of pick-up electrodes arranged into a closely spaced two-dimensional grid. An example of CKC-based sEMG decomposition is illustrated in Figure 15.6.

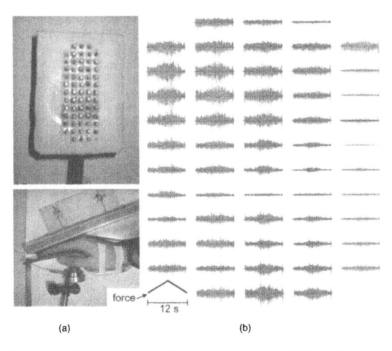

force

12 s

(a)   (b)

Figure 15.6. Experimental setup for detection of surface EMG signals from abductor pollicis brevis muscle. (a) Matrix of 64 surface electrodes, arranged into 13 lines and 5 columns and with the four corner electrodes missing (upper panel), and the isometric braces measuring the force produced during the abduction of the thumb (lower panel). (b) Example of recorded surface EMG signals (positions of the signals reflect the spatial organization of the pick-up electrodes)

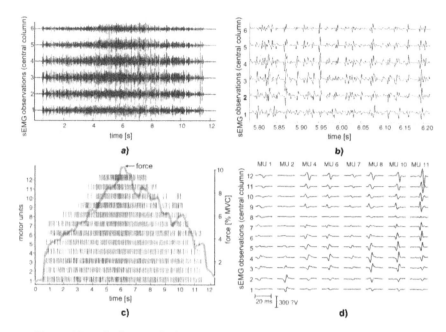

Figure 15.7. Surface EMG signals, recorded during a 6 s ramp-up (from 0% to 10% of maximum contraction level, MVC) and 6 s ramp-down (from 10% to 0% MVC) contraction of abductor pollicis brevis muscle and their decomposition into contributions of different motor units. (a) Surface EMG signals detected by the first six electrodes of the central column (Figure 15.6). (b) The same as in (a) with the portion of the signal zoomed in. (c) Discharge patterns of 12 identified motor units and their dependence on the exerted muscle force. Each vertical mark corresponds to a single motor unit discharge. (d) MUAP templates of eight different motor units, as reconstructed by a spike-triggered averaging of the sEMG signals from the central electrode column (Figure 15.6), taking the identified discharge patterns as triggers.

APPENDIX

Eigenvalue Decomposition

A vector $\mathbf{v}_i$ that changes length but not direction when operated upon by a matrix $\mathbf{A}$ is said to be an eigenvector of $\mathbf{A}$. The length-scale factor is called the eigenvalue of $\mathbf{A}$. Eigenvalues λ_i and eigenvectors (directions) $\mathbf{v}_i$ of a matrix $\mathbf{A}$ with dimensions $M \times M$ are defined by

$$\mathbf{A}\mathbf{v}_i = \lambda_i\mathbf{v}_i \qquad (15.A1)$$

where λ_i are scalars and $\mathbf{v}_i$ are M-dimensional vectors. A matrix $\mathbf{A}$ can be represented in Jordan form as

$$\mathbf{V}^T\mathbf{A}\mathbf{V} = \mathrm{diag}(\mathbf{J}_1, \ldots, \mathbf{J}_r) \qquad (15.\mathrm{A2})$$

where $\mathbf{V} = [\mathbf{v}_1, \ldots, \mathbf{v}_M]$, diag is a block diagonal matrix, r is the number of independent eigenvectors of $\mathbf{A}$, and $\mathbf{J}_i$ indicates the Jordan block associated with the ith eigenvalue:

$$\mathbf{J}_i = \begin{bmatrix} \lambda_i & 1 & 0 & \cdots & 0 \\ 0 & \lambda_i & 1 & \ddots & \vdots \\ \vdots & \ddots & \ddots & \ddots & 0 \\ \vdots & \ddots & \ddots & \ddots & 1 \\ 0 & \cdots & \cdots & 0 & \lambda_i \end{bmatrix} \qquad (15.\mathrm{A3})$$

The dimension of the Jordan block is the multiplicity of the correspondent eigenvalue. If matrix $\mathbf{A}$ has M independent eigenvectors, the Jordan representation simplifies to a diagonal form:

$$\mathbf{\Lambda} = \begin{bmatrix} \lambda_1 & 0 & \cdots & 0 \\ 0 & \lambda_2 & \ddots & \vdots \\ \vdots & \ddots & \ddots & 0 \\ 0 & \cdots & 0 & \lambda_m \end{bmatrix} \qquad (15.\mathrm{A4})$$

Singular Value Decomposition

A rectangular matrix $\mathbf{B}$ with dimensions $M \times N$ can be represented as

$$\mathbf{V}^T\mathbf{B}\mathbf{U} = \mathbf{\Lambda} \qquad (15.\mathrm{A5})$$

where $\mathbf{U}$ ($N \times N$ matrix) and $\mathbf{V}$ ($M \times M$ matrix) are the matrices of the right and left eigenvectors, respectively, defined as

$$\mathbf{B}\mathbf{u}_i = \sigma_i \mathbf{v}_i \qquad (15.\mathrm{A6})$$

where $\sigma_1, \ldots, \sigma_p$ are the singular values (i.e., the square root of the eigenvalues of $\mathbf{B}^T\mathbf{B}$). $\mathbf{\Lambda}$ is $M \times N$ matrix with the singular values σ_i on the diagonal and zero elsewhere. The left eigenvectors $\mathbf{v}_i$ are also the eigenvectors of matrix $\mathbf{B}\mathbf{B}^T$. In tensorial notation, matrix $\mathbf{B}$ can be represented as a sum of dyadic forms:

$$\mathbf{B} = \sum_{k=1}^{m} \sigma_i \mathbf{v}_i \mathbf{u}_i^T \qquad (15.\mathrm{A7})$$

The pseudoinverse matrix of **B** is defined as

$$\mathbf{B}^\# = \sum_{k=1}^{p} \frac{1}{\sigma_i} \mathbf{u}_i \mathbf{v}_i^T \qquad (15.A8)$$

Consider matrix **B** as the mixing matrix with $M > N$. The problem of identification of the N sources $\mathbf{s}(t)$ from the M mixtures $\mathbf{x}(t)$ requires the overdetermination of the system, so that a solution of the problem

$$\mathbf{Bs}(t) = \mathbf{x}(t) \quad \text{or} \quad \mathbf{Bs}(t) - \mathbf{x}(t) = 0 \qquad (15.A9)$$

does not exist in general, as there are more independent conditions than unknowns [the independence of the conditions comes from the noise, which is always superimposed on the observations $\mathbf{x}(t)$]. For a solution to exists, a weaker definition of the solution is introduced, that is, the function $s_d(t)$ minimizing the squared error:

$$\mathbf{s}_d(t) = \min_{\mathbf{s}} \arg \left\| \mathbf{Bs}(t) - \mathbf{x}(t) \right\|^2 \qquad (15.A10)$$

The theorem of projections for Hilbert spaces implies that

$$\mathbf{Bs}_d(t) - \mathbf{x}(t) \in Im(\mathbf{B})^{\perp} = Ker(\mathbf{B}^T) \qquad (15.A11)$$

where $Im(\mathbf{B})$ and $Ker(\mathbf{B})$ are the image and the kernel of the matrix **B** and the symbol $\perp$ indicates the orthogonal space. Thus, we get

$$\mathbf{B}^T \mathbf{Bs}_d(t) = \mathbf{B}^T \mathbf{x}(t) \qquad (15.A12)$$

and

$$\mathbf{s}_d(t) = (\mathbf{B}^T \mathbf{B})^{-1} \mathbf{B}^T \mathbf{x}(t) = \mathbf{B}^\# \mathbf{x}(t) \qquad (15.A13)$$

where the definition of Equation 15.A8 is used. The pseudoinverse multiplied by the vector of observations gives the sources $\mathbf{s}_d(t)$ that minimize the squared error (Equation 15.A10).

ACKNOWLEDGMENTS

This work was supported by the European Project CyberManS (contract no. 016712), by Marie Curie reintegration grant iMOVE (Contract No. 239216) and TREMOR EU project (Contract No. 224051) within the 7th European Community Framework Programme, and by Fondazione Cassa di Risparmio di Torino and Compagnia di San Paolo di Torino.

REFERENCES

Anemüller, J., and Gramss, T., On-Line Blind Separation of Moving Sound Sources, in Cardoso, J.F., Jutten, C., and Loubaton, P. (Eds.), *Proceedings of the First International Workshop on Independent Component Analysis and Blind Signal Separation,* Aussois, France, pp. 331–334, 1999.

Bar-Ness, Y., Bootstrapping Adaptive Interference Cancelers: Some Practical Limitations, in *The Globecom Conference,* Miami, Paper F 3.7, pp. 1251–1255, 1982.

Belouchrani, A., Abed-Meraim, K., Cardoso, J. F., and Moulines, E., A Blind Source Separation Technique Using Second-Order Statistics, *IEEE Trans. Signal Proc.,* Vol. 45, 434–443, 1997.

Belouchrani, A., and Amin, M. G., Blind Source Separation Based on Time-Frequency Signal Representations, *IEEE Trans. Signal Proc.,* Vol. 46, 2888–2897, 1998.

Belouchrani, A., Abed-Meraim, K., Amin, M. G., and Zoubir, A., Joint-Antidiagonalization for Blind Source Separation, in *Proceedings of ICASSP,* Vol. 5, 2789–2792, 2001.

Bousbiah-Salah, H., Belouchrani, A., and Abed-Meraim, K., Jacobi-Like Algorithm for Blind Source Separation, *Electronic Letters,* Vol. 37, No. 16, 1049–1050, 2001.

Bousbiah-Salah, H., Belouchrani, A., and Abed-Meraim, K., Blind Separation of Nonstationary Sources Using Joint Block Diagonalization, in *Proceedings of IEEE Workshop on Statistical Signal Processing,* pp. 448–451, 2001.

Cardoso, J. F., Sources Separation Using Higher Order Moments, in *Proceedings of Internat. Conference on Acoustic Speech Signal Processes,* Glasgow, pp. 2109–2112, 1989.

Cardoso, J. F., and Souloumiac, A., Jacobi Angles for Simultaneous Diagonalization, *SIAM J. Mat. Anal. Appl.,* Vol. 17, 161–164, 1996.

Comon, P., Independent Component Analysis, in *International Signal Processing Workshop on High-Order Statistics,* pp. 29–38, Elsevier, 1992.

Comon, P., Independent Component Analysis: A New Concept?, *Signal Processing,* Vol. 36, 287–314, 1994.

De Luca, C. J, Adam, A., Wotiz, R., Gilmore, L. D., and Nawab, S. H., Decomposition of Surface EMG Signals, *J. Neurophysiol.,* Vol. 96, 1646–1657, 2006.

Farina, D., Févote, C., Doncarli, C., and Merletti, R., Blind Separation of Linear Instantaneous Mixtures of Non-stationary Surface Myoelectric Signals, *IEEE Trans. Biomed. Eng.,* Vol. 51, No. 9, 1555–1567, 2004.

Fety, L., *Methodes de Traitement d'antenne Adaptèes aux Radiocommunications,* Doctorate Thesis, ENST, 1988.

Gaeta, M., and Lacounme, J.-L., Source Separation without a Priori Knowledge: The Maximum Likelihood Solution, in *Proceedings of European Signal Processing Conference,* pp. 621–624, 1990.

Garcia, G. A., Okuno, R., and Akazawa, K., A Decomposition Algorithm for Surface Electrode-Array Electromyogram: A Noninvasive, Three-Step Approach to Analyze Surface EMG Signals, *IEEE Eng. Med. Biol. Mag.,* Vol. 24, 63–72, 2005.

Giannakis, G. B., Inouye, Y., and Mendel, J. M., Cumulant Based Identification of Multichannel Moving Average Models, *IEEE Trans. Automat. Control,* Vol. 34, 783–787, 1989.

Holobar, A., and Zazula, D., A New Approach for Blind Source Separation of Convolutive Mixtures of Pulse Trains, in *Proceedings of BSI02*, pp. 163–166, Como, Italy, 2002.

Holobar, A., and Zazula, D., Surface EMG Decomposition Using a Novel Approach for Blind Source Separation, *Informatica Medica Slovenica*, Vol. 8, No. 1, 2–14, 2003.

Holobar A., and Zazula, D., Correlation-based Decomposition of Surface Electromyograms at Low Contraction Forces, *Med. Biol. Eng. Comput.*, Vol. 42, No. 4, 487–495, 2004.

Holobar, A., and Zazula, D., Multichannel Blind Source Separation Using Convolution Kernel Compensation. *IEEE Trans. Signal Process.*, Vol. 55, 4487–4496, 2007.

Holobar, A., Ojsteršek, M., and Zazula, D., Distributed Jacobi Joint Diagonalization on Clusters of Personal Computers. *Int. J. Parallel Program.*, Vol. 34, 509–530, 2006.

Hyvarinen, A., New Approximations of Differential Entropy for Independent Component Analysis and Projection Pursuit, in *Advances in Neural Information Processing Systems*, Vol. 10, pp. 273–279. MIT press, 1998.

Hyvarinen, A., Survey on Independent Component Analysis, *Neural Computing Surveys*, Vol. 2, 94–128, 1999.

Hyvarinen A., and Oja, E., Independent Component Analysis: Algorithm and Applications, *Neural Networks*, Vol. 13, Nos. 4–5, 411–430, 2000.

Hyvarinen, A., Karhunen, J., and Oja, E., *Independent Component Analysis*, Wiley, 2001.

Inouye, Y., and Matsui, T., Cumulant Based Parameter Estimation of Linear Systems, in *Proceedings of Workshop on Higher Order Spectral Analysis*, pp. 180–185, Vail, Colorado, 1989.

Jolliffe, I. T., *Principal Component Analysis*, Springer-Verlag, 1986.

Jones, M. C., and Sibson, R., What is Projection Pursuit? *J. of the Royal Statistical Society, Ser. A*, Vol. 150, No. 1, 36, 1987.

Jutten, C., and Herault, J., Blind Separation of Sources, Part I: An Adaptive Algorithm Based on Neuromimetic Architecture, *Signal Processing*, Vol. 24, 1–10, 1991.

Jutten, C., *Calcul Neuromimétique et Traitement du Signal: Analyse en Composante Indépendantes*, Ph.D. thesis, INPG, Univ. Grenoble, 1987.

Karhunen, J., Oja, E., Wang, L., Vigário, R., and Joutsensalo, J., A Class of Neural Networks for Independent Component Analysis, *IEEE Trans. on Neural Networks*, Vol. 8, No. 3, pp. 486–504, 1997.

Kleine, B. U., van Dijk, J. P., Lapatki, B. G., Zwarts, M. J., and Stegman, D. F., Using Two-Dimensional Spatial Information in Decomposition of Surface EMG Signals, *J. of Electromyogr. Kinesiol.*, Vol. 5, 535–548, 2007.

Koutras, A., Dermatas, E., and Kokkinakis, G., Blind Separation of Speakers in Noisy Environments: A Neural Network Approach, *Neural Network World*, Vol. 10, No. 4, 619–630, 2000.

Lacoume, J. L., A Survey of Source Separation, in *ICA '99*, pp. 1–6, Aussois, France, January 11–15, 1999.

Makeig, S., Bell, A. J., Jung, T.-P., and Sejnowski, T.-J., Independent Component Analysis of Electroencephalographic Data, in *Advances in Neural Information Processing Systems 8*, pp. 145–151, MIT Press, 1996.

Merletti, R., and Parker, P. A., *Electromyography: Physiology, Engineering, and Non-Invasive Applications,* IEEE Press/Wiley, 2004.

Nakamura, H., Yoshida, M., Kotani, M., Akazawa, K., and Moritani, T., The Application of Independent Component Analysis to the Multi-Channel Surface Electromyographic Signals for Separation of Motor Unit Action Potential Trains: Part I—Measuring Techniques, *J. Electromyog. Kinesiol.,* Vol. 14, 423–432, 2004.

Pham, D.-T., Garrat, P., and Jutten, C., Separation of a Mixture of Independent Sources through a Maximum Likelihood Approach, in *Proceedings of EUSIP-CO,* pp. 771–774, 1992.

Ruiz, P., and Lacoume, J. L., Extraction of Independent Sources from Correlated Inputs, in *Proceedings of Workshop on Higher Order Spectral Analysis,* pp. 146–151, Vail, Colorado, 1989.

Sorouchyari, E., Blind Source Separation, Part III: Stability Analysis, *Signal Processing,* Vol. 24, 21–29, 1991.

Souloumiac, A., and Cardoso, J.-F., Comparaison de Methodes de Separation de Sources, in *Proceedings of GRETSI,* Juan les Pins, France, 1991.

Vigário, R., Extraction of Ocular Artifacts from EEG Using Independent Component Analysis, *Electroenceph. Clin. Neurophysiol.,* Vol. 103, No. 3, 395–404, 1997.

Vigário, R., Jousmäki, V., Hämäläinen, M., Hari, R., and Oja, E., Independent Component Analysis for Identification of Artifacts in Magnetoencephalographic Recordings, in *Advances in Neural Information Processing 10* (Proceedings of NIPS'97), pp. 229–235, MIT Press, 1998.

Vigário, R., Särelä, J., Jousmäki, V., and Oja, E., Independent Component Analysis in Decomposition of Auditory and Somatosensory Evoked Fields, in *Proceedings of International Workshop on Independent Component Analysis and Signal Separation (ICA'99),* pp. 167–172, Aussois, France, 1999.

Thispageintentionallyleftblank

CHAPTER 16

HIGHER ORDER SPECTRA

Giovanni Calcagnini and Federica Censi

16.1. INTRODUCTION

The estimation of the power spectrum density has been shown to be a powerful tool not only for the definition of indexes and parameters from several biomedical signals, but also for the comprehension of the complex physiological mechanisms underlying the generation of such signals.

Examples of results obtained using the spectral approach can be found in applications such as the analysis of the predominant rhythms in the electroencephalographic signal, the interpretation of the baroreflex control loop and the spontaneous fluctuations of heart rate, as well as in the field of the neuromuscular signals.

Various methods for spectrum-density analysis have been developed with the aim of obtaining reliable estimates in view of the peculiar constraint posed by the biological time series; some of these methods rely on models of generation of the series.

Power spectrum density (PSD) provides the distribution over the frequency of the signal power, neglecting the phase relationships between the harmonic components. The information embedded in the PSD is the same as that of the autocorrelation function of the signal. Such information is sufficient for a statistical description of Gaussian signals, although in some cases it may be important to obtain information regarding the deviation from the Gaussianity or the phase relationships between harmonics. Higher order statistics and spectra do contain some of this information. A particular case of higher order spectra (HOS) is represented by the third-order spectrum, also called the bispectrum, which is defined as the Fourier transform of the third-order statistics of a stationary signal.

Figure 16.1 shows the relationships between the signal-statistics order and the corresponding spectra. The traditional PSDs belong to the class of higher order spectra called second-order spectra.

Advanced Methods of Biomedical Signal Processing. Edited by S. Cerutti and C. Marchesi   **411**
Copyright © 2011 the Institute of Electrical and Electronics Engineers, Inc.

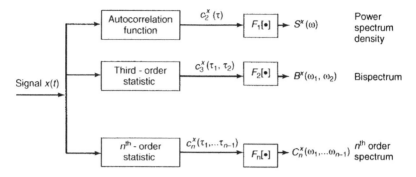

Figure 16.1. Classification of the higher order spectra of a signal $x(t)$. $F[\cdot]$ indicates the Fourier transform.

Potential application fields of the HOS include:

- Nonminimum-phase system identification
- Extraction and characterization of nonlinear features from signals and nonlinear system identification
- Analysis and quantification of deviation from Gaussianity
- Suppression of colored (nonwhite) additive Gaussian noise of unknown spectral density

16.2. HIGHER ORDER STATISTICS: DEFINITION AND MAIN PROPERTIES

The definition of the HOS takes into account the deterministic or stochastic nature of the phenomenon under investigation. For deterministic signals, HOS are defined as the Fourier transforms of the higher order moments, whereas for stochastic signals, HOS are defined as the Fourier transform of the higher order cumulants.

If $x(k)$, $k = 0, \ldots, N$, is a real, stationary, discrete-time signal and its moments up to the nth order exist, the nth-order moment can be written as

$$m_n^x(\tau_1, \tau_2, \ldots, \tau_{n-1}) = E[x(k)x(k + \tau_1)\ldots x(k + \tau_{n-1})]$$

where $E[\cdot]$ denotes the statistical expectation. If $x(t)$ is a real stationary signal, the nth-order moments depend only on the differences $\tau_1, \tau_2, \ldots, \tau_{n-1}$. The second-order moment $m_2^x(\tau_1)$ is the autocorrelation function of $x(k)$. The nth-order cumulant of a stationary random signal, for $n = 3, 4$, can be written as

$$c_n^x(\tau_1, \tau_1, \ldots, \tau_{n-1}) = m_n^x(\tau_1, \tau_1, \ldots, \tau_{n-1}) - m_n^G(\tau_1, \tau_1, \ldots, \tau_{n-1})$$

where $m_n^x(\tau_1, \tau_1, \ldots, \tau_{n-1})$ is the nth-order moment $x(k)$ and $m_n^G(\tau_1, \tau_1, \ldots, \tau_{n-1})$ is the nth-order moment of a Gaussian signal with the same mean value and autocorrelation sequence of $x(k)$. For Gaussian signals, $m_n^x(\tau_1, \tau_1, \ldots, \tau_{n-1}) = m_n^G(\tau_1, \tau_1, \ldots, \tau_{n-1})$ and, thus, $c_n^x(\tau_1, \tau_1, \ldots, \tau_{n-1}) = 0$. For Gaussian signals, $c_n^x(\tau_1, \tau_1, \ldots, \tau_{n-1}) = 0$ for any order.

To give a general definition of cumulants, the characteristic functions must be used. The first characteristic function, $\varphi(\omega)$, of a random variable x is defined as the Fourier transform of its probability density function $p_x(x)$. $\varphi(\omega)$ is also called a moment-generating function as the coefficients of its expansion into Taylor series give us the moments of x. The second characteristic function of x is defined as $\psi(\omega) = \ln[\varphi(\omega)]$, and is also called a cumulant generating function as cumulants are defined as the coefficients of the Taylor series expansion of $\psi(\omega)$ (Papoulis, 1991). The main properties of the cumulants are summarized in Table 16.1 (Nikias and Mendel, 1993). For a better comprehension of cumulants, it is useful to highlight their relationships with the moments:

Table 16.1. Main properties of moments (*Mom*) and cumulants (*Cum*)

1. Moments and cumulants of scaled quantities equal the product of the unscaled quantities:

$Mom[a_1x_1, a_2x_2, \ldots, a_nx_n] = a_1 \ldots a_n Mom[x_1 \ldots x_n]$

$Cum[a_1x_1, a_2x_2, \ldots, a_nx_n] = a_1 \ldots a_n Cum[x_1 \ldots x_n]$ where $a_1, a_1, \ldots, a_n$ are constant (nonrandom) values

2. Cumulants are blind to additive constants: $Cum[a + x_1, x_2, \ldots, x_n] = Cum[x_1, x_2, \ldots, x_n]$, where a is a constant

3. Moments and cumulants are invariant to any permutation of their arguments, for example, $Mom[x_1, x_2, x_3] = Mom[x_2, x_1, x_3] = Mom[x_3, x_2, x_1]$

4. Cumulants are additive in their arguments:

$Cum[x_1 + y_1, x_2, \ldots, x_n] = Cum[x_1, x_2, \ldots, x_n] + Cum[y_1, x_2, \ldots, x_n]$

5. Cumulants of a sum of statistically independent quantities equal the sum of cumulants of the individual quantities:

$Cum[x_1 + y_1, x_2 + y_2, \ldots, x_n + y_n] = Cum[x_1, x_2, \ldots, x_n] + Cum[y_1, y_2, \ldots, y_n]$

where $\{x_1, x_1, \ldots x_n\}$ are independent of the random variables $\{y_1, y_1, \ldots y_n\}$

6. If a set of random variable is jointly Gaussian, its higher order (greater than 2) cumulants are zero, and $\{x_1, x_1, \ldots x_n\}$ is a colored or white process.

7. $Cum[x_1, x_2, x_3] = Cum[x_2, x_1, x_3] = Cum[x_3, x_2, x_1]$. If a subset k of the n random variables $\{x_1, x_1, \ldots x_n\}$ is independent of the rest, then the n-order cumulant is zero, whereas, in general, the n-order moments are nonnull (Mendel, 1991).

Order 1: $c_1^x = m_1^x = E\{X(k)\}$

Order 2: $c_2^x(\tau_1) = m_2^x(\tau_1) - (m_1^x)^2 = m_2^x(-\tau_1) - (m_1^x)^2 = c_2^x(-\tau_1)$; the second-order cumulant is the covariance sequence; $m_2^x(\tau_1)$ is the autocorrelation sequence.

Order 3: $c_3^x(\tau_1, \tau_2) = m_3^x(\tau_1, \tau_2) - m_1^x(m_2^x(\tau_1) + m_2^x(\tau_2) + m_2^x(\tau_2 - \tau_1)] + 2 \cdot (m_1^x)^3$, where $m_3^x(\tau_1, \tau_2)$ is the third-order moment.

16.2.1. Observations

If $x(k)$ is a zero-mean process, second- and third-order cumulants are identical to second- and third-order moments, whereas the fourth-order cumulant is a function of the second and fourth-order moments. The following symmetries are a consequence of the properties of moments and the relationships between the third-order cumulant and moment:

$$c_3^x(\tau_1, \tau_2) = c_3^x(\tau_2, \tau_1) = c_3^x(-\tau_2, \tau_1 - \tau_2) = c_3^x(\tau_2 - \tau_1, -\tau_1)$$

$$= c_3^x(\tau_1 - \tau_2, -\tau_2) = c_3^x(-\tau_1, \tau_2 - \tau_1)$$

Thus, the third-order cumulant has six regions of symmetry (Figure 16.2). The knowledge of the cumulant in any of the six sectors (including the boundaries) allow one to find the entire third-order cumulant sequence (Papoulis, 1991).

By setting $\tau_1 = \tau_2 = \tau_3 = 0$ and assuming $m_1^x = 0$, we obtain:

$$\gamma_2^x = E\{X^2(k)\} = c_2^x(0) \qquad \text{(variance)}$$

$$\gamma_3^x = E\{X^3(k)\} = c_3^x(0,0) \qquad \text{(skewness)}$$

$$\gamma_4^x = E\{X^4(k)\} - 3 \cdot [\gamma_2^x]^2 = c_4^x(0,0,0) \qquad \text{(kurtosis)}$$

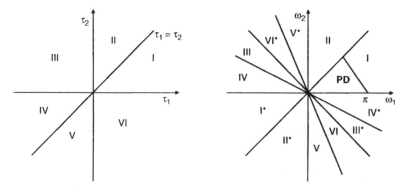

Figure 16.2. Regions of symmetry of third-order cumulant (left) and of the bispectrum (right). PD denotes the principal domain; * indicates the conjugate operator.

Throughout this chapter, we concentrate on the third-order spectra of real processes.

16.3. BISPECTRUM AND BICOHERENCE: DEFINITIONS, PROPERTIES, AND ESTIMATION METHODS

16.3.1. Definitions and Properties

Higher order spectra are defined as the n-dimensional Fourier transforms of the higher order statistics: spectra of cumulants and spectra of moments. As stated before, there are advantages to using cumulants for stochastic signals and moments for deterministic ones. In the following, we will refer to second- and third-order spectra and bispectra.

The power spectrum density is defined as

$$S^x(\omega) = \sum_{\tau=-\infty}^{+\infty} c_2^x(\tau)\exp\left\{-j(\omega\tau)\right\}$$

where $|\omega| \leq \pi$ and $c_2^x(\tau)$ is the second-order cumulant (covariance sequence). For real signals, the power spectrum is real nonnegative and symmetric with respect to the frequency. Note that for zero-mean signals, this definition of spectrum coincides with the usual one, since the second-order cumulant is the covariance sequence of the signal.

The bispectrum is defined as

$$B^x(\omega_1,\omega_2) = \sum_{\tau_1=-\infty}^{+\infty} \sum_{\tau_2=-\infty}^{+\infty} c_3^x(\tau_1,\tau_2)\exp\left\{-j(\omega_1\tau_1 + \omega_2\tau_2)\right\}$$

where $c_3^x(\tau_1, \tau_2)$ is the third-order cumulant. The bispectrum of a real stationary signal has twelve symmetry zones, descending from the property of invariance to the argument permutation of zone $c_3^x(\tau_1, \tau_2)$ (Figure 16.2). In addition, $B^x(\omega_1, \omega_2)$ is periodic in ω_1 and ω_2, with period 2π. Thus, knowledge of the bispectrum in the triangular region $\omega_1 \geq \omega_2$, $\omega_2 \geq 0$, $\omega_1 + \omega_2 \leq \pi$, also referred to as the principal domain (PD), is enough for a complete description of the bispectrum (Hinich, 1982; Hinich and Messer, 1995; Hinich and Wolinsky, 1988).

Whereas deterministic signals are better analyzed using cumulant spectra, for stochastic signals, there are certain advantages in using cumulant spectra, for the following reasons:

• For Gaussian signals only, all cumulant spectra of order greater than two are identically zero. Thus, nonzero cumulants provide information on deviation from Gaussianity.

- Cumulants provide an estimate of the degree of statistical dependence in time series.
- The cumulants of high-order white noise are multidimensional impulse functions, and their corresponding polyspectra are multidimensionally flat.
- The cumulant of a sum of two statistically independent processes equals the sum of the cumulants of the individual processes, whereas this does not hold true for moments.

A useful function can be obtained by normalizing the bispectrum with respect to the corresponding power spectrum. The bicoherence (or third-order coherence index) is defined as

$$BIC^x(\omega_1, \omega_2) \doteq \frac{B^x(\omega_1, \omega_2)}{\left[S^x(\omega_1) \cdot S^x(\omega_2) \cdot S^x(\omega_1 + \omega_2) \right]^{1/2}}$$

The bicoherence is particularly important in discriminating linear from nonlinear processes, as well as in detecting and characterizing nonlinearities in time series. A signal is said to be a linear non-Gaussian process of order 3 if the magnitude of the bicoherence is constant over all frequencies; otherwise, the signal is said to be a nonlinear process. This definition can be generalized to the nth-order coherence indexes.

16.3.2. Bispectrum Estimation: Nonparametric and Parametric Approaches

The use of the abovementioned definitions and the properties of higher order cumulants and spectra would require either the availability of an infinite-length signal or the knowledge of the higher order statistics of the underling process. In general, in a real application only a finite number of signal samples are available and, thus, appropriate methods of estimation should be employed. As in case of the power spectrum, there are also different ways to estimate higher order spectra: nonparametric (also known as conventional) and parametric methods.

Parametric methods are based on the identification of a model that is assumed to be able to generate the given data. The model is driven by a zero-mean, nth-order, non-Gaussian white noise ($\gamma_n^x \neq 0$ for $n = 2, 3, \ldots$). The bispectrum is then estimated from the model coefficients (i.e., from the model transfer function) and from the spectrum of the input signal. A detailed description of the parametric estimation of the spectrum can be found in Raghuveer and Nikias (1985).

As far as the differences between conventional and parametric methods, we can draw the same general conclusions usually drawn for spectral estimators: parametric methods (AR and ARMA) have higher resolution and, thus, are better detectors of nonlinearities and phase coupling of harmonics close in frequency, whereas conventional methods provide better quantification and estimates of the degree of phase coupling between harmonics.

Conventional estimation methods can be divided into direct and indirect ones. The indirect method of bispectrum estimation is based on the estimation of third-order moments and cumulants, as follows. The signal segment is divided into K subsegments of M samples and the subsegments are detrended. Third-order moment sequences of the subsegments are estimated using the following formula:

$$r^i(m,n) = \frac{1}{M} \sum_{l=s_1}^{s_2} x^{(i)}(l)x^{(i)}(l+m)x^{(i)}(l+n)$$

where

$$i = 1,2,...,k$$
$$s_1 = \max(-m,-n)$$
$$s_2 = \min(M-1, M-m-1, M-n-1)$$

The estimate for the third-order moment sequence of the whole signal segment is achieved by averaging:

$$\hat{c}_3^x(m,n) = \frac{1}{K} \sum_{i=1}^{K} r^{(i)}(m,n)$$

Note that since the subsegments where detrended, the estimate of the third-order cumulant sequence is equal to that of the third-order moment sequence. The bispectrum estimate can be calculated as the two-dimensional Fourier transform:

$$\hat{B}^x(\omega_1,\omega_2) = \sum_{m=-L}^{L} \sum_{n=-L}^{L} \hat{c}_3^x(m,n)W(m,n)\exp[-j(\omega_1 m + \omega_2 n)]$$

where $W(m,n)$ is a two-dimensional window function that can be used to improve the estimation (Raghuveer and Nikias, 1985).

The direct method of estimating the bispectrum is similar to the power-spectrum estimation based on the periodogram and contains the following steps: the signal segment is divided into K subsegments of M samples each ($N = K \times M$), the subsegments are detrended (average value is subtracted), and the Fourier transform is taken from each subsegment.

$$X^{(i)}(\omega) = \sum_{k=0}^{M-1} x^{(i)}(k)\exp(-j\omega k)$$

The third-order bispectrum estimate of the subsegments is calculated by

$$\hat{b}_i(\omega_1,\omega_2) = X^{(i)}(\omega_1)X^{(i)}(\omega_2)X^{(i)*}(\omega_1 + \omega_2)$$

where * denotes complex conjugate. Then the bispectrum estimate is obtained by averaging over the subsegment bispectra:

$$\hat{B}^{\gamma}(\omega_1,\omega_2) = \frac{1}{K}\sum_{i=1}^{K} \hat{b}_i(\omega_1,\omega_2)$$

The described methods for bispectrum estimation give, in general, different results. However, both methods have been shown to give asymptotically unbiased (the bias between the estimate and the true value of the bispectrum approaches zero as the length of the data segment increases) and consistent (the variance of the estimate goes to zero with increasing segment length) estimates.

16.4.   ANALYSIS OF NONLINEAR SIGNALS: QUADRATIC PHASE COUPLING

A particular nonlinear phenomenon that the bispectrum is capable of detecting and quantifying is quadratic phase coupling (QPC). This phenomenon occurs when the interaction between two harmonic components contributes to the power at their sum and/or difference frequencies. In this case, the relations between the signal phases are exactly the same as the frequency relations. Thus, in some circumstances we may be interested in assessing if spectral peaks at harmonically related positions are also phase coupled. Let us consider a signal containing six harmonics:

$$x(t) = \sum_{i=1}^{6} \cos(2\pi f_i t + \theta_i)$$

where $f_3 = f_1 + f_2$, $f_6 = f_4 + f_5$, $\theta_1, \theta_2, \ldots, \theta_5$ are independent, uniformly distributed random variables over $(-\pi, +\pi)$. QPC occurs between f_4 and f_5 if $\theta_6 = \theta_4 + \theta_5$.

Note that the existence of a relation between harmonics (the sum, in this example) is a necessary condition for the QPC. The power spectrum can detect the relations between the frequency, but since it suppresses all the phase relations, it cannot provide information on the phase relations (Figure 16.3, left panel). The bispectrum is capable of detecting the QPC since only the phase-coupled components contribute to the third-order cu-

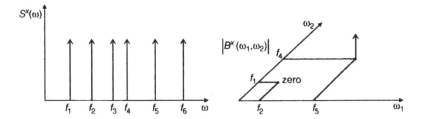

Figure 16.3. Quadratic phase coupling. Power spectrum density (left panel) and magnitude bispectrum (right panel) of a signal with six harmonics and a QPC between f_4 and f_5.

mulant. Thus, the bispectrum shows an impulse only at frequencies f_4 and f_5. (Figure 16.3, right panel).

A particular case occurs when a signal contains two harmonics f_1 and f_2 with $f_2 = 2f_1$, that is, $x(t) = \sin(2\pi f_1 t + \theta_1) + \sin(2\pi f_2 t + \theta_2)$ with $f_2 = 2f_1$. In this case, there is coupling between f_1 and f_2 if $\theta_2 = 2\theta_1$ and the bispectrum will show an impulse at (f_1, f_1).

When the bispectrum and the spectrum are estimated with nonparametric methods, the magnitude bicoherence allow the quantification of the QPC. A bicoherence value close to one at the phase-coupled frequency pair indicates a strong (almost 100%) quadratic interaction.

It is important to stress that the value of the coherence index should be considered only at those frequencies at which a phase coupling may occur, whereas elsewhere it is meaningless. (Raghuveer and Nikias, 1985). A method to discriminate between "true" and "spurious" peaks or, in other words, to set a threshold in the bicoherence function, in the case of poor signal-to-noise ratio can be found in the work of Pinhas et al. (2004).

16.5. IDENTIFICATION OF LINEAR SYSTEMS

The bispectrum can be used for the identification of linear systems (Papoulis, 1991). In general, if we have access to the input and output, $x(k)$ and $y(k)$, of a linear system with transfer function $H(\omega) = A(\omega)e^{j\varphi(\omega)}$, the identification of the system is achieved by using the second-order statistics (input/output cross correlation and input autocorrelation function) or, equivalently, using the input/output cross spectrum and the input autospectrum:

$$S^{xy}(\omega) = S^x(\omega) H^*(\omega)$$

In many applications, we cannot estimate the second-order statistics because we do not have access to the system input $x(k)$. In such cases, a possible solution is the assumption that the system is driven by a white noise. With this assumption we obtain

$$S^{y}(\omega)= S^{x}(\omega)\left|H(\omega)\right|^{2} = qA^{2}(\omega)$$

This relationship allows the estimation (within a constant factor) of $A(\omega)$ from $S^{y}(\omega)$. If the system is minimum phase, $H(\omega)$ is completely determined because $\phi(\omega)$ and $H(\omega)$ can be expressed in terms of $A(\omega)$. For arbitrary systems (nonminimum phase) the phase cannot be determined from the second-order statistics only (Giannakis and Mendel, 1989). It can, however, be determined if the third-order moment of $y(k)$ is known.

For a linear system driven by a noise process with constant bispectrum equal to Q, we have (Papoulis, 1991):

$$B^{y}\left(\omega_{1},\omega_{2}\right)=\left|B(\omega_{1},\omega_{2})\right|e^{j\theta(\omega_{1},\omega_{2})} = QH(\omega_{1})H(\omega_{2})H^{*}(\omega_{1}+\omega_{2})$$

where $|B(\omega_{1}, \omega_{2})|$ and $\theta(\omega_{1}, \omega_{2})$ are the amplitude and the phase of $B^{y}(\omega_{1}, \omega_{2})$, respectively. By equating amplitude and phase we obtain:

$$\left|B(\omega_{1},\omega_{2})\right| = QA(\omega_{1})A(\omega_{2})A(\omega_{1}+\omega_{2})$$

$$\theta\left(\omega_{1},\omega_{2}\right)= \phi(\omega_{1})+\phi(\omega_{2})-\phi(\omega_{1}+\omega_{2})$$

Thus, the amplitude and phase of the linear system can be determined by solving the two equations.

16.6.   INTERACTION AMONG CARDIORESPIRATORY SIGNALS

The respiration activity interacts with the complex mechanisms that control the heart rate and blood pressure (Akselrod, 1981; *Circulation*, 1996). Such interaction is reflected in the cardiovascular variability time series (blood pressure and heart rate) by a harmonic component synchronous with the respiratory activity (high frequency, HF, 0.15–0.40 Hz). In addition to this component, a lower frequency component (low frequency, LF, 0.04–0.15 Hz), an expression of the baroreflex control mechanism of the blood pressure, characterizes the cardiovascular signal of normal subjects. Power-spectrum analysis has been widely use to detect and quantify such components, and allowed the definition of clinical indexes able to quantify the cardiovascular neuroregulation function.

As mentioned before, due to the limitations of the power spectrum and of the associated linear modeling approach, the interactions (if any) between the LF and HF rhythms of the cardiovascular variability series cannot be thoroughly investigated. However, investigation with the power spectrum has indeed revealed some interactions. It was shown that the LF rhythm, whose frequency is generally centred around 0.10 Hz, could be shifted by appropriate breathing exercises (Censi et al., 2000; Kitney et al., 1982). In particular, a breathing frequency around 0.13 Hz can entrain the LF rhythm up to a complete synchronization to the respiration. Breathing at higher frequency can synchronize the LF rhythm at its subharmonic. Power-spectrum analysis can detect the interactions between the frequencies, whereas bispectrum and bicoherence analysis can detect and quantify the phase relations.

Figure 16.4 shows two examples of interaction between respiration and heart period, in a normal subject during controled breathing exercises at 12 and 15 breaths/min (0.20 Hz and 0.25 Hz, respectively).

During controlled breathing at 12 breaths/min, spectrum analysis detects a 2:1 frequency relation between the two harmonics characterizing the signal (LF at 0.10 Hz and HF at 0.20 Hz; Figure 16.4, left panel). Bispectrum analysis detects a phase relation between the two harmonics (bispectral peak at 0.10,0.10 Hz). A higher breathing rate (0.25 Hz) does not elicit frequency coupling with the LF rhythm, which is only barely detectable in the power spectrum (Figure 16.4, right panel). Indeed, bispectrum analysis detects a quadratic phase coupling between the respiration (at 0.25 Hz) and a harmonic component at 0.50 Hz, which suggests a nonlinear interaction of the respiration with the heart-period signal. Unfortunately, we can not discriminate whether this finding is the expression of a nonlinear interaction between the respiration and the heart-period fluctuations unless we analyze the respiration signal, its bispectrum, and the cross spectrum and cross bispectrum between the two signals. The quadratic phase coupling between respiration and heart rate was investigated by Jamsek et al. (2004) using time-phase bispectral analysis in adults during spontaneous and paced breathing, and in neonates by Schwab et al. (2006) using a time-variant parametric estimation of the bispectrum.

16.7. CLINICAL APPLICATIONS OF HOS: BISPECTRAL INDEX FOR ASSESSMENT OF ANAESTHESIA DEPTH

Anaesthesia is a complex phenomenon aimed at controlling various physiological components: consciousness, analgesia, autonomic responses, and muscle relaxation. Depth of anaesthesia is difficult to measure accurately, but attempts have been made to measure it and to reduce the incidence of all causes of awareness. A lot of research has been aimed at finding appropriate parameters to monitor the depth of anaesthesia. Historically, this was measured using pulse, blood pressure, sweating, lacrimation, and skin conductance. However, these parameters have several limitations: they are subjective and operator dependent, and other drugs may interfere with the signs (e.g., atropine causes tachycardia and papillary dilatation) (Agarwal

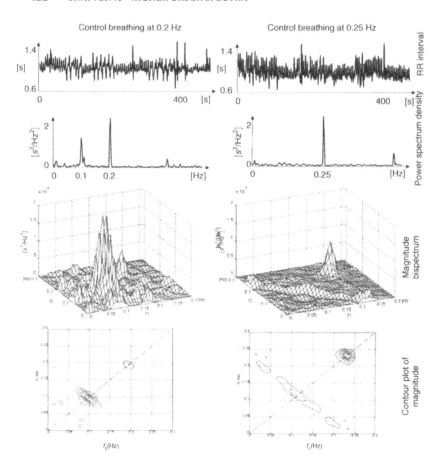

Figure 16.4. Upper panels: heart period signals. Middle panels: power spectrum density. Lower panels: amplitude bispectrum. Signals were recorded from a normal subject during controlled breathing exercises at 0.20 Hz (left) and 0.25 Hz (right).

and Griffiths, 2004). New tools are EEG, bispectral index (BIS), and evoked potentials. It is well known that the EEG signal is characterized by five main frequency bands whose powers are related to the degree of awareness/attention of the patients (Table 16.2).

Power-spectrum analysis has been widely used to quantify the relative power of the bands so as to extract descriptors of the level of consciousness for various types and dosage of anaesthetic drugs (Levy, 1986; Rampil and Matteo, 1987).

Time domain analysis has provided parameters able to quantify deep anaesthesia (Rampil and Laster, 1992; Rampil et al., 1988). During deep

Table 16.2. EEG rhythms

Rhythm	Frequency (Hz)	Amplitude (mV)	Subject states and physiological correlates
Delta	0.5–4	20–200	Pathological conditions (coma, cerebral ischemia), general anaesthesia
Theta	4–8	5–100	Deep sleep, dreaming
Alpha	8–13	10–200	Mental relaxation
Beta	13–22	1–20	Attention, cortical activation
Beta$_2$	>22	1–20	Attention, cortical activation, sensory stimulation

anaesthesia, head trauma, or brain ischemia the EEG may develop a peculiar pattern of activity characterized by alternating periods of normal-to-high voltage activity changing to low voltage or even isoelectric. The burst suppression ratio (BSR) quantifies the degree of burst suppression. Burst suppression is defined as those periods longer than 0.5 s during which the EEG does not exceed approximately 5 μV. The time in a suppressed state is measured and the BSR is calculated as the fraction of the epoch length in which EEG is suppressed (Rampil, 1998).

Although these variables somehow detect changes in the EEG caused by anaesthetic drugs, the patient-specific responses and the sensitivity to different EEG patterns induced by different drug significantly affect their performance as anaesthetic monitors.

The use of bispectrum analysis to analyze the EEG was suggested by Barnett et al. (1971), who investigated the phase coupling between the α and β rhythms. By estimating the bispectrum and the bicoherence of surface EEGs, they showed a significant quadratic coupling between α and β waves in awake subjects. They found that about 50% of the β activity could be attributed to harmonic coupling with α peaks. During sleep, the degree of interaction diminished significantly.

The combination of time-domain, frequency-domain, and high-order spectral parameters has been used by the Aspect Medical System to introduce a complex parameter to monitor the anaesthesia depth: the bispectral index (BIS) (Figure 16.5). The (proprietary) expression of the subparameters and their weights as a single variable was derived empirically on large databases (training set) and then clinically validated over large populations (test set). BIS decreases continuously with decreasing level of consciousness, ranging between 0 (no cerebral activity) and 100 (patient awake).

Figure 16.6 shows an example of BIS values during a surgical procedure. During the anaesthesia induction, the index decreases and it rises again at the end of the procedure when the patient is wakened. After 50 min, an increase in patient consciousness is observed, requiring an anaesthesia supplement.

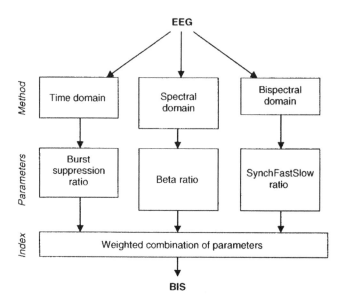

Figure 16.5. Algorithm for BIS index development from the EEG signal.

It is important to point out that automated and quantitative indexes for anaesthesia depth monitoring not only may allow better management of the patient in terms of drug type and dose, but also open the way for designing feedback systems for analgesia administration (closed-loop anaesthesia).

As the generation of the EEG signal is extremely complex, the phys-

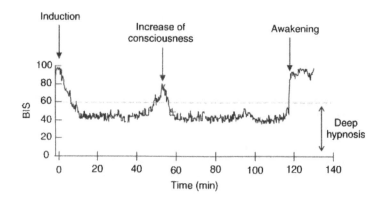

Figure 16.6. Example of BIS values during a general anaesthesia surgical procedure.

iological meaning of this parameter as well as its relationship with depth of anaesthesia are not known. It has been suggested that strong phase relationships relate inversely to the number of independent EEG generators. Since it is applied to the surface EEG, it has been hypothesized that the BIS somehow reflects a particular component of the anaesthesia that is linked to the level of consciousness and to the suppression of the cerebral mechanisms of elaboration of the pain stimuli arriving from the periphery (hypnosis).

REFERENCES

Agarwal, M., and Griffiths, R., Monitoring the Depth of Anaesthesia, *Anaesthesia & Intensive Care Medicine,* Vol. 5, No. 10, 343–344, 2004.

Akselrod, S., Power Spectrum Analysis of Heart Rate Fluctuation: A Quantitative Probe of Beat-to-Beat Cardiovascular Control, *Science,* Vol. 213, 220–222, 1981.

Barnett, T. P., Johnson, L. C., Naitoh, P., Hicks, N., and Nute, C., Bispectrum Analysis of Electroencephalogram Signals During Waking and Sleeping, *Science,* Vol. 172, No. 23, 401–402, 1971.

Censi, F., Calcagnini, G., Lino, S., Seydnejad, S. R., Kitney, R. I., and Cerutti, S., Transient Phase Locking Patterns Among Respiration, Heart Rate and Blood Pressure During Cardiorespiratory Synchronisation in Humans, *Med. Biol. Eng. Comput.,* Vol. 38, No. 4, 416–426, 2000.

Giannakis, G. B., and Mendel, J. M., Identification of Non-Minimum Phase Systems Using Higher-Order Statistics, *IEEE Trans. on Acoustics, Speech and Signal Processing,* Vol. 37, No. 3, 360 –377, 1989.

Hinich, M. J., Detecting a Transient Signal by Bispectral Analysis, *IEEE Trans. on Acoustic, Speech and Signal Processing,* Vol. 38, No. 7, 1277–1283, 1990.

Hinich, M. J., Testing for Gaussianity and Linearity of a Stationary Time Series, *J. Time Series Analysis,* Vol. 3, 169–176, 1982.

Hinich, M. J., and Messer, H., On the Principal Domain of the Discrete Bispectrum of a Stationary Signal, *IEEE Trans. on Signal Processing,* Vol. 43, No. 9, 2130–2134, 1995.

Hinich, M. J., and Wolinsky, M. A., A Test for Aliasing Using Bispectral Analysis, *J. American Statistical Association,* Vol. 83, No. 402, 449–502, 1988.

Jamsek, J., Stefanovska, A., and McClintock, P. V., Nonlinear Cardio-Respiratory Interactions Revealed by Time-Phase Bispectral Analysis, *Phys Med Biol.,* Vol. 49, No. 18, 4407–4425, 2004.

Kitney, R. I., Linkens, D., Sekman, A., and McDonald, A., The Interaction Between Heart Rate and Respiration: Part II—Non-Linear Analysis Based on Computer Modelling, *Automedica,* Vol. 4, 141–153, 1982.

Levy, W. J., Power Spectrum Correlates of Changes in Consciousness During Anesthetic Induction with Enflurane, *Anesthesiology,* Vol. 64, No. 6, 688–693, 1986.

Mendel, J. M., Cumulant-based Approach to Harmonic Retrieval and Related Problems, *IEEE Trans. on Signal Processing.* Vol. 39, No. 5, 1099–1109, 1991.

Nikias, C. L., and Mendel, J. M., Signal processing with higher-order spectra, *IEEE Signal Processing Magazine*, Vol. 10, No. 3, 10–37, 1993.

Papoulis, A., *Probability, Random Variables and Stochastic Processes*, McGraw-Hill, 1991.

Pinhas, I., Toledo, E., Aravot, D., and Akselrod, S., Bicoherence Analysis of New Cardiovascular Spectral Components Observed in Heart-Transplant Patients: Statistical Approach for Bicoherence Thresholding, *IEEEE Trans. Biomed. Eng.*, Oct., 51, 2004.

Raghuveer, M. R., and Nikias, C. L., Bispectrum Estimation: A Parametric Approach, *IEEE Trans. on Acoustics, Speech and Signal Processing*, Vol. 33, No. 5, 1213–1230, 1985.

Rampil, I. J., and Laster, M. J., No Correlation Between Quantitative Electroencephalographic Measurements and Movement Response to Noxious Stimuli During Isoflurane Anesthesia in Rats, *Anesthesiology*, Vol. 77, No. 5, 920–925, 1992.

Rampil, I. J., and Matteo, R. S., Changes in EEG Spectral Edge Frequency Correlate with the Hemodynamic Response to Laryngoscopy and Intubation, *Anesthesiology*, Vol. 67, No. 1, 139–142, 1987.

Rampil, I. J., Weiskopf, R. B., Brown, J. G., Eger, E. I., 2nd, Johnson, B. H., Holmes, M. A., and Donegan, J. H., I653 and Isoflurane Produce Similar Dose-Related Changes in the Electroencephalogram of Pigs, *Anesthesiology*, Vol. 69, No. 3, 298–302, 1988.

Rampil, I. J. A Primer for EEG Signal Processing in Anesthesia, *Anesthesiology*, Vol. 89, No. 4, 980–1002, 1998.

Schwab, K., Eiselt, M., Putsche, P., Helbig, M., and Witte, H., Time-Variant Parametric Estimation of Transient Quadratic Phase Couplings Between Heart Rate Components in Healthy Neonates, *Med. Biol. Eng. Computing*, Vol. 44, No. 12, 1077–1083, 2006.

Task Force Heart Rate Variability. Standard of Measurement, Physiological Interpretation and Clinical Use, *Circulation*, Vol. 93, 1043–1065, 1996.

PART VI

INFORMATION PROCESSING OF MOLECULAR BIOLOGY DATA

Thispageintentionallyleftblank

CHAPTER *17*

MOLECULAR BIOENGINEERING AND NANOBIOSCIENCE
Data Analysis and Processing Methods

Carmelina Ruggiero

17.1 INTRODUCTION

In the past few years, a growing interest has arisen in using nanotechnolo-
gy in biomedical applications, particularly molecular bioengineering,
whereas microscale technology is often used in cellular engineering. Contributions to nanobioscience and molecular bioengineering are made, at
times independently and at times collaboratively, by scientists with various backgrounds: physicists, chemists, biologists, engineers, mathematicians, and computer scientists.

Molecular and subcellular aspects of bioengineering and the related
technology at the nanometer scale have been considered extensively. The
recent developments in microscopy, measurement methods, nanostructure
fabrication, and informatics have fueled this interest in nanometer-scale
research. Specifically, atomic force microscopy has led to the ability to
measure nanometer-scale surface properties of biological samples such as
living cells, DNA, proteins, and biomaterials. Other methods, such as microarray-based analysis and spectroscopy for the characterization of biomolecular structures, have also achieved significant developments. A body
of knowledge is being established that provides a basis for measurement
methods and techniques at the nanometer scale.

The developments mentioned above in measurement methods and
techniques have been paralleled by fabrication methods and techniques for
the fabrication of ordered nanostructures, nanomaterials, and nanodevices
for a wide variety of applications. The biological, biotechnological, and

Advanced Methods of Biomedical Signal Processing. Edited by S. Cerutti and C. Marchesi
Copyright © 2011 the Institute of Electrical and Electronics Engineers, Inc. **429**

medical ones are relevant parts of these applications. Such methods include self-assembly, electron-beam lithography, and nanocontact printing. Some related areas in the biomedical field are biosensors for experimental laboratory work, clinical purposes, and environmental applications (Vo-Dinh and Cullum, 2000; Liu et al., 2003; Castillo et al., 2004; Kissinger, 2005; Rodriguez-Mozaz et al., 2005; Mohanty and Kougianos, 2006; Pastorino et al., 2006a; Erickson et al., 2008; Borisove and Wolbeis, 2008); the fabrication of implants to be inserted in the human body for tissue engineering and regenerative medicine; the biocompatibility of materials at the nanoscale level; the design and fabrication of surfaces by assembly of ensembles of molecules; and surface patterning to guide specific interactions with molecules and cells (Brown, 2000; Desai, 2000; Evans, 2001; Wang et al., 2001; Tyroen-Toth et al., 2001; Wilkinson et al., 2002; Sinani et al., 2003; Wang and Lineaweaver, 2003; Belkas et al., 2004; Li et al., 2005; Pastorino et al., 2006b; Soumetz et al., 2008). Most interestingly, cells have been found to be capable of perceiving details on surfaces at the nanoscale; therefore, the nanotopography of surfaces (and, specifically, order and symmetry) impacts cell adhesion onto surfaces and cell–cell interactions in general (Curtis, 2004; Ruggiero et al., 2005).

The design of molecules, especially in genomics and proteomics, is closely related to the areas described above, both as relates to experimental methods and techniques and as relates to bioinformatics, which plays a key role in the present postgenomic period. Some significant examples are software tools for data mining, workflows, and integrative bioinformatics; methods and techniques for the sequence-structure analysis of proteins; molecular dynamics tools; docking modeling and simulation; and DNA microarray and protein microarray data processing tools.

A further related area is molecular electronics, which focuses on the use of molecules to perform tasks that are at present performed by semiconductor-based elements in electronic circuits. Molecules have much smaller dimensions than those that can be obtained today by semiconductor technology, and their use would allow reducing dimensions and, therefore, increasing potential applications, such as computational power.

The convergence of molecular oriented approaches by different scientific communities has lead to a need for interdisciplinary and more coherent approaches. In this respect, as relates to biomedical data and signal analysis, it may be observed that often methods and techniques that are employed are common to more than one scientific community involved in molecular engineering. Specifically, some methods that are being used for the study of DNA and proteins have also been used for the analysis of data sequences of various kinds, and methods and techniques used in biomedical signal analysis have been successfully used in genomics and proteomics.

Since the 1980s, Fourier analysis has been applied along with biomedical signal analysis for the prediction of protein structures, specifically (but not only) for the study of the hydrophobicity of sequences. At the same time, work based on Fourier analysis has been carried out for DNA sequencing, both for the identification of structural patterns in the double helix and for singling out variations in the double helix in relation to specific nucleotides. Since then, Fourier-analysis-based methods have been successfully used for the analysis of various aspects of DNA and protein sequencing.

At the end of the 1980s, the wavelet transform was introduced. Whereas in Fourier analysis the basic functions are localized in frequency but not in time, the basic functions in wavelet analysis are local both in the time domain and in the frequency domain. For several kinds of signals, wavelet-transform-based analysis allows one to obtain a more compact representation than Fourier analysis. This is the case for signals with local peaks, such as some biomedical signals, in which the presence of such peaks in the time domain originates many components in the frequency domain. Wavelet-transform analysis has been successfully applied to various kinds of signals and images, including biomedical signals and images, and in sequence and image analysis for genomics (Arneodo et al., 1995; Audit et al., 2002; Audit and Ouzounis, 2003; Chain et al., 2003; Aggarwal et al., 2005; Touchon et al., 2005; Haimovich, 2006; Kwan et al., 2006; Thurman et al., 2007). The recent achievements in genome sequencing and in DNA microarray technology have originated a significant amount of data analysis and processing work. A variety of methods have been used. In genomics, especially as relates to microarrays and to the problem of inferring significance from microarray data, clustering methods and other statistical methods are often used, as well as machine learning methods and data mining methods. Such methods have been used for many years, prior to the advent of microarrays, for biomedical signal and image analysis. Simple clustering methods have been used since the 1960s for biomedical signal analysis, such as evoked potentials, ECGs, and other signals. More recently, artificial neural networks (whose main feature is the capability of learning from examples, that is, of generalizing and extracting knowledge from data) have been successfully applied for the solution of many types of pattern recognition, prediction, and classification problems both for biomedical signals and images, and for genomics and proteomics.

In the following, analysis and processing methods that play a key role in genomics and proteomics are described.

17.2 DATA ANALYSIS AND PROCESSING METHODS FOR GENOMICS IN THE POSTGENOMIC ERA

In the past several years since the time when the sequencing of the human genome was completed, rapid progress in the understanding of the human genome and the genetic basis of disease has been achieved. Moreover, high-throughput technologies related to microarrays have brought about the rapid generation of large-scale datasets focused on genes and gene products. The great increase of the available information on DNA sequences and the development of methods and techniques for the use of this knowledge are the main aspects of a most significant transition phase in biomedical research. The recent achievements are opening up the way to the identification of further topics and to the solution of related problems. Using sequence-coupling techniques, it is possible to identify sequences that are related to diseases, to locate such sequences in the human genome, and to identify specific genes. However, in spite of the constant increase in these data, many genes are not known, and the same applies to most functions of the genes that have been discovered, to the processes leading to proteins, and to the regulatory mechanisms that control such processes.

The production of data by centers that obtain sequences has increased to a great extent in the last two decades, whereas the analysis of sequences and genomes has not progressed at the same rate.

The continuing improvements in gene sequencing and the continuing increases in sequence databases have led to a demand for the functional analysis that follows sequencing. Comparing complete genomes is the next step in solving problems such as coding region or regulation-signal identification (Xie and Hood, 2003; Frazer et al., 2004; El-Sayed et al., 2005; Notredame, 2007; Zhou et al., 2008). The main requirements for such analyses are sequence comparison, visualization, and analysis.

17.2.1 Genome Sequence Alignment

Sequence alignment has provided one of the main tools in sequence analysis and has led to the development of a great number of informatic and statistical genome-oriented tools (Margulies, 2008; Huang et al., 2007). Moreover, Web-based tools have been developed that provide shared databases and data mining and processing software (Carmona-Saez et al., 2007; Brudno, 2007; Brudno et al., 2007; Bruford et al., 2008; Karolchik et al., 2008). A variety of alignment algorithms are available (Li, 1997; Bradley et al., 2008; Kapustin, 2008). They are based on scoring all possible alignments according to similarity/identity parameters for each residue, followed by alignment optimization. The early algorithms relate to DNA sequences containing one gene only, whereas longer alignments bring about computational speed problems. Comparing complete genomes requires solving a great number of problems, such as dealing with repeated elements, including or eliminating elements, and reorganizations. Some of these problems have been solved using algorithms that find exact length correspondences, and start from minimum length correspondences and form contiguous, more extended correspondences.

On the basis of the observation of similar aspects of analogous genes, it is possible to obtain insights into the possible regulation and functions of such genes and on their evolutionistic history.

17.2.2 Genome Sequence Analysis

DNA sequences can be represented as symbolic strings relating to the nucleotides adenine, cytosine, guanine, and thymine (A, C, G, T), similar to the 20 amino acid strings for proteins. The correlation structure of strings can be completely characterized starting from the possible nucleotide–nucleotide correlation functions or from their corresponding power spectra. More generally, the statistical analysis of DNA sequences plays an important role in the understanding of the structure and function of genomes.

The knowledge and understanding of the correlation among bases in DNA sequences has been of great interest for a long time. Before the human genome project, long and continuous DNA sequences were not available; therefore, only short correlations were considered. Since then, long DNA sequences have become easily available and it has been possible to obtain more complete characterizations of correlations among base couples both for short distances and for long distances (Herzel and Grope, 1997; Choe et al., 2000).

From a molecular biology point of view, long-distance correlations are not surprising, since the complex organization of genomes involves relations at very diverse distances. For example, it has been experimentally proved that fragments containing up to 104 base couples exhibit rather large variances in the content of the sum of guanine plus cytosine, and this cannot be explained by considering fluctuations relating to short distances [49]. More recently, pronounced fluctuations in the content of the sum of guanine plus cytosine with a period of about 105 couples of bases have been found (Choe et al., 2000).

A DNA sequence can be regarded as a string of characters whose correlation structure can be characterized by all possible base–base functions or by their corresponding power spectra. The correlation structure of DNA sequences can be evaluated by analyzing various estimators such as Fourier spectra and the wavelet transform (the latter has been found to be a very useful for the study of the heterogeneity of DNA sequences (Herzel and Grope, 1997). Other algorithms that can be used are based on machine learning methods, such as artificial neural networks, which provide the most powerful tools for pattern recognition problems and have, therefore, been employed to extract information from DNA sequences. Some examples are the identification of the genome regions that code proteins, predicting mRNA donor and acceptor sites for the DNA sequence (Abe et al., 2003), and the use of the oligonucleotide frequency in order to distinguish genomes (Wu, 1997).

Other applications relate to gene identification, which can be achieved by two complementary approaches: search by content (which takes into account the protein coding potential of sequences) and search by signal (based on the identification of sequences that limit coding regions) (Oakley and Hanna, 2004). Artificial neural networks have been successfully used for the identification and analysis of sites (for example, regulation sites and transcription sites), for sequence classification, for the identification of significant sequence features, and for the understanding of biological rules that guide the structure and regulation of genes (Oakley and Hanna, 2004).

17.2.3 DNA Microarray Data Analysis

DNA microarrays, which are very frequently used for genome analysis, provide the most relevant contributions for the understanding of DNA sequences. Data mining tools (such as Bayesian networks, clustering algorithms, genetic algorithms, Markov models, and artificial neural networks) are being extensively used in DNA microarray data analysis (Bertone and Gerstein, 2001; Valafar, 2002; Greer and Khan, 2007; Kim and Cho, 2008; Tan et al., 2008).

Microarray technology is based on the immobilization of fragments of oligonucleotides with known sequences on matrixes and on their hybridization by exposure to DNA markers. The signals corresponding to hybridized fragments are quantified and originate an image that is the result of the simultaneous examination of thousands of genes.

The analysis of microarray data includes the search for genes that have similar or related expression patterns. In order to understand and interpret data deriving from microarray technology, specifically tailored computational methods have been developed (Salzburg, 1998), even though their basis lies in statistical and computational intelligence methods (such as genome data analysis in a broader sense). Data analysis methods aim to identify correlations between the microarray data and an underlying function or biological condition.

For a specific function or biological condition, the question to be answered is whether, when this function or condition is present, the expression levels of genes or gene sequences change significantly with respect to the case in which it is absent. A simple example of such analysis is the t-test, which compares averages of observations. Another example is principal-component analysis, a linear technique that finds basis vectors (principal components) that expand the space of the problem (the genic expression space). A principal component can be regarded as a relevant pattern in the ensemble of the genic expression data. Other statistical methods, such as Bayesian analysis, can take into account aspects relating to noise and to the typical variability of microarray data. Clustering methods, both k-means and hierarchical clustering, originate simple and easy to set up tools that have been successfully applied to genic expression data. For example, k-means analysis can be applied to microarray data forming a fixed number of groups with similar expression patterns. All data can be randomly assigned to each of the groups. Subsequently, the distances among the data and the averages of each group and the distances among the averages of each group are calculated. At this point, distances within each group and distances among groups are maximized. Hierarchical clustering algorithms can be used for microarray data, establishing similarities among groups and using them as a basis to form new groups.

Artificial neural networks provide a further, most suitable analysis method for the classification of DNA microarray data. Self-organizing neural networks have features that are to some extent similar to k-means algorithms, but each group is represented by a node to which a specific weight is associated. Weights and positions of nodes are updated during a learning process in which relations among groups are obtained. When it is possible to establish the number of classes, the most appropriate method is supervised learning by backpropagation training, in which training takes place starting from a dataset for which the classification is known. The following phase classifies the data for which the classification is not available.

17.3 FROM GENOMICS TO PROTEOMICS

The analysis of DNA sequences is mostly aimed at charatecterizing sequences and genes. For proteins, the analysis of the amino acid sequences is mostly aimed at secondary and tertiary structure prediction, the identification of sequence patterns related to functional domains, and the prediction of the function of molcules or of domains of proteins (Tyers and Mann, 2003). The most frequently used methods are based on homologies with known molecular structures and on analyses starting from basic principles (using knowledge of fundamental atomic interactions and energy-based approaches).

The term proteomics originated within the work carried out on the genome. This term relates to the study of the proteins that are present in the cell, to structural descriptions of proteins and their interactions, to the description of protein complexes, and to protein structure modifications needed to change protein structure. Work on such aspects has great prospects for improving the understanding of cellular functions and drug design.

Proteomics can be regarded as an area that complements functional genomic aspects, such as genic expression profiles based on microarrays and phenotypic profiles at the cellular and organism levels (Baker and Sal, 2001). Proteomics is based on recent results in genomics, which elucidated aspects relating to genes that brought about specific analyses of the related proteins. Genome sequencing discovers sequences of amino acids. However, in order to understand the biological role of the corresponding proteins it is necessary to know their structure, which determines their function. Functional genomics, which uses experimental and algorithmic methods to characterize protein sequences, is focused on this aim. Various approaches have been adopted, ranging from focusing on folded structure only to analyzing all proteins that are present in one genome, which is taken as a model (Pei, 2008).

17.4 PROTEIN STRUCTURE DETERMINATION

After many years of work, the determination of protein structure remains one of the key goals of computational biology. Protein structure knowledge is particularly important for functional and structural genomics. Experimental methods (especially X-ray crystallography and NMR spectroscopy) have determined the structure of a great number of proteins. Often, such methods are complemented by computer-based structure-prediction methods, which play a key role in cases in which experimental determination is not possible or difficult.

For proteins whose structure is similar to the structure of known proteins, structure prediction can be carried out by locating similar structure parts and aligning them with the unknown sequence. This can be achieved by simple methods that have been available for quite some time if the structure identity with a known protein is greater than 25–30%. The most effective methods that are presently available use sequence alignment, prediction algorithms without use of known structures, and algorithms based on conformational energy (Lim, 1974a, b; Dumas and Ninio, 1982; Keskin et al., 2005; Floudas et al., 2006; Katzman et al., 2008; Viklund and Elofsson, 2008).

Computer-based methods for protein structure prediction have been used since the 1970s. The most frequently used methods are the ones for the prediction of secondary structure elements. The first methods that were set up are based on simple stereochemical principles (Lim, 1974b; Fasman, 1989) that take into account structural features such as compactness and the presence of an internal hydrophobic, tightly compressed part and an external, polar part. Another very early method (Garnier et al., 1978) focuses on the frequency of occurrence of each of three conformations (alpha helix, beta sheet, and coil) in the residues present in a set of proteins. Yet another method uses parameters based on the frequency of occurrence of an amino acid in one protein, on its presence in each type of secondary structure, and on the percentage of amino acids in that structure type, together with empirical rules. A further early method is based on the observation that the conformation of an amino acid depends on the amino acid that surrounds it in the sequence.

For amphipatic structure prediction, the hydrophobic pattern has been recognized as a key element, so effective prediction methods are based on it. Fourier analysis of the hydrophobic profile has the great advantage of taking into account cooperativity among amino acids in protein folding. This aspect is difficult to take into account in methods based on the use of databases (Garnier et al., 1978).

Starting from the late 1980s, artificial neural networks have been used to a significant extent, and better results have been obtained with respect to previously used methods. Moreover, encouraging results have been obtained even in some tertiary-structure prediction cases (Oakley and Hanna, 2004). It can be noticed that, in general, the use of neural networks has given very good results for many kinds of molecular sequence analysis problems. Moreover, the association of neural-net-based methods and of other methods such as de novo structure-prediction methods have given most promising results (Pei, 2008). The latter are based on the assumption that the native state of a protein corresponds to the minimum of the free energy. Free-energy-based methods have limitations deriving from the great number of variables that are involved, from the uncertainties regarding the formulae of the terms that represent the energy, and from the fact that many conformations exist that correspond to local minima of the global potential energy. Using fragments of known structures allows one to reduce such limitations to a great extent.

17.5 CONCLUSIONS

The methods described above play a key role in the achievement of a body of knowledge relating to molecular bioengineering and nanobioscience. These methods include statistical ones, databases, mathematical modeling, and machine learning, and relate to several existing disciplines.

REFERENCES

Abe, T., Kanaya, S., Kinouchi, M., Ichiba, Y., Kozuki, T., and Ikemura, T., Informatics for Unveiling Hidden Genome Signatures, *Genome Research,* 693–702, 2003.

Aggarwal, A., Leong, S. H., Lee, C., Kon, O. L., and Tan, P., Wavelet Trans- formations of Tumor Expression Profiles Reveals a Pervasive Genome-Wide Imprinting of Aneuploidy on the Cancer Transcriptome, *Cancer Res.,* Vol. 65,186–194, 2005.

Arneodo, A., Bacry, E., Graves, P. V., and Muzy, J. F., Characterizing Long-Range Correlations in DNA Sequences from Wavelet Analysis, *Phys. Rev.Lett.,* Vol. 74, 3293–3296, 1995.

Audit, B., and Ouzounis, C. A., From Genes to Genomes: Universal Scale-invariant Proprties of Microbial Chromosome Organisation, *J. Mol. Biol.,* Vol. 332, 617–633, 2003.

Audit, B., Bacry, E., Muzy, J. F., and Arneodo, A., Wavelet-Based Estimators of Scaling Behavior, *IEEE Transactions on Information Theory,* Vol. 48, No. 11, 2002.

Baker, F. and Sali A., Protein Structure Prediction and Structural Genomics, Science, Vol. 294, No. 5540, 93–96, 2001.

Belkas, J. S., Shoichet, M. S., and Rajiv, M., Peripheral Nerve Regeneration Through Guidance Tubes, *Neurological Research.,* Vol. 26, 151–160, 2004.

Bertone, P., and Gerstein, M., Integrative Data Mining: The New Direction in Bioinformatics, *IEEE Engineering in Medicine and Biology,* July/August, 2001.

Borisov, S. M., and Wolfbeis, O. S., Optical Biosensor, *Chem. Rev.,* Vol. 108, No. 2, 423–461, 2008.

Bradley, R. K., Pachter, L., and Holmes, I., Specific Alignment of Structured RNA: Stochastic Grammars and Sequence Annealing, *Bioinformatics*, Vol. 24, No. 23, 2677–2683, 2008.

Brown, R. A., Bioartificial Implants: Design and Tissue Engineering, In *Structural Biological Materials: Design And Structure—Property Relationships*, Elices, M. (Ed.), pp. 105–160, Pergamon, 2000.

Brudno, M., An Introduction to the Lagan Alignment Toolkit, *Comparative Genomics*, Vol. 395, 205–219, 2007.

Brudno, M., Poliakov, A., Minovitsky, S., Ratnere, I., and Dubchak, I., Multiple Whole Genome Alignments and Novel Biomedical Applications at the VISTA Portal, *Nucleic Acids Res.*, Vol. 35 (Web Server issue), W669–674,

Bruford, E. A., Lush, M. J., Wright, M. W., Sneddon, T. P., Povey, S., and Birney, E., The HGNC Database in 2008: A Resource for the Human Genome, *Nucleic Acids Res.*, Vol. 36 (Database issue), D445–448, 2008.

Carmona-Saez, P., Chagoyen, M., Tirado, T., Carazo, J. M., and Pascual-Montano, A., GENECODIS: a web-based tool for finding significant concurrent annotations in gene lists, *Genome Biology*, Vol. 8, R3, 2007.

Castillo, J., Gáspár, S., Leth, S., Niculescu, M., Mortari, A., Bontidean, I., Soukharev, V., Dorneanu, S. A., Ryabov, A. D., and

Csöregi, E., Biosensors for Life Quality: Design, Development and Applications, *Sensors and Actuators B, Chemical*, Vol. 102, 79, 2004.

Chain, P., Kurtz, S., Ohlebush, E., and Slezak, T., An Applications-Focused Review of Comparative Genomics Tools: Capabilities, Limitations and Future Challenges, *Briefings in Bioinformatics*, Vol. 4, No. 2, 105–123, 2003.

Choe, W., Ersoy, O. K., and Bina, M., Neural Network Schemes for Detecting Rare Events in Human Genomic DNA, *Bioinformatics*, Vol. 16, No. 12, 1062–1072, 2000.

Curtis, A., Tutorial on the Biology of Nanotopography, *IEEE Trans Nanobioscience*, Vol. 3, No. 4, 293–295, 2004. Database: 2008 Update, *Nucl. Acids Res.*, Vol. 36, D773–D779, 2008.

Desai, T. A., Micro- and Nanoscale Structures for Tissue Engineering Constructs, *Med. Eng. Phys.*, Vol. 22, 595–606, 2000.

Dumas, J., and Ninio, J., Efficient Algorithms for Folding and Comparing Nucleic Acid Sequences, *Nucleic Acids Res.*, Vol. 10, 197–206, 1982.

El-Sayed, N.M., Myler, P. J., Blandin, G., Berriman, M., Crabtree, J., Aggarwal, G., Caler, E., Renauld, H., Worthey, E. A., Hertz-Fowler, C., Ghedin, E., Peacock, C., Bartholomeu, D. C., Haas, B. J. Tran, A., Wortman, J. R., Alsmark, U. C. M., Angiuoli, S., Anupama, A., Badger, J., Bringaud, F., Cadag, E., Carlton, J. M., Cerqueira, G. C., Creasy, T., Delcher, A. L., Djikeng, A.,Embley, T. M., Hauser, C., Ivens, A. C., Kummerfeld, S. K., Pereira-Leal, J. B., Nilsson, D., Peterson, J., Salzberg, S. L., Shallom, J., Silva, J. C., Sundaram, J., Westenberger, S., White, O., Melville, S. E., Donelson, J. E., Andersson, B., Stuart, K. D., and Hall, N., Comparative Genomics of Trypanosomatid Parasitic Protozoa, *Science*, Vol. 309, No. 5733, 404, 2005.

Erickson, D., Mandal, S., Yang, A. H. J., and Cordovez, B., Nanobiosensors:

Optofluidic, Electrical and Mechanical Approaches to Biomolecular Detection at the Nanoscale, *Microfluid Nanofluid.*, Vol. 4, 33–52, 2008.

Evans, G. R., Peripheral Nerve Injury: A Review and Approach to Tissue Engineered Constructs, *Anatomical Record*, Vol. 263, 396–404, 2001.

Fasman, G. D., *Prediction of Protein Structure and the Principles of Protein Conformation*, Plenum Press, 1989.

Floudas, C. A., Fung, H. K., et al. Advances in Protein Structure Prediction and De Novo Protein Design: A Review, *Chemical Engineering Science*, Vol. 61, No. 3, 966–988, 2006.

Frazer, K. A., Pachter, L., Poliakov, A., Rubin, E. M., and Dubchak, I., VISTA: Computational Tools for Comparative Genomics, *Nucl. Acids Res.*, Vol. 32, W273–W279, 2004.

Garnier, J., Osguthorpe, D. J., and Robsonk B., Analysis of the Accuracy and Implications of Simple Methods for Predicting the Secondary Structure of Globular Proteins, *J. Mol. Biol.*, Vol. 120, 97–120, 1978.

Greer, B., and Khan, J., Online Analysis of Microarray Data Using Artificial Neural Networks, *Microarray Data Analysis*, Vol. 377, 61–73, 2008.

Haimovich, A. D., Byrne, B., Ramaswamy, R., and Welsh, W. J., Wavelet Analysis of DNA Walks, *Journal of Computational Biology*, Vol. 13, No. 7, 1289–1298, 2006.

Herzel, H., and Grope, I., Correlations in DNA Sequences: The Role of Protein Coding Segments, *Physical Review E*, Vol. 55, No. 1, 1997.

Huang, D. W., Sherman, B. T., Tan, Q., Collins, J. R., Alvord, W. G., Roayaei, J., Stephens, R., Baseler, M. W., Lane, H. C., and

Lempicki, R. A., The DAVID Gene Functional Classification Tool: A Novel Biological Module- Centric Algorithm to Functionally Analyze Large Gene Lists, *Genome Biology*, Vol. 8, R183, 2007.

Kapustin, Y., Souvorov, A., Tatusova, T., and Lipman, D., Splign Algorithms for Computing Spliced Alignments with Identification of Paralogs, *Biol. Direct.*, Vol. 3, 20, 2008.

Karolchik, D., Kuhn, R. M., Baertsch, R., Barber, G. P., Clawson, H., Diekhans, M., Giardine, B., Harte, R. A., Hinrichs, A. S.,

Hsu, F., Kober, K. M., Miller, W., Pedersen, J. S., Pohl, A., Raney, B. J., Rhead, B., Rosenbloom, K. R., Smith, K. E., Stanke, M., Thakkapallayil, A., Trumbower, H., Wang, T., Zweig, A. S., Haussler, D., and Kent, W. J., The UCSC Genome Browser

Katzman, S., C. Barrett, et al. PREDICT-2ND: A Tool for Generalized Protein Local Structure Prediction, *Bioinformatics*, Vol. 24, No. 21, 2453–2459, 2008.

Keskin, O., Nussinov, R., and Gursoy, A., Prism: Protein–Protein Interaction Prediction by Structural Matching, *Functional Proteomics*, Vol. 484, 505–521, 2005.

Kim, K.-J., and Cho, S.-B., An Evolutionary Algorithm Approach to Optimal Ensemble Classifiers for DNA Microarray Data Analysis, *IEEE Transactions on Evolutionary Computation*, Vol. 12, No. 3, 377–388.

Kissinger, P. T., Biosensors—A Perspective, Biosensors and Bioelectronics, Vol. 20, 2512, 2005.

Kwan, B. Y. M., Kwan, J. Y. Y., and Kwan, H. K., Wavelet Analysis of the

Genome of the Model Plant *Arabidopsis thaliana,* in *TENCON 2006. 2006 IEEE Region 10 Conference.* 14–17 Nov. 2006.

Li, M., Mills, D. K., Cui, T., and McShane, M. J., Cellular Response to Gelatin- and Fibronectin-Coated Multilayers Polyelectrolyte Nanofilms, *IEEE Transactions on Nanobioscience,* Vol. 4, No. 2, 170–179, 2005.

Li, W., The Study of Correlation Structures of DNA Sequences—A Critical Review, *Computers & Chemistry,* Vol. 21, No. 4, 257–272, 1997.

Lim V.I., Algorithms for Prediction of _-Helices and _-Structural Regions in Globular Proteins, *J. Mol. Biol.,* Vol. 88, 873–894, 1974b.

Lim, V. I., Structural Principles of the Globular Organization of Protein Chains: A Stereochemical Theory of Globular Protein Secondary Structure, *J. Mol. Biol.,* Vol. 88, 857–872, 1974a.

Liu, Y., Yu, X., Zhao, R., Shangguan, D. H., Bo, Z. Y., and Liu, G. Q., Quartz Crystal Biosensor for Real-Time Monitoring of Molecular Recognition Between Protein and Small Molecular Drug, *Biosensors and Bioelectronics,* Vol.19, 9, 2003.

Margulies, E. H., Confidence in Comparative Genomics, *Genome Res.,* Vol. 18, 199–200, 2008.

Mohanty, S. P., and Kougianos, E., Biosensors: A Tutorial Review, *IEEE Potentials,* March/April, 35–40, 2006.

Notredame, C., Recent Evolutions of Multiple Sequence Alignment Algorithms, *PLoS Computational Biology,* Vol. 3, No. 8, 2007.

Oakley, B. A., and Hanna, D. M., A Review of Nanobioscience and Bioinformatics Initiatives in North America, *IEEE Transactions on NanoBioscience,* Vol. 3, 1, 74–84, 2004.

Pastorino, L., Soumetz, F. C., and Ruggiero, C., Nanofunctionalisation for the Treatment of Peripheral Nervous System Injuries, *IEE Proceedings Nanobiotechnology,* Vol. 153, No. 2, 16–20, 2006.

Pastorino, L., Soumetz, F. C., Giacomini, M., and Ruggiero, C., Development of a Piezoelectric Immunosensor for Paclitaxel Measurement, Journal of Immunological Methods, Vol. 313, 119–198, 2006a.

Pei, J., Multiple Protein Sequence Alignment, *Current Opinion in Structural Biology,* Vol. 18, No. 3, 382–386, 2008.

PNAS, Vol. 102, 9836–9841, 2005.

Rodriguez-Mozaz, S., López de Alda, M. J., Marco, M. P., and Barceló, D., Biosensors for Environmental Monitoring: A Global Perspective, *Talanta,* Vol. 65, 291, 2005.

Ruggiero, C., Mantelli, M., Curtis, A., and Rolfe, P., Protein–Surface Interactions: An Energy-Based Mathematical Model, *Cell Biochem Biophys.,* Vol. 43, No. 3, 407–417, 2005.

Salzburg, S. L., Searls, D. B., and Kash, S., *Computational Methods in Molecular Biology,* Elsevier, 1998.

Sinani, V. A., Koktysh, D. S., Yun, B. G., Matts, R. L., Pappas, T. C., Motamedi, M., Thomas, S. N., and Kotov, N. A., Collagen Coating Promotes Biocompatibility of Semiconductor Nanoparticles in Stratified LBL Films, *Nano Letters,* Vol. 3, No. 9, 1177–1182, 2003.

Soumetz, F. C., Pastorino, L., and Ruggiero, C., Human Osteoblast-Like Cells Response to Nanofunctionalised Surfaces for Tissue Engineering, *Journal of Bio-*

medical Materials Research—Part B, Applied Biomaterials, Vol. 84B, No. 1, 249–255, 2008.

Tan, M. P., Smith, E. N., Broach, J. R., and Floudas, C., A., Microarray Data Mining: A Novel Optimization-Based Approach to Uncover Biologically Coherent Structures, *BMC Bioinformatics,* Vol. 9, 268, 2008.

Thurman, R. E., Day, N., Noble, W. S., and Stamatoyannopoulos, J. A., Identification of Higher-Order Functional Domains in the Human ENCODE Regions, *Genome Res.,* Vol. 17, 917–927, 2007.

Touchon M., Nicolay, S., Audit, B., Brodie of Brodie, E.-B., d'Aubenton- Carafa, Y., Arneodo, A., and Thermes, C., Replication-Associated Strand Asymmetries in Mammalian Genomes: Toward Detection of Replication Origins,

Tyers, M., and Mann, M., From Genomics to Proteomics, *Nature,* Vol. 442,193–197, 2003.

Tyroen-Tóth, P., Vautier, D., Haikel, Y., Voegel, J., Schaaf, P., Chluba, J., and Ogier, J., Viability, Adhesion, and Bone Phenotype of Osteoblast-Like Cells on Polyelectrolyte Multilayer Films, *J. Biomed. Mater. Res.,* Vol. 60, 657–667, 2002.

Valafar F., Pattern Recognition Techniques in Microarray Data Analysis: A Survey, *Annals of New York Accademy of Sciences,* Vol. 980, 41–64, December 2002.

Viklund, H., and Elofsson, A., OCTOPUS: Improving Topology Prediction by Two-Track ANN-Based Preference Scores and an Extended Topological Grammar, *Bioinformatics,* Vol. 24, No. 15, 1662–1668, 2008.

Vo-Dinh, T., and Cullum, B., Biosensors and Biochips: Advances in Biological and Medical Diagnostics, *Fresenius J. Anal. Chem.,* Vol. 366, 540–551, 2000.

Wang, H., and Lineaweaver, W., Nerve Conduits for Nerve Reconstruction, *Operative Techniques in Plastic and Reconstructive Surgery,* Vol. 9, 59–66, 2003.

Wang, Y., Du, W., Spillman, W. B., and Claus, R. O., Biocompatible Thin Film Coatings Fabricated Using the Electrostatic Self-Assembly Process, *Proc. SPIE,* Vol. 4265, 142–151, 2001.

Wilkinson, C. D. W., Riehle, M., Wood, M., Gallagher, J., and Curtis, A. S. G., The Use of Materials Patterned on a Nano- and Micro-Metric Scale in Cellular Engineering, *Mater. Sci. Eng., C,* Vol. 19, 263–269, 2002.

Wu, C. H., Artificial Neural Networks for Molecular Sequence Analysis, *Computers & Chemistry,* Vol. 21, No. 4, 237–256, 1997.

Xie, T., and Hood, L., ACGT—A Comparative Genomics Tool, *Bioinformatics,* Vol. 19, No. 8, 1039–40, 2003.

Zhou, J., Zhu, T., Hu, C., Li, H., Chen, G., Xu, G., Wang, S., Zhou, J., and Ma, D., Comparative Genomics and Function Analysis on Bl1 Family, *Computational Biology and Chemistry,* Vol. 32, No. 3, 159–162, 2008.

Thispageintentionallyleftblank

CHAPTER 18

MICROARRAY DATA ANALYSIS
General Concepts, Gene Selection, and Classification

Riccardo Bellazzi, Silvio Bicciato, Claudio Cobelli,
Barbara Di Camillo, Fulvia Ferrazzi, Paolo Magni,
Lucia Sacchi, and Gianna Toffolo

18.1 INTRODUCTION

Discoveries from the genome sequencing projects facilitated the development of novel techniques able to screen thousands of molecules in parallel and identify sets of potentially interesting sequences associated with physiological/pathological conditions. As a consequence, high-throughput, large-scale experimental methodologies, combined with bioinformatics analysis of DNA, RNA, and protein data projected biological sciences into the so-called post-genomic functional genomics era. The exploration of all genes or proteins at once, in a systematic fashion, represents a sort of revolution, shifting molecular biology and medicine research from a reductionistic, hypothesis-driven approach toward deciphering how genes and their products work, how they interact in pathways within the cells, and what roles they play in health and disease (Chipping Forecast I, 1999; Chipping Forecast II, 2002). Oligonucleotide and cDNA microarrays for transcriptional profiling (Lockhart et al., 1996; Schena et al., 1995) allow measuring such interaction patterns, thus representing an unprecedented opportunity to boost the identification of diagnostic and therapeutic targets (Brown et al., 1999).

The principle of a microarray for gene expression analysis is basically that of the classical northern blot extended to the whole genome level. Specifically, mRNA from a given cell line or tissue is labeled with a fluorescent dye and hybridized to a large number of DNA sequences, immobi-

Advanced Methods of Biomedical Signal Processing. Edited by S. Cerutti and C. Marchesi **443**
Copyright © 2011 the Institute of Electrical and Electronics Engineers, Inc.

lized on a solid surface (i.e., a glass slide or a silica wafer) in an ordered array. In such a way, tens of thousands of transcript species can be detected simultaneously, exploiting the highly specific complementarity of nucleic acids. Indeed, mRNA molecules (targets) will couple with the corresponding complementary probe immobilized on the array and the fluorescence emission of any spot in the array will represent a quantification of the expression level of each target.

Although in the last decade academic groups and commercial suppliers have developed many different microarray platforms, the most commonly used technologies can be divided into two groups, according to the type of probe (i.e., complementary DNA or oligonucleotides) and the probe deposition system (spotting or in-situ synthesis of the probe). Probes for cDNA arrays are usually products of the polymerase chain reactions (PCRs) generated from cDNA or clone libraries and deposited onto glass slides as spots at defined locations. Spots are typically 100–300 μm in size and are spaced about the same distance apart. Using this technique, arrays containing more than 30,000 cDNAs can be fitted onto the surface of a conventional microscope slide. Probes for oligonucleotide arrays can be short 20–25mers synthesized in situ onto silicon wafers by photolithography (GeneChip® technology from Affymetrix) or longer presynthesized oligonucleotides (50–75mers) printed onto glass slides. Both spotted and in-situ-synthesized oligonucleotide arrays allow monitoring the expression profile of an entire genome, although they present peculiarities in terms of preparation, experimental design, and signal analysis.

Microarrays are the technology of choice to investigate global changes in gene expression that might be peculiar to a given phenotype or characterize the physiopathological state of a cell. Then, the expression patterns can be used as a basis to formulate hypotheses on the networks of molecular interactions or can be turned into a diagnostic/prognostic tool providing a molecular portrait of diseases. Not surprisingly, parallel to these technological advances has been the development of bioinformatics methods able to turn the data generated by this new kind of experiment into accurate and robust biological hypotheses or mechanisms. Indeed, high-throughput profiling faces researchers with the challenge to evaluate, analyze, model, and interpret an overwhelming mass of information. Nowadays, the identification of candidate markers sharing peculiar profiles from huge matrices bearing values for tens of thousands of molecules and the translation of patterns and relationships into biologically and/or clinically useful information constitute one of the major objectives of bioinformatics. Data mining and statistical methods are the tools of choice to extract patterns of gene expression inherent in microarrays and to correlate numerical values of expression intensities to physiological states. Drawing from fields such as pattern recognition, statistics, artificial intelligence, and signal processing, data mining techniques are geared primarily toward uncovering patterns in the data and utilizing this knowledge to create empirical (data-driven) models that describe a system behavior. In general, computational approaches to the analysis of high-throughput data can be categorized into two main groups: unsupervised and supervised (Quackenbush, 2002; Ringner et al., 2002; Valafar, 2002; Shannon et al., 2003). In unsupervised methods, expression patterns are grouped based solely on expression levels. Unsupervised methods, for example, visual discovery and interpretation procedures (DeRisi et al., 1997), singular value decomposition and projection on principal component planes (Alter et al., 2000), Fourier analysis (Spellman et al., 1998), and clustering techniques (Eisen et al., 1998; Golub et al., 1999; Tamayo et al., 1999; Butte et al., 2000), are mostly applied to explore and visualize patterns or similarities at an early stage of investigation. Supervised methods represent a powerful alternative in all those experiments where some prior information or hypothesis about which samples or molecules are expected to group together is available. Supervised techniques represent the standard computational approach to extract potential targets from thousands of monitored molecules. The most widely used supervised approaches range from inclusion/exclusion criteria (i.e., a fold change threshold) and statistical measures (molecules are selected on the basis of confidence or P-values) to advanced machine learning tools, like support vector machines (SVMs) and artificial neural networks (ANNs) (Narayanan et al., 2002; Ringner and Peterson, 2003; Greer and Khan, 2004; Hanai and Honda, 2004).

The analysis of gene expression data starts with the quantification of gene expression level from the probe fluorescence intensities and with some preprocessing steps that allow turning a microarray experiment into a matrix of numerical values. In particular, a gene expression experiment is normally represented by a matrix in which each row corresponds to a probe in the array, each column to a different array, and each cell quantifies the expression level of a specific gene in a given sample (Figure 18.1).

One of the interesting features of microarray data is the fact that they contain information on a large number of genes (tens of thousands) from a small number of samples (tens or hundreds). Considering genes as variables, microarray data represent a dim scenario in which observations are in a thousands-dimensional space and the number of variables is enormously larger than the number of samples. This means that, a priori, an enormous amount of observations (microarrays) are needed to obtain a good estimate of gene function (for example, which genes have altered expression patterns in a specific physiological state). However, in microarray data the distinction between variable and samples can be switched depending on the issue that is to be investigated (transposable data). Specifically, if the objective of the analysis is the identification of groups of coexpressed genes, then the rows of the matrix will be the observations; other-

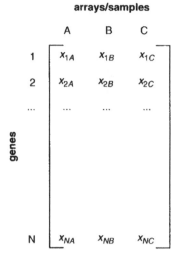

Figure 18.1. Matrix of gene expression data. Rows correspond to probes in the array (i.e., genes), columns to the different arrays (e.g., experimental conditions).

wise, if the aim is defining clusters of samples sharing the same expression pattern, the data matrix will be analyzed column-wise.

18.2   FROM MICROARRRAY TO GENE EXPRESSION DATA

Microarray data analysis starts with the acquisition of the array image (i.e., the fluorescence emission), the quantification of the expression levels, the preprocessing, and the normalization of the raw data. Although in this section these phases are discussed in brief, it is worthwhile noting that extreme caution needs to be taken in these preprocessing steps since any technique, from image acquisition to signal normalization, severely determines the final quality of the data and of the downstream analyses. Once quantified, all data, high or poor quality, are essentially fixed, and poor quality data will lead to a decrease in the power of the analysis.

18.2.1   Image Acquisition and Analysis

After performing all biological and hybridization experiments, the first step of data analysis is scanning the slide and extracting the raw intensity data from the images. Image acquisition and analysis can be divided into four basic steps: (1) scanning, (2) spot recognition, (3) segmentation, and (4) intensity extraction (Leung and Cavalieri, 2003). The first step requires setting the scanner parameters: laser power, photomultiplier tube (PMT) gain, and scanner resolution (see Leung and Cavalieri, 2003 for further details). Spot recognition is a routine task for most image analysis software; it allows one to adjust a grid on the microarray geometry (i.e., the spot positions) and flag low-quality spots. Segmentation is a process used to differentiate the foreground pixels (i.e., the true signal) in a spot from the background levels and can be performed by several algorithms that vary depending on the adopted microarray technology (spotted or in situ). After the segmentation process, the fluorescence intensities of the foreground and background areas in the image are averaged separately to give the foreground and background signals, respectively. Various intensity extraction methods, coded in commercial or freely available software, can be used, depending on the specific technology. In general, the spot intensity is calculated by subtracting the background intensity from the foreground intensity and/or summarizing these levels over multiple probes.

18.2.2 Preprocessing

The data extracted from image analysis need to be preprocessed to exclude poor-quality spots and remove systematic variations, artifacts, and errors that can arise from the starting material or from the sample manipulation. This task comprises a variety of methods for data cleaning and quality verification that depend on the type of microarray and are coded in several commercial software programs (e.g., Affymetrix GCOS or Silicon Genetics GeneSpring) as well as shareware applications (e.g., dChip, Bioconductor). The final step of preprocessing is the log transformation of the data so that levels of up- and down-regulation are of the same scale and comparable.

18.2.3 Normalization and Data Warehousing

The process of normalization aims to remove systematic errors by balancing the fluorescence intensities of the various probes or, as in the case of the spotted microarray, of the two labeling dyes.

Systematic errors and biases can be due to different incorporation efficiencies of the fluorescence dyes, different settings of the scanning systems, and varying hybridization rates in various zones of the microarray or in different microarrays. Some commonly used methods for calculating the normalization factor include global normalization, which scales all values on the array so that the median (or mean intensity) is the same over multiple arrays; housekeeping genes or invariant-set normalization, which use constantly expressed housekeeping/invariant genes; and internal controls normalization, which uses a known amount of exogenous control genes added during hybridization. Unfortunately, these normalization methods are inadequate to cope with nonlinear biases (i.e., errors whose amplitudes depend on the signal intensity) and, for this reason, techniques for nonlinear normalization have been developed. These strategies comprise Loess normalization, cyclic normalizations, and Q-Q diagnostic plots, which can be applied to either raw image intensities or expression levels for the normalization within an experiment and between multiple experiments (Quackenbush, 2002).

Finally, along with the spread of the technology there has been a pressing need to define methods for accurate, reliable standardization among multiple experiments and technologies. The definition of standards and normalized experimental procedures are critical issues for establishing a central, public-domain repository of transcriptional information that acts like GenBank or SwissProt do for sequenced genomes/proteomes and integrates with them. This problem has been addressed in several databases, making it possible to search for published microarray data that have undergone uniform processing and filtering and providing links to the original publications for more detailed information. Specifically, the Microarray Gene Expression Data (MGED) Society provides guidelines, formats, and tools for the verification and integration of data coming form multiple sources, and promoting the construction of public data warehouses for high-throughput information, such as Gene Expression Omnibus, Array-Express, and the Stanford Microarray Database (Moreau et al., 2003; Stevens and Doerge, 2005).

18.2.4   Technical and Biological Variability in Gene Expression Data

Gene expression data are affected by not only systematic errors that can be corrected by the normalization methods, but also by variability due to all the different steps of microarray fabrication (surface handling, probe synthesis, and deposition) and utilization (cell purity, RNA extraction, dye labeling, and hybridization conditions). Such sources of technical variability have been extensively discussed in the literature (Parmigiani et al., 2003; Zakharkin et al., 2005) and, at least for the Affymetrix high-density oligonucleotide technology, their contributions result in a level of random noise that is deemed to be low for many practical purposes, although that depends on the intensity of the expression signal [i.e., high at low expression levels and decreasing at higher signals (Figure 18.2)].

Biological variability is mostly due to intrinsic genomic differences among the samples, such as different individuals or tissue types, but may also depend on stochastic effects on the control mechanisms of the transcriptional machinery that differently affect different genes. Overall, the greater the experimental variability, the less is the power of the downstream analytical methods for the identification of differentially expressed genes (Nadon and Shoemaker, 2002). Thus, a major priority in designing

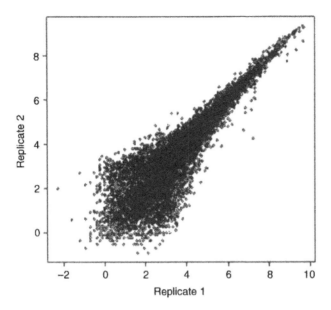

Figure 18.2. Expression levels in logarithmic scale of all probes in two replicates of the same experiment measured by Affymetrix microarrays.

an experiment involving the use of microarrays is to minimize the possible sources of variation and, possibly, to plan a number of replicates that allow quantifying both technical and biological variability. Obtaining replicates from different biological samples (e.g., patients) increases error due to biological variability but produces results that have better external validity and broader applicability (wider inference). A minimum of three or four replicates per group or experimental condition has been suggested to account for random variation and to provide good sensitivity, although there is no general law to determine the minimum number of replicates.

In general, a robust experimental design to accurately identify differentially expressed genes over multiple conditions or on a meaningful temporal window has to include a large number of replicates for each condition. Unfortunately, this request cannot always be satisfied in particular settings, as in time-course microarray experiments, in which replicates at each time point and experimental condition are often not feasible because of budget limitations or scarcity of biological material.

18.2.5 Microarray Data Annotation

Both cDNA/oligo and Affymetrix platforms provide a series of identification/accession numbers for each represented probe that could be used to integrate and match the data. However, the probes in the various types of microarrays have different lengths (~25/70mer oligonucleotides for the oligo and Affymetrix chips versus ~300/500 bp for the cDNA chip) and different matching sequences, making it difficult to identify different gene isoforms. Thus, the use of the probe identification number in some cases can be misleading. Different matching approaches have been proposed in the literature with ambiguous and discordant results. As an example, the matching of probes from different platforms can be addressed first, annotating each probe in terms of unique accession numbers (i.e., Entrez-GeneID) and, in case of discrepancies, using sequence comparison approaches (BLAST of probe sequences), as described in Romualdi et al. (2006) and Severgnini et al. (2006).

With regard to Affymetrix data, it is also worth noting that its selection of probes relies on earlier genome and transcriptome annotation, which is significantly different from current knowledge. Under the assumption that current genome and transcriptome databases are more accurate than those used for gene chip design, Affymetrix probes can be reorganized into gene-, transcript-, and exon-specific probe sets in light of up-to-date genome and cDNA/EST clustering. This reannotation step can be based both on chip definition files (Dai et al., 2005) or on the use of the GeneAnnot database (Ferrari et al., 2007a).

18.3 IDENTIFICATION OF DIFFERENTIALLY EXPRESSED GENES

A major challenge in interpreting microarray gene expression data is the development of proper computational methods for feature reduction, gene selection, and false-positive control that can cope with the small number of samples and the large number of molecules, and correlate numerical values of expression intensities to physiological states.

18.3.1 The Fold-Change Approach

Traditionally, differentially expressed genes have been identified using a fixed cut-off approach, that is, setting a threshold on the increase or decrease of the expression level or expression ratio. A typical threshold is one that considers a two-fold change in the expression signal of a gene between two experimental conditions. Specifically, a gene i will be identified as differentially expressed if its expression levels x_{iC} and x_{iT} in conditions C (e.g., control) and T (e.g., treated), respectively, satisfy the inequality

$$\left| \log\left(x_{iC}\right) - \log\left(x_{iT}\right) \right| > 2 \qquad (18.1)$$

Although extensively used, the fold-change method is statistically ineffi-
cient, due to the numerous systemic and biological variations that occur
during a microarray experiment. Indeed, using a fixed threshold to infer
significance increases the proportion of false positives (due to highly variable signals at low expression levels) and of false negatives (due to stable
genes at high expression levels).

18.3.2 Approaches Based on Statistical Tests

A more appropriate approach for inference of differentially expressed
genes includes the calculation of a statistic based on replicate-array data,
ranking genes according to the significance of their differential expression,
and the selection of a threshold value for rejecting the null hypothesis that
a gene is not differentially expressed. As such, replication of a microarray
experiment is essential to estimate the variation of the transcriptional signal for statistics calculation. Statistical methods such as Student's t-test
and its parametric and nonparametric variants and ANOVA can be used to
rank the genes (from replicated data) in the comparison of two or multiple
populations of samples, respectively. Parametric tests require that the data
fulfill some assumptions in order to be eligible for the statistical analysis.
First of all, the data should be normally distributed and the variances of the
compared groups should be similar. Although the technical and biological
variability can, in most cases, be assumed to be Gaussian noise, the limited
number of replicates often hampers a robust estimation of the experimental
variability and its independence from the signal intensity (Irizzary et al.,
2003; Tu et al., 2002). A further complication is represented by the choice
of the significance cutoff to obtain an optimal balance between false positives (Type I error) and false negatives (Type II error). This issue is rendered more severe by the thousands of tests that have to be simultaneously
performed to identify differentially expressed genes (multiple testing problem). Indeed, the more analyses are performed on a dataset, the more the
results will meet the conventional significance level by chance alone. For
example, in an experiment with an array containing $N = 100$ probes in
which the significance level is set at $\alpha = 0.01$, the probability of wrongly
selecting a specific gene as differentially expressed is 1%, whereas the
probability of wrongly selecting at least one gene as differentially expressed is given by $\alpha' = 1 - (1 - \alpha)^N = 1 - (0.99)^{100} = 63\%$. Therefore, using a p-value of 0.01 is likely to exaggerate Type I errors (Dudoit et al.,
2002). The multiple-hypothesis-testing problem is conventionally tackled
by correcting the p-value threshold through approaches that control the
family-wise error rate (FWER), the probability of having at least one false
positive among all testing hypotheses. One of these methods is the Bonferroni correction, where the original p-value is adjusted by the number of
comparisons (e.g., α'/N) to create a new corrected p-value against which

those comparisons will be tested. However, controlling the FWER can be too stringent and limits the power to identify significantly differentially expressed genes, although it could be acceptable to have a few false positives if the majority of true positives are chosen. Therefore, it might be more practical to control the false discovery rate (FDR), the expected proportion of false positives among the number of rejected hypotheses, instead of the chance of any false positives as the Bonferroni correction does. The FDR threshold is calculated from the distribution of observed p-values, but FDR is not interpreted in a similar way to p-value. For example, a p-value of 0.2 is in most cases unacceptable, whereas, if a hundred genes with an FDR of 0.2 are identified, then 20 of them should be correctly selected.

Among the variants of the standard t-test, the method called SAM (statistical analysis of microarray) addresses both the problem of variance estimation and multiple testing control. Let X be the matrix of normalized expression levels x_{ij} for gene i in sample j ($i = 1, 2, \ldots, G; j = 1, 2, \ldots, n$) and Y a response vector y_j ($j = 1, 2, \ldots, n$) for the n samples. The statistic d_i is based on the ratio of change in gene expression r_i to the standard deviation in the data set s_i for each probe set i, as defined by Tusher et al. (2001):

$$d_i = \frac{r_i}{s_i + s_0} \qquad (18.2)$$

where the estimates of gene-specific variance over repeated measurements are stabilized by a fudge factor s_0 (see Tusher et al., 2001 and SAM technical manual (www-stat.stanford.edu/~tibs/SAM/sam.pdf) for details).

The quantities r_i and s_i have different formulations in different experimental designs: two- and multiclass problems, paired data, quantitative responses, time-course experiments, and survival analyses. Specifically, in the two-class unpaired case, $y_j = 1$ or $y_j = 2$. Considering C_k the set of indices for the n_k samples in group k, $C_k = \{j: y_j = k, k = 1, 2\}$, $\bar{x}_{i1} = \sum_{j \in C_1} x_{ij}/n_1$ and $\bar{x}_{i2} = \sum_{j \in C_2} x_{ij}/n_2$, then r_i and s_i can be computed as follows:

$$r_i = \bar{x}_{i2} - \bar{x}_{i1} \qquad (18.3)$$

$$s_i = \left[(1/n_1 + 1/n_2) \left\{ \sum_{j \in C_1} \left(x_{ij} - \bar{x}_{i1} \right)^2 + \sum_{j \in C_2} \left(x_{ij} - \bar{x}_{i2} \right)^2 \right\} / \left(n_1 + n_2 - 2 \right) \right]^{1/2}$$
$$(18.4)$$

and d_i represents a two-sample t-like statistic with variance stabilization.

Finally, SAM uses a permutation strategy for calculation of empirical p-values. First, the t-like statistic for the original dataset is calculated, then the sample labels, denoting the group the individual samples belong to, are randomized several times (10,000). The t-like statistic is recalculated for each of the randomized datasets, and the empirical p-value is the percentage of randomized datasets that got a larger t-like statistic than the original dataset. The genes having the smallest empirical p-values best discriminate the groups from each other.

18.3.3 Analysis of Time-Course Microarray Experiments

Although hundreds of studies fully demonstrated the relevancy of microarrays in describing the transcriptional status of different physiological/pathological conditions, to access and reconstruct complex interaction pathways it is necessary to overcome the intrinsic limitation associated with static expression experiments and analyze dynamics of gene expression. A time-course design can be applied to investigate periodical processes or to monitor the differential response to treatment of a cell or tissue, for example, after drug administration. In both cases, transcript abundance is monitored over time, thus implying a longitudinal correlation among the observations at different time points (Guo et al., 2003; Tai and Speed, 2004).

Selection of differentially expressed genes in time-series experiments deserves a separate discussion. In time-series experiments, gene expression is often monitored over time during a transition between two stationary states, induced by appropriate cell treatment/perturbation. The objective of selection procedures is to identify genes whose expression significantly changes in time with respect to a baseline condition. Often, experiments are designed to monitor significant dynamical changes in the pattern of expression of treated cultures with respect to a control culture and thus to isolate the effect of treatments from other processes that take place in the cell simultaneously, but are not induced or inhibited by the treatments. The computational analysis of these experimental setups poses the challenge to detect transcripts whose expression level is differentially modulated by physiological state over time or in treated versus control culture in two or more populations. A straightforward approach to this problem is to mimic any of the various methods for detecting differentially expressed genes in static experiments and somehow combine information obtained at single time points over the entire temporal window.

ANOVA or ANOVA-based procedures (Park et al., 2003) have been proposed for this purpose. However, as in static differential studies, a robust experimental design to accurately identify differentially expressed genes over a meaningful temporal window would require large amounts of microarrays, and time-course experiments can be quite costly; thus, replicates are often sacrificed to the possibility of testing more conditions in more time points. Therefore, ANOVA tests are seldom applicable. For this reason, differentially expressed genes in time-series experiments are often selected using an empirical-constant fold-change threshold. For example, in a study on human fibroblasts exposed to UV radiation (Gentile et al., 2003), cells were exposed to no (control), medium, or high doses of UV radiation, and samples were collected at 6, 12, 18, and 24 hours after the beginning of the treatment. Differentially expressed genes were selected using a fold change equal to 3. This is far from ideal, since it is based on an arbitrary choice (e.g., FC = 3), which does not take into account the characteristics of the measurement error (Figure 18.2).

Moreover, analyzing each time point in isolation would result in cumbersome combinatorial analyses and in the loss of the longitudinal correlation of time-course samples. Thus, several ad-hoc statistical frameworks have been developed, which are based on analysis of variance (Kerr et al., 2000; Wolfinger et al., 2001) and estimating equation techniques (Guo et al., 2003), Bayes statistics (Tai and Speed, 2004), hidden Markov models (Yuan et al., 2003), correspondence analysis (Tan et al., 2004), and regression models (Xu et al., 2002; DeCook et al., 2004, Storey et al., 2005) are based on fitting the time-series expression profiles by polynomial and spline models, respectively; time series are then compared based on model parameters and goodness of fit. These latter methods do not require replicates; however, the effect of the number of available time samples on their performance is not clear. Recently, a method was proposed based on the analysis of measurement error (Di Camillo et al., 2007). First, the deviation of expression of gene X in treated (T) and control (C) samples is calculated for each sample t_k as

$$d(t_k) = x^T(t_k) - x^C(t_k) \tag{18.5}$$

The area A bounded by the two expression profiles T and C (Figure 18.3) is then calculated for each gene X as the sum of the contributions of partial areas from consecutive pairs of samples:

$$A = \sum_{k=1}^{M-1} A_k \tag{18.6}$$

where each contribution A_k is calculated from the deviation of expression in T and C (Equation 5), as

$$A_k = \frac{\left[\left|d(t_{k+1})\right| + \left|d(t_k)\right|\right] \times (t_{k+1} - t_k)}{2} \tag{18.7}$$

if $d(t_k)$ and $d(t_{k+1})$ have the same sign; if not, Equation 18.7 has to be modified as follows, to quantify the area within the two time-series expression profiles:

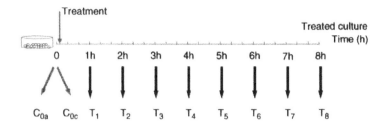

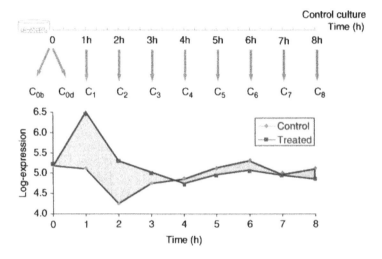

Figure 18.3. Case study providing a typical example of experimental design in time-series gene expression studies. Samples are collected from treated and control cultures. The expression level measured in the two cultures for a single gene is shown in the lower part of the figure. The area A bounded by the two expression profiles T and C is colored in gray.

$$A_k = \frac{\left(\left| d\left(t_{k+1}\right) \right| \cdot \frac{\left| d\left(t_{k+1}\right) \right|}{\left| d\left(t_{k+1}\right) \right| + \left| d\left(t_k\right) \right|} + \left| d\left(t_k\right) \right| \cdot \frac{\left| d\left(t_k\right) \right|}{\left| d\left(t_{k+1}\right) \right| + \left| d\left(t_k\right) \right|} \right) \cdot \left(t_{k+1} - t_k\right)}{2}$$

$$(18.8)$$

Gene X is considered to be differentially expressed in T versus C if the area A (Equation 18.6) exceeds a threshold θ_A:

$$A > \theta_A \qquad (18.9)$$

where θ_A is determined, in correspondence to a significance level α, based on the null hypothesis distribution of A, that is, the distribution of A derived from experimental replicates. This corresponds to the distribution of areas between two profiles of not differentially expressed genes whose differences are due to biological and technical variability. At least two replicates for each time sample would be necessary to derive the distribution of A from the data under the null hypothesis. Since they are often not available, a Monte Carlo procedure is used to derive the null distribution of area A, as detailed in Di Camillo et al. (2007), by propagating errors affecting microarray measurements.

In Affymetrix chips, it is well known that errors have an intensity-dependent distribution (Tu et al., 2002), as suggested by Figure 18.2, which clearly indicates that low-intensity measurements are affected by higher noise than high-intensity measurements. In particular, analysis of technical replicates of the Affymetrix Human chip has shown that the standardized variable $s(t_k)$ [obtained by dividing $d(t_k)$ by its intensity-dependent standard deviation] has an intensity-independent distribution (Tu et al., 2002). Therefore, in the case of data showing intensity dependency of the variable $d(t_k)$, it is convenient to calculate the values of A and A^{H0} using $s(t_k)$ instead of $d(t_k)$.

18.4   CLASSIFICATION: UNSUPERVISED METHODS

Clustering techniques are routinely applied as part of the standard bioinformatics pipeline in the analysis of DNA microarray data. The main goals of cluster analysis are: (1) finding groups of genes with similar expression profiles over different experimental conditions, including different time points or different patients; (2) finding groups of experimental conditions (patients, toxic agents) that are similar in terms of their genome-wide expression profiles (Thomas et al., 2001).

Without loss of generality,* in the following we will refer to the first goal, which is typical of functional genomics, that is, the study of genes, their resulting proteins, and the role played by the proteins in the body's biochemical processes. In this case, we are interested in finding clusters of genes (the rows of the matrix in Figure 18.1) on the basis of their expression values (the columns of the matrix). The main hypothesis underlying the application of clustering methods in functional genomics studies is that genes with similar expression patterns, that is, coexpressed genes, are involved in the same cellular processes (Brown and Botstein, 1999). We will be interested in particular in experiments in which gene expression is collected over time.

All clustering approaches aim at finding a partition of a set of examples (genes) on the basis of a number of measurements (gene expression

---

*The methods presented in this chapter can be also applied to the problems of clustering experimental conditions (Alizadeh et al., 2000)

values); the partition corresponds to natural groups in the data, or clusters. Usually, clustering algorithms search for partitions that satisfy two main criteria:

1. Internal cohesion: the examples of a cluster should be similar to each other.
2. External separation: the examples of one cluster should be very different from the examples of another cluster.

Among the different computational strategies proposed in the literature, we can distinguish four main classes of algorithms: (1) distance-based methods, (2) model-based methods, (3) template-based methods, and (4) density-based methods. In the following, we will briefly describe methods 1 to 3, which have been widely applied in gene expression analysis, whereas for density-based approaches we refer the reader to the relevant literature (Ester et al., 1996).

18.4.1 Distance-Based Methods

Clustering methods based on similarity are the most widely used approaches in the bioinformatics pipeline. These methods rely on the definition of a distance measure between gene expression profiles and group together genes with a short distance (or high similarity) between each other. Given m examples (genes), each one characterized by a vector of n measurements, the distance is computed in the n-dimensional space of the available measurements. The first step needed by these methods is, therefore, the choice of the distance measure. In functional genomics, the most widely used measures are the Euclidean distance and Pearson correlation function.

Given a distance measure, distance-based approaches differ from each other in the way they build up the clusters and in the strategy they apply to compute distances between clusters. The two main classes of distance-based methods are partitional clustering and hierarchical clustering.

In partitional clustering, the n-dimensional measurement space is divided into g regions, corresponding to the clusters; the number of regions is often defined in advance by the data analyst. The different partitional clustering methods, including k-means, k-medoids, and self-organizing maps (Hastie et al., 2001), have been largely applied to the analysis of gene expression data.

Hierarchical clustering algorithms are divided into agglomerative and divisive ones. Agglomerative clustering starts with m groups of one element each, corresponding to the m examples (genes), and then, through m − 1 consecutive steps, progressively clusters the data into groups with a larger number of examples, until a single cluster with m examples is obtained (Quaglini, 1985; Hastie et al., 2001). Divisive clustering starts with one group of m genes and progressively partitions the data into smaller clusters, until m clusters of one example are obtained (Jiang et al., 2003).

Agglomerative hierarchical clustering is the most widely used method in functional genomics (Eisen, 1998; see also http://rana.lbl.gov/ and http://cmgm.stanford.edu/pbrown/). When the data collected are time series, the Pearson correlation coefficient is used as the similarity metric. As a matter of fact, the (standardized) correlation between two gene profiles well describes the biological notion of coexpressed, and maybe coregulated, genes (Eisen et al., 1998); two genes are similar if their temporal profiles have the same "shape," even if the absolute values are very different. Moreover, the correlation similarity may also allow clustering of counterregulated genes. The result of agglomerative clustering is depicted with a binary tree known as a dendrogram; the leaves of the dendrogram are the initial clusters with one example, whereas internal nodes represent the clusters obtained by grouping the examples that correspond to the child nodes.

One of the main reasons for the success of the clustering techniques is the joint visualization of the dendrogram and of a color map (known as a heath map) of the gene expression levels in the different experimental conditions, as shown in Figure 18.4.

This figure reports the results obtained with the hierarchical clustering algorithm of Eisen and coworkers (1998) on a set of gene expression time series measured during the cell division cycle in a human cancer cell line (Whitfield et al., 2002). This type of visualization clearly shows the homogeneity regions of the gene expression profiles, highlighting the natural clusters in the data and allowing the user to assess the quality of the clusters obtained by the algorithm

18.4.2 Model-Based Clustering

The use of distance-based methods for clustering gene expression time series may suffer from the failure of one the assumptions underlying distance computation: usually, the applied distance measures are invariant with respect to the order of the measurements, which are assumed to be independent from each other. This assumption is clearly not valid in the case of time series.

Several alternative approaches have been proposed to deal with this problem, ranging from a transformation of the original time series to an alternative definition of the distance function (Aach and Church, 2001). Given the very nature of the data, however, which are characterized by a small number of points (up to 40 measurements) and a small signal-to-noise ratio, an interesting solution is represented by a different class of clustering algorithms called model-based methods.

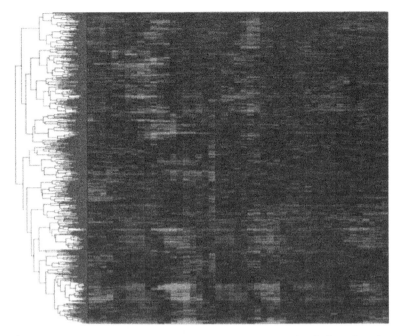

Figure 18.4. The results obtained by applying the hierarchical clustering algorithm (Eisen et al., 1998) on gene expression data measured during the cell division cycle in a human cancer-cell line (Whitfield et al., 2002). The dendrogram is reported on the left. Genes are grouped with a decreasing level of similarity from the right to the left. The height of each branch of the tree is directly proportional to the distance between the clusters that are grouped together; the more the clusters are similar, the smaller is the height of the branch, and vice versa. The image with the different gray levels shows the intensity of the gene expression levels at the different measurement time points.

The main assumption of model-based clustering is that the data are randomly extracted from a population made up of a number of subpopulations corresponding to the clusters, each one characterized by a different probability density function (Yeung et al., 2001). The subpopulations and their density functions are the model generating the data. The clustering problem is thus transformed into a model-selection problem, which can be solved by relying on probability and statistical modeling techniques.

In time-series analysis, each cluster is modeled by a different stochastic process, which is supposed to have generated the data. Usually, it is assumed that all the time series can be described by the same class of stochastic processes, and that the clusters differ from each other only because of different parameter values, such as a different autocovariance functions. Denoting with Y the set of available examples (in our case the m

× n matrix of DNA microarray data), with M the model of the data and with θ the parameters of the stochastic model generating the data, there are two main approaches for model selection that have been proposed in the literature (Baldi and Brunak, 1998; Kay, 1993; Ramoni and Sebastiani, 1999):

1. The maximum-likelihood approach, which searches the model that maximizes the likelihood function $P(Y|\theta, M)$, that is, the probability of the data given the model M and the parameters θ.
2. The Bayesian approach, which searches the model that maximizes $P(M|Y)$, that is, the posterior probability of the model M given the data Y.

Although both approaches have been applied to gene expression analysis, in this chapter we will describe in more detail the Bayesian method, which has some advantages in terms of model parsimony with respect to the maximum-likelihood strategy.

The Bayesian approach relies on Bayes' theorem, which computes the posterior probability of a model given the data as

$$P(M\mid Y) = \frac{P(Y\mid M)P(M)}{P(Y)} \qquad (18.10)$$

The posterior distribution $P(M|Y)$ is proportional to the product of the marginal likelihood, $P(Y|M)$ and of the prior distribution $P(M)$ ($P(Y)$ is constant if the data Y are known).

The prior distribution $P(M)$ is the estimate of the probability of each model M before having observed the data.

The marginal likelihood $P(Y|M)$ is computed as a function of θ and M as follows:

$$P(Y\mid M) = \int P(Y\mid \theta, M)P(\theta\mid M)d\theta \qquad (18.11)$$

Although the Bayesian approach allows one to exploit prior information on the clusters and the model complexity, usually all the models are considered a priori equally likely and, thus, the marginal likelihood is maximized, since $P(M)$ is constant and $P(M|Y)$ is proportional to $P(Y|M)$.

If the number of available data $N = m \times n$ is high, maximizing the marginal likelihood is equivalent to finding a compromise between the likelihood of a model and the number of its parameters.

A theorem proved by Schwartz (1978) shows that, if $N \rightarrow \infty$, then

$$\log(P(Y\mid M)) = \log[P(Y\mid \theta_M, M)] - \frac{1}{2}\log(N)\dim(M) + O(1) \qquad (18.12)$$

where dim(M) are the degrees of freedom of the model M and θ_M is the estimate of the model parameters. If we choose θ_M as the maximum likelihood estimate, the Bayesian approach looks for models with high likelihood and low dimensionality.

Model-based clustering has been successfully applied to cluster gene-expression time series. The CAGED (cluster analysis of gene expression dynamics; Ramoni et al., 2002) software is one of the most interesting tools for clustering time series in functional genomics (Svrakic et al., 2003; Tomczak et al., 2004). CAGED assumes that the time series are generated by an unknown number of autoregressive stochastic processes (AR). This assumption, together with a number of hypotheses on the probability distribution of the AR model parameters and of the measurement error, allows one to compute in closed form the integral of (11), that is, the marginal likelihood, for each model M and, thus, for each possible clustering of the data. Since it is computationally unfeasible to generate and compare all possible models, it is necessary to couple marginal likelihood computation with an efficient search strategy in the cluster space. To this end, similarly to hierarchical clustering, CAGED exploits an agglomerative procedure. The time series are iteratively clustered, selecting at each step the aggregation that maximizes the marginal likelihood. In this way, CAGED is able to select the optimal number of clusters, by ranking the marginal likelihood of each level of the hierarchy. Finally, the results can be shown in the same way as hierarchical clustering, with a dendrogram coupled with a heath map.

CAGED was evaluated on gene expression time series measured in serum-stimulated fibroblast cells (Iyer et al., 1999; Ramoni et al., 2002). In the original work of Iyer and coworkers, 10 clusters were found by visual inspection, but CAGED selected four clusters. Those groups seem to be able to better cluster the dynamics of the data, avoiding overfitting problems often related to a visual analysis of the dendrogram. Recently, different methods have been proposed to improve the CAGED approach by relaxing some of its hypotheses, such as the stationarity of the process generating the data or the regular sampling-time grid. In particular, more general stochastic processes have been applied, such as random walks (Ferrazzi et al., 2005) hidden Markov models (Schliep et al., 2003), or more flexible models, such as polynomial ones (Sebastiani et al., 2006).

18.4.3 Template-Based Clustering

Gene expression time series are usually characterized by a small number of time points. A recent review has shown that more than the 80% of the time series available in the Stanford Microarray Database have a number of points that is smaller than or equal to 8. The main reasons are related to the high cost and high complexity of those experiments. Since the data are also noisy, alternative clustering strategies have been investigated. One of those strategies is to group time series on the basis of the matching of the series with a pattern or a template, which may have qualitative characteristics, such as the presence of an increasing or decreasing trend, of an up-and-down behavior.

If the templates are already available, the clustering problem becomes a pattern-matching one, which can be also carried out with qualitative templates (Sacchi et. al., 2005; Shahar, 1997). In the majority of cases, the templates are not available and the template-based clustering approaches must automatically find the qualitative templates in the data.

A method widely used by researchers was proposed by Ernst and coworkers (2005) and implemented in the software STEM (short time-series expression miner) (Ernst and Bar-Joseph, 2006). STEM works with up to 10 time points. The method starts by enumerating all possible qualitative patterns of a gene profile of n time points, given the parameter c, which represents the possible unit changes of each gene from one time point to the next one. For example, if $c = 2$ each gene may increase or decrease by one or two qualitative units from one point to the next (or to remain steady). This allows one to generate $(2c + 1)(n - 1)$ qualitative templates. The second step of the algorithm reduces the number of such templates to a number m predefined by the user. The reduction is performed by clustering the qualitative profiles on the basis of their mutual distance. After this step, the original time series are assigned to the m clusters with a nearest neighbor strategy. The Pearson correlation is used as the similarity function. Finally, the number of clusters is further reduced by: (1) computing the statistical significance of each group, through a permutation-based test; and (2) aggregating the remaining clusters that are closer than a predefined threshold.

The validation of the STEM approach has been performed by comparing its performance with CAGED on a set of "short" time series; in the paper (Ernst and Bar-Joseph, 2006), STEM showed a better capability of obtaining more homogeneous groups in terms of their biological functions.

Another template-based approach was proposed by Sacchi and colleagues (Sacchi et al., 2005) who modeled the time-series data as set of consecutive trend temporal abstractions, that is, intervals in which one of the trend templates (increasing, decreasing, or steady) is verified. Clustering is then performed in an efficient way at three different levels of aggregation of the qualitative labels. At the first level, gene-expression time series with the same sequence of increasing or decreasing patterns are clustered together. At the second level, time series with the same sequence of increasing, steady, or decreasing patterns are grouped, whereas at the third level the time series sharing the same labels on the same time intervals are clustered together. The results of this method, known as TA-clustering, can be visualized as a three-level hierarchical tree and, as such, are easily interpreted. The authors demonstrated the utility of the proposed algorithm on a set of two simulated datasets and on yeast gene expression data.

Finally, an interesting knowledge-based template clustering has been presented by Hvidsten and colleagues (Hvidsten et al., 2003). In their work, the main goal of which was to find descriptive rules about the behavior of functional classes, they grouped and summarized the available gene expression time series by resorting to template-based clustering. They first enumerated all possible subintervals in the time series and labeled them as increasing, decreasing, and steady with a temporal abstraction-like procedure. Then, they clustered together genes matching the same templates over the same subintervals. In this way, a single gene may be present in more than one cluster. The overall system has been evaluated on the data published by Cho and colleagues (Cho et al., 2001) on the cell cycle in human fibroblasts.

Rather interestingly, the different clustering approaches can be now applied in an integrated way thanks to new software tools such as TimeClust, a freely downloadable software that allows the clustering of gene expression time series with distance-based, model-based, and template-based methods (Magni et al., 2008).

18.5 CLASSIFICATION: SUPERVISED METHODS

Supervised classification methods have been widely applied for the mining of DNA microarray data in functional genomics and deriving new prognostic and diagnostic models, in particular in cancer research and in pharmacology.

In functional genomics, it is possible to build a training set with a number of gene expression profiles with known biological or molecular function. The training set is used to learn a set of decision rules that allow one to classify genes with unknown function on the basis of their expression values. For example, Brown and coworkers (2000) successfully applied support vector machines to the analysis of yeast gene expression data.

As previously mentioned, DNA microarrays have been widely applied in clinical applications to perform molecular classification. In this case, the classes are a certain number of mutually exclusive diseases. The classification problem is then to find a decision rule that could correctly diagnose the patient's disease on the basis of the DNA microarray data. From a methodological viewpoint, this problem suffers from the small n, large m problem, that is, a small number of cases (few patients, on the order of hundreds) and a large number of classification attributes (many genes, on the order of tens of thousands). It is then usually necessary to apply gene selection and dimensionality reduction algorithms, such as principal component analysis or independent component analysis (Hastie et al., 2001). After gene selection and dimensionality reduction, many algorithms have been proposed to perform molecular diagnosis, ranging from decision stumps (Golub et al., 1999) to random forests. Support vector machines and random forests are nowadays considered to be the state-of-the-art approach to deal with this class of problems.

Supervised classification algorithms can be applied to derive prognostic models from DNA microarrays, that is, a prognosis on the outcomes of a certain disease on the basis of the molecular information coming from a certain patient. Many papers have been published in cancer research, although, due to the dimensionality problems previously mentioned, the model proposed has poor generalization properties and cannot be easily applied in routine clinical settings (Van't Veer et al., 2002).

Another area of great interest from an application viewpoint is pharmacology and chemobioinformatics, with particular reference to the oncology field. For example, the lymphoma leukemia project (Rosenwald et al., 2002) has developed a method to predict survival after chemotherapy for diffuse, large B-cell lymphoma. In this study, the gene expressions of 160 patients treated with antracycline chemotherapy were used to build a Cox survival model. The model was then tested on 80 patients, showing good performance in predicting five-year survival and providing interesting hypotheses on the patients who are good therapy responders.

18.6 CONCLUSIONS

The availability of high-throughput gene expression data, coupled with bioinformatics tools for their analysis, represented a scientific breakthrough in the quest for understanding and reconstructing biological mechanisms. A plethora of computational methods have been developed in the last ten years to analyze and upgrade the information content of microarray data, proving the effectiveness of expression signatures, for example, in cancer diagnosis. However, the massive accumulation of high-quality structural and functional annotations of the genomes imposes the need for computational frameworks able to analyze, model, and interpret enormous amount of data and information.

In this chapter, an overview of microarray analysis techniques has been presented, from preprocessing to selection of differentially expressed genes and supervised/unsupervised classification of genes and experimental conditions. Data mining, bioinformatics, and bioengineering methods allow identification of genes and expression patterns that represent a fingerprint of molecular processes under study, but there is a tendency to leave to the final user the task of discovering and interpreting the biological processes that may underlie the expression patterns. Although lists of modulated genes can be limited in size, as compared to raw databases, the translation of statistically significant signals into new hypotheses about the underlying biology is still the rate-limiting process, and reaching a biologically meaningful end point requires tools for annotation, mining, pathway analysis, and data integration. Several public consortia are currently focusing on the annotation and ontological description of gene-specific functional data allowing the retrieval of a wealth of information about every single transcript. These resources provide exceptional depth and coverage of any data available for a given gene, but are not designed to effectively summarize the biological knowledge associated with hundreds of genes in parallel. Most of the time, the final result is simply an explosion of the information content associated with each element of the list without any integration of the modulated genes into networks of cellular events or any formulation of new hypotheses about the examined physiological state. Computational methods should aim at identifying, and quantitatively correlating into networks of interactions, how the structural elements of genomes impact the molecular mechanisms of functional utilization (Bortoluzzi et al., 1998; Coppe et al., 2006; Callegaro et al., 2006). Lately, some tools have been designed for aiding human cognition to detect and organize lists of modulated genes, not by inspection alone but along multiple lines of conceptual similarity, and to combine functionally descriptive records with intuitive graphical displays (e.g., Genomatix Suite, Ingenuity Pathway, and Cytoscape).

The integration of different types of genomic data (gene sequences, transcriptional levels, and functional characteristics) is a fundamental step in the identification of gene function, which will allow turning genomic research into accurate and robust biological hypotheses. Recent studies on the relationships between gene structure and gene function in eukaryotic genomes showed how groups of physically contiguous genes are characterized by similar, coordinated transcriptional profiles (Caron et al., 2001; Versteeg et al., 2003). In particular, Caron and coworkers (2001) illustrated how whole chromosome views reveal a higher order organization of the genome, as there is a strong clustering of expressed genes with most chromosomes presenting large regions of highly transcribed genes interspersed with regions where gene expression is low. Several other studies reinforced these findings, highlighting that coexpressed and tissue-specific genes are often grouped in distinct chromosomal regions (Fukuoka et al., 2004; Bortoluzzi et al., 1998; Yamashita et al., 2004; Vogel et al., 2005; Ferrari et al., 2007b).

A further step to elucidate gene function and molecular mechanisms is the identification of networks of molecular interactions. To do that, it is necessary (1) to monitor dynamic expression profiles during opportune experimental stimulus and (2) to use and develop methodologies able to detect cause–effect relationships from gene expression profiles.

The set of methods able to reconstruct a regulatory network from its observed dynamic output is known as reverse engineering. Reverse engineering methods, which will be presented in the next chapter, are assuming relevance as more data become available in regard to both gene and protein expression data, and information on gene sequences, protein structure, and molecules functional characteristics.

REFERENCES

Aach, J., and Church, G., Aligning Gene Expression Time Series with Time Warping Algorithms, *Bioinformatics.* Vol. 17, 495–508. 2001.

Alizadeh, A. A., et al. Distinct Types of Diffuse Large B-Cell Lymphoma Identified by Gene Expression Profiling, *Nature.* Vol. 403, No. 3, 503–511, 2000.

Alter, O., Brown, P. O., and Botstein, D., Singular Value Decomposition for Genome-Wide Expression Data Processing and Modeling, *P. Natl. Acad. Sci. USA.* Vol. 97, No. 18, 10101–10106, 2000.

Baldi, P., and Brunak, S., *Bioinformatics: The Machine Learning Approach.* MIT Press, 1998.

Bortoluzzi, S., Rampoldi, L., Simionati, B., Zimbello, R., Barbon, A., d'Alessi,, F., Tiso, N., Pallavicini, A., Toppo, S., Cannata, N., Valle, G., Lanfranchi, G., and Danieli, G. A., A Comprehensive, High-Resolution Genomic Transcript Map of Human Skeletal Muscle, *Genome Res.,* Vol. 8, No. 8, 817–825, 1998.

Brown, M., Grundy, W., Lin, D., Cristianini, N., Sugnet, C., Furey, T., Ares, M., Jr., and Haussler, D., Knowledge-Based Analysis of Microarray Gene-Expression Data by Using Support Vector Machines, *PNAS,* Vol. 97, No. 1, 262–267, 2000.

Brown, P. O., and Botstein, D., Exploring the New World of the Genome With DNA Microarrays, *Nat. Gen.,* Vol. 21 (supplement), 33–37, 1999.

Butte, A. J., Tamayo, P., Slonim, D. K., Golub, T. R., and Kohane, I. S., Discovering Functional Relationships Between RNA Expression and Chemotherapeutic Susceptibility Using Relevance Networks, *Proc. Natl. Acad. Sci. USA,* Vol. 97, No. 22, 12182–12186, 2000.

Callegaro, A., Basso, D., and Bicciato, S., A Locally Adaptive Statistical Procedure (LAP) to Identify Differentially Expressed Chromosomal Regions, *Bioinformatics,* Vol. 22, No. 21, 2658–2666, 2006.

Caron, H., van Schaik, B., van der Mee, M., Baas, F., Riggins, G., van Sluis, P., Hermus, M. C., van Asperen, R., Boon, K., Voute, P. A., Heisterkamp, S., van Kampen, A., and Versteeg, R., The human transcriptome map: clustering of highly expressed genes in chromosomal domains, *Science,* Vol. 16, No. 291, 1289–1292, 2001.

Chipping Forecast 1999, The Chipping Forecast. Special Supplement. *Nature Genet.* Vol. 21, Jan. 1999, http://www.nature.com/cgi-taf/DynaPage.taf?file=/ng/journal/v21/n1s/index.html.

Cho, R. J., Huang, M., Campbell, M. J., Dong, H., Steinmetz, L., Sapinoso, L., Hampton, G., Elledge, S. J., Davis, R. W., and Lockhart, D. J., Transcriptional

Regulation and Function During the Human Cell Cycle, *Nat. Genet.*, Vol. 27, 48–54. 2001.

Coppe, A., Danieli, G. A., and Bortoluzzi, S., REEF: Searching REgionally Enriched Features in Genomes, *BMC Bioinformatics*, Vol. 7, Oct. 16, 453, 2006.

Dai, M., Wang, P., Boyd, A. D., Kostov, G., Athey, B., Jones, E. G., Bunney, W. E., Myers, R. M., Speed, T. P., Akil, H., Watson, S. J., and Meng, F., Evolving Gene/Transcript Definitions Significantly Alter the Interpretation of GeneChip Data, *Nucleic Acids Res.*, Vol. 33, No. 20, 175, 2005.

DeCook, R., Nettleton, D., Foster, C. M., and Wurtele, E., Differentially Expressed Genes in Unreplicated Multiple-Treatment Microarray Timecourse Experiments, *Comput. Stat. Data An., 2004.*

DeRisi, J. L., Iyer, V. R., and Brown, P. O., Exploring the Metabolic and Genetic Control of Gene Expression on a Genomic Scale, *Science,* Vol. 278, No. 5338, 680–686, 1997.

Di Camillo, B., Toffolo, G. S. K., Nair, Greenlund, L. J., and Cobelli, C., Significance Analysis of Microarray Transcript Levels in Time Series Experiments, *BMC Bioinformatics,* 8 (Suppl 1), S1, 1–13, 2007.

Dudoit, S., Shaffer, J. P., and Boldrick, J. C., Multiple Hypothesis Testing in Microarray Experiments, Technical Report Tech. Report # 110, U.C. Berkeley Division of Biostatistics, Working Paper Series, 2002.

Eisen, M. B., Spellman, P. T., Brown, P. O., and Botstein, D., Cluster Analysis and Display of Genome Wide Expression Patterns, *Proc. Natl. Acad. Sci. USA,* Vol. 95, No. 25, 14,863–14,868, 1998.

Ernst, J., and Bar-Joseph, Z., STEM: A Tool for the Analysis of Short Time Series Gene Expression Data, *BMC Bioinformatics,* Vol. 7, Apr. 5, 191, 2006.

Ernst, J., Nau, G. J., and Bar-Joseph, Z., Clustering short time series gene expression data, *Bioinformatics,* Suppl. 1, June 21, 159–168, 2005.

Ester, M., Kriegel, H. P., Sander, J., and Xu, X., A Density-Based Algorithm for Discovering Clusters in Large Spatial Databases with Noise, in *Proceedings of 2nd International Conference on Knowledge Discovery and Data Mining (KDD P96),* Portland, Oregon, AAAI Press, 1996.

Ferrari, F., Bortoluzzi, S., Coppe, A., Basso, D., Bicciato, S., Zini, R., Gemelli, C., Danieli, G. A., and Ferrari, S., Genomic Expression During Human Myelopoiesis, *BMC Genomics,* Vol. 8, Aug. 3, 264, 2007.

Ferrari, F., Bortoluzzi, S., Coppe, A., Sirota, A., Safran, M., Shmoish, M., Ferrari, S., Lancet, D., Danieli, G. A., and Bicciato, S., Novel Definition Files for Human GeneChips Based on GeneAnnot, *BMC Bioinformatics,* Vol. 8, Nov. 15, 446, 2007.

Ferrazzi, F., Magni, P., and Bellazzi, R., Random Walk Models for Bayesian Clustering of Gene Expression Profiles, *Appl. Bioinformatics,* Vol. 4, No. 4, 263–276, 2005.

Fukuoka, J., Fujii, T., Shih, J. H., Dracheva, T., Meerzaman, D., Player, A., Hong, K., Settnek, S., Gupta, A., Buetow, K., Hewitt, S., Travis, W. D., and Jen, J., Chromatin Remodeling Factors and BRM/BRG1 Expression as Prognostic Indicators in Non-Small Cell Lung Cancer, *Clin. Cancer Res.,* Vol. 10, No. 13, 4314–4324, 2004.

Gentile, M., Latonen, L., and Laiho, M., Cell Cycle Arrest and Apoptosis Provoked by UV Radiation-Induced DNA Damage are Transcriptionally Highly

Divergent Responses, *Nucleic Acids Research*, Vol. 31, No. 16, 4779–4790, 2003.

Golub, T., Slonim, D., Tamayo P., Huard, C., Gaasenbeeh, M., Mesirov J., Coller, H., Loh, M., Downing J., Caligiuri, M., Bloomfield, C., and Lander, E., Molecular Classification of Cancer: Class Discovery and Class Prediction by Gene Expression Monitoring, *Science*, Vol. 286, No. 5439, 531–537, 1999.

Greer, B. T., and Khan, J., Diagnostic Classification of Cancer Using DNA Microarrays and Artificial Intelligence, *Ann. NY Acad. Sci.*, Vol. 1020, May, 49–46, 2004.

Guo, X., Qi, H., Verfaillie, C. M., and Pan, W., Statistical Significance Analysis of Longitudinal Gene Expression Data, *Bioinformatics*, Vol. 19, No. 13, 1628–1635, 2003.

Hanai, T., and Honda, H., Application of Knowledge Information Processing Methods to Biochemical Engineering, Biomedical and Bioinformatics Fields, *Adv. Biochem. Eng. Biotechnol.*, Vol. 91, 51–73, 2004.

Hastie, T., Tibshirani, R., and Friedman, J., *The Elements of Statistical Learning: Data Mining, Inference and Prediction*, Springer, 2001.

Hvidsten, T. R., Laegreid, A., and Komorowski, J., Learning Rule-Based Models of Biological Process from Gene Expression Time Profiles Using Gene Ontology, *Bioinformatics*, Vol. 19, 1116–1123, 2003.

Irizarry, R. A., Hobbs, B., Collin, F., Beazer-Barclay, Y. D., Antonellis, K. J., Scherf, U., and Speed, T. P., Exploration, Normalization and Summaries of High Density Oligonucleotide Array Probe Level Data, *Biostatistics*, Vol. 4, No. 2, 249–264, 2003.

Iyer, et al., The Transcriptional Program in the Response of Human Fibroblasts to Serum, *Science*, Vol. 283, 83–87, 1999.

Jiang, D., Pei, J., and Zhang, A. DHC: A Density-based Hierarchical Clustering Method for Time Series Gene Expression Data, in *Proceedings of 3rd IEEE International Symposium on BioInformatics and BioEngineering (BIBE 2003)*, pp. 393–400, 2003.

Kay, S. M., *Fundaments of Statistical Signal Processing: Estimation Theory*, Prentice-Hall, 1993.

Kerr, M. K., Martin, M., and Churchill, G. A., Analysis of Variance for Gene Expression Microarray Data, *J. Comput. Biol.*, Vol. 7, No. 6, 819–837, 2000.

Leung, Y. F., and Cavalieri, D., Fundamentals of cDNA Microarray Data Analysis, *Trends Genet.*, Vol. 19, No. 11, 649–659, 2003.

Lockhart, D. J., Dong, H., Byrne, M. C., Follettie, M. T., Gallo, M. V., Chee, M. S., Mittmann, M., Wang, C., Kobayashi, M., Horton, H., and Brown, E. L., Expression Monitoring by Hybridization to High-Density Oligonucleotide Arrays, *Nat. Biotechnol.*, Vol. 14, No. 13, 1675–1680, 1996.

Magni, P., Ferrazzi, F., Sacchi, L., and Bellazzi, R., TimeClust: A Clustering Tool for Gene Expression Time Series, *Bioinformatics*, Vol. 24, No. 3, 430–432, 2008.

Moreau, Y., Aerts, S., Moor, B. D., Strooper, B. D., and Dabrowski, M., Comparison and Meta-Analysis of Microarray Data: From the Bench to the Computer Desk, *Trends in Genetics*, Vol. 19, No. 10, 570–577, 2003.

Nadon, R., and Shoemaker, J., Statistical Issues with Microarrays: Processing and Analysis, *Trends Genet.*, Vol. 18, No. 5, 265–271, 2002.

Narayanan, A., Keedwell, E. C., and Olsson, B., Artificial Intelligence Techniques for Bioinformatics, *Appl. Bioinformatics*, Vol. 1, No. 4, 191–222, 2002.

Parmigiani, G., Garrett, E. S., Irizarry, R. L., and Zeger, S. L., The Analysis of Gene Expression Data: An Overview of Methods and Software, In Parmigiani, G., Garrett, E. S., Irizarry, R. L., Zeger, S. L., (Eds.), *The Analysis of Gene Expression Data,* Springer, 2003.

Park, T., Yi, S. G., Lee, S., Lee, S. Y., Yoo, D. H., Ahn, J. I., and Lee, Y. S., Statistical Tests for Identifying Differentially Expressed Genes in Time-Course Microarray Experiments, *Bioinformatics,* Vol. 19, 694–703, 2003.

Quackenbush, J., Microarray Data Normalization and Transformation, *Nature Genetics,* Vol. 32, 496–501, 2002.

Quaglini, S., Tecniche per l'Analisi Esploratoria di Osservazioni Multivariate, In C. Berzuini, *Il Calcolatore Nella Pratica Clinica,* Patron Editore, pp. 124–158, 1985.

Ramoni, M., and Sebastiani, P., Bayesian Methods for Intelligent Data Analysis, In Berthold, M., and Hand, D. J., (Eds.), *Intelligent Data Analysis: An Introduction,* Springer, 1999.

Ramoni, M., Sebastiani, P., and Kohane, I. S., Cluster Analysis of Gene Expression Dynamics, *Proc. Natl. Acad. Sci. USA,* Vol. 99, No. 14, 9121–9126, 2002.

Ringner, M., Peterson, C., and Khan, J., Analyzing Array Data Using Supervised Methods, *Pharmacogenomics,* Vol. 3, No. 3, 403–415, 2002.

Ringner, M., and Peterson, C., Microarray-Based Cancer Diagnosis with Artificial Neural Networks, *Biotechniques,* Supplement, March, 30–35, 2003.

Romualdi, C., De Pittà, C., Tombolan, L., Bortoluzzi, S., Sartori, F., Rosolen, A., and Lanfranchi, G., Defining the Gene Expression Signature of Rhabdomyosarcoma by Meta-Analysis, *BMC Genomics,* Vol. 7, Nov. 7, 287, 2006.

Rosenwald, A., Wright, G., Chan, W. C., Connors, J. M., Campo, E., Fisher, R. I., Gascoyne, R. D., Muller-Hermelink, H. K., Smeland, E. B., Giltnane, J. M., Hurt, E. M., Zhao, H., Averett, L., Yang, L., Wilson, W. H., Jaffe, E. S., Simon, R., Klausner, R. D., Powell, J., Duffey, P. L., Longo, D. L., Greiner, T. C., Weisenburger, D. D., Sanger, W. G., Dave, B. J., Lynch, J. C., Vose, J., Armitage, J. O., Montserrat, E., López-Guillermo, A., Grogan, T. M., Miller, T. P., LeBlanc, M., Ott, G., Kvaloy, S., Delabie, J., Holte, H., Krajci, P., Stokke, T., Staudt, L. M., Lymphoma/Leukemia Molecular Profiling Project, *New England Journal of Medicine,* Vol. 346, No. 25, 1937–1947, 2002.

Sacchi, L., Bellazzi, R., Larizza, C., Magni, P., Curk, T., Petrovic, U., and Zupan, B., TA-Clustering: Cluster Analysis of Gene Expression Profiles Through Temporal Abstractions, *Int. J. Med. Inform.,* Vol. 74, 505–517, 2005.

Schena, M., Shalon, D., Davis, R. W., and Brown, P. O., Quantitative Monitoring of Gene Expression Patterns with a Complementary DNA Microarray, *Science,* Vol. 270, No. 5235, 467–470, 1995.

Schliep, A., Schonhuth, A., and Steinhoff, C., Using Hidden Markov Models to Analyze Gene Expression Time Course Data, *Bioinformatics,* Vol. 19, 255–263, 2003.

Schwartz, G., Estimating the Dimension of a Model, *The Annals of Statistics,* Vol. 6, No. 2, 461–464, 1978.

Sebastiani, P., Xie, H., and Ramoni, M. F., Bayesian Analysis of Comparative Microarray Experiments by Model Averaging, *Bayes. Anal.,* Vol. 1, No. 4, 707–732, 2006.

Severgnini, M., Bicciato, S., Mangano, E., Scarlatti, F., Mezzelani, A., Mattioli, M., Ghidoni, R., Peano, C., Bonnal, R., Viti, F., Milanesi, L., De Bellis, G., and Battaglia, C., Strategies for Comparing Gene Expression Profiles from Different Microarray Platforms: Application to a Case-Control Experiment, *Anal. Biochem.*, Vol. 353, No. 1, 43–56, 2006.

Shahar, Y., A Framework for Knowledge-Based Temporal Abstraction, *Artificial Intelligence*, Vol. 90, 79–133, 1997.

Shannon, W., Culverhouse, R., and Duncan, J., Analyzing Microarray Data Using Cluster Analysis, *Pharmacogenomics*, Vol. 4, No. 1, 41–45, 2003.

Spellman, P. T., Sherlock, G., Iyer, V. R., Zhang, M., Anders, K., Eisen, M. B., Brown, P. O., Botstein, D., and Futcher, B., Comprehensive Identification of Cell Cycle-Regulated Genes of the Yeast Saccharomyces Cerevisiae by Microarray Hybridization, *Mol. Biol. Cell.*, Vol. 9, 3273–3277, 1998.

Stevens, J. R., and Doerge, R. W., Combining Affymetrix Microarray Results, *BMC Bioinformatics*, Vol. 6, No. 1, 57, 2005.

Storey, J. D., Xiao, W., Leek, J. T., Tompkins, R. G., and Davis, R. W., Significance Analysis of Time Course Microarray Experiments, *Proc. Natl. Acad. Sci. USA*, Vol. 102, No. 36, 12837–12842, 2005.

Svrakic, N. M., Nesic, O., Dasu, M. R. K., Herndon, D., and Perez-Polo, J., R., Statistical Approach to DNA Chip Analysis, *Recent Prog. Horm. Res.*, Vol. 58, No. 1, 75–93, 2003.

Tai, Y. C., and Speed, T. P., Technical Report #667, Department of Statistics, University of California, Berkeley, 2004.

Tamayo, P., Slonim, D. K., Mesirov, J., Zhu, Q., Kitareewan, S., Dmitrovsky, E., Lander, E. S., and Golub, T. R., Interpreting Patterns of Gene Expression with Self-Organizing Maps: Methods and Application to Hematopoietic Differentiation, *P Natl Acad Sci USA*, Vol. 96, No. 6, 2907–2902, 1999.

Tan, Q., Brusgaard, K., Kruse, T. A., Oakeley, E., Hemmings, B., Beck-Nielsen, H., Hansen, L., and Gaster, M., Correspondence Analysis of Microarray Time-Course Data in Case-Control Design, *J Biomed Inform.*, Vol. 37, No. 5, 358–365, 2004.

Thomas, R., Rank, D., Penn, S., Zastrow, G., Hayes, K., Pande, K., Glover, E., Silander, T., Craven, M., Redddy, J., Jovanovich, S., and Bradfield, C. Identification of Toxicologically Predictive Gene Sets Using cDNA Microarrays, *Molecular Pharmacology*, Vol. 60, No. 6, 1189–1195, 2001.

Tomczak, K. K., Marinescu, V., D., Ramoni, M. F., Sanoudou, D., Montanaro, F., Han, M., Kunkel, L. M., Kohane, I. S., and Beggs, A. H., Expression Profiling and Identification of Novel Genes Involved in Myogenic Differentiation, *FASEB, J.*, Vol. 18, No. 2, 403–405, 2004.

Tu, Y., Stolovitzky, G., and Klein, U., Quantitative Noise Analysis for Gene Expression Microarray Experiment, *P. Natl. Acad. Sci. USA*, Vol. 99, No. 22, 14031–14036, 2002.

Tusher, G., T., Tibshirani, R., and Chu, G., Significance Analysis of Microarrays Applied to the Ionizing Radiation Response, *Proc. Natl. Acad. Sci. USA*, Vol. 98, No. 9, 5116–5121, 2001.

Valafar, F., Pattern Recognition Techniques in Microarray Data Analysis: A Survey, *Ann. NY Acad. Sci.*, Vol. 980, 41–64, 2002.

Van't Veer, L., et al., Gene Expression Profiling Predicts Clinical Outcome of

Breast Cancer, *Nature*, Vol. 415, No. 6871, 530–536, 2002.

Versteeg, R., van Schaik, B. D., van Batenburg, M. F., Roos, M., Monajemi, R., Caron, H., Bussemaker, H. J., and van Kampen, A. H., The Human Transcriptome Map Reveals Extremes in Gene Density, Intron Length, GC Content, and Repeat Pattern for Domains of Highly and Weakly Expressed Genes, *Genome Res.*, Vol. 13, No. 9, 1998–2004, 2003.

Vogel, J. H., von Heydebreck, A., Purmann, A., and Sperling, S., Chromosomal Clustering of a Human Transcriptome Reveals Regulatory Background, *BMC Bioinformatics*, Vol. 6, 230, 2005.

Whitfield, M. L., Sherlock, G., Saldanha, A. J., Murray, J. I., Ball, C. A., Alexander, K. E., Matese, J. C., Perou, C. M., Hurt, M. M., Brown, P. O., and Botstein, D., Identification of Genes Periodically Expressed in the Human Cell Cycle and their Expression in Tumors, *Mol. Biol. Cell.*, Vol. 13, No. 6, 1977–2000, 2002.

Wolfinger, R. D., Gibson, G., Wolfinger, E. D., Bennett, L., Hamadeh, H., Bushel, P., Afshari, C., and Paules, R. S., Assessing Gene Significance from cDNA Microarray Expression Data Via Mixed Models, *J. Comput. Biol.*, Vol. 8, No. 6, 625–637, 2001.

Xu, X. L., Olson, J. M., and Zhao, L. P., A Regression-Based Method to Identify Differentially Expressed Genes in Microarray Time Course Studies and its Application in an Inducible Huntington's Disease Transgenic Model, *Hum. Mol. Genet.*, Vol. 11, No. 17, 1977–1985, 2002.

Yamashita, T., Honda, M., Takatori, H., Nishino, R., Hoshino, N., and Kaneko, S., Genome-Wide Transcriptome Mapping Analysis Identifies Organ-Specific Gene Expression Patterns Along Human Chromosomes, *Genomics*, Vol. 84, No. 5, 867–875, 2004.

Yeung, K. Y., Fraley, C., Murua, A., Raftery, A. E., and Ruzzo, W. L., Model-Based Clustering and Data Transformations for Gene Expression Data, *Bioinformatics*, Vol. 17, No. 10, 977–987, 2001.

Yuan, M., Kendziorski, C., Park, F., Porter, J. R., Hayes, K., and Bradeld, C. A., Technical Report # 178, Department of Biostatistics and Medical Informatics, University of Wisconsin, Madison, 2003.

Zakharkin, S. O., Kim, K., Mehta, T., Chen, L., Barnes, S., Scheirer, E., Parrish, R. S., Allison, D. B., and Page, G. P., Sources of Variation in Affymetrix Microarray Experiments, *BMC Bioinformatics*, Vol. 6, Aug., 214, 2005.

INTERNET RESOURCES

Chipping Forecast II. Special Supplement. *Nature Genet.* Vol. 32, Dec. 2002, http://www.nature.com/cgi-taf/dynapage.taf?file=/ng/journal/v32/n4s/index.html.

Bioconductor: http://www.bioconductor.org/

Bioinformatics: http://bioinformatics.oupjournals.org/

Interesting functional genomics links at http://ihome.cuhk.edu.hk/%7Eb400559/

Eisen, M. B., *ScanAlyze*. Available at http://rana.stanford.edu/software

Thispageintentionallyleftblank

CHAPTER *19*

MICROARRAY DATA ANALYSIS
Gene Regulatory Networks

Riccardo Bellazzi, Silvio Bicciato, Claudio Cobelli,
Barbara Di Camillo, Fulvia Ferrazzi, Paolo Magni,
Lucia Sacchi, and Gianna Toffolo

19.1 INTRODUCTION

Cellular processes involve million of molecules playing a coherent role in the exchange of matter, energy, and information, both among themselves and with the environment. These processes are regulated by proteins whose expression is controlled by a tight network of interactions between genes, proteins, and other molecules. There is evidence that some pathologies of major social impact, such as cancer and diabetes, involve groups of genes and proteins that are functionally related pathways rather than the expression of a single gene or protein. Therefore, it is important to investigate the global modifications of a specific regulatory pathway rather than the expression of a single gene. This is a major goal of systems-biology approaches, devoted to the elucidation of the complex network of interacting DNA sequences, RNAs, and proteins regulating and controlling gene expression.

Today, high-throughput technologies such as microarray and mass spectrometry measure the cellular molecular expression in a given instant, thus making possible, at least in principle, the reconstruction of the regulatory network from its observed output through reverse engineering approaches. Unfortunately, microarray technology cost restricts the number of samples (on the order of 10^1–10^2) available for each experiment with respect to the number of monitored genes (on the order of 10^4). Mass spectrometry techniques have some limitations as well, since at present they are not able to provide a precise quantification of protein expression and require one to identify the original protein from the spectrum of its fragments by mining specific databases. For these reasons, reverse engineering approaches are usually limited to very general and abstract models in which RNA expression is considered as a proxy for protein expression in controlling gene transcription. Several reverse engineering approaches based on this model have been proposed in the last few years to infer gene regulatory networks from microarray gene expression data. Among them are Boolean models, models based on differential equations, Bayesian networks, and methods based on pair-wise gene expression correlation.

19.2 BOOLEAN MODELS

Boolean models are interesting tools to approach reverse engineering problems. They require a preliminary quantization of gene expression levels, which are continuous in nature. Although Boolean models typically resort to data binarization, a discretization in three levels is often used in relation to differential expression (1 overexpressed, 0 nondifferentially expressed, −1 underexpressed). Boolean methods aim to identify, in the logic rules and-or-not space, a rule able to represent gene expression level x as a logic function of the levels of other genes. For instance, the rule $q = x$ and y and $\neg z$ means that gene q is expressed if and only if genes x and y are simultaneously expressed and gene z is underexpressed (Figure 19.1).

If gene expression time series are available, either instantaneous or synchronous regulation models can be adopted: gene x expression at time t is modeled in terms of the expression of other genes measured at time t in the former case, at time $t + \Delta t$ in the latter. Both deterministic and probabilistic models are available, originally proposed in Somogyi and Sniegoski (1996) and Shmulevich and coworkers (2002), respectively.

x	y	z	q	x AND y AND ¬ z = q
1	1	1	0	
1	1	0	1	
1	0	1	0	
1	0	0	0	
0	1	1	0	
0	1	0	0	
0	0	1	0	
0	0	0	0	

Figure 19.1. Gene q is regulated by genes x, y, and z according to the rule the rule $q = x$ and y and $\neg z$. To reconstruct the rule without uncertainty, all the $2^3 = 8$ possible input states must be observed.

A prototype of Boolean models applied to gene expression data is the algorithm REVEAL (Reverse Engineering ALgorithm) (Liang et al., 1998), which aims at identifying a minimum set of input genes able to predict the behavior of the output gene, building on the concepts of entropy (H) and mutual information (M). Entropy (Shannon and Weaver 1963) is a measure of information content defined as the probability to observe a frequency of occurrence in the predefined set of possible events:

$$H(x) = -\sum_{i=1}^{L} P(x \in l_i) \cdot \log_2 P(x \in l_i) \qquad (19.1)$$

In Equation 19.1, L is the number of levels l_i used to discretize gene expression, and $P(x \in l_i)$ is the probability that gene x assumes values within the interval l_i. Entropy definition can be extended to measure the joint entropy $H(x, y)$ of two genes x and y:

$$H(x, y) = -\sum_{l=1}^{L} \sum_{l'=1}^{L} P(x \in l, y \in l') \cdot \log_2 P(x \in l, y \in l') \qquad (19.2)$$

and the conditional entropy $H(x|y)$, representing a measure of the information of gene x that is not shared by gene y:

$$H(x/y) = H(x,y) - H(y) \qquad (19.3)$$

Similarly,

$$H(y/x) = H(x,y) - H(x) \qquad (19.4)$$

Finally, mutual information $M(x, y)$ is defned as the information of gene x after removing the information not shared with y:

$$\begin{aligned} M(x,y) &= H(y) - H(y/x) \\ &= H(x) - H(x/y) \qquad (19.5) \\ &= H(x) + H(y) - H(x,y) \end{aligned}$$

A functional relationship $y \rightarrow x$, that is, y univocally determines x, is inferred when mutual information between x and y equals the entropy of gene x:

$$M(x, y) = H(x) \qquad (19.6)$$

For example, Equation 19.6 is verified when $y = (-1, -1, 0, 0, 0, 1, 11, -1)$ and $x = (1, 1, 00, 0, 0, -1, -1, 1)$. Analogously, it is possible to consider the gene expression profile of gene x with respect to two genes y and z, thus inferring an interaction of order $k = 2$, where x is univocally determined by the pair (y, z) if $M[x, (y, z)] = H(x)$, analogously for $k = 3$, and so on.

The REVEAL algorithm searches for each gene x all the possible interactions of order $k = 1$, that is, all genes y for which Equation 19.6 is satisfied. If no genes satisfy this condition, the research is extended to $k = 2$ interactions and so forth. In practice, computational cost limits the search to $k = 3$. This also limits the number of false positive regulations deriving from the low number of available samples as compared to the high number of monitored genes.

The most immediate implementation of entropy and mutual information is associated with two or three levels of discretization. Relevance networks, based on a discretization in ten levels, were proposed by Butte and coworkers (2000). A relationship between genes x and y (no direction is specified with this method) is identified if the two genes are characterized by a high value of mutual information, that is, exceeding a threshold equal to the highest value of mutual information observed on randomized data.

More recently, ARACNe (Algorithm for the Reconstruction of Accurate Cellular Networks) was proposed (Basso et al., 2005; Margolin et al., 2006), in which entropy definition is extended to the continuous domain, in order to avoid data discretization. The threshold on mutual information is calculated based on the false discovery rate, which compromises between false positive and false negative interactions. A further innovative feature of the method is an additional pruning step on the inferred network topology aimed at eliminating those connections between genes x and y that can be explained in terms of their interaction with a third common gene z.

19.3 DIFFERENTIAL EQUATION MODELS

Methods based on differential equations describe the derivative of gene expression i at time t as a function of all other genes and, possibly, external outputs. This function can be linear or not and its complexity depends on the model assumed for regulation (D'Haeseleer et al., 1999; Chen et al., 1999; Weaver et al., 1999).

A simple yet general model of regulation assumes that the rate of transcription of gene i as a function of a linear combination of the expression of other genes ($\Sigma w_{ij} x_j$) and can be represented, at time t, by the following differential equation:

$$\frac{dx_i(t_k)}{dt} = R_i \times f\left(\sum_{j=1}^{N} w_{ij} \times x_j(t_k) + \sum_{s=1}^{S} v_{is} \times c_s(t_k) + B_i\right) \quad (19.7)$$

where t_k is the sampling time, x_i is the observed gene expression value for gene i ($i = 1, \ldots, N$) with N number of genes, R_i is the activation constant for gene i, c_s is the concentration of the external input s, and B_i is the basal activation level of gene i. w_{ij} and v_{is} are control parameters, assumed to be time independent, which are positive, negative, or null values depending on the positive, negative or null control that gene j or output s exerts on gene i. Finally, f is the activation function, assumed to be linear (D'Haeseleer et al., 1999) or sigmoidal (Weaver et al., 1999), depending on the model hypothesis.

Parameters w_{ij} and v_{is} are unknown and must be identified from gene expression data available at different time samples. To solve the identification problem, the number of available samples must be at least equal to the number of parameters to be identified for each gene i. This condition is seldom satisfied since the number of arrays is usually much lower than the number of analyzed genes. Therefore, even after a gene selection step, the system of model equations (19.7) is usually undetermined and it is necessary to resort to heuristics to solve it. For example, cluster analysis was often applied to diminish the number of profiles to be analyzed, or a fitting followed by a resampling procedure was adopted to augment the number of samples. In one of the first applications to microarray data (D'Haeseleer et al., 1999), gene regulation during the development of the central nervous system in the rat was studied by measuring the expression of 65 preselected genes in 28 time samples. Each gene expression profile was smoothed by cubic splines and resampled with a five-fold higher frequency; least squares methods were then applied to solve the differential equation system.

Other strategies solve Equation 19.7 by constraining the weight w_{ij} matrix to be sparse; the observation that biological networks are scale-free suggests that the number of nonnull weights w_{ij} for each gene is less than 10–12. As an example, in a study on simulated networks, Yeung and coworkers (2002) used singular value decomposition (SVD) to obtain a set of solutions consistent with data and selected the optimal one based on a criterion of network sparseness.

More recently, Cosentino and coworkers (2007) proposed an approach based on dynamical linear systems identification theory. A novel algorithm, based on linear matrix inequalities, is devised to infer the interaction network. This approach allows taking into account, within the optimization procedure, the a priori available knowledge of the biological system.

As for Boolean networks, differential-equation-based models can be extended to probabilistic models by accounting for the probability of observing a certain expression level for each gene at time t. These models, however, require a higher number of observations and can be considered as a special case of Bayesian models, which will be described in the following section.

19.4   BAYESIAN MODELS

Bayesian networks (BNs; Pearl, 1988; Friedman et al., 2000; Friedman, 2004) are general formalisms for the representation and use of probabilistic knowledge employed in a wide range of fields, including artificial intelligence, automatic reasoning, epidemiology, and genetics. In functional genomics, BNs have been successfully applied to gene regulatory network modeling, as they are able to take into account the uncertainty about gene relationships inferred from experimental data (Friedman, 2004). BNs are indeed flexible and easily interpretable models that allow the representation of multivariate probabilistic relationships both at qualitative and quantitative levels. At the qualitative level, the structure of BNs describe the relationships between variables as conditional independences represented in a graph, whereas at the quantitative level, these relationships are expressed by conditional probability distributions. The application of BNs to gene regulatory network modeling is not without problems. In particular, as the goal is the discovery of gene relationships from data, it is necessary to infer both the network structure and the conditional probability distributions on the basis of the available observations. This can be nontrivial because of the presence of datasets with an extremely high number of variables (genes) and few observations (gene expression values collected over time). In the following, we will describe BNs and an extension of BNs that is particularly suited to analyze time series—dynamic Bayesian networks (Friedman et al., 1998; Ong et al., 2002; Ferrazzi et al., 2007). We will thus introduce methods to learn conditional probability distributions and the structure of gene regulatory networks. Finally, we will present some methodological variants that have been proposed for the analysis of DNA microarray data.

A BN is a directed acyclic graph in which nodes represent probabilistic variables and arcs conditional dependence relationships. If an arc from A to B is present, A is a parent of B (and B is a child of A). A BN is completely specified if all the conditional probability distributions of children given their parents are known. For example, the BN in Figure 19.2 is specified by the probability distributions $P(A)$, $P(B|A)$, and $P(C|A; B)$.

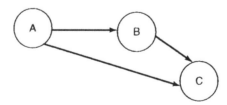

Figure 19.2. A simple Bayesian network.

A very interesting property of BNs is their ability to express complex joint probability distributions by means of local models. Given a BN with m random variables $X_1, \ldots, X_m$ and calling $pa(X_i)$ the set of parents of variable X_i, the following equality holds (called the chain rule):

$$P(X_1, \ldots, X_n) = \prod_{i=1}^{m} P[X_i \mid pa(X_i)] \qquad (19.8)$$

where $P(X_1, \ldots, X_m)$ is the joint probability distribution. In the case of a gene regulatory network, a BN allows us to express the joint probability distribution of the expression levels of a set of genes (or the entire genome), as well as the probabilistic relationships between the values taken by the different genes (for example, if A is overexpressed then B is underexpressed with probability $= 0.2$).

In the literature, different algorithms exist to calculate the marginal posterior distributions of a node once some evidence is available (Lauritzen and Spiegelhalter, 1988). For example, in the case of the network in Figure 19.2 it is possible to calculate $P(A|B = B^*, C = C^*)$, where C^* and B^* are values taken by the variables B and C. This allows the generation of hypotheses about the most probable values of unknown variables and, in general, it allows probabilistic inference.

Yet, in the representation of gene regulatory networks (and, in general, any dynamic system), the absence of cycles in the formalism of BNs can constitute a problem, as in biological processes in which the state of a variable often depends on its previous values. Dynamic Bayesian networks (DBNs) are a extensions of BNs to take into account the dynamics of a system. A DBN represents the evolution of a system in a finite number n of time points $(t_1, \ldots, t_n)$ called time slices. Each node of the network is repeated at every time point and arcs can express both instantaneous dependencies (within the same time slice) and delayed dependencies (between one time slice and another). Figure 19.3 shows a possible dynamic version of the BN in Figure 19.2.

The most relevant problem in the case of gene regulatory networks is learning BNs and DBNs from data (Buntine, 1994; Heckerman et al., 1995; Heckerman, 1998). This problem can be decomposed into two subproblems: (1) learning the conditional distributions given the structure and (2) learning the structure. In the following, we will deal with both topics in detail.

19.4.1 Learning Conditional Probability Distributions

A probability distribution is typically characterized by a set of parameters θ that can be learned from data by following different strategies. For example, if n samples of a Gaussian random variable with mean μ and variance σ^2 are available, it is known that the maximum likelihood estimates

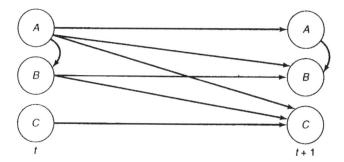

Figure 19.3. A possible dynamic Bayesian network compatible with the Bayesian network in Figure 19.2. With respect to Figure 19.2, the link between *A* and *B* expresses both an instantaneous and a delayed dependence, whereas the dependences between *A* and *B* and *A* and *C* are only delayed.

for μ and σ^2 are the sample mean and variance. In the case of discrete variables with multinomial distributions, the parameters are easily estimated by counts. For example, the probability of success of a binary (binomial) variable has a maximum likelihood estimate equal to the number of observed successes divided by the total number of considered cases. In all cases, it is possible to derive a Bayesian estimate of probability distribution parameters. For example, in the case of discrete variables, it is assumed that the multinomial distribution parameters have a suitable prior distribution, such as a Dirichlet distribution (or a Beta distribution for binary variables). In this case, the resulting estimate is a compromise between the maximum likelihood estimate and the prior opinion (Spiegelhalter and Lauritzen, 1990).

Learning conditional distributions in BNs can be treated as the estimation of probability distribution parameters thanks to two fundamental assumptions: (1) it is assumed that the probability distribution parameters of different nodes are independent (global independence) and (2) it is assumed that probability distribution parameters of the same node given a configuration of the parent nodes are independent (local independence). This allows the independent treatment (and, thus, estimation) of the parameters during learning. The most well-known algorithms for learning conditional probability distributions have been described (Buntine, 1994; Lauritzen, 1995) and also allow obtaining estimates in the presence of missing data.

19.4.2 Learning the Structure of Bayesian Networks

A more complex problem is represented by learning the structure of BNs from data (Murphy and Mian, 1999). This problem is usually solved by resorting to a model-selection approach. The objective is to learn the structure M of a BN on the basis of a set of data D, that is, finding the structure M that maximizes $P(M|D)$. Using Bayes' theorem, we obtain $P(M|D) \propto P(D|M)P(M)$ and, therefore, assuming that all models are a priori equally probable, the problem can be approached as the search for a structure that maximizes the marginal likelihood:

$$P(D \mid M) = \int_{\Theta} P(D \mid \theta, M) P(\theta \mid M) d\theta \qquad (19.9)$$

The problem thus consists of two steps: calculating the score function (the marginal likelihood) and searching for the maximum of the score:

1. A closed-form solution of the maximums-likelihood integral is possible only in particular cases, such as in the cases of multinomial or linear Gaussian variables. Yet, some approximations exist, such as the Bayesian information criterion (BIC), which allows the use of more general models. The BIC is defined as

$$\log(f(D \mid M)) \approx \log(f(D \mid \theta_M, M)) - \frac{1}{2}\log(N)\dim(M)$$

where θ_M is the estimate of model parameters (for example, the maximum likelihood estimate), N is the number of data, and $\dim(M)$ is the total number of model parameters. Thanks to the chain rule and to conditional independence assumptions, the first term, which is the log likelihood of data with respect to the model and the parameters estimate, becomes the sum of the log likelihoods of single nodes.
2. The search for the maximum of the marginal likelihood is a NP-problem, as the number of possible networks is exponential in the number of nodes. For this reason, various heuristic strategies have been proposed, the most well-known being the K2 method by Cooper and Herskovits (1992). Yet this strategy requires an ordering of the nodes that allows a significant reduction of the search space for possible parents of a network node. Alternatively, hill-climbing methods and variants of the REVEAL method previously described have been used. Recently, in the case of dynamic networks with constraints on the graph connectivity, the use of Monte-Carlo-based methods has been proposed, in which the structure is searched by means of an iterative stochastic procedure (Husmeier, 2003).

It is necessary to emphasize that in the literature other approaches have been presented that are able to learn networks with hidden nodes. In this case, the various proposed strategies include optimization methods such as gradient descent (Bindler et al., 1997) or Gibbs sampling (Bellazzi et al., 1997) when the structure is known, or of a technique known as structural EM when the structure is unknown (Friedman, 1998).

To conclude, it is necessary to note that all the above presented approaches have high computational complexity. Their application thus requires filtering data beforehand and/or modeling only subportions of the regulatory network under study, assuming that gene interactions are modular (Ong et al., 2002). In order to deal with these problems, new methodologies have been proposed, which we will briefly describe in the following.

19.4.3   Module Networks

A particularly interesting method to learn gene regulatory networks with BNs is represented by module networks (MNs; Segal et al., 2005). MNs were employed to analyze yeast expression data and expression profiles of different tumor types by E. Segal and colleagues (Segal et al., 2003, Segal et al., 2004). MNs start from the assumption that the dynamic behaviors of coexpressed genes are not distinguishable and, thus, these genes have to be grouped into a unique variable. Moreover, it is assumed that the temporal profile of each coexpressed gene can be considered an independent realization of the same stochastic process. On this basis, MNs partition the variables space into modules; each module is characterized by a certain number of parent modules and a unique conditional probability distribution. The search for a module network requires solving a learning problem with three components: (1) estimating the probability distributions in a generic module, (2) searching for the number of modules and the variables they contain, and (3) searching for the best structure of the module network. In order to solve problem (1), the proposed approach assumes that the variables in a module express the same quantities. In the case of a gene regulatory network, this assumption is natural, as variables correspond to gene expression values on a continuous scale or discretized into the same number of values. This allows treating all variables belonging to the module as unique variables and estimating the probability distributions as if the data observed for the module variables were all independent samples of the same variable. In this way, the available samples are increased. Steps (2) and (3) are dealt with by extending in a natural way what is done for BN learning. The procedure first requires the definition of a Bayesian score to evaluate the proposed models M, which include both the structure and the partition into modules. The employed principle is again the maximization of the posterior probability, $P(M \mid D) = P(S, A \mid D) \propto P(D \mid S; A)$ $P(S \mid A) P(A)$, where A and S, respectively, indicate the partition into modules and the structure of the module network. As previously described, the problem of score calculation is reflected in the calculation of the marginal likelihood $P(D \mid S; A)$, which requires the evaluation of an integral similar to that of Equation 19.9. Once the score is defined, the algorithm starts the partition into modules through a clustering procedure, which groups together genes with similar expression profiles. When the modules are known, it is possible to use a standard algorithm for BN learning. Once the network structure is learned, which is optimal with respect to the employed partition into modules, the search for an alternative partition into modules is performed by performing local variations in the modules. In particular, whether the score increases if a node is moved to another module is evaluated and the procedure is iterated for each module. Steps (2) and (3) are repeated until neither the structure nor the module division changes.

19.4.4 Integrating Prior Knowledge

The BN framework allows one to introduce background knowledge in several different ways. Imoto and coworkers (Imoto et al., 2004) represent the prior probability of each model $p(G)$ as a function of the available background knowledge, which is modeled with the Gibbs distribution:

$$p(G) = Z^{-1} \exp\{-\lambda E(G)\} \qquad (19.10)$$

where $E(G)$ is the energy of the network, Z is a normalizing constant, and λ is a suitable hyperparameter. $E(G)$ can be decomposed by exploiting the locality property of BNs in order to take into account the strength of prior evidence for an arc, expressed in terms of the energy of the arc itself:

$$E(G) = \sum_{i=1}^{n} E_i = \sum_{i=1}^{n} \sum_{j \in pa(x_i)} E_{ij} \qquad (19.11)$$

where E_{ij} is the energy of the arc going to the ith gene from its jth parent. The problem of specifying prior knowledge is then transformed into the specifications of a number of hyperparameters through a set of algebraic transformations. The search process is then guided by the prior knowledge using the usual Bayesian scoring metric. The method was tested on simulated data and yeast data. A similar approach was applied by Tamada and coworkers (2003) to infer gene networks by combining gene expression data and the structure of a gene promoter region.

Later on, the same group (Nariai et al., 2005) proposed an algorithm for the joint learning of gene-regulatory networks and protein–protein interaction networks in terms of Bayesian and Markov (undirected) networks. Given DNA microarray data (X) and interaction networks data (Y), the posterior probability of the gene-regulatory network (G_r) and of the protein interaction network (G_p) was calculated as

$$p(G_r, G_p \mid X, Y) \propto P(X \mid G_r)P(X \mid G_p)P(G_r \mid G_p)P(G_p) \qquad (19.12)$$

The described approach was tested on a mutant expression dataset (Hughes et al., 2000) and protein–protein interaction data (Gavin et al., 2002) and, additionally, relied on the background knowledge contained in the MIPS functional-category database. The approach has been evaluated through a comparison with an external reference-knowledge source (KEGG). The results showed an increase in accuracy in recovering protein–protein interactions of about 10% with respect to learning without gene expression information, and an improvement in recovering correct regulatory interactions with respect to learning without protein–protein interaction information.

Le Phillip and colleagues (Le Phillip et al., 2004) described the effect of using background knowledge on the learning of a BN from a set of simulated data describing glucose homeostasis. The effect of the inclusion of prior knowledge was evaluated by clamping a number of arcs in the gene-regulation network to the correct ones as obtained from background knowledge, and then evaluating the sensitivity, that is, the proportion of true edges extracted by the network over the total number of true edges. The results pointed out that even a relatively small proportion of clamped edges may improve the sensitivity of the algorithm and, at the same time, reduce the number of expression profiles required for learning.

Bernard and Hartemink (2005), exploited a transcription-factor database to derive a model for defining the prior probability $p(G)$ of a DBN. In particular, they derived the probability of an arc connecting two genes in the network by analyzing the data on transcription-binding sites. By assuming that the evidence that a transcription factor regulates a gene is expressed through a p-value, they computed the prior probability as the composition of local models, each one expressing the probability of an edge being present given its p-value. Since the scoring metric can be decomposed into local models, the search procedure is easily modified by taking into account different priors for each different gene model. The effectiveness of the algorithm has been shown on simulated data on the gene-regulatory network of the yeast cell cycle, producing similar findings as Le Phillip and coworkers (Le Phillip et al., 2004).

19.5 CONCLUSIONS

In this chapter, reverse engineering methods have been outlined as an interesting instrument for network reconstruction from gene expression data. These algorithms use different approaches; however, it is not clear which methods perform best and in which experimental situations. Since no biological network is understood well enough to serve as a standard, quantitative assessment of reverse engineering methods is usually accomplished in vitro (Mendes et al., 2003; Van den Bulcke et al., 2006). To this purpose, a novel gene-network simulator was recently presented (Di Camillo et al., 2007), which resembles some of the main features of transcriptional regulatory networks, related to topology, interaction among regulators of transcription, and expression dynamics. Based on that, the ability of four reverse engineering methods (linear and nonlinear dynamic Bayesian networks, ARACNe, and graphical Gaussian models) to reconstruct network topology and to detect hubs was assessed in a variety of situations with a wide range of system complexity and experimental design (Corradin et al., 2008). Results suggest that reliable hypotheses on gene interactions are only possible with a limited number of genes (on the order of ten), with their expression monitored during multiple experiments so as to excite different states of the system.

As regards the biological meaning of these interactions, they should be attributed to functional relationships rather than direct regulatory actions, since RNA undergoes a series of posttranscriptional modifications before being translated into proteins, and proteins themselves undergo a series of posttranslational transformations before performing their regulatory action. Therefore, to infer direct regulatory actions it is necessary to integrate gene expression data with both protein data and information on physical/chemical interactions. Information on protein–protein and protein–DNA interactions either derives from a priori knowledge or can be inferred from the analysis of protein structures (and, thus, on their possibility to interact) or DNA sequencess (looking for characteristic motifs that are the site of protein interaction). Given this experimental evidence, the standard methodologies for the analysis of gene expression profiles seem to be severely limited. Bioinformatics and computational biology need to overcome this limitation and develop approaches that allow the integration of high-throughput transcriptional data with any gene structural and functional information.

Current systems biology approaches try to combine different types of data both to improve the performance of reverse engineering applications and to get the deepest biological insight. Development of data-integration strategies and dynamic aspects of regulation constitute the current challenge in gene network reconstruction

REFERENCES

Basso, K., Margolin, A. A., Stolovitzky, G., Klein, U., Dalla-Favera, R., and Califano, A.. Reverse Engineering of Regulatory Networks in Human B Cells, *Nat Genet.*, Vol. 37, No. 4, 382–390, 2005.

Bellazzi, R., Magni, P., and De Nicolao, G., Dynamic Probabilistic Networks for Modelling and Identifying Dynamic Systems: A MCMC Approach, *Intelligent Data Analysis*, Vol. 1, No. 4, 245–262, 1997.

Bernard, A., and Hartemink, A. J., Informative Structure Priors: Joint Learning of Dynamic Regulatory Networks from Multiple Types of Data, in *Pacific Symposium on Biocomputing,* pp. 459–470, Morgan Kaufman, San Francisco, 2005.

Bindler, J., Koller, D., Russel, J., and Kanazawa, D., Adaptive Probabilistic Networks with Hidden Variables, *Machine Learning,* Vol. 29, 213–244, 1997.

Buntine, W., Operations for Learning with Graphical Models, *J. of Artificial Intelligence Research,* Vol. 2, 159–225, 1994.

Butte, A. J., Tamayo, P., Slonim, D., Golub, T. R., and Kohane, I. S., Discovering Functional Relationships Between RNA Expression and Chemotherapeutic Susceptibility Using Relevance Networks, *Proc. Natl. Acad. Sci. USA,* Vol. 97, No. 22, 12182–12186, 2000.

Chen, T., Hongyu, L. H., and Church, G. M., Modelling Gene Expression with Differential Equations, in *Pacific Symposium on Biocomputing,* Vol. 4, pp. 29–40, Morgan Kaufman, San Francisco, 1999

Cooper, G., F., and Herskovits, E., A Bayesian Method for the Induction of Probabilistic Networks from Data, *Machine Learning,* Vol. 9, 309–347, 1992.

Corradin, A., Di Camillo, B., Toffolo, G., and Cobelli, C., In Vitro Assessment of Four Reverse Engineering Algorithms: Role of Network Complexity and Multi-Experiment Design in Network Reconstruction and Hub Detection, in *EN-FIN—DREAM Conference Assessment of Computational Methods in Systems Biology,* April 28–29, Madrid, 2008.

Cosentino, C. W., Curatola, F., Montefusco, M., Bansal, D., di Bernardo, and F., Amato, Linear Matrix Inequalities Approach to Reconstruction of Biological Networks, *IET Systems Biology,* Vol. 1, No. 3, 164–173, 2007.

D'Haeseleer, P., Wen, X., Fuhrman, S., and Somogyi, R., Linear Modeling of mRNA Expression Levels During CNS Development and Injury, in *Pacific Symposium on Biocomputing,* Vol. 4, pp. 41–52, Morgan Kaufman, San Francisco, 1999.

Di Camillo, B., Toffolo, G., and Cobelli, C., (2007). A Gene Network Simulator to Assess Reverse Engineering Algorithms, in *DREAM2: Second Dialogue for Reverse Engineering Assessments and Methods,* December 3 and 4, New York Academy of Sciences, New York, 2007.

Ferrazzi, F., Sebastiani, P., Ramoni, M. F., and Bellazzi, R., Bayesian Approaches to Reverse Engineer Cellular Systems: A Simulation Study on Nonlinear Gaussian Networks, *BMC Bioinformatics,* Vol. 8, Suppl. 5, S2, 2007.

Friedman, N., Murphy, K., and Russel, S., Learning the Structure of Dynamic Probabilistic Networks, in *Proceedings of the Fourteenth Conference on Uncertainty in Artificial Intelligence,* pp. 139–147, Morgan Kaufman, 1998.

Friedman, N., The Bayesian Structural EM, in *Proceedings of Fourteenth Conference on Uncertainty in Artificial Intelligence (UAI '98),* pp. 129–138, G. F. Cooper and S. Moral (Eds.), Morgan Kaufmann, 1998.

Friedman, N., Inferring Cellular Networks Using Probabilistic Graphical Models, *Science,* Vol. 303, 799–805, 2004.

Friedman, N., Linial, M., Nachman, I., and Pe'er, D., Using Bayesian Networks to Analyze Expression Data, *J. Comput. Biol.,* Vol. 7, Nos. 3–4, 601–620, 2000.

Gavin, A. C., Bosche, M., Krause, R., Grandi, P., Marzioch, M., Bauer, A., Schultz, J., Rick, J. M., Michon, A. M., Cruciat, C. M., Remor, M., Hofert, C., Schelder, M., Brajenovic, M., Ruffner, H., Merino, A., Klein, K., Hudak, M.,

Dickson, D., Rudi, T., Gnau, V., Bauch, A., Bastuck, S., Huhse, B., Leutwein, C., Heurtier, M. A., Copley, R. R., Edelmann, A., Querfurth, E., Rybin, V., Drewes, G., Raida, M., Bouwmeester, T., Bork, P., Seraphin, B., Kuster, B., Neubauer, G., and Superti-Furga, G., Functional Organization of the Yeast Proteome by Systematic Analysis of Protein Complexes, *Nature*, Vol. 415, 141–147, 2002.

Heckerman, D., A Tutorial on Learning With Bayesian Networks, in, M. I. Jordan (Ed.), *Learning in Graphical Models*, Kluwer, 1998.

Heckerman, D., Geiger, D., and Chickering, D., M. Learning Bayesian Networks: The Combination of Knowledge and Statistical Data, *Machine Learning*, Vol. 20, 197–243, 1995.

Hughes, T. R., Marton, M. J., Jones, A. R., Roberts, C. J., Stoughton, R., Armour, C. D., Bennett, H. A., Coffey, E., Dai, H., He, Y. D., Kidd, M. J., King, A. M., Meyer, M. R., Slade, D., Lum, P. Y., Stepaniants, S. B., Shoemaker, D. D., Gachotte, D., Chakraburtty, K., Simon, J., Bard, M., and Friend, S. H., Functional Discovery via a Compendium of Expression Profiles, *Cell*, Vol. 102, 109–126, 2000.

Husmeier, D., Sensitivity and Specificity of Inferring Genetic Regulatory Interactions from Microarray Experiments with Dynamic Bayesian Networks, *Bioinformatics*, Vol. 19, No. 17, 2271–2282, 2003.

Imoto, S., Higuchi, T., Goto, T., Tashiro, K., Kuhara, S., and Miyano, S., Combining Microarrays and Biological Knowledge for Estimating Gene Networks via Bayesian Networks, *J. Bioinform. Comput. Biol.*, Vol. 2, 77–98, 2004.

Lauritzen, S. L., The EM Algorithm for Graphical Association Models with Missing Data, *Computational Statistics and Data Analysis*, Vol. 19, 191–201, 1995.

Lauritzen, S. L., and Spiegelhalter, D. J., Local Computations with Probabilities on Graphical Structures and Their Application to Expert Systems (with discussion), *Journal of the Royal Statistical Society, Series B*, Vol. 50, 157–224, 1988.

Le Phillip, P., Bahl, A., and Ungar, L. H., Using Prior Knowledge to Improve Genetic Network Reconstruction from Microarray Data, *In Vitro Biol.*, Vol. 4, 335–353, 2004.

Liang, S., Fuhrman, S., and Somogyi, R., REVEAL, a General Reverse Engineering Algorithm for Inference of Genetic Network Architectures, *Pacific Symp. Biocomp.*, Vol. 98, No. 3, 18–29, 1998.

Margolin, A A, Nemenman, I., Basso, K., Wiggins, C., Stolovitzky, G., Dalla Favera, R., and Califano, A. ARACNE: An Algorithm for the Reconstruction of Gene Regulatory Networks in a Mammalian Cellular Context, *BMC Bioinformatics*, Vol. 7, Suppl. 1, S7, 2006.

Mendes, P., Sha, W., and Ye, K., Artificial Gene Networks for Objective Comparison of Analysis Algorithms, *Bioinformatics*, Vol. 19, Suppl. 2: 122–129, 2003.

Murphy, K., and Mian, S., Modeling Gene Expression Data Using Dynamic Bayesian Networks, Technical Report, University of California, Berkeley http://www.cs.ubc.ca/~murphyk/Papers/ismb99.pdf, 1999.

Nariai, N., Tamada, Y., Imoto, S., and Miyano, S. Estimating Gene Regulatory Networks and Protein-Protein Interactions of *Saccharomyces Cerevisiae* from Multiple Genome-Wide Data, *Bioinformatics*, Vol. 21, Suppl. 2, ii206–ii212, 2005.

Ong, I. M., Glasner, J. D., and Page, D., Modelling Regulatory Pathways in *E. Coli*

From Time Series Expression Profiles, *Bioinformatics,* Vol. 18, No. 1, 241–248, 2002.

Pearl, J., *Probabilistic Reasoning in Intelligent Systems: Networks Of Plausible Inference.* Morgan Kaufmann, 1988.

Segal, E., Shapira, M., Regev, A., Pe'er, D., Botstein, D., Koller, D., and Friedman, N., Module Networks: Identifying Regulatory Modules and their Condition-Specific Regulators from Gene Expression Data, *Nat. Genet.,* Vol. 34, No. 2, 166–176, 2003.

Segal, E., Friedman, N., Koller, D., and Regev, A., A Module Map Showing Conditional Activity of Expression Modules in Cancer, *Nature Genetics,* Vol. 36, No. 10, 1090–8, 2004.

Segal, E., Pe'er, D., Regev, A., and Koller, D., Friedman, N. Learning Module Networks, *Journal of Machine Learning Research.* Vol. 6, 557–88, 2005.

Shannon, C. E., and Weaver, W., The Mathematical Theory of Communication. University of Illinois Press, 1963.

Shmulevich, I., Dougherty, E. R., Kim, S., and Zhang, W., Probabilistic Boolean Networks: A Rule-based Uncertainty Model for Gene Regulatory Networks, *Bioinfomatics,* Vol. 18, 261–274, 2002.

Somogyi, R., and Sniegoski, Modeling the Complexity of Gene Networks Understanding Multigenetic and Pleiotropic Regulation, *Complexity.* Vol. 1, 45–63, 1996.

Spiegelhalter, D. J., and Lauritzen, S. L., Sequential Updating of Conditional Probabilities on Directed Graphical Structures, *Networks,* Vol. 20, 579–605, 1990.

Tamada, Y., Kim, S., Bannai, H., Imoto, S., Tashiro, K., Kuhara, S., and Miyano, S., Estimating Gene Networks from Gene Expression Data by Combining Bayesian Network Model with Promoter Element Detection, *Bioinformatics,* Vol. 19, Suppl. 2, II227–II236, 2003.

Van den Bulcke, T., et al., SynTReN: A Generator of Synthetic Gene Expression Data for Design and Analysis of Structure Learning Algorithms, *BMC Bioinformatics,* Vol. 7, 43, 2006.

Weaver, D. C., Workman, C. T., and Stormo, G. D., Modeling Regulatory Networks with Weight Matrices, in *Pacific Symposium on Biocomputing.* Vol. 4, pp. 112–123, Morgan Kaufman, San Francisco, 1999.

Yeung, M. K. S., Tegnér, J., and Collins, J., Reverse Engineering Gene Networks Using Singular Value Decomposition and Robust Regression, *Proc. Natl. Acad. Sci. USA.* Vol. 99, No. 9, 6163–6168, 2002.

CHAPTER 20

BIOMOLECULAR SEQUENCE ANALYSIS

Linda Pattini and Sergio Cerutti

20.1 INTRODUCTION

The analysis of biological data using mathematics, statistics, and, in general, computational approaches, is a necessary complement to research activities in molecular biology. The sequences of the informational macromolecules of the cell (DNA, RNA, and proteins) and the huge amount of data at the subcellular scale, which are provided by the current technologies, require different processing techniques. This chapter provides a survey and examples of nucleic acid and protein sequences analyzed through methods borrowed from signal processing and information theory. In some cases, the use of such techniques allows one to extract information about the physical and experimental reality and to establish a relation between the information content of the sequence and the structural and/or functional aspects of the molecule; in other cases, visualization and data management serve as useful tools of investigation.

20.2 CORRELATION IN DNA SEQUENCES

20.2.1 Coding and Noncoding Sequences

The application of methods generally used in the processing of numerical series requires an appropriate transcodification of the DNA sequence. This transformation can be obtained according to different criteria privileging different aspects of information (for example, the occurrence of the purinic or pyrimidinic bases, or the occurrence of bases characterized by a strong bond with the complementary base, or simply the occurrence of a single base in the sequence). In Table 20.1, the main kinds of transcodifications for a nucleotidic sequence are reported; they allow one to obtain from a

Table 20.1. Mapping rules from a nucleotidic sequence n_i to a numerical series u_i

Purine–pyrimidine	if n_i = A or G, u_i = 1; if n_i = C or T, u_i = –1
Hydrogen bond energy rule	if n_i = C or G, u_i = 1; if n_i = A or T, u_i = –1
Hybrid rule	if n_i = A or C, u_i = 1; if n_i = G or T, u_i = –1
A–notA	if n_i = A, u_i = 1; if n_i ≠ A, u_i = –1
C–notC	if n_i = C, u_i = 1; if n_i ≠ C, u_i = –1
G–notG	if n_i = G, u_i = 1; if n_i ≠ G, u_i = –1
T–notT	if n_i = T, u_i = 1; if n_i ≠ T, u_i = –1

string of characters a binary series whose cumulative sum, also known as a *DNA walk*, may be processed through numerical methods.

A study of about one thousand coding sequences and as many non-coding sequences (Buldyrev et al., 1995), through the computation of the Fourier transform and of the power spectrum of nonoverlapping subsequences of fixed length, revealed the different characteristics of the two kinds of DNA regions. Coding and noncoding sequences show, as depicted in Figure 20.1, different behaviors of the spectral averages. The plots

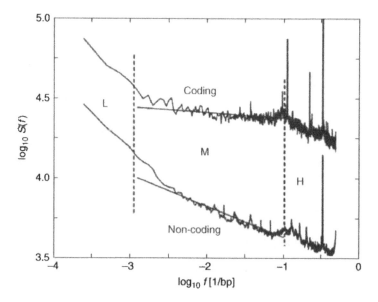

Figure 20.1. The power spectrum of the DNA walk can discriminate coding from noncoding sequences, particularly in the range of intermediate frequencies from 1/10 to 1/1000 bp. The exponent β, which vanishes in the case of white noise, for coding sequences has the value 0.00 ± 0.04. Conversely, for noncoding sequences, the mean value of β (0.16 ± 0.05), significantly greater than 0, shows the presence of long-range correlation. (From Buldyrev et al., 1995, reprinted with permission from the American Physical Society.)

may be subdivided, as indicated, into three spectral regimes—H, M, and L—corresponding, respectively, to high-frequency, mid-frequency, and low-frequency regions. In the region of high frequencies, two main peaks at 1/3 bp^{-1} and at 1/9 bp^{-1} are particularly evident in the coding sequences, due to the codon structure. Their presence, even in the noncoding sequences, may be motivated by the presence of ancient coding sequences that lost their functional role during evolution. The region L is not particularly informative because of artifacts. The intermediate region M shows in the log–log plot a linear behavior for both kinds of sequences. For the spectrum that assumes the expression $S(f) \sim f^{-\beta}$, a value for the slope β is found very near to 0 in the coding sequences and significantly greater than 0 for the noncoding sequences. These results, therefore, are characterized by the presence of long-range correlation. Evidently, the information is redundant with respect to the coding sequences that are similar to white noise. These properties are found even when other methods of physical statistics are exploited to implement algorithms for gene finding in genomic sequences (Stanley et al., 1999; Uberbacher et al., 1991).

20.2.2 DNA Sequence–Structure Relationship

Analyses of the assessment of long-range correlation were also made on sequences of genomic DNA (Audit et al., 2002). In this case, the presence of autosimilarity could be related to the structural organization of DNA, which in the nucleus is densely packed to form chromatin. Audit and coworkers used the wavelet transform of DNA walks to compute the coefficient of long-range correlation H (Hurst coefficient). Values of H greater than one-half, that is, the value that corresponds to random sequences not correlated, indicate that the sequence has this kind of property. The estimation of H is obtained according to the expression $\sigma(a) \propto a^H$, where a represents the scale and $\sigma(a)$ is the root mean square of the coefficients of the transform at a given scale a. The use of the wavelets in this kind of analysis is particularly useful, since it allows the assessment of the scale invariance even in the case of evident trends in the DNA walk due to dishomogeneity in the nucleotidic composition of the different regions. Figure 20.2 shows the DNA walk correspondent to a human sequence according to the purine–pyrimidine rule. Three intervals with a different nucleotidic content are visible. For this sequence, the wavelet transform was computed with two different mother wavelets shown in the panels below: the derivatives of the Gaussian functions of the first and second order (the latter known as "sombrero" for its characteristic shape). Panels (c), (d), and (e) show the profiles of the first transform at the scales $a = a_1 = 32$ and $a = a_2 = 512$, and the profile of the second transform at the scale a_2. The latter does not appear to be influenced by the macroscopic trend of the considered sequence.

The analysis was accomplished on the whole genome of the first eu-

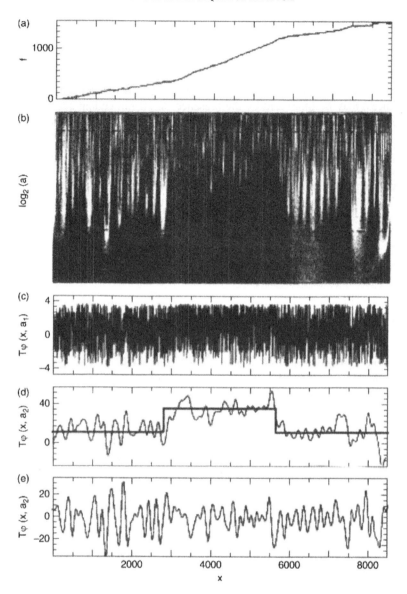

Figure 20.2. Wavelet analysis of a human DNA sequence (coding for desmo-plakin) transcoded according to the purine–pirimidine rule. (a) DNA walk of the sequence. (b) Scalogram of the wavelet transform with the first derivative of the Gaussian function as mother wavelet. (c) Transform profile at the scale $a = a_1 = 32$. (d) Transform profile at the scale $a = a_2 = 512$. (e) Behavior of the transform, with second-derivative Gaussian wavelet, at the scale $a = a_2 = 512$. This profile is not influenced by the fluctuations in the nucleotidic composition of the sequence. (Reprinted from Arneodo et al., 1996 with permission from Elsevier.)

karyote to be sequenced, *Saccharomyces cerevisiae*. The 16 chromosomes were transcodified according the rule A–notA and the results of the first estimation of H is reported in Figure 20.3. For values of scale a between 20 and 200 bp, $H = 0.59 \pm 0.02$; for $200 \leq a \leq 5000$, $H = 0.82 \pm 0.01$. Two other kinds of conversion of the DNA string, which concern the mechanical properties of the molecule experimentally observed, were considered: the codification Pnuc, obtained according to the nucleosome positioning, and the DNase I rule, which is another scale of flexibility. For both of them, a numerical value is associated to each triplet of nucleotides and a cumulative series is obtained (Table 20.2). In Figure 20.3, the plots of $\sigma(a)$ are reported also for these two transcodifications. The analysis of the flexibility profile Pnuc provides results comparable to those obtained on the "textual" profile. On the contrary, with the DNase rule, a sensible reduction of the estimated H is apparent. A characteristic scale emerges between 100 and 200 bp, which separates two different regimes of long-range correlation likely related to the structural organization of DNA around the nucleosomes, since the strand that wraps a single nucleosome has approximatively that length.

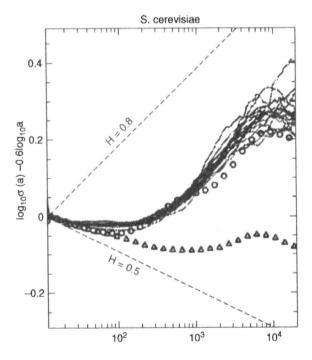

Figure 20.3. Estimation of $\sigma(a)$. $Log_{10}\ \sigma(a) - 0.6\ log_{10}\ a$ is plotted versus $log_{10}\ a$. (——) A–notA rule; (○) PNuc rule; (△) DNase rule. (Reprinted from Audit et al., 2001 with permission from the American Physical Society.)

Table 20.2. PNuc and DNase I scales

	Pnue	Dnase I
AAA/TTT	0.0	0.1
AAC/CTT	3.7	1.6
AAG/CCT	5/2	4.2
AAT ATT	9.7	0.0
ACA/TGT	5.2	5.8
AGC/GGT	5.4	5.2
ACT/AGT	5.8	2.0
AGA/TCT	3.3	5.5
AGC/GCT	7.5	6.3
AGG/CCT	5.4	4.7
ATA/TAT	2.8	9.7
ATC/GAT	67	8.7
CAA/TTG	3.3	6.2
CAC/GTG	6.5	6.8
CAG/CTG	4.2	9.6
CCA/TGG	5.4	0.7
CCC/GGG	6.0	5.7
CCG/CGG	4.7	3.0
CGA/TCG	8.3	5.8
CGC/GCG	7.5	4.3
CTA/TAG	2.2	7.8
CTC/GAG	5.4	6.6
GAA/TTC	3.0	5.1
GAC/GTC	5.4	5.6
GCA/TGC	6.0	7.5
GCC/GGC	10.0	8.2
GGA/TCC	3.8	6.2
GTA/TAC	3.7	6.4
TAA/TTA	2.0	7.3
TCA/TGA	5.4	10.0

20.3   SPECTRAL METHODS IN GENOMICS

For the analysis of the frequency content of DNA sequences, color spectrograms were proposed that allow the visualization of characteristic patterns in an immediate manner (Anastassiou, 2000).

The spectrogram of a numerical sequence can be obtained through the application of the short-time Fourier transform (STFT). The transcodification of the string of characters can be obtained by means of four binary indicators $u_A[n]$, $u_C[n]$, $u_G[n]$, and $u_T[n]$, one for each of the four bases, that for the position n may assume the value 1 or 0 according to the presence or the absence of the correspondent bases. The so obtained four series are not linearly independent; three of them are sufficient to rebuild the original DNA sequence. Applying an opportune transformation, it is possible to obtain three independent series, each of which can be associated with one color of the RGB spectrum. In this way, the superimposition of the spectrograms corresponding to the three different channels creates a color image. The series has the following expressions:

$$x_r[n] = a_r u_A[n] + t_r u_T[n] + c_r u_C[n] + g_r u_G[n]$$

$$x_g[n] = a_g u_A[n] + t_g u_T[n] + c_g u_C[n] + g_g u_G[n]$$

$$x_b[n] = a_b u_A[n] + t_b u_T[n] + c_b u_C[n] + g_g u_G[n]$$

where the coefficients are chosen opportunely (Sussillo et al., 2004), assigning each DNA base to the four vertices of a regular tetrahedron whose center is positioned at the origin of a Cartesian system and projecting the vertices on the axes r, g, and b:

$a_r = 0$	$a_g = 0$	$a_b = 1,$
$t_r = 0.911$	$t_g = -0.244,$	$t_b = -0.333$
$c_r = 0.244$	$c_g = 0.911$	$c_b = -0.333$
$g_r = -0.817$	$g_g = -0.471$	$g_b = 0.471$

These coefficients allow the association of a characteristic color to each bases (blue for adenine, red for thymine, green for cytosine, yellow for guanine), further improving the readability of the image. In the obtained spectrogram, the horizontal axis represents the position in the sequence and the vertical axis refers to the frequency of the STFT. The corresponding period is N/k, where k is the value of the vertical axis and N is the length of the epoch used in the STFT.

Figure 20.4 shows the spectrogram obtained through this technique for the entire chromosome of *E. coli* of about 4.6 Mbp in length, on windows of length $N = 10,000$. The three-period ($k = 10000/3$) characteristic of the coding sequences is apparent, consistent with the fact that prokaryotes have compact genomes, that is, they lack noncoding regions.

This kind of representation for DNA sequences may reveal other patterns and local characteristics. CpG islands, for example, are DNA regions rich in cytosine and guanine and characterized by a high frequency, with respect to the expected value, of the dinucleotide CG, denoted as CpG ("p" refers to the phosphate group that binds the two nucleotides). Their importance derives from the their localization; in vertebrates, particularly in mammals, they often are present in promoter regions of genes frequently expressed in cells (housekeeping) and, thus, they represent an im-

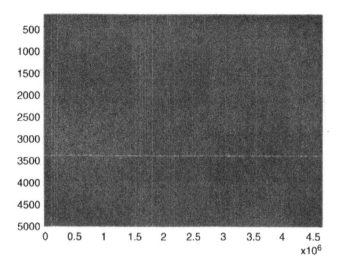

Figure 20.4. Spectrogram for the entire genome of *E. Coli* K12. The light line reveals the 3 bp periodicity ($k = 10,000/3$) along the whole sequence. (Reprinted from Sussillo et al., 2004 with permission.)

portant indication for the localization of such genes. Moreover, they represent a discriminant from a phylogenetic point of view since the CpG content decreases in higher organisms. These regions may be easily identified and represented by means of the described technique; in a color spectrogram, they are represented as regions in which the green color is dominant.

20.4   INFORMATION THEORY

20.4.1   Analysis of Genomic Sequences through Chaos Game Representation

Information theory provides a further approach to analyzing, assessing, and comparing the complexity of nucleotidic sequences. We present here a method that allows both representation and extraction of quantitative indices of complexity, discriminating between sequences coming from different species. This kind of representation is based on a technique borrowed from chaos theory and provides a fractal image of the sequence, revealing statistical and informational properties (Oliver et al., 1993; Almeida et al., 2001). A diagram is obtained by assigning to each of the four vertices of a square one of the four DNA bases and iteratively adding a point for each position in the sequence by starting from the center of the square and adding, for each character of the string, the median point of the segment between the vertex of the corresponding base and the current position. In Figure 20.5a, the image obtained through this technique for a genomic sequence belonging to human chromosome 11 (Genbank Accession: U01317) is shown. Differently from a random sequence, which would give a uniform diagram, the representation for a real DNA sequence reveals structured information. Such distribution of points in continuous space is unique and each point allows one to reobtain the whole sequence up to the beginning base. Thus, it is sufficient to know the position of the last point of the diagram to build the entire string. From a quantitative point of view, it can be observed that subdividing the image into 4^m subquadrants, each of them corresponds univocally to the possible different subsequences of length m. Consequently, the density of points in a subquadrant represents the frequency of occurrence of the correspondent string of m characters within the entire sequence. It is possible at this step, for each value of m, to compute the histogram of the number of quadrants showing a certain density of points with respect to the density. Normalizing the occurrences according to the total number of 4m quadrants, the probabilities $P_{k,m}$ of finding quadrants containing k points at resolution m may be estimated. Then it is possible to apply the Shannon entropy operator to the histrogram relative to each value of resolution m:

$$H_m = -\sum P_{k,m} \cdot \log_2 P_{k,m}$$

Repeating this computation for different values of m, an entropy profile may be obtained. The comparison among profiles extracted from different sequences can give indications about possible differences in their complexity. In Figure 20.5b, the profiles obtained for sequences retrieved from

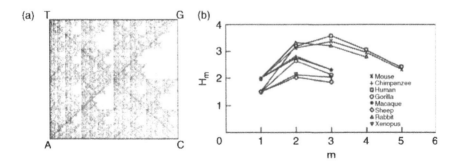

Figure 20.5. (a) Chaos game representation for a sequence of the human genome (Genbank Accession: U01317) of length 73,308 bp. (b) Variability of the entropic profiles obtained for genomic sequences belonging to different organisms. (Reprinted from Oliver et al., 1993 with permission from Elsevier.)

different genomes are reported. Their variability could be correlated to the different constraints acting on composition, and nucleotidic heterogeneity acting on different species.

20.5 PROCESSING OF PROTEIN SEQUENCES

20.5.1 Codification of Amino Acid Sequences

The extraction of information on structural and, consequently, functional characteristics of a peptide from its amino acid sequence is a primary issue in computational biology. As for nucleotide sequences, for amino acid sequences there are also codifications that allow the extraction from the string of characters of a numerical series that can be analyzed through signal processing methods. One of the possible codifications consists of the substitution of each amino acid with the correspondent electron–ion interaction potential (EIIP), which represents the average energy states of all valence electrons (Veljkovic and Slavic, 1972):

$$\langle \vec{k} + \vec{q} | w | \vec{k} \rangle = 0.25 Z \, \sin(\pi \cdot 1.04 Z) / 2\pi$$

where $\vec{q}$ is the change of momentum $\vec{k}$ of the delocalized electron in the interaction with potential w and

$$Z = \left(\sum_i Z_i \right) / N$$

where Z_i is the number of valence electrons of the ith component of each amino acid and N is the total number of atoms in the amino acid.

However, the data that are usually considered are the hydrophobicity properties (Cornette et al., 1987) and the relative accessible surface area (Lee and Richards, 1971) of residues that belong to the sequence. Both kinds of information may provide useful indications about spatial conformation and functional characteristics of the protein.

As concerns hydrophobicity, the polar/apolar character of single amino acids can be quantified through experimental scales. The assignment of the corresponding value to each residue allows one to obtain from the sequence a quantitative profile of the hydropathic character of the protein along its primary structure. From the literature, many hydrophobicity scales may be retrieved. They can be subdivided into two groups: those based upon experimental measurements about chemicophysical charcteristics of each amino acid such as as solubility and energy transfer between organic solvent and water, and those obtained from statistical evaluations of the spatial positions of each kind of amino acid in peptides whose structure is known. Among the most used scales are the Kyte–Doolittle and

Eisenberg hydrophobicity scales. The accessibility scale refers to the accessibility in the space of the surface of each residue for a molecule of solvent, generally the water molecule, which is modeled as a sphere of 1.4 Å.

The analysis and identification of patterns, periodicities, or characteristic behaviors of these profiles may provide useful information on the final spatial structure of the protein and on the functional aspects strictly connected to it.

20.5.2 Characterization and Comparison of Proteins

The EIIP codification is used for the application of the so-called resonance recognition model (RRM) (Cosic and Nesic, 1987). The method consists of the application of the DFT to the numerical series obtained from the sequence to identify characterizing information in its frequency content. The distance between samples is conventionally taken as 1; thus, the maximum spectral frequency remains unchanged and the length of the protein sequence determines the spectral resolution. This kind of approach may prove useful in the evaluation of similarities between amino acid sequences. For the exctraction of possible common characteristics in the frequency content of a pair of amino acid sequences, it is possible to apply the following cross-spectral function to the respective transcodifications $x(m)$ and $y(m)$:

$$S_n = X_n Y_n^* \qquad n = 1, 2, \ldots, N/2$$

where X_n are the coefficients of the DFT of the sequence $x(m)$, and Y_n^* are the complex conjugate DFT coefficients of the series $y(m)$.

In case of multiple sequences, the following cross-spectral function is applied:

$$|M_n| = |X1_n||X2_n| \ldots, |XM_n| \qquad n = 1, 2, \ldots, N/2$$

The cross-spectral analysis may highlight possible characteristic frequencies that may be related to the structural and/or functional characteristics that are shared by the compared sequences. For each frequency, a peak signal-to-noise ratio (S/N) is computed as the ratio between the signal intensity at a determined frequency and the average value over all the spectrum (in general, a S/N value above 20 is considered significant); it represents a similarity measure between the compared sequences (Cosic, 1994).

The method may be improved through the application of the discrete wavelet transform (DWT) to obtain a position-dependent frequency analysis. Each numerical series obtained through the EIIP conversion of the RRM is normalized to zero mean and unit standard deviation and zero-padded to have identical sequence length. The signal is decomposed by the

DWT to M levels of detail and an approximation at level M. For each level, the cross-correlation coefficients are computed to assess the similarity between the two compared protein sequences:

$$\rho^{12}(j) = \frac{r^{12}(j)}{\frac{1}{N}\left[\sum_{n=0}^{N-1} s_1^2(n) \sum_{n=0}^{N-1} s_2^2(n)\right]^{\frac{1}{2}}} \qquad j = 0, \pm 1, \pm 2, \pm 3, \ldots$$

where N is the length of the signal, j is the lag, and $r^{12}(j)$ represents the cross-covariance estimated as

$$r^{12}(j) = \frac{1}{N}\sum_{n=0}^{N-1} s_2(n)s_1(n-j)$$

The maximum absolute value of the correlation coefficient for each of the $M + 1$ signals is considered, comprising a vector of $M + 1$ elements, as a similarity measure at the different scales (de Trad et al., 2002). In Figure 20.6, the plots depict the behavior of the cross-correlation between the wavelet decompositions to four levels for the human and horse hemoglobin α-chains. The similarity vector that is obtained for these two homologous sequences is [0.92, 0.83, 0.90, 0.89, 0.89] (the components correspond, respectively, to the decompositions [A4, D4, D3, D2, D1]), revealing a strong correlation at all the resolutions, consistent with what could be expected. Conventionally, pairs of sequences showing a cross-correlation coefficient greater than ±0.7 are considered strongly correlated, and weakly correlated for values between ±0.5 and ±0.7. An interesting example concerns the comparison between the proteins chymotrypsin and subtilisin, two enzymes with a very low sequence identity that do not show a significant similarity according to the conventional sequence alignment methods. Using the described technique, a similarity vector of [0.35, 0.60, 0.42, 0.35, 0.18] is found. Indeed, the two proteins, for a convergent evolution event, share some functional aspects and the multiscale analysis, in contrast to traditional alignment methods, recognizes the presense of a weak correlation at one level of detail (D4).

20.5.3 Detection of Repeating Motifs in Proteins

The modularity of protein structures often implies the existence of repeating motifs (patterns of secondary structure elements) and domains (three-dimensional portions characterized for structure and function) in which peptide chains are organized. In the work of Murray and coworkers (2002), some interesting appplications are reported in which the frequency analysis of accessibility and hydrophobicity profiles, accomplished by means of a wavelet transform, allows the detection of motifs that charac-

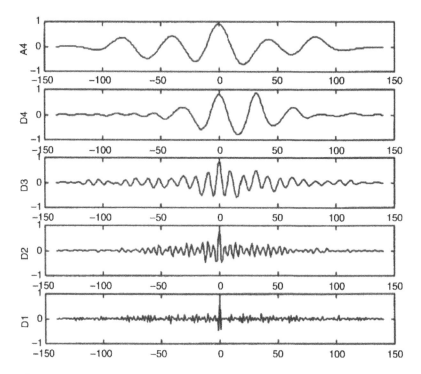

Figure 20.6. Cross-correlation functions between the signals obtained through the wavelet decomposition (wavelet Bior 3.3) of the sequences of the human and horse hemoglobin α-chains. The abscissa is the relative shift in terms of positions along the amino acid sequences. The similarity vector is [0.92, 0.83, 0.90, 0.89, 0.89], where the elements correspond, respectively, to the signals [A4, D4, D3, D2, D1], showing a high degree of correlation at all the resolutions for the two homologous sequences. (From de Trad et al., 2002, reprinted with permission from Oxford University Press.)

terize the final folding. One of the proposed examples concerns the propeller domain at the C-terminus of the protein emopexin (Protein Data Bank, PDB code: 1hxn) in the rabbit. This domain is composed of four repetitions of the same module, constituted by four or five beta sheets radially disposed around the central tunnel, as depicted in Figure 20.7a. Then the accessibility and hydrophobicity series are reported (Figure 20.7b), from which it is not possible to detect the presence of repeated motifs, whereas the corresponding spectra (Figure 20.7c) show a dominant frequency of approximately 50 residues (mean length of the propeller motif). The Fourier analysis reflects the overall frequency content without the spatial localization of the contributions. The scalogram of the wavelet transform of the accessibility profile highlights the repeating pattern along the

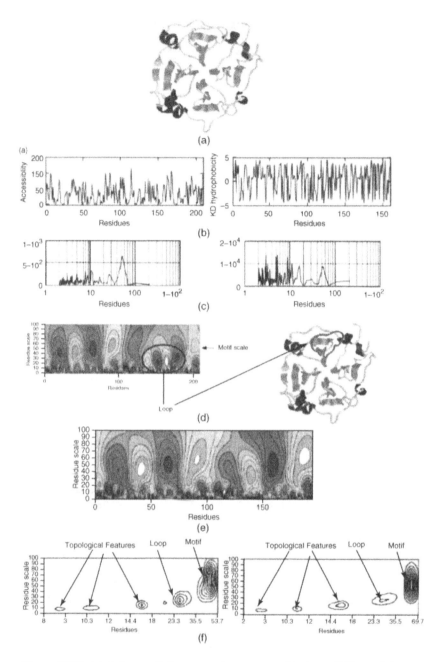

Figure 20.7. (a) Scheme of the emopexin structure. (b) Accessibility and hydrophobicity profiles. (c) Spectra of the accessibility and hydrophobicity profiles. (d) Scalogram of the wavelet transform of the hydrophobicity profile of the original sequence containing a hydrophilic loop. (e) Scalogram of the wavelet transform of the hydrophobicity profile of the sequence after the removal of the hydrophilic loop. (f) Frequency analysis of the wavelet coefficients obtained for the accessibility and hydrophobicity profiles, scale by scale. (From Murray et al., 2002, reprinted with permission from Elsevier.)

sequence (Figure 20.7d). At positions 124–185, centred on a scale of 26 residues, an irregularity appears in the pattern because of the insertion of a wide hydrophilic loop. Removing the amino acids corresponding to this region, the dominant periodicities evidenced by the scalogram are centered at 50 residues along the whole sequence (Figure 20.7e).

For an even more compact representation of the periodicities identified by the wavelet transform, it is possible to apply the Fourier transform to the coefficients of the transform at each scale:

$$\hat{P}(\omega,a) = \int_{-\infty}^{+\infty} T(a,b)e^{-2i\omega b}db$$

obtaining a normalized spectrum in which each scale has the same energy content:

$$C(\omega,a) = \frac{\left|\hat{P}(\omega,a)\right|^2}{N^{-1}\sum_{\omega}\left|\hat{P}(\omega,a)\right|^2}$$

In Figure 20.7f, such spectra are reported, computed both for the accessibility and the hydrophobicity series. It is apparent that there is a notable correspondence between the diagrams; they show components relative to internal periodicities of each motif, components correspondent to the hydrophilic loop and the evident trace of the repeating motifs.

20.5.4 Prediction of Transmembrane Alpha Helices

The wavelet transform was applied successfully even for the prediction of transmembrane helices. These represent a very common structure, since there are a lot of membrane proteins and they have a strong hydrophobic character that allow them to remain anchored in the double phospholipidic bilayer of the cell membrane. In the work of Liò and Vannucci (2000), a discrete wavelet transform is applied to the hydrophobicity profile and to a "propensity" profile obtained by assessing the occurrence of each kind of amino acid in a database of transmembrane helices. The wavelet coefficients are then filtered by means of a "change point" technique; for each scale, a statistical test is made to establish whether the behavior of the coefficients is equivalent to white noise or there is an informative signal. In this way, the largest coefficients are selected; the filtering depends on the confidence level of the test. Both the profiles, filtered through this technique, allow one to identify in the analyzed sequences the transmembrane domains, whose occurrence is indicated by the local maxima of the profiles.

20.5.5   Prediction of Amphiphilic Alpha Helices

The amphiphilic alpha helix is a particular element of the secondary structure of a protein. It has one side more polar than the opposite one, since hydrophilic and hydrophobic amino acids tend to segregate, remaining on different sides of the helix. The period corresponding to this state is equal to the physical period of the alpha helix, that is, 3.6 amino acids (eighteen residues for every five turns). Analyzing the frequency content of the hydrophobicity profile, it is possible to identify characteristic periodicities. The amphiphilicity of a peptide may be quantified through the computation of the hydrophobic moment $\mu(\delta)$:

$$\mu(\delta) = \left\{ \left[\sum_{n=1}^{N} H_n \sin(\delta n) \right]^2 + \left[\sum_{n=1}^{N} H_n \cos(\delta n) \right]^2 \right\}^{\frac{1}{2}} =$$

$$= \left| \sum_{n=1}^{N} H_n e^{j\delta n} \right| \qquad \delta = \frac{2\pi}{m}$$

where $\delta = 2\pi/m$ (m is the number of residues per turn) is the angle in radians that separates the projections along the axis of the residues that are disposed around the helix and H_n is the value of the hydrophobicity profile at the nth position (Phoenix and Harris, 2002). For the alpha helix ($m = 3.6$) that shows an amphipathic character, the hydrophobic moment should show a maximum at $\delta = 100°$, as in the example reported in Figure 20.8, where the plot of

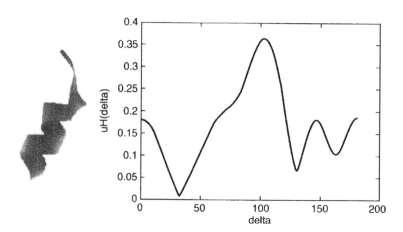

Figure 20.8. Three-dimensional representation of a synthetic model of the amphiphilic alpha helix 1AL1 (PDB code) and correspondent hydrophobic moment plot.

$\mu(\delta)$ for a model of an amphiphilic alpha helix (PDB code 1AL1) is shown. It has absolute maximum at $\delta = 100°$; other minor peaks are present because of the finiteness of the sequence. It is apparent that the hydrophobic moment is equivalent to the application of the DFT to the hydrophobicity series.

Even in this case, the analysis may be accomplished through the application of a wavelet transform, exploiting the greater flexibility of this operator. To identify the regions of a sequence with a high potential to assume the conformation of an amphiphilic alpha helix, a method was proposed (Pattini and Cerutti, 2004) based on the use of a continuous wavelet transform, with the Morlet function as mother wavelet, of the hydrophobicity profile, obtained according to the Eisenberg scale (Eisenberg et al., 1982). Selecting the scales corresponding to the critical 3.6 periodicity, it is possible to identify the portions of the sequence that are more likely to assume this kind of secondary structure. Figure 20.9 shows an example of prediction of the sequence of a protein containing two amphiphilic alpha helices (PDB code 1G83). It is possible to visualize that the transform coefficients, selected for the informative scales, are greater at the positions occupied by the alpha helices that are experimentally known.

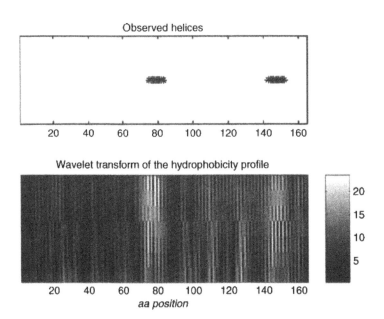

Figure 20.9. The wavelet transform highlights the periodicities of the hydrophobicity profile. In this example (protein PDB ID 1G83), the coefficients at the selected scales show greater values in the positions correspoding to the amno acids belonging to the amphiphilic alpha helices experimentally known (indicated by the stars in the upper panel).

This method, which, with respect to the traditional hydrophobic moment, shows a better accuracy, was utilized in the implementation of a predictor of mitochondrial proteins. As is known, many of these proteins have an N-terminal sequence characterized by a positive net charge and by the presence of an amphiphilic alpha helix (Roise and Theiler, 1988).

These peculiarities are responsible for the mechanism of targeting of the proteins that are synthesized in the cytoplasm by nuclear genes, but that are destined to the mitochondrial compartment. Even this last application, concerning the so-called protein sorting, represents an interesting application of signal processing to the field of computational biology, proving that its contribution may lead to useful applications.

BIBLIOGRAPHY

Audit, B., Vaillant, C., Arneodo, A., d'Aubenton-Carafa, Y., and Thermes, C., Long-Range Correlations Between DNA Bending Sites: Relation to the Structure and Dynamics of Nucleosomes, *J. Mol. Biol.*, Vol. 316, 903–918, 2002.

Almeida, J. S., Carriço, J. A., Maretzek, A., Noble, P. A., and Fletcher, M., Analysis of Genomic Sequences by Chaos Game Representation, *Bioinformatics*, Vol. 17, No. 5, 429–437, 2001.

Anastassiou, D., Frequency-Domain Analysis of Biomolecular Sequences, *Bioinformatics*, Vol. 16, No. 12, 1073–1081, 2000.

Arneodo, A., d'Aubenton-Carafa, Y., Bacry, E., Graves, P. V., Muzy, J. F., and Thermes, C., Wavelet Based Fractal Analysis of DNA Sequences, *Physica D*, Vol. 96, 291–320, 1996.

Audit, B., Thermes, C., Vaillant, C., d'Aubenton-Carafa, Y., Muzy, J. F., and Arneodo, A., Long-Range Correlations in Genomic DNA: A Signature of the Nucleosomal Structure, *Physical Review Letters*, Vol. 86, No. 11, 2471–2474, 2001.

Bruckner, I., Sanchez, R., Suck, D., and Pongor, S., Sequence-Dependent Bending Propensity of DNA as Revealed by Dnase I: Parameters for Trinucleotides, *EMBO J.*, Vol. 14, 1812–1818, 1995.

Buldyrev, S. V., Goldberger, A. L., Havlin, S., Mantegna, R. N., Matsa, M. E., Peng, C.-K., Simons N., and Stanley, H. E., Long-Range Correlation Properties of Coding and Noncoding DNA Sequences: Genbank Analysis, *Physical Review E*, Vol 51, No. 5, 5084–5091, 1995.

Cornette, J. L., Kemp, B. C., Margalit, H., Spouge, J. L., Berzofsky, J. A., and De Lisi, C., Hydrophobicity Scales and Computational Techniques for Detecting Amphipathic Structures in Proteins, *J.Mol.Biol.*, Vol. 195, 659–685, 1987.

Cosic, I., Macromolecular Bioactivity: Is it Resonant Interaction Between Macromolecules?—Theory and Applications, *IEEE Trans. Biomed. Eng.*, Vol. 41, No. 12, 1101–1114, 1994.

Cosic, I., and Nesic, D., Prediction of "Hot Spots" in SV40 Enhancer and Relation with Experimental Data, *Eur. J. Biochem.*, Vol. 170, 247–252, 1987.

de Trad, C. H., Fang, Q., and Cosic, I., Protein Sequence Comparison Based on the Wavelet Transform Approach, *Protein Eng.*, Vol. 15, No. 3, 193–203, 2002.

Eisenberg, D., Weiss, R. M., and Terwilliger, T. C., The Hydrophobic Moment Detects Periodicity in Protein Hydrophobicity, *Proc. Natl. Acad. Sci. USA,* Vol. 81, 140–144, 1982.

Gabrielian, A., and Pongor, S., Correlation of Intrinsic DNA Curvature with DNA Property Periodicity, *FEBS Letters,* Vol. 393, 65–68, 1996.

Lee, B., and Richards, F. M., The Interpretation of Protein Structures: Estimation of Static Accessibility, *J.Mol.Biol.,* Vol. 55, 379–400, 1971.

Liò, P., Wavelets in Bioinformatics and Computational Biology: State of Art and Perspectives, *Bioinformatics,* Vol. 19, No. 1, 2–9, 2003.

Liò, P., and Vannucci, M., Wavelet Change-Point Prediction of Transmembrane Proteins, *Bioinformatics,* Vol. 16, No. 4, 376–382, 2000.

Murray, K. B., Gorse, D., and Thornton, J. M., Wavelet Transforms for the Characterization and Detection of Repeating Motifs, *J.Mol.Biol.,* Vol. 316, 341–363, 2002.

Oliver, J. L., Bernaola-Galvàn, P., Guerriero-Garcìa, J., and Romàn-Roldàn, R., Entropic Profiles of DNA Sequences through Chaos-Game Derived Images, *J. Theor. Biol.,* Vol. 160, 457–470, 1993.

Pattini, L., and Cerutti, S., Hydrophobicity Analysis of Protein Primary Structures to Identify Helical Regions, *Meth. Inf. Med.,* Vol. 43, No. 1, 102–105, 2004.

Phoenix, D. A., and Harris, F., The Hydrophobic Moment and its Use in the Classification of Amphiphilic Structures, *Mol. Membr. Biol.,* Vol. 19, 1–10, 2002.

Pigorova, E., Simon, G. P., and Cosic, I., Investigation of the Applicability of Dielectric Relaxation Properties of Amino Acid Solutions within the Resonant Recognition Model, *IEEE T. Nanobiosci.,* Vol. 2, No. 2, 63–69, 2003.

Roise, D., and Theiler, F., Amphiphilicity is Essential for Mitochondrial Presequence Function, *EMBO J.,* Vol. 7, 649–653, 1988.

Simons M., Stanley H.E., Long-Range Correlation Properties of Coding and Noncoding DNA Sequences: Genbank Analysis, *Physical Review E* Vol. 51, No. 5, 5084–5091, 1995.

Stanley, H. E., Buldyrev, S. V. Goldberger, A. L., Havlin, S., Peng, C.-K., and Simons, M., Scaling Features of Noncoding DNA, *Physica A ,* Vol. 273, 1–18, 1999.Sussillo, D., Kundaje, A., and Anastassiou, D., Spectrogram Analysis of Genomes, *EURASIP Journal on Applied Signal Processing,* Vol. 1, 29–42, 2004.

Uberbacher, E. C., and Mural, R. J., Locating Protein-Coding Regions in Human DNA Sequences by a Multiple Sensor-Neural Network Approach, *Proc. Natl. Acad. Sci. USA,* Vol. 88, 11261–11265, 1991.

Veljkovic, V., and Slavic, I., Simple General-Model Pseudopotential, *Phys. Rev. Lett.,* Vol. 29, 105–108, 1972.

Zhang, X.-Y, Chen, F., Zhang, Y.-T, Agner, S. C., Akay, M., Lu, Z.-H., Waye, M. M. Y., and Tsui, S. K.-W., Signal Processing Techniques in Genomic Engineering, *P. IEEE,* Vol. 90, No. 12, 1822–1833, 2002.

Thispageintentionallyleftblank

PART VII

CLASSIFICATION AND FEATURE EXTRACTION

Thispageintentionallyleftblank

CHAPTER *21*

SOFT COMPUTING IN SIGNAL AND DATA ANALYSIS
Neural Networks, Neuro-Fuzzy Networks, and Genetic Algorithms

Giovanni Magenes, Francesco Lunghi, and Stefano Ramat

21.1 INTRODUCTION

The term "soft computing" was first coined by Zadeh, in his works concerning fuzzy logic (1965) and fuzzy approaches to complex systems (1973), to denote a series of methods allowing one to deal with problems whose knowledge is inaccurate, uncertain, or partial, and to find approximate solutions. This term is used in opposition to traditional computation (hard computing), which is based on analysis, numerical simulation, and search for exact solutions. Soft computing (SC) methods exploit uncertain and imprecise information to reach treatable, resiliant, low-computational-cost solutions as well as high machine intelligence, contrary to traditional computing, in which uncertainty and inaccuracy are undesired properties.

According to Zadeh, the way a human brain thinks, by intrinsically dealing with an enormous amount of qualitative and incomplete information, is the optimal model to inspire soft computing methods.

Nowadays, the most important SC techniques are fuzzy logic, neural networks theory, and probabilistic reasoning, including probabilistic networks, genetic algorithms, chaos theory, and learning theory. It must be said that SC is not a mixture of the above-mentioned methods, but more a collection of them, whereby each one supplies a solution to a specific domain of problems. From this point of view, the different methodologies are not competing among each other but are instead complementary, allowing complex problems to be approached by combining multiple techniques. A clear example of such interaction is provided by neuro-fuzzy systems.

This chapter contains an introduction to some of the most commonly used SC techniques: neural networks, neuro-fuzzy networks, and genetic algorithms. Other methodologies have already been described in depth in other chapters of this book. It is worth noting that the authors have here emphasized the discussion of neural and neuro-fuzzy networks by considering them as part of the larger topic of adaptive networks.

An adaptive network is basically a mathematical instrument, able to approximate a function transforming a vector from the input space into a vector in the output space. From a computational perspective, the use of adaptive networks allows the development of a series of models in order to mimic the working principles of complex and highly nonlinear natural processes and/or to solve classification or clustering problems as well as prediction and control tasks.

In neurosciences and in cognitive sciences, there is a strong interest in the development of adaptive networks because, besides being good models for classification and processing of biological signals, they are able to intrinsically synthesize (within their parameters) a form of knowledge representation that is different from the classical symbolic approach typical of artificial intelligence (connectionism, symbolism, and dualism).

We favor a methodological approach over an applicative one focusing on the processing of biomedical signals. The main reason for this choice lies in the desire to sketch out a description as general as possible of the theoretical context while keeping the treatment terse. Hence, this coverage cannot be considered exhaustive, if compared to the extent of the topic, nor particularly deep in methodological terms, but can represent, in the authors' opinion, a reasonable compromise, allowing the reader to comprehend some of the principles and basic issues that can represent the starting point for further study. The interested reader will find complete treatments of the methods that here are only briefly described by referring to the publications cited in the references.

21.2 ADAPTIVE NETWORKS

An adaptive network is a structure composed of a set of nodes interconnected by directional links. Each node corresponds to a single computational unit and the connections represent causal relations among the nodes. The signal is propagated inside the network, from a set of input nodes, following the different links, and is transformed into the signal produced by the set of output nodes.

Adaptive networks are usually represented as oriented graphs, organized either in layers or in topological order. In the first approach, the network nodes are clustered in separate sets, each one corresponding to one layer; in the second approach, the network nodes are aligned in a sequence corresponding to the signal flow.

In the following, we will consider only the first approach to network representation since it has many advantages in terms of modularity and is the preferred representation in the scientific literature. In this representation, the input space is the layer in which the input signal is applied to the network, the output space is the one providing the output as the result of the network processing, and the other nodes make up the internal or so-called hidden layers.

Adaptive networks may be classified in two main categories based on their architecture: networks with only forward connections (feed-forward) and networks with recursive or fed-back connections (feed-back).

In feed-forward networks (Figure 21.1), each layer has only input connections coming from the preceding layer and output projections to the following one, whereas recursive networks have feed-back connections that create circular paths in the signal flow, providing the network with state dynamics.

Nodes constitute elementary units in which the signal is processed. The computation performed inside the node represents its transfer or activation function, which is determined by the functional relationship between the input to each node and its output. Such relationship depends both on the choice, at design time, of the activation function (i.e., linear or

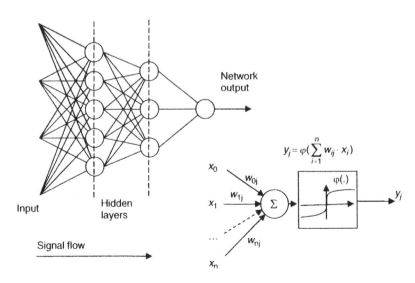

$$y_j = \varphi\left(\sum_{i=1}^{n} w_{ij} \cdot x_i\right)$$

Figure 21.1. An example of a feed-forward neural network and a schematic representation of an artificial neuron.

nonlinear; monotonic or not) and on a set of parameters characterizing its behavior.

In addition to the parameters characterizing the transfer functions of each node, the set of connection weights between nodes completes the set of free parameters of the adaptive network. The input to each node is either a linear combination of the inputs to the network or of the outputs of nodes belonging to the preceding layer.

Such weights represent the strength of the connection between two nodes and are to be optimized in order to adapt the behavior of the network (that is, to obtain the desired behavior).

21.3   NEURAL NETWORKS

Neural networks (NNs) are one of the most commonly studied and frequently applied solutions in the field of adaptive networks. Despite being a formal computational method, they are inspired by the working principles of networks of biological neurons and represent a tool for both understanding the functioning mechanism of the human brain and for mimicking its behavior in terms of the ability to process information. Similar to biological neural networks, the computational element is the neuron, which is, in this case, artificial.

An artificial neuron, similar to a biological one, is made up of:

- A set of synapses or connections, each one characterized by a strength (or weight)
- A concentrator or adder, conveying the sum of weighted signals coming from neuron's input connections
- An activation function, generating the output signal of the neuron

The great innovation introduced with the use of NNs in soft computing lies mainly in their being very different from traditional Von Neumann machines, which are the most representative exponents of hard computing systems. A Von Neumann machine implements the abstraction of a human processing system based on two different units for computation and storage, whereas neural networks are a metaphor of the high architectural parallelism of nervous systems, which do not make a distinction between those units.

For this reason, neural networks, considered as computational machines, can be easily compared to a multiprocessor system with very simple but fast processing elements, in which the "storage" function is performed by the structure itself, in terms of its synaptic weights.

Different examples of neural network architectures have been implemented electronically using the very large scale integration (VLSI) technology.

Considering the layered representation, connections between neurons can be of four types:

1. Feed-forward. The output signal of a neuron (or layer) is the input for one or more neurons of the following layer.
2. Feed-back. The signal is propagated in both directions (both to the following and preceding layers).
3. Lateral, that is, between neurons in the same layer. Usually, these connections are of the inhibitory kind and favor a supremacy principle between peers (winner-takes-all), considering as winner the neuron with the highest activation strength. These connections are also used to implement lateral–orthogonal networks, in which, by means of lateral connections, neurons belonging to the same layer are forced to extract orthogonal components of the input signal.
4. Delayed. The signal is processed by units introducing lags. They are manly used to produce dynamical models.

The following are among the most attractive characteristics of neural networks for their application to different topics in engineering:

- Adaptive learning—the ability to learn how to perform desired actions.
- Self-organization. A neural network can automatically create its internal representation of the information received as input.
- Real-time functioning. Being a collection of very fast parallel computational elements, NNs can synthesize very quick implementations, in spite of their great complexity.
- Fault tolerance. Redundancy of stored information inside a network keeps it operative even when facing partial destruction. Moreover, a damaged network, if further trained, can recover from the handicap.

The distinctive aspect of NNs is, thus, their ability to elaborate, adaptively and in parallel, highly nonlinear information. NNs are widely used in different applications from artificial vision to signal and image processing (especially in biomedical applications), to text and speech recognition, to robotics, and in scientific exploration in general.

In an attempt to sketch the set of high-level tasks in which the power of neural networks is best exploited, one should mention association, clustering, and classification; pattern recognition; regression or generalization; and optimization.

21.3.1 Association, Clustering, and Classification

In these categories of problems, static input patterns or time series must be classified and/or recognized. A classifier is "trained" in order to correctly recognize even distorted versions of the stimulus.

These applications may be further detailed as:

- *Auto- or hetero-association.* Auto-association is the ability to obtain the complete pattern starting from a partial description of the desired one. Hetero-association is the capability to retrieve a pattern belonging to a desired set starting from the processing of information concerning patterns that do not belong to the same set or class. Usually, the association is performed using the Hebb's learning rule (Hebb, 1949).
- *Nonsupervised clustering.* Network weights are constantly tuned as a consequence of new stimulations.
- *Supervised classification.* It performs an approximation or interpolation of the data. It is usually achieved (for example, in different language, text, or image recognition applications) by creating supervised networks trained on a set of stimulus–membership class pairs.

21.3.2 Pattern Completion

In many classification tasks, the goal is to complete a partial amount of information, in other words, recovering the desired pattern from an incomplete structure. There are two main typologies of completion: the temporal one based on sequences of stimuli, and the static one based on a single stimulus.

21.3.3 Regression and Generalization

Regression means to achieve a model (linear or not) adapting as well as possible to the training data. It is a matter of interpolation. Generalization is the ability to create a model able to respond to the stimuli used for training as well as to new inputs. This procedure is generally performed by determining the main characteristics of the studied process and by recognizing them in the inputs presented after the training phase.

21.3.4 Optimization

Neural networks can be exploited to solve optimization problems, consisting of researching a stationary point (minimum or maximum) of a cost functional, generally expressed in terms of an energy function. In general, NNs rapidly converge to stationary points but tend to be captured by stationary conditions. The implementation of a neural network for this purpose is, therefore, commonly faced with being restricted by local minima (or maxima).

21.4 LEARNING

The concept of learning has many definitions, often related to the different scientific contexts or to the nature of concerned topics. In the different fields of engineering, a learning machine is commonly described as a system that is able to:

- Learn: acquire knowledge through direct or indirect experience
- Memorize: store and internally represent this knowledge
- Reason: exploit the acquired knowledge in order to solve problems or, more generally, to adapt itself to the external environment

In NNs, similarly to the human brain, knowledge is represented in terms of synaptic weights, whereas the reasoning mechanism corresponds to the evolution (flow) of a stimulus through the network.

In general terms, we may consider that in an adaptive network learning is the process by which free parameters are tuned by means of a sequence of stimulations (inputs) produced by the environment in which the network belongs. When developing a NN, except in rare cases discussed in the following paragraphs, the synaptic weights are the only free parameters.

The learning process can be thus interpreted as a temporal sequence of changes in synaptic weights, suited to enhance the network performance, storing the knowledge provided by stimulations. Considering $w_{kj}(n)$ as the value at time n of the synapses between the kth and jth neurons, we may write $w_{kj}(n + 1) = w_{kj}(n) + \Delta w_{kj}(n)$, where $\Delta w_{kj}(n)$ is the variation of the synapse.

Neural networks are generally subdivided into two categories, depending on their learning paradigm: supervised and nonsupervised networks. Nonsupervised networks (Figure 21.2) calculate the updating of the free parameters based only on the experience acquired through previous stimulation and have no learning control mechanism referring to desired output of the system.

These networks extract features and regularities from the input space, grouping the received stimuli into classes of internal representa-

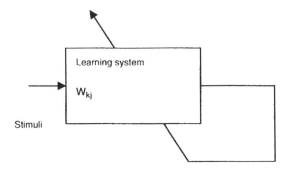

Figure 21.2. Nonsupervised learning.

tions, without any further interaction with the outer environment. This is the main reason why nonsupervised networks are usually preferred in problems of clustering and extraction of characteristics and similarities.

In supervised networks (Figure 21.3), the learning process is performed using examples and through the intervention of a teacher, that is, an algorithm able to provide information regarding the correctness of the system response and to iteratively correct it.

The examples used for training are, in general, desired input–output pairs available to the network designer, and the teacher is basically an algorithm that minimizes the error between the actual and desired output of the network.

21.4.1 Nonsupervised Learning

The two most commonly used paradigms in nonsupervised learning are Hebbian and competitive learning. Both are inspired by neurobiological considerations and partially reproduce learning principles observed in nature. There is a third kind of unsupervised learning, explained later in this chapter, named memory-based learning, which is mainly used in the construction of the so-called associative memory networks.

21.4.1.1 Hebbian Learning. Hebbian learning is derived from studies concerning the behavior of human neurons carried out by Hebb, a neuropsychologist of the twentieth century (Hebb, 1949). His learning theory postulates that when two connected neurons are activated at the same time, the strength of their connection is increased. This principle can be translated into two rules:

1. If two connected neurons are activated simultaneously, then the strength of their common synapse is increased.

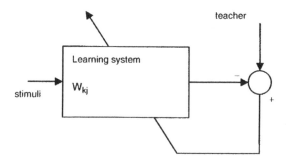

Figure 21.3. Supervised learning.

2. If they are activated in an asynchronous way, then their connection is weakened or eliminated.

The easiest way to mathematically represent the synaptic update is the following formula, where the update Δw_{kj} is expressed as a function of the pre- and postsynaptic activities (respectively, x_j and y_k):

$$\Delta w_{kj}(n) = \eta y_k(n) x_j(n)$$

where η is the learning rate.

In other words, the change in the amount of synaptic weight is proportional to the correlation between the input and output signals. As can be easily understood, such a rule could lead to saturation in the event of a repeated input x_j. Another way to represent Hebb's principle is to make the weight variation proportional to the covariance of the pre- and postynaptic activities (Sejnowski, 1977). To overcome the exponential growth of synaptic weights, the formula is completed with a forgetting term:

$$\Delta w_{kj}(n) = \eta y_k(n) x_j(n) - \alpha y_k(n) w_{kj}(n)$$

where α is the forgetting coefficient.

21.4.1.2 Competitive Learning. The competitive learning paradigm, as the name suggests, consists of the free competition between neurons belonging to the same group aimed at defining the only activated neuron.

Differently from the Hebbian learning, which envisages the event of multiple neurons contemporarily activated, this kind of rule provides for the activation of a single neuron at a time. This characteristic makes competitive learning extremely suitable for highlighting static features in order to correctly classify a set of input stimuli.

To realize a competitive learning machine, we need (Rumelhart and Zipser, 1985):

- A set of identical neurons connected with synapses having random weights (in order to differently react to the same stimulus)
- A limitation to the overall activation strength of each neuron (i.e., $\Sigma_j w_{kj} = 1 \; \forall k$)
- A competitive mechanism. The neuron winning the competition becomes the winner-takes-all neuron.

A competitive network usually includes excitatory forward connections and inhibitory lateral connections. Forward connections (feed-forward) generally implement the Hebbian rule: when an input neuron persistently participates in the activation of an output neuron, the activation strength of the synapses connecting them is increased. Lateral synapses provide the inhibition of competitors and, thus, determine the winner.

In a winner-takes-all scheme, the output unit with the highest excitation level (the winner) obtains its maximum value, whereas every other element is inhibited by means of lateral connections.

A neural model trained with this competitive–inhibitory mechanism is able to internally preserve a sort of topological order. Nonsupervised competitive networks are often based on a clustering, or grouping, strategy that forms groups of similar patterns. For this reason, these networks are best exploited in classification problems.

Thus, it is necessary to define a criterion quantifying similarity to be used for determining how close two patterns are. The most common methods are the inner product

$$<x_i, x_j> = x_i^T x_j = \|x_i\| \|x_j\| \cos(x_i, x_j)$$

and the Euclidean norm

$$d(x_i, x_j) = \Sigma_k (x_i(k) - x_j(k))^2$$

Many examples of unsupervised neural networks using the learning paradigms described above can be found in the literature. Here, we will mention and briefly describe only a few of the most common ones. In particular, we will describe the extraction of principal components as an implementation of the Hebbian rule, Kohonen maps (Kohonen, 1982) as implementations of competitive learning, and associative networks (Hopfield, 1982) as examples of fixed-weight, nonsupervised networks.

21.4.1.3 *Principal Components Analysis (PCA).*

One of the most important operations that need to be performed in a classification problem is certainly that of recognizing the characteristics (or input patterns) that have the most descriptive power or, in computer science terms, those having the highest information content. This task can be performed by a single-layer, one-output neural network trained with a slightly modified version of the Hebbian algorithm, allowing one to extract the principal components, namely the eigenvectors corresponding to the largest eigenvalues of the input correlation matrix. Such processing, called principal components analysis (Loève, 1963), maps the input space into a new space of extracted features.

Considering a p-dimensional vector x that we need to reduce to a vector u of $m < p$ dimensions while losing the least possible information content, the PCA algorithm will find the vector (u) for which the variance of the dataset projected over u is maximized. In doing so, the PCA will compute the correlation matrix of the data provided, and compute its eigenvalues. The starting p-dimensions are then reduced to m by discarding the $p - m$ components showing the lowest eigenvalues.

21.4.1.4 Self-Organizing Maps (Kohonen Maps).

The idea underlying self-organizing maps (SOMs) is that of including a notion of neighborhood (between neurons) and of time in a competitive learning rule. The introduction of a proximity criterion avoids the condition of neurons not participating in the learning process and favors the acquisition of topological properties that should be kept in the features' space.

Considering an input pattern with N features, represented by a vector x in a N-dimensional space, the network will map this pattern in a mono- or bidimensional space of output nodes, presenting a determined topological order (Figure 21.4). Kohonen (1982) proposed a method for making this kind of network self-train, hence the name SOM, and added lateral connection between the nodes in the output layer.

The most innovative contribution to the classic competitive learning paradigm is the introduction of the neighborhood principle, in other words, the extension of the excitatory impulse to a group of neurons in the proximity of the winning one.

The sequence of operation of this training phase can be summarized in the following steps:

1. Weights are randomly initialized
2. A sample x is chosen from the input distribution
3. The winning neuron $i(x)$ is selected using the minimum Euclidean distance criterion $i(x) = \arg\min_j (\|x - w_j\|)$

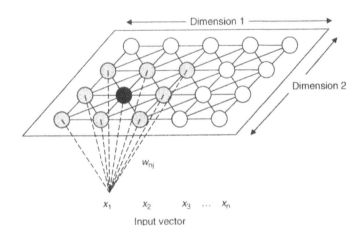

Figure 21.4. Schema of a self-organizing map (Kohonen map).

4. The weights are updated, not only for the winning neuron but considering also its neighbors:

$$\Delta w_{kj} = \eta(x - w_j) \text{ for each neuron } j \text{ belonging to } \Lambda i(x)$$
$$\text{(proximity function)}$$

$$\Delta w_{kj} = 0 \text{ otherwise}$$

where $\Lambda i(x)$ is a function for determinig the neighbors of neuron i. Such a function should become narrower as the number of iterations increases.

5. Loop to step 2

These maps are formally called topologically ordered computational maps and mimic the working principle of the brain, in which contiguous areas are stimulated by similar stimuli (e.g., the arrangement of sensory areas). They are topological maps in which the spatial position of a neuron (i.e., its coordinate in the mono- or bidimensional features space) corresponds to intrinsic characteristics of the input patterns.

Frequently when solving classification problems, SOMs are used in modular and composite networks as a first features extraction layer and their output is generally postprocessed by supervised linear classifiers.

21.4.1.5 Associative Memory Networks. This kind of network is characterized by a sharp temporal segregation of training and usage, during which weights are kept fixed and undergo no further tuning. The learning process, therefore, is performed only before using the network. Associative networks internally store a representation of the features of training samples and can then bind an input stimulus, albeit noisy or incomplete, with one of the internal classes.

An associative memory (AM) has the useful property of being able to recall one of the training patterns (stored in the memorization phase), starting from a distorted or sketchy version of it presented as input stimulus.

These kinds of memories can be used for problems of hetero- or autoassociation. In the first condition, the dimensions of the input and output space are identical; in the second, they can differ.

Two possible implementations of AM networks are the feed-forward associative memory networks (FAMN) and the feedback AMN. Their obvious distinction consists in the different signal flow: in the first the input signal is only propagated forward, whereas the second typology allows an iterative path.

Linear associative memory networks (LAMs) can be considered as examples of associative memories and are structurally identical to a single-layer feed-forward neural network with thresholded activation functions.

On the other hand, the Hopfield networks (Hopfield, 1982), are among the most famous implementations of recursive AMs.

Hopfield networks are completely interconnected recursive networks (except that they have no autoconnections, that is, no connections of a neuron with itself). These networks have two working modalities, corresponding to the two phases of memorization and recall:

1. Memorization. Weights are tuned by means of a Hebbian rule in the event of a continuous Hopfield network, or in a single computational step for discrete networks.
2. Recall. A Hopfield network implements a dynamic system in which stationary points or attractors are bound with the information to be extracted based upon the content. Each attractor is characterized by an attraction basin, a region of the state space whose trajectories converge to a stationary point. Each time the network is stimulated with an input pattern, namely, an initial state, it evolves, switching states, until it remains trapped in a stable one. Each stable condition corresponds to one of the classes of patterns in which the input training set was subdivided during the memorization phase.

21.4.2 Supervised Learning

This category contains every learning paradigm needing the intervention of an external element, called a teacher or supervisor, supporting the weights' adaptation process. This is undoubtedly the most studied and implemented category of neural networks. Within it, two different classes of algorithms can be considered according to the kind of information available to the supervisor and to the way the teacher exploits this information: approximation learning and reinforcement learning.

21.4.2.1 Error Correction Learning (Approximation Learning). In some scenarios, in addition to the training input dataset, a corresponding set of desired outcomes may be available. It is a common situation in classification tasks in which the membership class of a set of input samples is known or when dealing with physical systems allowing one to obtain a real-time response to the stimulation from the system itself.

In these conditions, the supervisor can use the correct input–output pairs and instantaneously compute information concerning the network's error:

$$e(k) = d(k) - y(k)$$

that is, the error at time k is the difference between the desired (d) and the actual output (y) of the network.

Approximation learning is intended to be a modification of the free parameters of the network in order to obtain the reduction of a cost function, which is usually represented by the mean-squared error, defined as the mean (sampled mean, in absence of statistical information) of the sum of the quadratic errors.

Among the most commonly used methods for approximation learning is certainly the backpropagation algorithm, implementing a correction of synaptic weights by propagating the error backward, from the output nodes to the input layer, flowing through the hidden layers (Rumelhart et al., 1986a, b).

21.4.2.2   *Reinforcement Learning.*

Another possible intervention by the supervisor is that of providing qualitative, rather than quantitative, information concerning the network error. In other words, the supervisor may provide a series of evaluations related to the learning progress. This criterion is, therefore, a sort of corrective learning, acquiring information about misinterpretations of the input data. The basic principle of this technique is that if an action is followed by an improvement, then the trend to reproduce that action must be favoured (reinforced). Otherwise, this trend must be inhibited.

It is therefore necessary to choose a function for evaluating the performance (Sutton et al., 1991). Considering a learning system described by a discrete time dynamic system with a finite set of possible states X, being in state $x(n)$ at time n, the system will choose to perform the action $a(n)$ and, independently of the past, switch to a state y, with probability $P_{xy}(a)$ (the probability of transition between x and y, given action a). After this transition, the system will obtain a reinforcement $r(n + 1)$ as a function of $x(n)$ and $a(n)$. The aim of the reinforcement learning is to find a policy for selecting a sequence of optimal actions, supposing that state $x(n + 1)$ only depends on state $x(n)$ and, therefore, the states sequence is a Markov chain. A possible evaluation of the performance is (Barto et al., 1990)

$$J(x) = E[\Sigma_k \gamma^k r(k + 1) \mid x(0) = x]$$

where γ is the discount-rate parameter with $0 \leq \gamma < 1$, x is the initial state, and $E[]$ represents the expectation function.

The above summation is called the cumulative discounted reinforcement (cumulative since it is summed, and discounted for the term γ). The basic idea of reinforcement learning is to learn the $J(x)$ function in order to predict the cumulative discounted reinforcement that will result from every possible state change.

An example of reinforcement learning is given by networks using a critic block, like the adaptive heuristic critic (Barto et al., 1983), receiving both the reinforcement signal and the state of the system as input and elaborating an enhanced reinforcement based on estimating the reinforcement that may be obtained from executing the undergoing actions. This predictor, by computing data coming from k previous state transitions, is able to estimate $\hat{J}_n(k)$, which is the prediction of the reinforcement at time n, and, consequently, provide a heuristic reinforcement $\hat{r}(n + 1)$.

A duality between the two supervised learning techniques—approximation and reinforcement—is apparent: in the first, the teacher provides local information concerning the estimate of the gradient of the error surface (useful to compute the convergence to a minimum); in the second, only temporal data are available (commonly referred to as a delayed reward) of a trial-and-error nature, representing a technique performing the exploration of the error surface. The duality is thus between identification and control, that is, between the wish to exploit the knowledge acquired and that of acquiring more information by means of exploration (and, thus, susceptible to errors).

In the following paragraphs, some of the most well-known implementations of the supervised approximation learning paradigm will be examined, where the availability of a set of desired input–output pairs is assumed.

21.4.2.3 Perceptron and Adaline. The artificial neuron, in the form represented in Figure 21.1, is commonly called a perceptron, from the pioneering work by Rosenblatt (1958). The perceptron, provided with a thresholded activation function (McCulloch and Pitts, 1943), is able to linearly separate the input space. Rosenblatt showed that if a perceptron is trained using examples taken from two linearly separable classes, then the algorithm converges and places the decision surface as a hyperplane between the two classes.

A perceptron made out of only one neuron can identify only two classes but increasing the number of neurons allows one to recognize elements belonging to more classes, as long as they are linearly separable. Obviously, a single-layer network of perceptrons is of little use, since it is unable to solve the simplest nonlinear problems (the most famous example in the literature is the XOR function). As we will see, it is necessary to combine multiple layers of perceptrons in order to develop classifiers for the category of nonlinearly separable problems.

Another implementation of the artificial neuron with an adaptive learning algorithm is the adaptive linear element (adaline), suggested by Widrow and Hoff (1960). It is basically a version of the perceptron without a nonlinear activation function, in which the network output is simply a linear combination of its inputs. The network is, therefore, a linear model whose solution could be determined by using a least mean squares (LMS) regression method.

The algorithm proposed by the authors, instead, named the delta rule, exploits the instantaneous value of the error $E(w) = 1/2e^2(n)$ as a cost function, which is considered as an approximation of the mean squared error.

By differentiating, we get

$$\frac{\partial E(w)}{\partial w} = e(n)\frac{\partial e(n)}{\partial w}$$

Since $e(n) = d(n) - w^T(n)x(n)$, it follows that

$$\frac{\partial e(n)}{\partial w(n)} = -x(n)$$

and thus,

$$\frac{\partial E(w)}{\partial w(n)} = -x(n)e(n) = \hat{g}(n)$$

The update of the synaptic weights in the delta rule is therefore

$$\Delta w(n) = -\eta\hat{g}(n) = \eta x(n)e(n)$$

with η being the learning coefficient. Actually, this is an LMS method that does not follow a gradient descent path over the surface error but moves with a trajectory comparable to Brownian motion.

In order to solve the XOR problem, the authors proposed a couple of interconnected adalines, but this solution also appeared to be of little interest. Indeed, the resulting madaline [multiple adaline (Widrow, 1962)] was affected, like the perceptron, by all the limitations of being a single-layer network.

21.4.2.4 MLP Multilayer Perceptron. The low approximation and classification power of single-layer neural networks led to the development of multilayer perceptrons (MLP). These are networks with multiple neurons organized in layers that are connected in cascades. Neurons generally have sigmoid activation functions (tansig or logsig) to allow nonlinear relationships between inputs and outputs (Rumelhart et al., 1986).

With the introduction of MLP networks, learning becomes a nonlinear parametric estimation problem, which is solved using a sum of squared errors (SSE) minimization criterion. The most frequently used algorithm for solving this problem is the backpropagation algorithm. The name backpropagation derives from a term by Rosenblatt (1962) in his attempts to generalize the perceptron learning mechanism to the multilayer case.

In the 1960s and 1970s, various attempts were made to develop a learning algorithm for optimizing the weights of a neural network based on evaluating its output error, but only in 1982 (Rumelhart et al., 1986b) was a real implementation achieved.

The backpropagation algorithm, namely the method of backward propagation of the error, is a gradient descent method in which the error of the network is propagated from the output nodes up to the input layer. In this context, we will not prove the convergence of the backpropagation algorithm but we will analyze the main steps for computing the update of the synaptic weights. This algorithm derives from the delta rule but does estimate the contribution of each neuron to the final error (including the hidden ones) by computing the partial derivative of the error surface with respect to each single network weight. This is interpreted as a measure of sensitivity, determining the search direction in the weights space. For each neuron belonging to a hidden layer, where a desired output is not available, the error is recursively determined from the errors of all the neurons to which it projects.

The backpropagation algorithm requires the activation function of every neuron to be differentiable (although nonlinear) and this is the main reason for the choice of sigmoid functions instead of thresholds.

The algorithm is itself subdivided into two main phases: the forward propagation of the input signal and the backward propagation of the error. In the first step, the stimulus is applied to the input of the network and the corresponding output is computed. Then the error is calculated as the difference from the desired (known) result, $e_j(n) = d_j(n) - y_j(n)$, where j is the jth output neuron and d and y are, respectively, the desired and actual outputs.

In the second step, such error is backward propagated through the network, starting form the output layer up to the input neurons, and only in this latter step are the connection weights tuned.

Considering the following as functional cost to be minimized,

$$E(n) = \frac{1}{2} \sum_{j \in C} e^2(n)$$

where C is the set of the output neurons, and evaluating the outputs corresponding to all the training set, that is, the stimuli used for the training phase,

$$E_{\text{Mean}} = \frac{1}{N} \sum_{n=1}^{N} E(n)$$

the transfer function of the single neuron can be described as

$$y_j(n) = \varphi[v_j(n)]$$

where

$$v_j(n) = \sum_{i=1}^{m} w_{ji}(n)y_i(n)$$

is the weighted sum of inputs of the jth neuron, corresponding to the outputs of the neurons belonging to the preceding layer connected to the jth neuron multiplied by the weights of their relative connections.

The weight correction performed by the backpropagation algorithm is proportional to the error derivative $\partial E(n)/\partial w_{p_{ij}}(n)$, which may be expanded as

$$\frac{\partial E(n)}{\partial w_{ji}(n)} = \frac{\partial E(n)}{\partial e_j(n)}\frac{\partial e_j(n)}{\partial y_j(n)}\frac{\partial y_j(n)}{\partial v_j(n)}\frac{\partial v_j(n)}{\partial w_{ji}(n)}$$

It is easy to verify how

$$\frac{\partial E(n)}{\partial w_{ji}(n)} = -e_j(n)\varphi_j'[v_j(n)]y_j(n)$$

The weight correction w_{ji} with respect to the nth input, following the delta rule, is then

$$\Delta w_{ji}(n) = -\eta\frac{\partial E(n)}{\partial w_{ji}(n)}$$

In other words, the direction chosen for descent is opposite to that of the gradient in the weights space.

Thus,

$$\Delta w_{ji}(n) = -\eta\delta_j(n)y_j(n)$$

where $\delta_j(n)$ is the local gradient.

If the jth neuron lies in the output layer, the calculation of this gradient is simple because we have the desired output value and, thus, the error $e_j(n)$ can be directly computed.

If the neuron is in a hidden layer, the desired output is not available and the error must be computed from the information obtained from other neurons, recursively from the output up to all other neurons receiving projections from the selected one (generally belonging to the next layer).

Rewriting the term

$$\delta_j(n) = -\frac{\partial E(n)}{\partial y_j(n)}\varphi_j'[v_j(n)]$$

and differentiating the cost functional $E(n)$ with respect to $y_j(n)$, we obtain

$$\frac{\partial E(n)}{\partial y_j(n)} = \sum_k e_k \frac{\partial e_k(n)}{\partial y_j(n)}$$

that is, expanding the partial derivatives,

$$\frac{\partial E(n)}{\partial y_j(n)} = \sum_k e_k \frac{\partial e_k(n)}{\partial v_k(n)} \frac{\partial v_k(n)}{\partial y_j(n)}$$

Since $e_k(n) = d_k(n) - y_k(n) = d_k(n) - \varphi_k[v_k(n)]$, with output neuron k, differentiating the local field of neuron k with respect to $y_j(n)$, we obtain

$$\frac{\partial v_k(n)}{\partial y_j(n)} = w_{kj}(n)$$

Substituting in the previously calculated gradient formula gives

$$\frac{\partial E(n)}{\partial w_j(n)} = -\sum_k e_k(n) \varphi_k'[v_k(n)] w_{kj}(n) = -\sum_k \delta_k(n) w_{kj}(n)$$

The backpropagation rule is then

$$\delta_j(n) = -\varphi_j'[v_j(n)] \sum_k \delta_k(n) w_{kj}(n)$$

for a hidden layer j.

By analyzing the above equation, it may be noted that the term $\delta_k(n) w_{kj}(n)$ only depends on the errors made by the k neurons in the layer directly following the one of the selected jth neuron and on the weights of the corresponding connections. This piece of information is available since the algorithm is proceeding backward from the output to the input. To conclude, once the local gradient, depending on the layer a neuron belongs to, is calculated, the weights update is performed by applying the delta rule.

It must be remembered that Cybenko (1989) and Funahashi (1989) independently proved how a MLP network with at least one hidden layer is a universal approximator: it can approximate in an arbitrarily precise way every continuous function, as long as a sufficient number of neurons is employed. The theorem is valid also when the output neurons have a linear activation function.

21.4.2.5 *Radial Basis Function Networks (RBFs).* RBFs are generally three-layered networks. The first layer is a sensorial input (generally lacking weights); the second is the only hidden layer, with a large number of

neurons, applying the nonlinear transformation; and the last is a linear output layer, often implemented as a simple summation element.

Such a structure implements a high-dimensionality intermediate block, which may be explained as an application of the Cover theorem (Cover, 1965); a complex nonlinear pattern classification problem on a highly dimensional space is more likely to be linearly separable, that is, separable by a hyperplane, than in a lower dimensional space.

The activation function of the operative units in this kind of neural network is a radial function of the form

$$f(x) = h(\|x - u\|)$$

where h is a bell-shaped Gaussian or logistic function and u is the center of the radial function.

Considering, for simplicity, that the output is a simple summation, the overall transfer function of the network will be

$$g(x) = \Sigma_k w_k \, h(\|x - u_k\|)$$

where w_k assumes the meaning of the amplitude of the radial function.

The main difference from the neural networks mentioned so far is that RBFs also have free parameters to be tuned in the activation functions, inside the computational units (neurons).

It can be shown that RBF networks can also act as universal approximators. Moreover, once the coordinates of the centers of the activation functions are known, the optimization problem becomes linear in the parameters and a linear optimization algorithm such as, for example, the LMS, can be employed. The basic idea of the learning process of a RBF is indeed to break the training into two distinct phases:

1. Place the centers
2. Compute the w_k values

In the first phase, an a priori strategy can be followed, placing the centers in a uniform grid and choosing a random subset of the input data by means of a clustering algorithm (K-means), or using a stepwise regression.

A crossvalidation algorithm can be used for selecting the number of centers and, thus, neurons in the hidden layer.

21.5   STRUCTURAL ADAPTATION

In the first part of this chapter, we cosidered that the structure of the network (in terms of number of neurons, number of layers, etc.) is a priori fixed after the initial design, corresponding to the typology and dimensionality of the problem to solve. However, there are many cases in which a neural network requires a sort of structural adaptation to the task. This can be obtained through two main categories of intervention, causing, respectively, network growth or reduction (pruning), both acting on the number of neurons and on their connections.

A typical example of network growth is the cascade correlation learning architecture (Fahlman and Lebiere, 1990). Starting with a minimal network without neurons in the hidden layer and training it with LMS methods, a hidden neuron is added at each iteration. It will be connected to every input neuron and to any other in the hidden or output layers. For each added neuron, the weights corresponding to synapses to the input layer are frozen and only the neuron's output connections are trained. The iterative procedure continues until a satisfactory level is reached.

More frequently, one can use implementations of pruning methods, which start from an overdimensioned network and reduce it, trying to optimize a cost function. Among them, we can mention the complexity regularization, consisting of estimates of the penalization caused by the network size (it uses complexity-related measures of the model, such as the minimum description length).

Another pruning method is based on the computation of the Hessian matrix of the error surface, providing a local model of the error surface, in order to predict the effect on the network of a perturbation of its synaptic weights. To this purpose, the method searches for the set of parameters whose exclusion will cause the least error increase. The most famous pruning algorithm using the Hessian matrix is the optimal brain damage procedure (LeCun et al., 1990), which eliminates (by zeroing their connection weights) the largest number of neurons in trying to minimize the error variation.

However, in many classification problems there is the need to optimize not only the accuracy of the result (reduction of the final error) but also the network structure. The support vector machines (SVM) (Boser et al., 1992; Cortes and Vapnik, 1995; Vapnik, 1995, 1998), whose theoretical foundation can be found in statistical learning theory (Vapnik, 1995), are algorithms able to obtain a structural optimization of the network. Synthetically, they allow one to transform any classification problem into a linear one of higher dimensionality and to retrieve the optimal separation hyperplane (Cover theorem).

21.5.1 Statistical Learning Theory

Consider two sets of random variables, $x \in X \subseteq R^d$ and $y \in Y \subseteq R$, connected by an unknown probabilistic relation $P(x, y)$ and having a set of l examples, that is (x_i, y_i) pairs with $i = 1, 2, \ldots, l$. The learning task consists of building an estimator, formally a function $f: X \to Y$, able to correctly evaluate the y corresponding to each x belonging to X. The solution to the learning problem, as seen so far, can be approached by determining a cost function $L[y, f(x)]$, also known as a risk function, providing a measurement of the error made by f when attempting to estimate y from x, and by searching for a f_0 belonging to the space of F functions from X to Y, with the minimum mean risk, in other words, an optimal function minimizing the expected risk:

$$E\left\{L[y, f(x)]\right\} = R[f] = \int_{X,Y} L[y, f(x)]P(x, y)dx\, dy$$

Since $P(x, y)$ is unknown, according to the empirical risk minimization principle, only an estimate of f_0 minimizing the risk function can be found. The expected risk is indeed approximated from the available data, obtaining the so-called empirical risk, mathematically based on a sample mean:

$$R_{emp}[f] = \frac{1}{l}\sum_{i=1}^{l} L[y_i, f(x_i)]$$

As already seen in other cases, we can take the squared error $[y - f(x)]^2$ as the cost function L.

The minimization of the empirical risk in F can lead to suboptimal solutions but, unfortunately, it can also cause an overfitting situation. The overfitting, namely, a surplus of approximation, is a classical example of the dilemma between generalization and approximation. The more a model tends to correctly approximate the dataset upon which it is optimized, the less it will generalize the underlying $P(x, y)$ that generated those data and which is the working mechanism of the studied process.

The statistical learning theory (SLT) adds probabilistic bounds to the distance between the empirical and the expected risk. These boundaries obviously depend on the number of available samples but also on an h factor characterizing the capacity of the functions space, corresponding to a complexity measure of the F space. One of the most used parameters for evaluating this complexity-related measure is the Vapnik–Chervonenkis (VC) dimension (Vapnik and Chervonenkis, 1971).

The expected risk is therefore bounded:

$$R[f] \le R_{emp}[f] + \phi\left(\frac{\ell}{h_k}\right)$$

where the second term is related to the network structure and, thus, implicitly to its generalization ability (structural risk minimization).

It is remarkable that the above inequality also fits the classical neural networks (MLP or RBF) theory, where the second term, contributing to the expected risk, is fixed and depends only on the designed architecture.

Minimizing the expected risk means, then, to find a compromise between the capacity of the machine, defined as the ability of a function to learn well (memorize) any training set and the smoothness (in terms of precision) of the approximation (not strictly following the points to interpolate but trying to infer a global trend). (See Figure 21.5.)

21.5.2 SVM Support Vector Machines

A classification problem can always be reduced to a dichotomy, that is, the choice between two classes, without losing its generality. For this reason, the following description will focus on the issue of separating two classes, starting from a set of examples of correct classifications (training set). First, the case of linearly separable classes will be examined.

Considering a a set of pairs,

$$(y_1, x_1), ..., (y_n, x_n), x \in R^n, y \in \{-1, +1\}$$

separated by the hyperplane

$$(w \cdot x) + b = 0$$

the input vector set is optimally separable if the hyperplane divides the dataset without misclassifications, maximizing the separation margin, de-

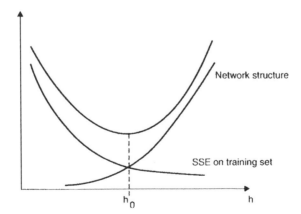

Figure 21.5. The plot highlights the expected risk minimization principle, composed of two factors: one related to the error against training data and the other depending on $\Phi(l/h)$.

fined as the minimum distance between a sample x and the hyperplane (Figure 21.6). A support vector machine is a method to find that hyperplane, trying to fit the training set, and to generalize the example vectors. The optimization task can be reduced to a constrained optimization of

$$\Phi(w) = \frac{1}{2}(w \cdot w)$$

under the constraints

$$y_i[(w \cdot x_i) + b] \geq 1 \qquad i = 1, 2, 3, \ldots, 1$$

The x points verifying the constraints with an equal relationship (vectors lying over the separation edge) are called support vectors.

If the training set is separable, then the solution of the optimization problem exists in closed form and is based on the constrained optimization of the quadratic function by means of the Lagrange multipliers and the duality theorem (Bertsekas, 1995).

The algorithm, if applied to nonlinearly separable data, cannot find any acceptable solution, since the objective functional (the dual-problem Lagrangian) can infinitely grow. To avoid this phenomenon, the constraints are loosened, introducing slack variables and allowing the data to trespass the edge bounds.

Cortes (1995) introduced a nonnegative variable $\xi > 0$ (slack variable) and a penalty function allowing one to generalize the searching mechanism of the optimal hyperplane.

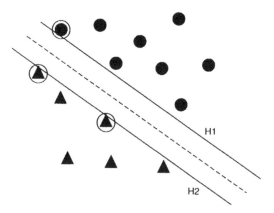

Figure 21.6. An example of linearly separable classes and (dotted line) the optimal separation. H1 and H2 are the lines (more generally, hyperplanes) maximizing the distance. The circles highlight the support vectors.

The optimization problem is then

$$\Phi(w,\xi) = \frac{1}{2}(w \cdot w) + C\sum_{i=1}^{l}\xi_i$$

with C a fixed value. The constraints are

$$y_i\left[(w \cdot x_i) + b\right] \geq 1 - \xi_i$$

Lagrange multipliers will then range between 0 and C.

The C parameter is a priori chosen and generally reflects the amount of presumed noise in the data. For large C values, the SVM will tend to overfit data; higher generalization and lower accuracy over training data will instead occur for smaller C. This fixed parameter, usually called the regularization parameter, must be chosen as a compromise between the machine complexity and the number of nonseparable samples. The choice is usually made in a former design phase, through experiments (crossvalidation) or analytically, by estimating the VC dimension of the model and using as a constraint (there are formulas in the literature that estimate it from the data samples). In this case, the search for the optimal C can be included in the optimization process of the SVM.

21.5.2 Generalizing to the Nonlinear Case.
So far, only data linearly separable by means of a hyperplane have been considered. If, on the other hand, the sampled vectors cannot be linearly separated, the Cover theorem can be exploited. This theorem allows one to map each input vector x in a $\mathcal{H}$ space, using a nonlinear transformation. The idea is to move the input vectors into a new space of higher dimensionality. If the function performing the mapping is nonlinear and the target space dimension is high enough, there is a high probability that the mapped configuration of the starting vectors in the new domain will become linearly separable. In other words, instead of building a nonlinear classifier in the input space, the nonlinear SVM searches for a transformation that maps data into a higher dimensional space where a linear classifier is sufficient to discriminate. The procedure can be explained as the extraction of features from data to build a linear classifier in the features space, instead of in the examples space.

Mathematically, data are mapped in the $\mathcal{H}$ space (of arbitrary dimension, usually very high), using a function Φ:

$$\Phi : \mathcal{R}^d \rightarrow \mathcal{H}, x \rightarrow \Phi(x) = [\Phi_1(x), \Phi_2(x), \Phi_3(x), \ldots, \Phi_N(x)]$$

where Φ_i are chosen functions. After that, a linear classifier is built in the $\mathcal{H}$ space. The space has a high dimensionality (theoretically infinite), so the Φ functions increase their computational weight and they become difficult to handle in numerical algorithms. This problem is solved in the SVM theory by starting from the observation of the SVM learning algorithm, where training-set data always appear in formulas computing the dot products $x_i \cdot x_j$.

By transforming x in $\Phi(x)$, the training process would then use the dot product of the mapped data $\Phi(x_i) \cdot \Phi(x_j)$. Then, choosing a function (called kernel) K, which performs the direct dot product mapping,

$$K(x_i, x_j) = \Phi(x_i)\Phi(x_j)$$

this function could be directly used in the learning procedure, without distinctly managing the Φ mapping.

It is remarkable how the replacement of $x_i \cdot x_j$ with $K(x_i, x_j)$ in the training algorithm will generate a SVM acting in a space with arbitrary dimension. That result will be computationally the same as if obtained in the original space.

The choice of the kernel becomes determinant and defines different kinds of SVMs (Gunn, 1998; Burges, 1998; Vapnik, 1995). Below are summarized some of the alternatives.

Polynomial Kernel

$$K(\bar{x}, \bar{y}) = (\bar{x} \cdot \bar{y})^d$$
$$d = 1, 2, \ldots$$

RBF Kernel

$$K(\bar{x}, \bar{y}) = \exp\left(-\frac{(\bar{x} - \bar{y})^2}{2\sigma^2}\right)$$

MLP Kernel. This definition is valid only for some values of b and α [Mercer's theorem (Mercer, 1908; Courant and Hilbert, 1970)]:

$$K(\bar{x}, \bar{y}) = \tanh\left[\alpha(\bar{x} \cdot \bar{y}) + b\right]$$

The names RBF and MLP were chosen in analogy with the corresponding neural networks and, under certain conditions, they show a functional equivalence with their counterparts.

Adaptive networks so far analyzed share a frequentist approach to the learning problem because they realize the process of knowledge acquisition just by means of experimental data.

21.6   NEURO-FUZZY NETWORKS

In the field of biomedical engineering, the problems are often characterized by a noticeable amount of uncertainty added to a high complexity,
which makes it difficult to create models of the observed phenomena in
simple mathematical terms. It is sometimes hard to exactly describe systems starting from a partial or noisy knowledge of them. Neural networks
are a a first approach for overcoming this issue but they are limited by being unable to integrate any a priori knowledge of the studied system,
which is fragmented and sometimes qualitative.

In real-world problems, it is common to join numerical data with
qualitative information derived from observation or deduced by evaluation
of similarity to known mechanisms, by human experience or intuition. In
this perspective, the fuzzy inference systems have been proposed as a
method for solving complex problems starting from a set of observations
and an amount of a priori information. These models are based on the
fuzzy logic theory and substitute numerical variables with linguistic ones.
(Zadeh 1965).

In his pioneering work, Zadeh (1965) stated that "as the complexity
of a system increases, our ability to make precise and, at the same time,
significant statements about its behavior reduces up to a threshold. Beyond
that limit, precision and significance become almost mutually exclusive
features." For this reason, numerical variables, assuming only definite values (crisp) and belonging to sharp bounded sets (e.g., "the age of the patient is < 25 years"), are substituted by linguistic variables in the fuzzy logic. These variables can describe uncertainty and confusion and belong to
hazy (fuzzy) sets. Hence, they are called fuzzy variables (e.g., "the patient
is young").

A fuzzy set is characterized by a membership function (MF) associating every element with a degree of membership, in analogy to the concept of probability, ranging from 0 to 1.

A fuzzy inference system (FIS) is a collection of rules of the form if
x is A then y is B, where A and B are linguistic values belonging to fuzzy
sets and x and y are linguistic variables.

Usually, a fuzzy rule contains intersection and union operators, such as
x is A and x is B. These operators are called triangular norm and conorm (T-
norm and T-conorm), respectively, and can be implemented in different ways
(Mamdani and Assilian, 1975; Takagi and Sugeno 1985; Tsukamoto, 1979).
Various FIS models differ according to the choice of composition methods of
the if–then rules and to the operators, including the implication operator. The
FIS described so far presents two heavy limitations (Jang, 1993):

1. There are no standard methods to transform human knowledge into
   fuzzy rules.

2. There is no effective way to model the membership functions in order to minimize some error measures or maximize a performance index.

Trying to overcome these limitations, since an FIS can be considered as an adaptive network, Jang proposed a representation of an inference system based on fuzzy logic called ANFIS (adaptive network-based fuzzy inference system). An ANFIS network is a feed-forward multilayer adaptive network. Historically, the Sugeno and Tsukamoto models are the most used implementations of ANFIS networks. In this chapter, we will focus on the Sugeno typology but it has to be remembered that the differences between the models are minimal. A Sugeno ANFIS is a feed-forward network composed by five layers that can be described separately as follows:

1. Every node in the first layer is an adaptive (with tuneable parameters) function transforming the input (crisp) in a membership value to a fuzzy set. This membership function is usually a bell-shaped curve, like a Gaussian or sigmoid, but it is only required to be continuous and differentiable (C^1). This layer is said to fuzzify the crisp input into a linguistic (fuzzy) signal.
2. Nodes in this layer are fixed (nonparametric) and produce as output the product of their input signals. In the Sugeno model, this represents the intersection operation, that is, the AND operator, thus chosen to be implemented as a product.
3. These nodes also are nonparametric and normalize the "activation strengths" (to continue with the neural network jargon). The ith node will then have as output the value of the corresponding ith node of the preceding layer, divided by the sum, of the exits of the previous layer
4. In this fourth layer, the actual inference is performed and the signal coming from the premise is turned into the consequence. In the Sugeno model, this operation is done through the linear interpolation of inputs, casting a linguistic variable into a crisp value. By means of a weighted sum, the layer defuzzifies the signal. Nodes in this layer are parametric with a polynomial transfer function, whose input variables are the normalized activation strengths and whose parameters are called consequence parameters.
5. The last layer completes the defuzzifying process by summing the output signals of the previous layer into the real exit of the inference system.

It is remarkable how, working in the data domain which is necessarily crisp, both the input and the output of an ANFIS must be numerical variables.

Provided there are a sufficient number of rules, it can be shown that an ANFIS is a universal approximator. Moreover, it is quite easy to demonstrate that, under certain hypotheses, an ANFIS network is functionally equivalent to a radial basis neural network (Jang and Sun 1993).

21.6.1 ANFIS Learning

Belonging to the adaptive network category, an ANFIS network can adopt learning mechanisms such as the already mentioned (error) backpropagation. It is easy to show that the optimization of the ANFIS model is nonlinear for the premise parameters but linear in the consequence ones. This allows us to apply, in analogy with radial basis neural networks, a hybrid learning procedure. The algorithm is, therefore, built up by combining the backpropagation with a linear optimization method. The two steps are computed in sequence. First (forward step), the consequence parameters are optimized by fixing the premise parameters and applying a least-squares linear estimate, using as signals the outputs of the nodes. Then a second (backward) step is computed, propagating the error signal from the exit toward the input layer. Likewise, the consequence parameters are kept fixed during this phase, in order to optimize the premise parameters through the backpropagation algorithm. Since the computational complexity of the least-squares optimization is much higher than the backpropagation's, it is often preferred to use one-step gradient-descent methods or approximated algorithms that linearize the premise, such as the Kalman filter or the Levemberg–Marquardt algorithm (Jang and Mizutani 1996).

21.6.1.1 CANFIS. The limitations of having a linear consequence and a single output can be eliminated by adopting a generalized neuro-fuzzy networks class, called CANFIS (CoActive Neuro-Fuzzy Inference System) (Mizutani and Jang 1995). Networks belonging to this typology, although derived by the ANFIS theory, share much more structural and functional aspects with common neural networks. Typically, a CANFIS network will have nonlinear (usually sigmoidal) rules both in the premise and in the consequence. Its structure can be represented, in a connectionist logic, as a modular neural network, where two blocks respectively elaborate the premise and infer the consequence. For this reason, they inherit many of the properties owned by neural networks, including their learning mechanism through backpropagation. CANFIS networks are basically an intermediate model class between a neuro-fuzzy and a neural approach.

21.6.1.2 Neuro-Fuzzy Spectrum. The compromise between approximation accuracy and system interpretability in a fuzzy perspective is well depicted by the neuro-fuzzy spectrum notion. Neuro-fuzzy models allow one to synthesize some a priori knowledge and offer some interpretability of the resulting model after training. On the other hand, neural networks, and especially MLP trained with backpropagation, are very effective in mapping input–output pairs, but they are black-box models, since they do not provide any information concerning the nature of the studied process. At the two extreme sides of the neuro-fuzzy spectrum, we can find, therefore, a completely understandable FIS and a neural network whose weights after training have no explanatory meaning.

The application of more and more advanced optimization methods to neuro-fuzzy networks allows one to get better and more accurate computational results but, on the other hand, it leads to a reduction in the symbolic meaning of their fuzzy rules. This is usually referred to as the dilemma between accuracy and interpretability.

Different methods try to resolve this dilemma, among which are the interpretation of ANFIS and CANFIS networks in a probabilistic perspective (such as Bayesian networks), the choice of different shapes and constraints to their membership functions, and the use of hybrid learning methods in order to preserve the logical meaning of rules.

21.6.2   Fuzzy Modeling

The design of a fuzzy model is the development of an inference system in order to predict or explain the working mechanism of an unknown process, described by examples. So far, the parametric identification process, on which the empirical risk minimization is based, has been analyzed, but, in order to complete the modeling procedure, the structure of the system must be determined as well. As we already reported, this is commonly known as the structural risk minimization.

In the construction of rules for an inference system, the premise defines regions inside the input space, whereas the consequence describes the behavior of the system within those regions. This suggests partitioning the input space, trying to identify the most meaningful zones and correspondingly minimizing the number of rules. To this purpose, one of the following strategies can be applied:

1. Grid partitioning. This is a method used for restricted input sets, since it needs a limited number of membership functions per input. Unfortunately, the number of rules grows exponentially with the number of inputs; in other words, this method is affected by the "curse of dimensionality."
2. Tree partitioning. The input space is subdivided into nonoverlapping regions of different dimensions. This method, generally implemented using the CART (classification and regression tree) algorithm (Breiman et al., 1984), allows reduction of the growth of model complexity but requires a larger number of functions per input, consequently causing a loss in their linguistic significance (interpretability).

3. Scattered partitioning. Probabilistic or clustering methods are applied in order to group the inputs in sets. Its effectiveness strongly depends on the available data.

21.7 GENETIC ALGORITHMS

Genetic algorithms (GAs) are search algorithms using operators based on mechanisms of natural selection and genetics, combining the principle of survival of the fittest with a structured, though stochastic, rule for exchanging information. GAs were first suggested by John Holland and his group in the mid nineteen-seventies (Holland, 1975), yet the scientific community was skeptical of the potential of the suggested approach. GAs became popular only over ten years later, when David E. Goldberg published a new detailed presentation of their structure and applications (Goldberg, 1989).

The scientific interest in GAs was due to the attempt to develop robust artificial systems, which may be able to deal with changes in both environmental and working conditions. Robustness is a compromise between efficacy and efficiency and is an extremely interesting property for artificial systems as it allows one to reduce the need for external intervention when facing changes in the environment or malfunctioning of some components of the system. Traditional artificial intelligence tools (based on a systems approach or on traditional AI principles) can hardly be endowed with such properties, whereas biological systems offer levels of robustness and adaptability that represent a long-desired goal for artificial ones. For this reason, Holland and colleagues decided to take inspiration from nature, where the evolutional principles described by Charles Darwin have led to biological systems being perfectly adapted to different environmental conditions and able to survive when faced with lesions or with environmental changes.

In nature, living beings survive and reproduce if their genotype encodes the characteristics that allow the development of a phenotype having somatic and behavioral traits that are adapted to the environment in which they live. The genotype of an individual is made up of one or more chromosomes; these are, in turn, composed of genes that encode (either singularly or as a set) the traits of the individual in terms of their values (alleles). If, for instance, we were to consider a gene that encodes the color of the eyes of a person, then the possible values (e.g., brown, green, or blue) represent its alleles. In biological systems, evolution is guided by the principle of survival of the fittest. Therefore, the individuals whose genotype developed into the most adapted phenotype for its environmental conditions will have a higher probability of reproducing and, thus, of transmitting their genetic characteristics to the following generation. The process of breeding that yields a new generation can be summarized in three main steps: selection of the individuals of the current generation that will become parents of offspring for the new generation, breeding of offsprings having a genotype combining genes from the two parents through a stochastic process, and stochastic change in one or more genes of the offspring, perhaps due to a copying error, that may introduce new alleles different from those in either parent.

GAs implement a metaphor of natural evolution by considering a population of candidate solutions that evolve based on the principle of survival of the fittest in order to produce increasingly fit individuals that are, therefore, better solutions to the problem at hand. Within such a metaphor, each candidate solution is coded in terms of a fixed-length string (e.g., binary) representing its chromosome. In a typical GA, an initial population of chromosomes is created randomly. The quality of each chromosome is quantified using a score (fitness) based on the performance of its phenotype as a solution. The selection of the parents for a new generation operates so that chromosomes with a higher fitness are more likely to be selected. Each pair of parents breeds one or more offsprings. These are chromosomes whose genotype comprises genes from both parents, inherited through a gene exchange operator usually called the crossover operator. Each offspring will thus be a string of the same length as the parent strings whose single genes are copied from one of the two parents. Each gene of the offspring may then undergo, with a low probability, a random change of its allele through the mutation operator mimicking an error in the gene copying process. The selection, crossover, and mutation operators iteratively create a new population of chromosomes of the same size as the initial one, representing a new generation. A new population fitness will then be computed and a new iteration of the breeding loop begins.

To better understand the operation of a GA, we will consider the problem of finding the minimum of the function $f(x) = [x - 16]^2$ and work through some steps of its solution. One of the first steps needed to set up a GA is the selection of a proper coding of the parameters involved in the optimization problem that needs to be faced. These are usually coded in tems of a finite-length string of symbols of some alphabet. In the context of our example problem, we may choose to limit the sought space of possible solutions to that of integers in the range [0, 31] and, therefore, consider solutions coded using five symbols drawn from the binary alphabet (i.e., 5 bits). A first population of four chromosomes is generated randomly as follows:

C1: 10101
C2: 00100
C3: 10010
C4: 01011

Next, we need to select a fitness function to evaluate each chromosome in the population in order to assess their quality as a solution to our problem. Since we are facing a minimization problem and a GA tends to find solutions that maximize the fitness, we can consider the fitness function fit$(x) = 1/[f(x) + 1]$. Evauating the fitness of each chromosome in the population, we will obtain. fit$(C1) = 0.04$, fit$(C2) = 0.0069$, fit$(C3) = 0.25$, and fit$(C4) = 0.04$.

The next step is that of selecting, based on these fitness scores, those chromosomes that will be parents for the chromosomes of the following generation so that the individuals with a higher fitness will have a higher probability of being selected. A commonly used selection operator is that called roulette wheel. In such an approach, an imaginary wheel is divided into as many sectors as there are chromosomes in the population, and the amplitude of each one of them is proportional to the fitness of the corresponding chromosome with respect to the total fitness of the population (0.3369 in our example). (See Figure 21.7.) A random number $r \in [0,1]$ will then be drawn to select each needed parent, and if $r \in [0,0.12)$ the selected parent will be C1, C2 if $r \in [0.12,0.14)$, C3 if $r \in [0.14,0.88)$, and C4 if $r \in [0.88,1]$.

Each chromosome may be selcted more than once. If we hypothesize that the four random numbers are 0.0579, 0.3529, 0.8132, and 0.0099, then the selected parents will be C1, C3, C3, and C1, respectively. Each pair of selected parents will then produce two offspring for the new generation. These may either be produced by applying a crossover operator, which

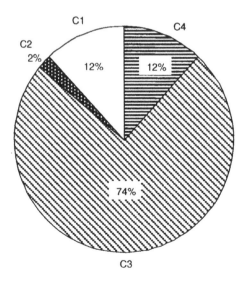

Figure 21.7. The roulette wheel corresponding to the current population.

usually occurs with a high probability (e.g., 0.7), or they may be exact copies of the parents. Many crossover operators may be found in the scientific literature; they are generally variations of the single-point crossover that we will consider here. Single-point crossover produces two offspring from each pair of parents. The genes in each offspring will be taken from each of the parents up to a randomly selected index, whereas the rest will come from the other chromosome. For instance, if we hypothesize that both our pairs (C1, C3) and (C3, C1) need to undergo the crossover operator and that the randomly selected indices are 2 and 4, respectively, then the four offsprings will be produced as follows:

C1: 10101	C1′: 10010
C3: 10010	C2′: 10101
C3: 10010	C3′: 10011
C1: 10101	C4′: 10100

As a last step to create the new generation, each gene of each offspring will be exposed to the mutation operator with a low probability (e.g., 0.01). We may, therefore, hypothesize that by drawing 20 random numbers for the 20 genes it turns out that the mutation operator needs to be applied only to the fourth gene in C3′, which will therefore become C3′ = 10001.

These steps have thus allowed us to produce a new generation, and the fitness corresponding to the new chromosomes is $\text{fit}(C1′) = 0.25$, $\text{fit}(C2′) = 0.04$, $\text{fit}(C3′) = 1$, and $\text{fit}(C4′) = 0.0625$. Note how both the maximum fitness and the total population fitness have increased and the best candidate solution in the second generation is now $x = 17$ (it was $x = 18$ in the first one). Repeating these steps to produce the following generations, the GA will tend to produce new populations with increasingly fit chromosomes, which will, therefore, be closer to the solution of the problem.

There are four main characteristics that set apart a GA from the other optimization techniques such as iterative descent and neural networks, or stochastic mechanisms such as random walks:

1. GAs operate on some coding of the problem parameters and not on the parameters themselves.
2. GAs search the solution space starting from a population of points and not from a single one.
3. GAs use a fitness function to evaluate the quality of each solution and no additional information relative to the problem at hand (e.g., error derivatives)
4. GAs use stochastic transiton rules to go from one state to the next instead of deterministic rules.

Thus, GAs are search algorithms producing increasingly fit solutions that do not necessarily converge to a globally optimal solution (fitness maximum) but are very efficient in finding a good approximation to such solution.

REFERENCES

Barto, A. G., Sutton, R. S., and Anderson, C. W., Neuronlike Adaptive Elements that can Solve Difficult Learning Control Problems, *IEEE Trans. Syst., Man, and Cybern.*, Vol. 13, 834–846, 1983.

Barto, A. G., Sutton, R. S., and Watkins, C. J. C. H., Learning and Sequential Decision Making, in *Learning and Computational Neuroscience*, M. Gabriel and J. W. Moore (Eds.), MIT Press, 1990.

Bertsekas, D. P., *Dynamic Programming and Optimal Control*, Vols. 1 and 2, Athenas Scientific, 1995.

Bertsekas, D. P., *Nonlinear Programming*, Athenas Scientific, 1995.

Boser, B., Guyon, I., and Vapnik, V. N., A Training Algorithm for Optimal Margin Classifiers, in *Fifth Annual Workshop on Computational Learning Theory*, Morgan Kaufmann, 1992.

Breiman, L., Friedman, J. H., Olshen, R. A., and Stone, C. J., *Classification and Regression Trees*, Wadsworth, 1984.

Brown, M., and Harris, C., *Neurofuzzy Adaptive Modeling and Control*, Prentice-Hall, 1994.

Burges, C.J.C., A Tutorial on Support Vector Machines for Pattern Recognition, *Data Mining and Knowledge Discovery*, Vol. 2, No. 2, 121–167, 1998.

Chauvin, Y., and Rumelhart, D. E., *Backpropagation: Theory, Architectures, and Applications*, Lawrence Erlbaum Associates, 1995.

Chen, S., Cowan, C. F. N., and Grant, P. M., Orthogonal Least Squares Learning Algorithm for Radial Basis Function Networks, *IEEE Trans. Neural Networks*, Vol. 2, 302–309, 1991.

Chen, V. C., and Pao, Y. H., Learning Control with Neural Networks, in *Proceedings of International Conference on Robotics and Automation*, 1448–1453, 1989.

Chen, Y.-Y., and Tsao, T.-C., A Description of the Dynamic Behaviour of Fuzzy Systems, *IEEE Trans. Syst. Man. and Cybern.*, Vol, 19, 745–755, 1989.

Chen, Y.-Y., A self-learning fuzzy controller, in *Proceedings of IEEE International Conference on Fuzzy Systems*, March 1992.

Chiu, S. L., Fuzzy Model Identification Based on Cluster Estimation, *J. Intell. and Fuzzy Syst.*, Vol. 2, No. 3, 267–278, 1994.

Cortes, C., and Vapnik, V. N., Support Vector Machines, *Machine Learning*, Vol. 20, 1995.

Courant, R., and Hilbert, D., *Methods of Mathematical Physics*, Vols. 1 and 2, Wiley–Interscience, 1970.

Crick, F. H. C., The Recent Excitement About Neural Networks, *Nature*, Vol. 337, No. 6203, 129–132, 1989.

Cover, T. M., Geometrical and Statistical Properties of Systems of Linear Inequalities with Applications in Pattern Recognition, *IEEE Trans. on Electronic Computers,* Vol. EC-14, 326–334, 1965.

Cybenko, G., *Approximation by Superpositions of a Sigmoidal Function,* University of Illinois Press, 1989.

Dubois, D., and Prade, H., *Fuzzy Sets and Systems: Systems: Theory and Applications,* Academic Press, 1980.

Fahlman, S. E., and Lebiere, C., The Cascade-Correlation Learning Architecture, *Advances in Neural Information Processing Systems,* Vol. 2, Morgan Kaufmann, 1990.

Funahashi, K., On the Approximate Realization of Continuous Mappings by Neural Networks, *Neural Networks,* Vol. 2, 183–192, 1989.

Goldberg, D. E., *Genetic Algorithms in Search, Optimization and Machine Learning.* Addison-Wesley, 1989.

Grossberg, S., Nonlinear Neural Networks: Principles, Mechanisms and Architectures, *Neural Networks,* Vol. 1, 17–61, 1988.

Gunn, S. R., Support Vector Machines for Classification and Regeression, Technical Report , Dept. of Electronics and Computer Science, Univ. of Southampton, 1–52, 1998.

Hayikin, S., *Neural Networks—A Comprehensive Foundation,* 2nd edition, Prentice-Hall, 1999.

Hebb, D. O., *The Organization of Behavior: A Neuropsychological Theory.* Wiley, 1949.

Holland, J. H., *Adaptation in Natural and Artificial Systems,* Universityof Michigan Press, Ann Arbor, 1975.

Holland, J. H., *Adaptation in Natural and Artificial Systems: An Introductory Analysis with Applications to Biology, Control and Artificial Intelligence,* MIT Press, 1998.

Hopfield, J. J., Neural Networks and Physical Systems with Emergent Collective Computational Abilities, *Proceedings of the National Academy of Sciences USA,* Vol. 79, 2554–2558, 1982.

Hopfield, J. J., Artificial Neural Networks, *IEEE Circuits and Devices Magazine.* Vol. 4, No. 5, pp. 3–10, 1988.

Hornik, K., Stinchcombe, M., and White, H., Multilayer Feedforward Networks are Universal Approximators, *Neural Networks,* Vol. 2, 359–366, 1989.

Hush, D. R., and Horne, B. G., Progress in Supervised Neural Networks, *IEEE Signal Processing Magazine,* Vol. 10, No. 1, 8–39, 1993.

Jang, J.-S. R., and Sun, C.-T., Functional Equivalence Between Radial Basis Function Networks and Fuzzy Inference Systems, *IEEE Trans. Neural Networks.* Vol. 4, 156–159, 1993.

Jang, J.-S. R., and Mizutani, E., Levemberg-Marquardt Method for ANFIS Learning, in *Proceedings of the International Joint Conference of The North American Fuzzy Information Processing Society Biannual Conference,* Bekeley, California, 1996.

Jang, J.-S. R., ANFIS: Adaptive-Network-Based Fuzzy Inference System, *IEEE Transactions on Systems, Man and Cybernetics,* Vol. 23, No. 3, 665–685, 1993.

Jang, J.-S. R., Self-Learning Fuzzy Controllers Based on Temporal Back Propagation, *IEEE Transactions on Neural Networks,* Vol. 3, No. 5, 142–148, 1992.

Jang, J.-S. R., Fuzzy Modeling Using Generalized Neural Networks and Kalman

Filter Algorithm, in *Proceedings of the Ninth National Conference of Artificial Intelligence* (AAAI-911), pp. 762–7G7, July, 1991.

Jang, J.-S. R., Sun, C.-T., and Mizutani, E., *Neuro-Fuzzy and Soft Computing,* Prentice-Hall, 1997.

Khanna, T., *Foundations of Neural Networks,* Addison-Wesley, 1990.

Kohonen, T., Clustering Taxonomy, and Topological Maps of Patterns, in *Proceedings of the 6th International Conference on Pattern Recognition,* Munich, Germany, 1982.

Kohonen, T., Self-Organized Formation of Topologically Correct Feature Maps, *Biological Cybernetics,* Vol. 43, No. 1, 59–69, 1982.

Kohonen, T., The Self-Organizing Map, *Proceedings of the IEEE,* Vol. 78, No. 9, 1990.

Kosko, B., *Neural Networks for Signal Processing,* Prentice-Hall International Editions, 1992.

Kumar, S. R., and Jayati, G., Neuro Fuzzy Approach to Pattern Recognition, *Neural Networks,* Vol. 10, No. 1, 161–182, 1997.

Lecun, Y., Denker, J. S., and Solla, S. A., Optimal Brain Damage, in *Advances in Neural Information Processing Systems,* Vol. 2, Touretzky, D. S. (Ed.), Morgan Kaufmann, 1990.

Li, H.-X., and Chen, C. L. P., The Equivalence Between Fuzzy Logic Systems and Feedforward Neural Networks, *IEEE Transactions on Neural Networks,* Vol. 11, No. 2, 2000.

Li, H.-X., Multifactorial Functions in Fuzzy Sets Theory, *Fuzzy Sets Syst.,* Vol. 35, 69–84, 1990.

Lippmann, R., A Critical Overview of Neural Network Classifiers, in *Neural Networks for Signal Processing: Proceedings of the 1991 IEEE Workshop,* Juang, B. H., Kung, S. Y., and Kamm, C. A. (Eds.), pp. 266–278, IEEE Press, 1991.

Lippmann, R. P., An Introduction to Computing with Neural Nets, *IEEE Acoustics, Speech and Signal Processing Magazine,* Vol. 4, No. 2, 4–22, 1987.

Loève, M., *Probability Theory,* 3rd edition, Van Nostrand, 1963.

Mamdani, E. H., and Assilian, S., An Experiment in Linguistic Synthesis with a Fuzzy Logic Controller, *Int. Journal of Man–Machine Studies,* Vol. 7, No.1, 1–13, 1975.

McCulloch, W. S., and Pitts, W., A Logical Calculus of the Idea Immanent in Nervous Activity, *Bulletin of Mathematical Biophysics,* Vol. 5, 115–133, 1943.

Mercer, J., Functions of Positive and Negative Type, And Their Connection with the Theory of Integral Equations, *Trans. of the London Philosophical Society (A),* Vol. 209, 415–446, 1909.

Mizutani, E., and Jang, J.-S. R., Coactive Neural Fuzzy Modeling, in *Proceedings of the International Conference on Neural Networks,* 1995.

Mizutani, E., Coactive Neuro-Fuzzy Modeling: Toward Generalized ANFIS, in *Neuro-Fuzzy and Soft Computing: A Computational Approach to Learning and Machine Intelligence,* Roger Jang, J.-S., Sun, C.-T., and Mizutani, E. (Eds.), pp. 369–400, Prentice Hall, 1997.

Muller, K. R., Mika, S., Ratsch, G., Tsuda, K., and Scholpov, B., An Introduction to Kernel Based Learning Algorithmns, *IEEE Trans. on Neural Networks,* Vol. 12, No 2, 181–201, 2001.

Osuna, E., Freund, R., and Girosi, F., An Improved Training Algorithm for Support Vector Machines, in *IEEE Workshop on Neural Networks and Signal Processing,* Amelia Island, FL, September 1997.

Rosenblatt, F., The Perceptron: A Probabilistic Model For Information Storage And Organization in the Brain, *Psychological Review,* Vol. 65, No.6, 386–408, 1958.

Rumelhart, D. E., Hinton, G. E., and Williams, R. J., Learning Internal Representation by Error Propagation, in *Parallel Data Processing,* Rumelhart, D. E., and McClelland, J. (Eds.), Vol. 1, pp. 318–362, MIT Press, 1986a.

Rumelhart, D. E., Hinton, G. E., and Williams, R. J., Learning Representations by Back Propagating Errors, *Nature* (London), Vol. 323, 533–536, 1986b.

Rumelhart, D. E., and Zipser, D., Feature Discovery By Competitive Learning, *Cognitive Science,* 1985.

Schölkopf, B., and Smola, A., *Learning with Kernels,* MIT Press, 2002.

Schölkopf, B., Burges, C. J. C., and Smola, A. J., *Advances in Kernel Methods—Support Vector Learning,* MIT Press, 1999.

Scholkopf, B., Sung, K.-K., Burges, C. J. C., Girosi, F., Niyogi, P., Poggio, T., and Vapnik, V., Comparing Support Vector Machines with Gaussian Kernel to Radial Basis Function Classifiers, *IEEE Transactions on Neural Networks,* Vol. 45, No. 11, 2758–2765, 1997.

Sejnowski, T. J., Statistical Constraints on Synaptic Plasticity, *Journal of Theoretical Biology,* Vol. 66, 385–389, 1977.

Sejnowski, T. J., Strong Covariance with Nonlinearly Interacting Neurons, *Journal of Mathematical Biology,* Vol. 4, 303–321, 1977.

Shepherd, G. M., The Significance of Real Neural Architectures for Neural Network Simulations, in *Computational Neuro Science,* Schwartz, E. L. (Ed.), MIT Press, 1990.

Stork, D., Is Back Propagation Biologically Plausible?, in *International Joint Conference on Neural Networks,* Vol. 2, Washington D.C., 1989.

Sugeno, M., Fuzzy Measures and Fuzzy Integrals: A Survey, in *Fuzzy Automata and Decision Processes,* Gupta, M. M., Saridis, G. N., and Gaines, B. R. (Eds.), pp. 89–102, North-Holland, 1977.

Sutton, R. S., Barto, A. G., and Williams, R. J., Reinforcement Learning is Direct Adaptive Optimal Control, in *Proceedings of the American Control Conference,* Boston, 1991.

Sutton, R. S., *Temporal Credit Assignment in Reinforcement Learning,* Ph.D. Dissertation, University of Massachussets, Amherst, MA, 1984.

Takagi, T., and Sugeno, M., Fuzzy Identification of Systems and its Applications to Modeling and Control, *IEEE Trans. Syst., Man, Cybern.,* Vol. SMC-15, 1–116, 1985.

Tsukamoto, Y., An Approach to Fuzzy Reasoning Method, in Gupta, M. M., Ragade, R., and Yager, R. R. (Eds.), *Advances in Fuzzy Set Theory and Applications,* North-Holland, 1979.

Vapnik, V., *The Nature Of Statistical Learning Theory,* Springer-Verlag, 1995.

Vapnik, V., An Overview of Statistical Learning Theory, *IEEE Transactions on Neural Networks,* Vol. 10, No. 5, 988–999, 1999.

Vapnik, V. N., *Statistical Learning Theory,* Wiley, 1998.

Vapnik, V. N., and Chervonenkis, A. Y., On the Uniform Convergence of Relative

Frequencies of Events to Their Probabilities, *Theory of Probability and its Applications,* Vol. 16, No. 2, 264–280, 1971.

Weigend, A. S., Rumelhart, D. E., and Hubeman, B. A., Back-Propagation, Weight Elimination and Time Series Prediction, in *Proceedings of the 1990 Connectionist Models Summer School,* pp. 65–80. Morgan Kaufmann, 1990.

White, H., Artificial Neural Networks, *Approximation and Learning Theory,* Blackwell, 1992.

Widrow, B., and Hoff, M. E. Jr., Adaptive Switching Circuits, *IRE WESCON, Convention Record,* 1960.

Widrow, B., ADALINE and MADALINE—1963, in *Proceedings IEEE 1st International Conference on Neural Networks,* Vol. 1, pp. 143–157, 1987.

Widrow, B., Generalization and Information Storage in Networks of Adaline neurons, in *Self-Organizing Systems,* Yovitz, M. C., Jacobi, G. T. and Goldstein, G. D. (Eds.), Sparta, 1962.

Yamakawa, T., Uchino, E., Miki, T., and Kusanagi, H., A Neo Fuzzy Neuron and its Application to System Identification and Prediction of the System Behaviour, in *Proceedings of the 2nd International Conference on Fuzzy Logic and Neural Networks,* pp. 477–483, 1992.

Zadeh, L. A., Fuzzy Sets, *Information and Control,* Vol. 8, 338–353, 1965.

Zadeh, L. A., A Fuzzy-Set-Theoretic Interpretation of Linguistic Hedges, *J. Cybernet.* Vol. 2, 4–34, 1972.

Zadeh, L. A., Outline of a New Approach to the Analysis of Complex Systems and Decision Processes, *IEEE Trans. Syst. Man, and Cybern.,* Vol. 3, 28–44, 1973.

Zadeh, L. A., Fuzzy Logic, *Computer,* Vol. 1, 83–93, 1988.

Thispageintentionallyleftblank

CHAPTER *22*

INTERPRETATION AND CLASSIFICATION OF PATIENT STATUS PATTERNS

Matteo Paoletti and Carlo Marchesi

T HIS CHAPTER CONCERNS the classification and interpretation processes of biosignals. It is divided into three main sections: a brief introduction to classification according to the Bayes' approach, cluster analysis
techniques and clinical applications, and interactive exploration and
graphic biomedical data analysis. The development of the first section is
explained almost completely from a theoretical point of view, but the second and the third sections describe some applicative examples showing results obtained by analyzing real clinical data acquired during patient-mon-
itoring sessions.

A biomedical signal processing system often consists of two main
structures. The first is an algorithmic subsection, defined by automatic
procedures and operations that produce numerical results representing parameters and series derived by extraction from the original signals. The
second structure represents a more subjective step and consists of interpretation and classification processes comprising methods for the integration
of heterogeneous data; hardware and software devices for human–computer interaction; and sophisticated techniques of artificial intelligence, which
help in the interpretation and, thus, the definition of decision-making support applications (Bruce, 2001).

In our opinion, the main role of classification is to link the two structures. This role makes the whole process contiguous, sometimes overlapping with traditional medical expertise, and this is probably the main cause
of uncertainty in the use of such terms as cluster analysis, feature extraction, pattern recognition, intelligent data analysis, learning from data, data
mining, and knowledge discovery. In addition to this, the uncertainty that

characterizes the different investigation methods that help us to reach the final result makes it arbitrary to determine these discipline boundaries. Taking advantage of this flexibility, and also referring to Chapter 2 in this book, we will focus on some aspects of the data classification process that may be of interest in biomedical applications.

22.1   THE CLASSIFICATION PROCESS

The classification process is one of the most universal approaches to clinical diagnostics and therapeutics. The multitude of applications and needs of different disciplines have enriched the number of solutions and methods based on the Bayesian decision theory (Duda et al., 2001).

It seems appropriate to recall here some principles related to the classification process and introduce the well-known and widely applied Bayesian approach. Then we will introduce some clinical applications based on nonparametric classifiers (cluster analysis).

We can introduce here an informal definition of classification: the discipline that studies the description of objects, events, and patterns, referring them to classes identified by one or more cognitive models through the mediation of chosen descriptive variables.

The aim of classification is basically to assign an individual to a group (class) using parameters interpreted by a decision rule.

In practice, each class generates a subspace in the parameters space. Elements in the space of measures are derived from these subspaces. The space of measures is the only accessible one and it is the space of the primary variables that can be observed.

During the process, the classification algorithms assign an individual to a class according to the measurement of the primary variables. The process that allows one to link an individual in the measures space (measured pattern) to one in the parameters space is usually called feature extraction.

22.1.1   Classification Principles

In the statistical classification process, we assume that each individual, represented in parameters space through the vector $X = [x_1, \ldots, x_d]^T$, is the realization of a random vector $\mathbf{X} = [\mathbf{x}_1, \ldots, \mathbf{x_d}]^T$ of size d and the transformation between the ith class cl_i, and the subspace is defined by a probability density function (pdf):

$$p_i(X) = p(X \,|\, cl_i)$$

Moreover, it is assumed that the number of classes L is known as the a priori probability of classes P_i. The classifiers are based on a decision rule by which the different individuals are assigned to a class with a risk of error equal to R.

In general, after having defined a set of discriminant functions, $fd_i(X)$ $i = 1, \ldots, L$, the decision rule is expressed by

$$fd_k(X) \overset{cl_k}{>} fd_i(X)$$

The geometric meaning of discriminatory functions is that space is divided into regions of the parameters for a decision so that if an individual belongs to one of the regions, it is then assigned to the class.

The problem of designing a classifier becomes the problem of planning the discriminant functions, as follows.

Given a random vector X with pdf $p(X)$ and conditional pdfs $p_i(X)$, the expected value is defined as $\mu = E\{X\} = \int_{R_n} Xp(X)dX$, variance is $\Sigma = E\{(X - M)(X - M)^T\}$, expected value of cl_i is $M_i = E\{X|cl_i\} = \int_{R_n} Xp_i(X)dX$, and variance of cl_i $\Sigma_i = E\{(X - M_i)(X - M_i)^T|cl_i\}$.

The density of probability Gaussian or normal is defined by

$$N(X, M, \Sigma) = \frac{1}{(2\pi)^{\frac{n}{2}} |\Sigma|^{\frac{1}{2}}} \exp\left[-\frac{1}{2}(X - M)^T \Sigma^{-1}(X - M)\right]$$

where $(X - M)^T \Sigma^{-1}(X - M_i)$ is known as the Mahalanobis distance, the distance carrier that depends on X (expected value) rather than M (normalized covariance matrix). This metric becomes the Euclidean distance if $\Sigma = 1$.

22.1.2 Error and Risk During Classification

A classifier does not always provide a perfect classification; there is a certain probability that an individual is assigned to a wrong class. Since our goal is to minimize the error probability, this is a measure of the performance of the classifier.

If we define $q_i(X) = P(cl_i|X)$, the a posteriori probability of the cl_i class given X, the conditional a priori probability to classify $X \in cl_i$ given X is

$$\varepsilon_i(X) = \sum_{j \neq i} q_j(X) = 1 - q_i(X)$$

To determine the a posteriori probabilities, the Bayes theorem can be applied:

$$q_i(X) = \frac{P_i p_i(X)}{p(X)}$$

The denominator in the second member is defined by the probability density function of X. This is difficult to determine unless the assumptions of the theorem of total probabilities can be applied. In this case,

$$p(X) = \sum_{i=1}^{L} P_i p_i(X)$$

Now imagine dividing the space of the parameters in regions Ω_i, $i = 1, \ldots, L$, so that individuals belonging to Ω_i are allocated to the class cl_i. The probability of error is

$$\varepsilon = \int_{\Omega_1} \varepsilon_1(X) p(X) dX + \ldots + \int_{\Omega_L} \varepsilon_L(X) p(X) dX$$

To assign an individual to one class rather than another can have different consequences; in particular, the concept of risk can be introduced.

Given c_{ij}, the risk of deciding $X \in cl_i$ when $X \in cl_j$, the conditional risk $X \in cl_i$ given X is

$$r_i(X) = \sum_{j=1}^{L} c_{ij} q_j(X)$$

and, as in the case of error,

$$r = \int_{\Omega_1} r_1(X) p(X) dX + \ldots + \int_{\Omega_L} r_L(X) p(X) dX$$

22.2 THE BAYES CLASSIFIER

The Bayes method introduces a decision strategy to minimize the expected value of the total classification cost (Duda et al., 2001).

Let us introduce a matrix cost L. The generic element $L(i, j)$ is the cost of the wrong decision of assigning an object to the class Ci whereas it would actually be a member of the Cj class:

$$L_x(k) = \sum_{j=1}^{g} L(k, j) p(j|X)$$

$Lx(k)$ represents the average cost associated with the decision "X is assigned to class Ck."

The expected value of the total classification cost is the minimum one in which each decision cost is minimized. The Bayes rule is:

X is assigned to the class Ck that minimizes $Lx(k)$

In the symmetric cost case, $L(i, j) = 1 - di, j$, the rule is:

$$X \text{ is assigned to } Ci \text{ if } P(i|X) \geq P(j|X) \ \forall j = 1, \ldots, g$$

The rule minimizes the probability of misclassification. The a posteriori probability may be computed by means of the a priori probability $P(i)$ and the probability densities $P(X|i)$.

The Bayes theorem is

$$P(i|X) = \frac{p(X|i)P(i)}{p(X)}$$

where $p(X)$ is the nonconditioned probability density and is defined as

$$p(X) = \sum_{i=1}^{g} p(X|i)P(i)$$

The decision rule is now

$$X \text{ is assigned to } Ci \text{ if } p(X|i)P(i) \geq p(X|j)P(j) \ \forall j = 1, \ldots, g$$

If all the classes show the same probability, then $P(i) = 1/g$ and the rule can be written as

$$X \text{ is assigned to } Ci \text{ if } p(X|i) \geq p(X|j) \ \forall j = 1, \ldots, g$$

The rule introduces the maximum likelihood concept and the pattern is assigned to the class showing the higher probability density. The Bayes decision rule can be written as

$$l(i|j) = \frac{p(X|i)}{p(X|j)} \geq \frac{P(j)}{P(i)}$$

The term $l(i|j)$ is called the likelihood ratio and the decision consists in a comparison of this value with a decision threshold t_{ij}. If the following condition is true,

$$t_{ij} \neq P(j)/P(i)$$

the decision will not be the "optimum" by means of the Bayesian method. However, these suboptimal classifications may be useful when we are not interested in minimizing a global cost but in introducing useful margins between success and error probabilities for each class.

In the end, the Bayes decision rule defines the following regions:

$$R_k = \{X: P(k)p(X|k) \geq P(j)p(X|j); j = 1, \ldots, g\}$$

and the discriminant functions can be chosen according to:

$$D_k(X) = \ln[P(k)p(X|k)]$$

Here, we list some useful remarks regarding statistical classification methods:

- A set of significative measures for a type of classifier may not be useful for another method.
- The initial variables set should be iteratively revised according to the chosen classification method.
- The group composed by the most n discriminant measures individually considered is not the best group of dimension n.

In biomedical applications, it is often difficult to estimate the distributions of the various probability functions. In many cases, the application of Bayes' theorem may be difficult and trivial. In the sections that follow, exploratory and unsupervised analysis methods are described; in particular, we will present some clustering methods that can be used to explore large clinical datasets when parametric models are missing.

22.3   A DIFFERENT APPROACH TO INTERPRET (AND CLASSIFY) DATA: CLUSTER ANALYSIS

Cluster analysis is a set of methodologies for automatic grouping of samples, or cases, into classes of similar objects. Clustering is a different kind of classification; in fact, the process is unsupervised and there is no target variable. With clustering methods, we do not try to classify, estimate, or predict the value of a target variable, but to segment the entire dataset into relatively homogeneous subgroups or clusters, where the similarity of the records within the cluster is maximized and the similarity to records outside this cluster is minimized (Han and Kamber, 2006). In other words, those methods aim to construct groups of records so that the between-cluster variation is large compared to the within-cluster variation. Examples of clustering tasks in business and research include target marketing (Larose, 2006), segmentation of financial behavior into benign and suspicious categories, dimension-reduction tools, and gene expression clustering. Clustering has often been used in some areas of medicine to try to identify whether there are different subtypes of diseases lumped together under a single diagnosis.

In the classification literature, we usually find two kinds of clustering algorithms: partitioning and hierarchical methods.

A partitioning method classifies objects into a specified number of N groups that satisfy the following criteria:

- Each group must contain at least one object.
- Each object must belong to one group.

Hierarchical algorithms do not construct a single partition with N clusters, but they deal with all values of N in the interval $[1, n]$. There are two main categories of hierarchical clustering methods: (1) agglomerative clustering and (2) divisive clustering. In hierarchical agglomeration (Ward, 1963), each object is initially considered as a cluster, then, iteratively, two clusters are chosen according to some criterion and merged into a new cluster. The procedure is continued until all the objects belong to the same cluster. The number of iterations is equal to the number of objects minus 1.

In divisive clustering, first, all objects belong to the same cluster; then, iteratively, a cluster of the current partition is chosen according to a selection criterion and bipartitioned according to a local criterion. The procedure continues until all clusters comprise single objects.

The choice of one algorithm rather than another is related to many factors (Kaufman and Rousseeuw, 1990). In particular, the following properties are to be evaluated in relation to the characteristics of the data matrix:

- Ability to work with different data types (heterogeneous)
- Ability to discriminate nonconvex clusters (arbitrary shapes)
- Scalability
- Insensitivity to outliers and noise
- Ability to work with multidimensional data
- Computational requirements
- Results interpretation

22.4 APPLICATIONS TO BIOMEDICAL DATA

In this chapter, we do not present the various algorithms from a theoretical point of view; instead, some examples of practical applications will be shown. In particular, two examples referring to classification problems in the presence of homogeneous and heterogeneous data will be presented. The problem is in both cases that of defining the various classes through a "prototype" characterizing the different clusters through a vector of representative parameters with which one may then compare the different patterns in order to properly classify them. Normally, we consider some distance (i.e., Minkowsky) in order to associate the different patterns to the different centers (prototype centers). In general, the degree of similarity between patterns grows as the distance between them gets smaller. The various components (individuals) will then be associated with one cluster rather than to another according to their similarities with the various prototypes of various classes.

22.4.1   Homogeneous Dataset

In this example, we examine a dataset containing parameters extracted from Holter ECG recordings acquired from patients suffering from abnormal ECG rhythm. In order to characterize the patient and his/her profile, the various ECG beats are extracted to be analyzed and classified on a morphological basis. In particular, it is interesting to discover the existence of pathological clusters and to characterize them through their prototypes (centers). After having extracted the various QRS complexes from the successive beats, it is possible to create a data matrix whose rows contain the samples to be analyzed (see Figure 22.1 for data matrix construction). After QRS complexes extraction, we apply the Karhunen–Loève (KL) transform, which provides an efficient parameter space dimension reduction.

In our example, we consider only the first three components that provide a three-dimensional mapping of the different cardiac cycles. As

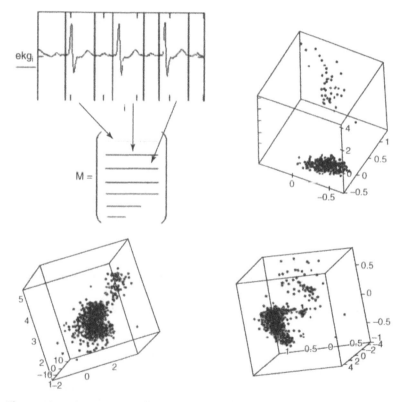

Figure 22.1. Construction of the data matrix and examples of projections in the principal components space. Each point represents a beat. Pathological beats or aberrant events are projected outside the main physiological cluster.

we can see in Figure 22.2, when exploiting the PCA reduced space it is often possible to enhance various clusters or subgroups, representing different beat morphologies (in Figure 22.2 some examples are visible; data provided by Physionet), suggesting the presence of different beat families to be characterized by cluster analysis techniques (Goldberger et al., 2000; Jolliffe, 2002).

22.4.1.1 KHM Clustering Method. To characterize different data groups (physiological and possible pathological clusters) observed in the dataset at hand, a recently introduced algorithm, called K-harmonic means (KHM) (Zhang et al., 2000), is exploited to calculate clusters centers. According to many authors, the well-known K-means method stands out among the many clustering algorithms developed in the past few years as

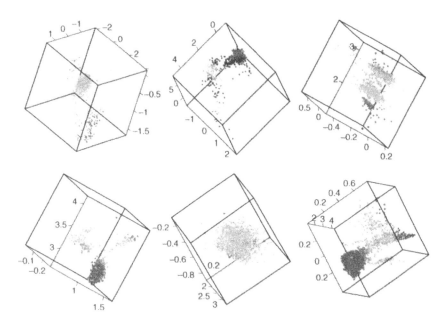

Figure 22.2. 3D graphs represent the projections of the ECG beats in the reduced PC space. Clusters represent different families of beats, characterized by different wave shapes (normal beats, PVC, ischemic events). Each point represents one beat. The first two examples are extracted from ECG recordings of patients suffering ischemic attacks. The last two graphs represent records containing ectopic beats. The various clusters are generated by different aberrant morphologies. The following graph was obtained from a control patient (not showing pathological events). In this case, we have only one physiological cluster. The last example was obtained from a recording of a patient suffering ischemic attack and ventricular tachycardia episodes. Data are provided by Physionet (Golberger et al., 2000).

one of the best methods, accepted by many application domains, even for classification of biomedical data. A major problem with this algorithm is that it is sensitive to the selection of the initial partition and may converge to a local minimum of the objective function if the initial centers are not properly chosen. The KHM algorithm, recently introduced by Zhang and coworkers (2000), solves this problem by replacing the minimum distance from a data point to the centers used in K-means by the harmonic averages of the distances from the data point to all centers. The harmonic average of N numbers is defined as

$$H(a_1, \ldots, a_N) = \frac{N}{\displaystyle\sum_{k=1}^{N} 1/a_k}$$

This quantity is small if at least one of the numbers is small. KHM performs better than K-means; in fact, the method uses the association provided by the harmonic means function to replace the winner-takes-all strategy of K-means.

The association of data points with the centers is distributed and the transition becomes continuous during convergence. According to Mac-Queen (1967), KHM has a built-in dynamic weighting function, which boosts the data that are not close to any center by giving them a higher weight in the next iteration. The recursive expression used to calculate center coordinates at each iteration is

$$c_k = \frac{\displaystyle\sum_{i=1}^{N} \frac{1}{d_{i,k}^3 \left(\sum_{l=1}^{K} 1/d_{i,l}^1\right)^2} x_i}{\displaystyle\sum_{i=1}^{N} \frac{1}{d_{i,k}^3 \left(\sum_{l=1}^{K} 1/d_{i,l}^1\right)^2}}$$

where $c_k, k = 0, \ldots, K$ are the center positions, x_i are the coordinates of the ith data point, and the terms $d_{i,l} = \|x_i - c_l\|$ represent the distances of data points from centers. The KHM algorithm starts with a set of initial positions of the centers, the terms $d_{i,l}$ are calculated, and then the new positions of the centers are obtained iteratively.

The algorithm has been extensively tested and compared with the well-known K-means (KM) and expectation maximization (EM) methods. With regard to the computational cost, Zhang and coworkers (2000) showed that the asymptotic computational complexity per iteration for KM, KHM, and EM are all $O(NKD)$, where N represents the number of data points, K indicates the number of clusters, and D indicates the dimension of the data vectors. For all three algorithms, since the costs are dominated (especially for highly dimensional data) by the distance calculations and there is exactly the same number of distances to be calculated, the coefficients of the cost term NKD of all three algorithms are very close. It is the convergence rate and the convergence quality that differentiate them in real-world applications. Space complexity of KHM is ND for data points, KD for the K centers, and $KD + 2K$ for temporary storage. The temporary storage requirement tends to be lower than KM because the latter needs a $O(N)$ temporary storage to keep the membership information and NK in real problems. It has been demonstrated, using benchmark datasets, that KHM outperforms KM and EM in terms of convergence rate and quality. Zhang and coworkers (2000) compare KM, EM, and KHM performances statistically, running the algorithms on 3600 pairs of dataset initializations to compare the statistical average and variation of the performance of these algorithms. The results are that KHM performs consistently better than KM and EM.

22.4.1.2 Applying the KHM Algorithm to Principal Components Space. In this example, we apply the KHM algorithm, explained above, to a 3 hour ECG recording acquired from a patient showing rhythm disorders (Goldberger et al., 2000). Figure 22.3 shows the projected beats in the space of the principal components (PC). As we can observe, three main subgroups representing different morphological families are visible. The largest cluster probably represents the physiological group, whereas the other two clusters are representative of aberrant beats. For simplicity, let us consider the projection of the beats patterns on the I and III principal

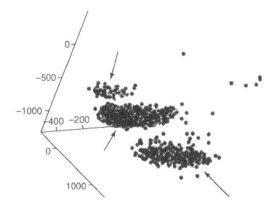

Figure 22.3. Projection into the principal components space of beats of 3 hour Holter recording of a patient with rhythm disorders (data from Goldberger et al., 2000).

components only (Jolliffe, 2002). To characterize the various clusters, three centers should be initialized at random and then the KHM algorithm should be run until convergence is reached. Figure 22.4 shows the evolution of the process within iterations. As we can see, after 10 iterations centers have reached their stable positions, featuring their clusters. Clusters 1 and 3, represented by C1 and C3, are collecting two families of ectopic beats, whereas cluster 2, represented by C2, is the set of physiological beats.

Combining the different patterns with centers found according to their similarities (distance), it is also possible to estimate the number of individuals belonging to the different families, providing medical information useful for diagnosis.

22.4.2 Heterogeneous Data

In this example, we present a cluster analysis performed on an array of heterogeneous data extracted from medical records relating to cardiopatic pa-

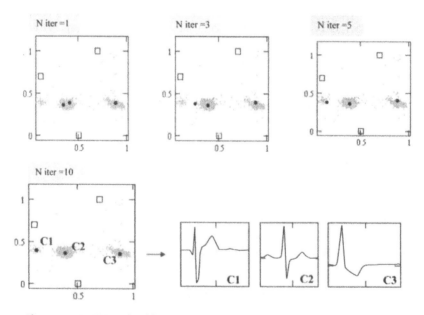

Figure 22.4. KHM algorithm convergence and identification of centers (C1, C2, C3). The ambulatory ECG signal was recorded during a Holter session on a patient suffering with rhythm disorder. Boxes represent the initial center positions; black circles indicate the center positions during the iterative process. Points are plotted in the I–II principal components plan. Data are provided by Physionet (Goldberger et al., 2000).

tients. The data come from the Heart Disease Database containing data collected at the Long Beach Medical Center and the Cleveland Clinic Foundation. Each of the 303 rows in data matrix represents a patient defined by 14 attributes (numeric and categorical). Table 22.1 summarizes the attributes and properties.

In these cases, unlike the previous example (homogeneous data), numerical algorithms cannot be directly applied to the data matrix because it consists of variables of different types (continuous, discrete, binary, and categorical). The clustering algorithms in these cases make calculations using variables derived from original data. In particular, the dissimilarity matrix is commonly calculated before applying numeric methods.

22.4.3 Dissimilarity Matrix

The dissimilarity matrix (Figure 22.5) is built from the data array and each term $d(i, j) = d(j, i)$ represents a dissimilarity measurement between the i and the j patterns. Each term indicates how much the two patterns are different. In order to compute dissimilarity between two patterns, a standard metric can be used in some cases (the distance in this case would represent a dissimilarity measurement) but a dissimilarity function is not a metric. As is typical for the dissimilarity calculation, functions that meet only the following axioms of a metric are used:

1. $d(i, i) = 0$
2. $d(i, j) >= 0$
3. $d(i, j) = d(j, i)$

A function often used to compute the dissimilarity matrix, in the case of a dataset consisting of **n** patterns characterized by **p** heterogeneous parameters, is

$$d(i, j) = \frac{\sum_{f=1}^{p} \delta_{ij}^{(f)} d_{ij}^{(f)}}{\sum_{f=1}^{p} \delta_{ij}^{(f)}} \in [0, 1]$$

The $\square i,j$ quantity is zero if $x_{i,f}$ or $x_{j,f}$ are missing data and in case $x_{i,f} = x_{j,f} = 0$. The $d_{i,f}$ term represents the contribution of each variable f depending on its own type:

- If f is binary (true/false) or categorical (variables that can take M possible values are not sorted): $d_{i,j} = 0$ if $x_{i,f} = x_{j,f}$, otherwise $d_{i,j} = 1$.
- If variables come from continuous measures on a almost linear scale (e.g., weight, temperature, height, and energy), we can estimate the

Table 22.1

Attribute	Property/Values	Description
Age	Numeric	
Sex	Value 0: Male Value 1: Female	
Cp	Value 1: typical angina Value 2: atypical angina Value 3: nonangina pain Value 4: asymptomatic	Chest pain type
TrestBps	Numeric	At-rest blood pressure
Cholesterol	Numeric	Serum cholesterol in mg/dl
FBS	1 = true; 0 = false	Fasting blood sugar > 120 mg/dl
Rest ECG	Value 0: normal Value 1: having ST-T wave abnormality (T wave inversions and/or ST elevation or depression of > 0.05 mV) Value 2: showing probable or definite left-ventricular hypertrophy by Estes' criteria	At-rest electrocardiographic results
Heart rate	Numeric	Maximum heart rate achieved during tests
ExAng	1 = yes; 0 = no	Exercise-induced angina
Oldpeak	Numeric	ST depression induced by exercise relative to rest
Slope	Value 1: upsloping Value 2: flat Value 3: downsloping	The slope of the peak exercise ST segment
Ca	0, 1, 2, 3	Number of major vessels (0–3) colored by fluoroscopy
Thal	Norm/Fix/Rev	X ECG test effects
NUM	0,1	Diagnosis of heart disease (angiographic disease status) Value 0: < 50% diameter narrowing Value 1: > 50% diameter narrowing (in any major vessel)

$$
\begin{bmatrix} x_{11} \cdots x_{1p} \\ \vdots \quad\;\; \vdots \\ x_{n1} \cdots x_{np} \end{bmatrix} \longrightarrow
\begin{bmatrix} 0 \\ d(2,1) \quad 0 \\ d(3,1)\; d(3,2)\; 0 \\ A \qquad A \qquad A \\ d(n,1)\; d(n,2) \cdots \cdots 0 \end{bmatrix}
$$

Figure 22.5. Data matrix (X) and dissimilarity matrix (D) computation.

contribution of the variable f with

$$
d_{ij}^{(f)} = \frac{|x_{i,f} - x_{j,f}|}{\max_h x_{hf} - \min_h x_{hf}}
$$

- If the f variable comes from measurements taken on an unknown or nonlinear scale (e.g., in the case of measures of exponential growth in bacterial populations), its contribution is computed using the following steps:

  1. Replace $x_{i,f}$ with the related rank $r_{i,f}\{1, \ldots, M_f\}$
  2. Transform the [0,1] scale through

$$
z_{if} = \frac{r_{if} - 1}{M_f - 1}
$$

  3. Continue as in the previous case

22.4.4   PAM (Partitioning Around Medoids) Algorithm

After having computed the dissimilarity matrix D, we can proceed with the cluster analysis using the PAM algorithm (MacQueen, 1967). It starts from the D matrix and computes k representative patterns, called medoids, which together determine the clustering of the data matrix. The number k of medoids must be the assigned before iteration, as we have seen with the KHM algorithm. Each pattern is then assigned to clusters corresponding to the closest medoids.

Applying the PAM algorithm to the dissimilarities matrix obtained from original data and specifying a two-cluster ($k = 2$) structure, we get a table detailing the medoids and linking each pattern to a cluster ID (in our case ID1 or ID2). Algorithm performances are generally evaluated by using a particular type of graphics called silhouette plots (Rousseeuw, 1987). This assigns to each pattern a score ranging from -1 (badly associated pattern) to $+1$ (well-associated pattern). The patterns with an intermediate score (score near zero) are not associated with any definitive medoid (intermediate data).

In Figure 22.6, the silhouette plot obtained with our data is shown.

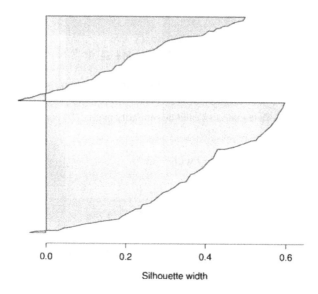

Silhouette width

Figure 22.6. Silhouette plot obtained from the PAM algorithm.

The average score of 0.35 indicates that the association between medoid and patterns in our case is very smooth (as expected, given the number and types of parameters) but still may provide useful information on the aggregation of data.

In the graph, the scores (sorted decreasingly) of the different patterns are drawn starting from the top to the bottom for cluster 1 and cluster 2. The low number of negative scores indicates that there are not many bad associated patterns, but many patterns remain in intermediate areas. Such an analysis could be used as a mean of distinguishing, for example, between healthy and sick patients when two well-defined clusters are found.

This type of analysis is also useful in all those cases in which the number of clusters is not known in advance. Running the algorithm several times with different values of k, it is possible to find the solution with the better average silhouette score and then choose the k number of clusters in structured data.

22.5 VISUAL EXPLORATION OF BIOMEDICAL DATA

When approaching a medical data analysis problem, we may already have some hypotheses (derived from the scientific knowledge in the specific application field or suggested by clinical practice) that we would like to test. In this case, we would use hypothesis-testing procedures to confirm our a priori hypotheses regarding the relationships between the variables in our dataset. The traditional statistical analysis literature provides several hypothesis-testing procedures, including methods for testing the following hypotheses: Z-tests for population mean, proportion, and difference in means for two populations; the t-tests for paired samples, population mean, slope of the regression line, and differences in means for two populations; tests for multinomial populations and independence of categorical variables; the F-test for the analysis of variance; and many other specific tests for time-series analysis, quality control, and nonparametric testing procedures.

However, we do not always have a priori notions of the expected relationships among the variables in the dataset or we do not want contradictory opinions or clinical debates, often regarding application fields in which a reference mathematical model is missing, to bias the classification process. In many cases, it is advisable to have before the various phases of a classification/interpretation study a visual exploratory data analysis (VEDA), instead of classic statistical methods, to delve into large unknown datasets, examining the interrelationships among the various attributes with a data-driven approach. During this phase, using appropriate graphics and visual inspections, it is possible to analyze the dataset at hand to identify interesting subsets of the observations and to assess the presence of separated clusters in our sample without any a priori hypotheses. Many two- or three-dimensional graphics are available to explore a dataset, for example: charts, bar plots, density plots, scatter plots, vector plots, and surface plots.

In this chapter, we will describe an application of a particular conditioned graph, the Trellis diagram, which seems to be very effective in analyzing multivariate data matrixes. We refer, in this example, to the same dataset described above.

Figure 22.7 represents the population distribution divided into healthy (0) and sick (1) subjects. To get a preliminary idea about the variables most affecting the output parameter (Out), we will try to condition the chart with other variables of the Cleveland Clinic database as introduced above. We can first try the blood cholesterol variable (upper header of each panel in Figure 22.8).

As we can observe, the healthy/sick populations seem to be related to the cholesterol value. For a value of 242 (panel 3) the inversion of the population can be noted. For lower values of cholesterol, the majority of both populations were negative (healthy).

Figure 22.9 shows a Trellis diagram of the population (conditioned variable) in respect to OldPeak (conditioning variable), representing the ST deviation achieved during XECG compared to resting ECG. Again, we can see how the parameter heavily conditions the output variable. In particular, we can observe how a value between 1.8 and 6.8 of OldPeak di-

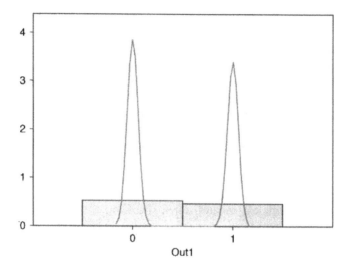

Figure 22.7. Healthy/sick population in the Cleveland Clinic database.

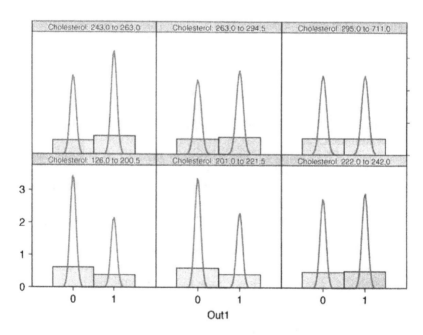

Figure 22.8. Cholesterol effects on the output variable distributions.

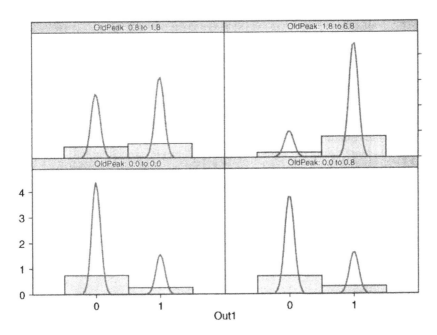

Figure 22.9. Trellis diagram on the conditioning variable OldPeak (ST deviation during the exercise).

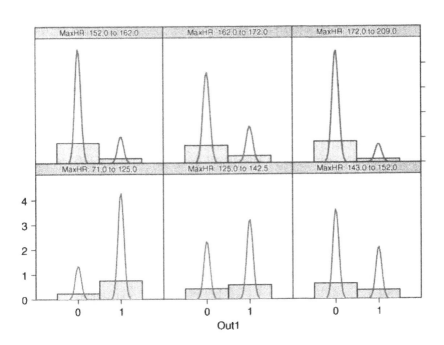

Figure 22.10. Trellis diagram of the conditioning variable MaxHR (maximum heart rate during exercise tests).

vides the population into two very unbalanced groups. Intervals (range OldPeak) of the various panels are calculated so as to collect the same number of individuals in the different modules. It is also possible to plot Trellis diagrams while maintaining the same range on all panels. The choice, of course, depends on the type of analysis we need and on the nature of the conditioning parameters.

It is interesting to consider another parameter, called maxHR, which represents the maximum heart rate achieved during the exercise test (Figure 22.10). We can see how higher values of MaxHR seem more common in healthy subjects, whereas maximum frequencies around 100–125 b/m are typical of sick individuals.

REFERENCES

Bruce, E. N., *Biomedical Signal Processing and Signal Modeling,* Proakis, 2001.

Duda, R. O., Hart, P. E., and Stork, D. G. (Eds.), *Pattern Classification,* Wiley, 2001.

Goldberger, A. L., Amaral, L. A. N., Glass, L., Hausdorff, J. M., Ivanov, P. Ch., Mark, R. G., Mietus, J. E., Moody, G. B. Peng, C. K., and Stanley, H. E., PhysioBank, PhysioToolkit, and PhysioNet: Components of a New Research Resource for Complex Physiologic Signals, *Circulation,* Vol. 101, No. 23, e215–e220, 2000.

Han, J., and Kamber, M., *Data Mining Concept and Techniques,* 2nd ed., Morgan Kaufmann, 2006.

Jolliffe, I. T., *Principal Component Analysis,* 2nd ed., Springer, 2002.

Kaufman, L., and Rousseeuw, P. J., *Finding Groups in Data: An Introduction to Cluster Analysis,* Wiley, 1990

Larose, D. T., *Data Mining Methods and Models,* Wiley, 2006.

MacQueen, J., Some Methods for Classification and Analysis of Multivariate Data, in *Fifth Berkeley Symposium,* Vol. 1, pp. 281–297, 1967.

Rousseeuw, P. J., Silhouettes: A Graphical Aid to the Interpretation and Validation of Cluster Analysis, *J. Comput. Appl. Math.,* Vol. 20, 53–65, 1987.

Ward, J. H., Hierarchical Groupings to Optimize an Objective Function, *Journal of the American Statistical Association,* Vol. 58, 236–244, 1963.

Zhang, B., Hsu, M., and Dayal, U., K-Harmonic Means, in *International Workshop on Temporal, Spatial and Spatio-Temporal Data Mining, TSDM2000,* Lyon, France, Sept. 12, 2000.

INDEX

IEEE Press Series in Biomedical Engineering

The focus of our series is to introduce current and emerging technologies to biomedical and electrical engineering practitioners, researchers, and students. This series seeks to foster interdisciplinary biomedical engineering education to satisfy the needs of the industrial and academic areas. This requires an innovative approach that overcomes the difficulties associated with the traditional textbooks and edited collections.

Series Editor: Metin Akay, University of Houston, Houston, Texas

1. *Time Frequency and Wavelets in Biomedical Signal Processing*
Metin Akay

2. *Neural Networks and Artificial Intelligence for Biomedical Engineering*
Donna L. Hudson, Maurice E. Cohen

3. *Physiological Control Systems: Analysis, Simulation, and Estimation*
Michael C. K. Khoo

4. *Principles of Magnetic Resonance Imaging: A Signal Processing Perspective*
Zhi-Pei Liang, Paul C. Lauterbur

5. *Nonlinear Biomedical Signal Processing, Volume 1, Fuzzy Logic, Neural Networks, and New Algorithms*
Metin Akay

6. *Fuzzy Control and Modeling: Analytical Foundations and Applications*
Hao Ying

7. *Nonlinear Biomedical Signal Processing, Volume 2, Dynamic Analysis and Modeling*
Metin Akay

8. *Biomedical Signal Analysis: A Case-Study Approach*
Rangaraj M. Rangayyan

9. *System Theory and Practical Applications of Biomedical Signals*
Gail D. Baura

10. *Introduction to Biomedical Imaging*
Andrew G. Webb

11. *Medical Image Analysis*
Atam P. Dhawan

12. *Identification of Nonlinear Physiological Systems*
David T. Westwick, Robert E. Kearney

13. *Electromyography: Physiology, Engineering, and Non-Invasive Applications*
Roberto Merletti, Philip Parker

14. *Nonlinear Dynamic Modeling of Physiological Systems*
Vasilis Z. Marmarelis

Printed and bound by CPI Group (UK) Ltd, Croydon, CR0 4YY

16/04/2025

14658461-0001